Amyloid
and Amyloidosis

Amyloid
and Amyloidosis

Edited by

Gilles Grateau
Robert A. Kyle
Martha Skinner

CRC Press
Taylor & Francis Group
Boca Raton London New York

CRC Press is an imprint of the
Taylor & Francis Group, an **informa** business

CRC Press
Taylor & Francis Group
6000 Broken Sound Parkway NW, Suite 300
Boca Raton, FL 33487-2742

© 2005 by Taylor & Francis Group, LLC
CRC Press is an imprint of Taylor & Francis Group, an Informa business

First issued in paperback 2019

No claim to original U.S. Government works

ISBN 13: 978-0-367-45425-8 (pbk)
ISBN 13: 978-0-8493-3534-1 (hbk)

Visit the Taylor & Francis Web site at
http://www.taylorandfrancis.com

and the CRC Press Web site at
http://www.crcpress.com

Library of Congress Cataloging-in-Publication Data

International Symposium on Amyloidosis (10th : 2004 : Tours, France)
    Amyloid and amyloidosis / edited by Gilles Grateau, Robert A. Kyle, Martha Skinner.
        p. cm.
    Includes bibliographical references and index.
    ISBN 0-8493-3534-5
    1. Amyloidosis-Congresses. 2. Amyloid--Congresses. I. Grateau, Gilles. II. Kyle, Robert A., 1928- III. Skinner, Martha, M.D., IV. Title.

    RC632.A5157 2004
    616.3'995--dc22
                                                                    2004059849

Preface

The Xth International Symposium on Amyloidosis was held in Tours, France, April 18–22, 2004. The meeting was organized by Dr. Gilles Grateau and colleagues, and was held in the Vinci Conference Center. The extended abstracts published in this book attest to the high quality scientific information that was presented. A new and well-received feature of the meeting was the workshops on diagnosis and typing of amyloidosis and on clinical management of patients with various systemic forms of amyloidosis.

The International Society of Amyloidosis was formally launched, and an election of officers took place. Robert Kyle (USA) became the first president of the Society. Other officers include: Vice President, Giampaolo Merlini (Italy); Secretary, Bouke Hazenberg (The Netherlands); Treasurer, Morie Gertz (USA); and three members-at–large, Robert Kisilevsky (Canada), Martha Skinner (USA), and Per Westermark (Sweden).

We are grateful to the Audrey and Martin Gruss Foundation, Neurochem Inc, and the International Society of Amyloidosis for support of this book of proceedings. We thank the presenters for submitting their papers promptly and in camera ready form. We appreciate the guidance and helpful format suggestions from Dr. Judith Spiegel of CRC Press. A very special thanks goes to Maura Brady for exceptional competence in collecting papers, working with the authors, and ensuring that each paper was correctly formatted.

Gilles Grateau, M.D.
Robert A. Kyle, M.D.
Martha Skinner, M.D.

Credits

Gellerman et al., p.17, Figure 1: From Schiller, J. et al. (1999) Lipid analysis by matrix-assisted laser desorption and ionization mass spectrometry: a methodological approach, *Anal. Biochem.* 267:46-56.

Seldin et al., p. 161 and 162, Figures 2 and 3: From Skinner, M. et al (2004) High-dose melphalan and autologous stem-cell transplantation in patients with AL amyloidosis: an 8-year study, *Ann. Intern. Med.* 140:85–93.

Hauck et al., pp 180–181, Tables 1–3: *Source:* Data on File, Neurochem Inc.

Omtvedt et al., pp. 209–210, Text and figure from Omtvedt, L. et al. (2004) Serum amyloid P component in mink, a non-glycosylated protein with affinity for phosphorylethanolamine and phosphorylcholine, *Amyloid: J. Protein Folding Disord.* 11(2):101–108. (http://www.tanaf.co.uk/journals)

Wein et al., p. 223, Table 1: *Source:* www.leica-microsystems.com

Stangou et al., p.313, Graph 1: *Source:* 2003 OPTN/SRTR Annual Report, Table 5.1.

Zeldenrust et al., pp. 329–331, Figures 1-4 and some text: From Bergen et al. (2004) Identification of transthyretin variants using sequential proteomic and genomic analysis, *Clin. Chem.* 50(9). With permission.

Zeldenrust et al., p. 331, Figure 5: From Nepomuceno, A.I. et al. (2004) Detection of genetic variants of transthyretin by LC dual electrospray ionization Fourier ion cyclotron resonance mass spectrometry, *Clin. Chem.* 50(9). With permission.

Haagsma et al., pp. 345–347, Text and figures from Haagsma, E.B. et al., (2004) Familial amyloidotic polyneuropathy with severe renal involvement in association with transthyretin Gly47Glu in Dutch, British, and American families, *Amyloid: J. Protein Folding Disord.* 11:44–49.

Röcken et al., p. 369, Table 1: Modified according to Benson, M.D. (2001) A new hereditary amyloidosis: the result of a stop-codon mutation in the apolipoprotein AII gene, *Geonomics,* 72:272–277.

Aucouturier et al., p. 377, Figure 1: From Ballerini et al., in preparation.

Aucouturier et al., p. 378, Figure 2: From Levavasseur et al., in preparation.

Motomiya et al., p. 427, Figures 1–3: From Motomiya et al. (in press) Studies on unfolded B2-microglobulin at C-terminal in dialysis related amyloidosis, *Kidney Int.* With permission.

Bartels et al., pp. 449–450, Text and figures from Bartels, H. et al. (2004) Laryngeal amyloidosis: localized versus systemic disease and update on diagnosis and therapy, *Ann. Otol. Rhinol.* 113(9):741–748.

Table of Contents

SECTION 3 — AA AMYLOIDOSIS

SECTION 4 — FAMILIAL AMYLOIDOSIS

SECTION 5 — CENTRAL NERVOUS SYSTEM AMYLOIDOSIS

SECTION 6 — LOCALIZED AMYLOIDOSIS

SECTION 7 — THERAPEUTICS

SECTION 8 — SATELLITE SYMPOSIUM, NEUROCHEM INC.

SECTION 1

AMYLOID FIBRILLOGENESIS

IN SILICO TO *IN VITRO* APPROACH FOR PROTEIN FOLDING AND MISFOLDING

D. Gilis, Y. Dehouck and M. Rooman

Genomic and Structural Bioinformatics, Free University of Brussels, Av. F. Roosevelt, 50 CP 165/61 1050 Brussels, Belgium.

Proteins are biopolymers that present the particularity to adopt a specific tridimensional structure (3D) dictated by their amino acid sequence. As they are generally only active in their folded form, it is important to get information about their 3D structure, whether by *in vitro* or *in silico* means. It has often been assumed that the 3D structure is unique, but the recent discovery that profound conformational changes may affect isolated proteins in native or non-native environments, or proteins interacting with molecules unrelated to their activity, has shaken this postulate. These conformational changes are at the basis of the so-called conformational diseases (1, 2), such as the Creutzfeld-Jakob disease provoked by the misfolding of the prion protein or those caused by serpin polymerization (3). We briefly survey the general techniques for predicting the 3D structure of a protein from its sequence, and present the *in silico* design of amino acid mutations that modify the misfolding propensities of α1-antitrypsin, a disease-causing serpin.

IN SILICO PREDICTION OF PROTEIN 3D STRUCTURE

Different approaches have been developed in view of predicting the 3D structure of a target protein. They can be classified in three categories, depending on the sequence similarity between the target and proteins of known structure.

Comparative modelling (4), presently the most accurate method for structure prediction, is based on the observation that similar sequences generally adopt similar folds. It is therefore restricted to target sequences that present significant similarity with proteins of known structures. In a first step, the protein structures whose sequences are most similar to the target are identified and defined as templates. In a second step, the target and template sequences are aligned, and the target sequence is mounted onto the template structures according to this alignment, so as to define a model of the target structure. Finally, this model is refined, essentially through an optimization of the side chain conformations (5-7).

A second approach is fold recognition (8), which is less accurate but can also be applied to target sequences that do not present marked sequence similarities with proteins of known structures. In this case, the sequence of the target protein is threaded through a library of known folds. The compatibility between the target sequence and the possible conformations is evaluated by means of an energy function and the best fit is retained. This method relies on the assumption that the target structure is included in the library or is part of a larger structure of the library (9,10).

Finally, the *ab initio* approach to structure prediction (11), so called because it uses the target sequence and an energy function as sole inputs, is the most demanding but also the most general method as it allows in principle to predict new folds. The target structure is assumed to correspond to the global minimum or a sufficiently deep local minimum in the free energy landscape. The conformational space is sampled and the compatibility between the target sequence and each conformation generated is evaluated using energy functions. The complexity of the system and the huge number of possible conformations require the use of efficient sampling algorithms combined with drastic simplifications in the protein representation and energy functions.

The efficiency of protein structure prediction thus depends strongly on the availability of similar sequences or folds: the more similar the target is to proteins of known structure, the easier is the prediction of its structure. Moreover, tackling protein structure prediction through a combination of different methods is generally a very rewarding approach. Note that every two years, an international contest (CASP) aiming at critically assessing the different prediction methods is held. During this contest, different groups submit blind protein structure predictions, which are then assessed according to their category, e.g. comparative modelling, fold recognition or *ab initio* (12-14).

MISFOLDING MODULATION: BLOCKING SERPIN POLYMERIZATION

Number of proteins are likely to adopt alternative, nonnative, structures that can be at the basis of conformational diseases. In particular, serine protease inhibitors (serpins) sometimes undergo dramatic conformational changes leading to inactive polymers. We focus here on α1-antitrypsin (α1-AT), a protein belonging to the serpin family. The mechanism by which α1-AT performs its biological function is usually compared to that of a mouse trap. In a first step, the reactive center loop (RCL) of α1-AT docks into the active site of a serine protease. This initiates the cleavage of the RCL and results in a dramatic conformational change within the serpin: the RCL inserts into the A β-sheet of the protein. During this process, the protease remains attached to α1-AT and is translated to the other end of the serpin. Its active site is thereby distorted and inactivated (15). Some variants of α1-AT, as well as, under certain conditions, the wild-type α1-AT, have been observed to polymerize by a propagated insertion of the uncleaved RCL of each α1-AT into the β-sheet of another α1-AT (16). The polymerization mechanism involves the formation of a partially folded state that self-associates (17).

These observations suggest that α1-AT polymerization could be prevented by stabilizing the native state and/or by destabilizing the polymerized state. To test this assumption, α1-AT variants that display different polymerization propensities were rationally designed *in silico* by means of the PoPMuSiC program (18, http://babylone.ulb.ac.be/popmusic), and their polymerization propensities were measured *in vitro* (19). PoPMuSiC proceeds by introducing *in silico* all possible single site mutations in a given protein structure and evaluating the resulting changes in folding free energy. It was applied to the X-ray structures of the active α1-AT form (PDB code 1QLP) and of α1-AT in complex with trypsin (PDB code 1EZX). The latter form presents its RCL inserted into its own A β-sheet and was used as an approximation of the α1-AT polymeric form. The aim of this *in silico* strategy was to identify mutations that simultaneously stabilize the native structure and destabilize the inserted form with the aim of decreasing the polymerization rate, and vice versa (19).

To test this strategy, five mutations selected by the *in silico* study were expressed *in vitro* and their polymerization propensities were measured. Among them, four were predicted to decrease the polymerization propensity (K331F, K331V, K331I and K331T) and one to increase it (S330R). The variants were incubated at 60°C. Aliquots were removed at different time points and analyzed so as to quantify the proportion of polymeric proteins (19).

A very good agreement between predictions and experiments was observed. The monomeric form of wild-type α1-AT was shown to disappear completely within 10 minutes of incubation due to the formation of polymers unable to enter the gel. For the mutation predicted to increase the polymerization propensity (S330R), the monomeric form disappeared completely within 5 minutes of incubation. For three of the four mutations predicted to decrease the polymerization propensity, monomers were still present after 20 minutes. Only the last mutant (K331T) polymerizes within the same timeframe as the wild-type protein.

CONCLUSIONS

The prediction of protein tertiary structure from the amino acid sequence is an eminently complex task, far from being totally settled at present time. However, significant progress is being made, and the models obtained by the different *in silico* prediction techniques are becoming more and more reliable (12-14). The level of reliability crucially depends on the sequence similarity between the target sequence and proteins of known structures.

Under appropriate conditions, some proteins undergo conformational changes leading from their native structure to alternative conformations. These changes often lead to the development of diseases. *In silico* approaches can contribute to understand this process and even to modulate it, for instance by identifying amino acid substitutions that tend to ease or block the conformational changes (19).

In conclusion, *in silico* approaches to protein folding and misfolding are becoming increasingly powerful tools for rationalizing, simulating, and understanding data obtained *in vitro*, and should lead eventually to rational protein design.

REFERENCES

1. Carrell R.W. and Lomas D.A. (1997), The Lancet, 350, 134.
2. Dobson C.M. (1999), Trends Biochem. Sci., 24, 329.
3. Whisstock J., Skinner S. and Lesk A.M. (1998), Trends Biochem. Sci., 23, 63.
4. Fiser A., Feig M., Brooks C.L. 3rd, and Sali A. (2002), Acc. Chem. Res., 35, 413.
5. Aszodi A. and Taylor W.R. (1996), Fold. Des., 1, 325.
6. Voorhoorst W.G.B., Warner A., de Vos W.M. and Siezen R.J. (1997), Protein Eng., 10, 905.
7. Gomar J., Sodano P., Ptak M. and Vovelle F. (1997), Fold. Des., 2, 183.
8. Godzik A. (2003), Methods Biochem Anal., 44, 525.
9. Skolnick J. and Kihara D. (2000), Proteins: Struct. Funct. Genet., 42, 319.
10. Panchenko A.R., Marchler-Bauer A. and Bryant S.H. (2000), J. Mol. Biol., 296, 1319.
11. Hardin C., Pogorelov T.V. and Luthey-Schulten Z. (2002), Curr. Opin. Struct. Biol., 12, 176.
12. Tramontano A. and Morea V. (2003), Proteins: Struct. Funct. Genet., 53, 352.
13. Kinch N.L., Wrabl J.O., Krishna S.S., Majumdar I., Sadreyev R.I., Qi Y., Pei J., Cheng H. and Grishin N.V. (2003), Proteins: Struct. Funct. Genet., 53, 395.
14. Aloy P., Stark A., Hadley C. and Russell R.B. (2003), Proteins: Struct. Funct. Genet., 53, 436.
15. Huntington J.A., Read R.J. and Carrell R.W. (2000), Nature, 407, 923.
16. Sivasothy P., Dafforn T. R., Gettins P. G. and Lomas, D.A. (2000), J. Biol. Chem., 275, 33663.
17. James E. L. and Bottomley S. P. (1998), Arch. Biochem. Biophys., 356, 296.
18. Kwasigroch J.M., Gilis D., Dehouck Y. and Rooman M. (2002), Bioinformatics, 18, 1701.
19. Gilis D., McLennan H.R., Dehouck Y., Cabrita L.D., Rooman M. and Bottomley S.P. (2003), J. Mol. Biol., 325, 581.

AMYLOIDOSIS *IN VIVO*: FROM MOLECULAR INTERACTIONS TO THERAPEUTIC TARGETS

R. Kisilevsky

Department of Pathology and Molecular Medicine, Queen's University, and The Syl and Molly Apps Research Centre, Kingston General Hospital, Kingston Ontario CANADA, K7L 3N6

INTRODUCTION

Over the last 15-20 years, based on *in vitro* studies, major progress has been made in understanding the mechanism of amyloid fibrillogenesis. These studies revealed that proteins with amylodogenic potential exist in a native conformation in equilibrium with forms that possess intermediate conformations. These in turn are in equilibrium with oligomers that may form frank amyloid fibrils. The sequences responsible for the amyloidogenic potential have been identified, as have aspects of the kinetics and thermodynamics of amyloid fibril assembly and, in many cases, the need for a nucleation event. The possible role of oligomers as the "cyto-toxic" species has emerged. Nevertheless, many of these studies have been performed at protein concentrations orders of magnitude greater than that which exists *in vivo*, in solvents that are not physiological and at pH's that are extreme. Furthermore, it has not yet been established that the fibrils generated *in vitro* are identical to those found *in situ*, nor do the *in vitro* studies address the tissue and anatomic specificity that characterize the deposition of the amyloid proteins *in vivo*. A critical need is the determination of the local tissue conditions, or factors, the microenvironment so to speak, that sets the stage for the appropriate kinetic and thermodynamics that allows fibrillogenesis to occur at specific sites. We must also address the question whether these additional constituents are simply cofactors, or are they just as important as the amyloidogenic proteins/peptides.

AA AMYLOIDOGENESIS AS A MODEL FOR STUDY

To address some of the issues indicated above we have used the rapid induction model of murine AA amyloidogenesis we devised many years ago (1). This model depends on two steps, 1) the induction of an acute inflammatory response by the subcutaneous injection of a small quantity of silver nitrate, which in turn induces the hepatic biosynthesis of serum amyloid A (SAA) the protein precursor to AA amyloid, and 2) the simultaneous intravenous injection of amyloid enhancing factor (AEF), which for our purposes is the nucleation factor for amyloid fibril assembly. Mice treated in this manner deposit substantial perifollicular splenic amyloid within 36 hr. Not only is the SAA deposited as amyloid, so is heparan sulfate (HS) as part of a HS proteoglycan (HSPG), laminin, collagen IV and entactin (2). With histochemical and immunostaining the HSPG can be shown to be part of the AA amyloid *in situ* (3). Furthermore, *in vitro* studies examining the influence of glycosaminoglycans (GAGs) on SAA conformation have indicated that HS, but no other GAG, induces a major increase in beta-sheet structure, the characteristic conformation of amyloids, in the specific SAA isoform that is amyloidogenic, but no other SAA isoform (4).

These studies suggested that amyloid *in vivo* is more than just a protein conformation problem involving only the amyloidogenic protein/peptide. In our view amyloid *in vivo* is composed of 2 sets of components. The first is the disease specific amyloidogenic protein, such as SAA (or Abeta in Alzheimer's disease), and the second is a group of extracellular constituents that includes serum amyloid P (SAP), HSPG, laminin, collagen IV, entactin and apoE. Studies with SAP and apoE gene knock-out mice have suggested that these two components are not absolutely essential for amyloidogenesis (5,6). They appear to play a role in the ease, and rate of initiation and deposition rather than being absolutely necessary requirements. Among the remaining components only HSPG has received significant attention.

STRATEGIES TO INTERFERE WITH HEPARAN SULFATE'S ROLE IN AA AMYLOIDOGENESIS

If one views *in vivo* AA amyloidogenesis, in part, as an interactive process between the amyloidogenic protein SAA and the common components, such as HSPG, then the identification of the complementary binding faces on these two molecules should provide information for the design and synthesis of molecules that will interfere with this interaction. Such studies, published in 1995 (7), illustrated that agents that interfere with the binding of SAA and HS proved to inhibit AA amyloidogenesis in the rapid induction model described above. They also showed that such agents, in addition to preventing AA amyloid deposition, caused regression of pre-existing splenic AA amyloid. Similarly, peptides derived from the HS binding site of SAA proved to be exceedingly effective in preventing AA amyloid deposition in a cell culture model.

Another possible route to achieve the same goal would be to interfere with HS biosynthesis so that SAA would have little ligand with which to interact. This required an understanding of the process of HS biosynthesis (8).

Heparan sulfate biosynthesis takes place in the Golgi where a serine residue(s) in the protein moiety of the proteoglycan serves as an attachment point for the synthesis of a tetrasaccharide composed of xylose-galactose-galactose-glucuronate, the xylose being linked to the serine. This is followed by alternating and sequential additions of multiple N-acetylglucosamine and glucuronate units which are then sulfated to varying degrees at the N-, and 6-O- positions of the glucosamine, and the 2-O- position of the uronate. The uronate may be either D-glucuronate or L-iduronate. The epimerization of the glucuronate to iduronate requires the successful N-deacetylation and sulfation of the adjacent glucosamine on the non-reducing side of the growing polysaccharide chain. The enzymatic addition of the N-acetylglucosamine to the glucuronate at the growing end of the polysaccharide chain takes place through an appropriate transferase that utilizes the UDP activated form of N-acetylglucosamine linking its C-1 hydroxyl to the C-4 hydroxyl of the glucuronate. Similarly, the enzymatic addition of the glucuronate to the N-acetylglucosamine at the growing end of the polysaccharide chain takes place through an appropriate transferase that utilizes the UDP activated form of glucuronate linking its C-1 hydroxyl to the C-4 hydroxyl of the N-acetylglucosamine. Thus an N-acetylglucosamine that contains a C-1 hydroxyl (the site of UDP activation) but lacks a C-4 hydroxyl, if incorporated into the growing polysaccharide chain, would terminate chain elongation. It was these considerations that prompted the design and synthesis of 4-deoxy analogues of N-acetylglucosamine. Such analogues proved to be potent inhibitors of AA amyloidogenesis in cell culture and *in vivo* (9).

Further evidence that heparan sulfate plays a critical role in amyloidogenesis in vivo was obtained from transgenic animals that over express human heparanase (10). One would predict that if heparan sulfate is a critical factor in AA amyloidogenesis that such transgenic mice would be resistant to AA amyloid induction. The results with such animals using the rapid murine induction protocol were both surprising and satisfying. Such mice developed splenic amyloid as rapidly and quantitatively equal to the wild-type controls. However, within the time frame examined, they failed to deposit AA amyloid in kidneys and liver, sites occupied by easily demonstrable AA amyloid in controls. Additional studies illustrated, with Northern blotting and immunohistochemistry, that the transgene was not expressed to any significant degree in the spleens of such mice but was expressed in the kidneys and livers. Analysis of the size of HS in these tissues demonstrated that the HS was fragmented in the liver and kidneys but not the spleen.. Thus there is an excellent correlation between the expression of tissue heparanase and a failure of such tissue to deposit AA amyloid.

SUMMARY

The *in vivo* evidence implicating HS and HSPG biosynthesis in AA amyloidogenesis is as follows:

1)   There is co-deposition of SAA and HS anatomically and temporally,

2)   HS and HSPG are part of the amyloid fibril *in situ*,

3)   Transcription of the mRNA for HSPG is up-regulated prior to, or coincident with amyloid deposition,

4)   Agents that inhibit SAA:HS binding inhibit AA amyloid deposition *in vivo* and in cell culture and promote regression of AA amyloid *in vivo*,

5)   Agents that truncate HS biosynthesis inhibit AA amyloid deposition *in vivo* and in culture, and

6)   Heparanase over-expression inhibits AA amyloid deposition but only in those tissues that over-express the gene.

CONCLUSIONS

1)   Heparan sulfate plays a critical role in AA amyloidogenesis, and by inference in other forms of amyloid such as Abeta and IAPP because HS is present in most, if not all, forms of amyloid.

2)   Heparan sulfate is a valid target for anti-amyloid therapy.

REFERENCES

1)   Axelrad, M.A., Kisilevsky, R., Willmer, J., Chen, S.J. and Skinner, M. *Lab. Invest.*, 47,139-146, 1982.

2)   Lyon, A.W., Narindrasorasak, S., Young, I.D., Anastassiades, T., Couchman, J.R., McCarthy, K., and Kisilevsky, R. *Lab. Invest.*, 64,785-790, 1991.

3)   Snow, A.D., Bramson, R., Mar, H., Wight, T.N. and Kisilevsky, R. *J. Histochem. Cytochem.*, 39,1321-1330, 1991.

4)   McCubbin, W.D., Kay, C.M., Narindrasorasak, S. and Kisilevsky, R. *Biochem. J.*, 256,775-783, 1988.

5)   Botto,M.; Hawkins,P.N.; Bickerstaff,M.C.M.; Herbert,J.; Bygrave,A.E.; Mcbride,A.; Hutchinson,W.L.; Tennent,G.A.; Walport,M.J.; Pepys,M.B. *Nat. Med.* 3,855-859, 1997.

6)   Hoshii,Y., Kawano,H., Cui,D., Takeda,T., Gondo,T., Takahashi,M., Kogishi,K., Higuchi,K., Ishihara,T. *Am.J.Pathol.*, 151,911-917, 1997.

7)   Kisilevsky, R., Lemieux, L.J., Fraser, P.E., Kong, X., Hultin, P., and Szarek, W.A. *Nat. Med.* 1,143-148, 1995.

8)   Salmivirta,M., Lidholt,K., Lindahl,U. *FASEB J.* 10,1270-1279,1996.

9)   Kisilevsky, R., Szarek. W.A., Ancsin, J.B., Elimova, E., Marone, S., Bhat, S., and Berkin, A. *Am. J. Pathol.* (in press).

10)   Li, J. P., Escobar-Galvis, M. L., Gong, F., Zhang, X., Zcharia, E., Metzger, S., Vlodavsky, I., Kisilevsky, R., and Lindahl, U. (in preparation).

BIOPHYSICAL CHARACTERIZATION OF MISFOLDED OLIGOMERIC PROTEIN STATES DURING THE VERY EARLY STAGES OF AMYLOID FORMATION

A.J. Modler[1] and K. Gast[2]

1. *Max-Delbrück-Centrum für Molekulare Medizin, Robert-Rössle-Str. 10, D-13125 Berlin, Germany.*
2. *Potsdam University, Physical Biochemistry, Karl-Liebknecht-Str. 24-25 Haus 25, D-14476 Golm b. Potsdam, Germany.*

We investigated the amyloid formation of phosphoglycerate kinase (PGK) (1), a pseudo wild-type of barstar (C40A/C82A/P27A) (2) and the Syrian hamster prion protein SHaPrP[90-232] (3) in vitro. The first two proteins are not related to any known disease or change of metabolic state of organisms. Changes in mass, size and morphology of assembling protein were monitored by static and dynamic light scattering, size exclusion chromatography (SEC) and electron microscopy. Transformations of secondary structure were probed by circular dichroism and Fourier transform infrared spectroscopy. Misfolding reactions were initiated by jumps to destabilizing solution conditions. In all cases low pH was applied and appropriate amounts of salt were added to screen intermolecular electrostatic repulsion between the protein molecules. Additionally, elevated temperatures were necessary to start amyloid formation of barstar, whereas misfolding of SHaPrP[90-232] could be achieved by addition of denaturants such as GuHCl.

Critical oligomeric states are involved in the earliest stages of amyloid formation

The initial state of PGK and SHaPrP[90-232], respectively, is a partially folded monomeric state with a predominant α-helical secondary structure. In both cases the conformation is expanded compared to the native state. In contrast, the initial state of barstar prior to assembly is an oligomer made of 16 monomers in a molten globule-like conformation, the so-called A-state of barstar. An ensemble of oligomers is buildt up by the first reaction stage of PGK and SHaPrP[90-232], respectively (see Figure 1). The oligomers of PGK have an average mass of ten monomers with a conformation in a predominant β-sheet structure. The oligomer distribution of SHaPrP[90-232] is very heterogeneous with an annular-shaped octamer as smallest product. The whole transition to β-sheet structure is already accomplished. During the second growth stage protofibrils are formed by coalescence of the oligomers delivered by the first stage. Barstar forms already protofibrils as product of the first reaction stage whereby it has to be kept in mind that this misfolding reaction starts already with an oligomeric state. The protofibrils of barstar could be further transformed to long, ribbon-like mature fibrils.

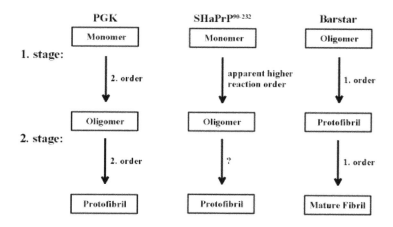

Figure 1. Schematic overview of intermediate states and reaction stages during amyloid formation of the investigated proteins.

In all three cases oligomeric states are involved which are long-lived and partially folded. They possess an intrinsic structural polarity. Thus we call this species critical oligomers.

Different Mechanisms of critical oligomer and protofibril formation

The reaction order derived from various biophysical methods is two for both growth stages of PGK. Populations of dimers, trimers etc. could be detected by SEC. Protofibril assembly of PGK can be described within the framework of a generalized diffusion-collision model (1).

Formation of critical oligomers of SHaPrP90-232 is an apparent two-state reaction between monomers and octamers. No oligomers being intermediate in size are detected by SEC. The reaction order varies between 2 and 4 depending on the used method (3,4). The reaction order of the second growth stage lies between 1 and 2. In case of SHaPrP$^{90\text{-}232}$, protofibrils are formed only by a small subpopulation of oligomers provided by the first growth stage. The annular shaped octamers are presumably an inert side-product not taking part in further assembly or transformation processes.

Protofibril formation of barstar is independent of initial protein concentration in the tested concentration range. The temperature dependence of the first growth stage follows an Arrhenius-like behavior with an activation energy of (19.2 ± 2.0) kcal/mol (2).

The time evolution of the progression curves derived from all methods can only be explained consistently by the assumption that critical oligomers do not interact with monomers to a considerable amount. This property is in contrast to a nucleus which is the defining hallmark of nucleation polymerization mechanisms (4). The critical oligomers represent a pre-aggregated state of proteins which has to be populated to allow further growth to fibrillar states. But the kinetics of their formation and their assembly to protofibrils are distinct. Especially the coupling between secondary and quaternary structure rearrangement is unique for each of the three proteins. The energetic barriers which have to be overcome and thus the detailed shape of the energy landscape to reach the common critical oligomeric state are different (4).

REFERENCES

1. Modler, Andreas J., Klaus Gast, Gudrun Lutsch, and Gregor Damaschun,. "Assembly of amyloid protofibrils via critical oligomers - A novel pathway of amyloid formation". *Journal of Molecular Biology* 325, (2003): 135-148.

2. Gast, Klaus, Andreas J. Modler, Hilde Damaschun, Reinhardt Kröber, Gudrun Lutsch, Dieter Zirwer, Ralf Golbik, and Gudrun Damaschun. "Effect of environmental conditions on aggregation and fibril formation of barstar". *European Biophysical .Journal* 32, (2003): 710-723.

3. Sokolowsk, Fabian, Andreas J. Modler, Ralf Masuch, Dieter Zirwer, Michael Baier, Gudrun Lutsch, David A. Moss, Klaus Gast, and Dieter Naumann, "Formation of critical oligomers is a key event during conformational transition of recombinant Syrian hamster prion protein". *Journal of Biological Chemistry* 278, (2003): 40481-40492.

4. Modler, Andreas J., Heinz Fabian, Fabian Sokolowski, Gudrun Lutsch, Klaus Gast, and Gregor Damaschun. "Polymerization of proteins into amyloid protofibrils shares common critical oligomeric states but differs in the mechanisms of their formation". To appear in *Amyloid – The Journal of Protein Folding Disorders* (2004).

STATUS OF CONGO RED STAINED AMYLOID IN POLARIZED LIGHT

Bart Kahr, Werner Kaminksy, Kacey Claborn, Miki Kurimoto, Lee-Way Jin

Departments of Chemistry and Pathology, University of Washington, Seattle WA 98195

New advances in polarized light microscopy have been employed to image Congo red (CR) stained amyloid plaques in sharp relief.[1] The rotating polarizer method has been used to produce separate false color maps of the transmission, linear birefringence (LB), orientation of the slow vibration direction, linear dichroism (LD), and orientation of the electric dipole transition moment. The consequences of these effects are typically convolved in an ordinary polarized light microscope. In this way, we show that the Alzheimer's plaques have disordered cores and radial distributions of CR.

INTRODUCTION

The principle diagnostic criterion of amyloidosis, established by Divry and Florkin in 1927, is the detection with a polarizing optical microscope of so-called "apple green" birefringence from CR stained tissue sections.[2] Despite the durability of this assay, the characterization of amyloid in polarized light has not progressed past this point. The birefringence is rarely quantified, a problem further confounded by the fact that CR does not stain amyloid consistently, and CR amyloid diagnosis is highly dependent on the skill of the investigator. Clearly, new optical contrast mechanisms are required for simple, reliable amyloid diagnosis.

 Here we show that recent advances in polarized light microscopy can be used to quickly quantify and refine our description of CR stained amyloid.[3] In particular, we have applied a newly developed imaging system to separate the optical transmission, refractive index anisotropy (LB), and optical extinction that are otherwise convolved to produce the ill-defined apple green birefringence. Moreover, we show that the system can be used to determine the absorption anisotropy (LD) and orientation of the electric dipole transition moment. The resulting micrographs that rely on sensitive CCD intensity measurements provide polarized light images of Congophilic amyloid with greater structural resolution than previously reported.

MICROSCOPY

The polarizing microscope, a prototype of the MetriPol System[4] now available from Oxford Cryosystems, is adapted with a stepper motor driven rotating polarizer, circular analyzer consisting of a linear analyzer and quarter wave plate aligned at 45°, an 8-bit monochrome CCD digital camera, and a PC. The modified polarizing microscope is operated in two modes, where the full optical path (both polarizer and circular analyzer) and reduced path (analyzer removed) are used to measure LB and LD, respectively. By modulating the intensity signal as a function of the polarizer angle α, the intensity ratio $I/I_o(\alpha)$ for each pixel is subject to a Fourier separation of the disparate optical contributions that are then displayed in false color images representing the overall transmission, the anisotropy (of refraction in LB mode and absorption in LD mode), and the orientation (of the slow vibration direction in LB mode and of the transition dipole moment in LD mode). The expressions for transmitted intensity with the full and reduced paths are given in expressions (1) and (2):

$$\frac{I}{I_O} = \frac{1}{2}\left[1 + \sin^2\left(\alpha - \varphi\right)\sin\left(2\pi L\,\frac{\Delta n}{\lambda}\right)\right] \tag{1}$$

$$\frac{I}{I_O} = \cosh\left(\frac{2\pi\Delta AL}{\lambda}\right) + \sinh\left(\frac{2\pi\Delta AL}{\lambda}\right)\cos\left(2\pi - 2\varphi'\right) \tag{2}$$

where L is the sample thickness, λ is the wavelength of light, Δn is the difference between the principal refractive indices or LB, ΔA is the scaled differential transmission along the eigenmodes of the sample or LD, φ and φ' are the orientations of the slow vibration direction and the most strongly absorbing direction, respectively, as measured counterclockwise from the horizontal axis. The resulting micrographs are recorded as functions of the phase shift of emergent light δ, where $\delta = 2\pi L\Delta n/\lambda$, the differential absorption coefficient ε, where $\varepsilon = 2\pi L\Delta A/\lambda$, the orientations φ and φ', and the overall intensity I/I_0. In this way, we are able to precisely quantify LB and LD, creating an optical matrix of the diseased tissue.

All measurements were made at 40x magnification. Three wavelengths, 547, 589, or 610 nm, were accessed with interference filters. In this work, dispersive effects were measured off resonance at 610 nm whereas dissipative effects were measured in the absorption band at 547 nm. The measurements were calibrated for a linear camera response, quarter wave plate alignment, and polarization bias of the light source, camera, and objective. Image precision is ~ 0.1% of the transmission, 0.05 nm of the retardance $L\Delta n$, and ~ 0.1° of the orientation.

RESULTS AND DISCUSSION

We applied the Oxford technique to visualize amyloid inclusions from the brains of deceased Alzheimer's disease (AD) patients. The images that we have thus far produced are far more detailed than those produced using conventional microscopies. Representative micrographs of a CR stained AD amyloid plaque are shown in Figure 1. Figure 1a is a map of the refractive index anisotropy (LB) plotted as the modulus of the sine of δ. Figure 1b shows the orientation of the slow vibration direction plotted counterclockwise from the horizontal axis. Virtually identical pictures are obtained with the reduced light path, which yield micrographs depicting the magnitude of absorption anisotropy (LD) and the orientation of maximum absorption direction. The correspondence of the two orientation images strongly indicates that the augmentation of the birefringence is a consequence of the anomalous dispersion in the absorption band of the dye.

While the round plaque seems to be homogeneous in transmission with radial order (Figure 1b), there is a disordered hole in the center as evidenced by the fact that $|\sin \delta| = 0$ in Figure 1a. The hole is mimicked in the map related to LD which is given as the hyperbolic tangent of ε.

Such structures resemble spherulites, aggregates built from concentric rings of small crystallites that many crystallographers often discard as products of failed crystallizations. Have the bio-pathological growth conditions changed such that the outer portion has begun to crystallize around a disordered core? Is the disordered plaque crystallizing from the outside in like a geode?   Surely, such questions regarding the mechanisms of crystal aggregate growth are of importance for designing strategies to inhibit the deposition of pathogenic amyloid plaques. Despite the fact that our inquiry is in its infancy, it is clear that optical micrographs of dyes associated with anisotropic structures reveal aspects of amyloid formation that are not discernable in electron or atomic force micrographs aimed at the identification of individual fibrils.

Nothing is known about the properties of amyloid plaques in circularly polarized light. Recently we described the a circular dichroism imaging microscope that we aim to use in the investigation of the chiroptical properties of CR stained amyloid.[5]

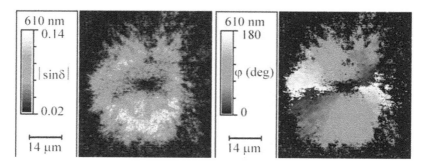

FIGURE 1. Linear birefringence of a cerebral Alzheimer's amyloid plaque stained with CR ($\times$ 600). (a) function of the retardation, $|\sin\delta|$, where $\delta = 2\pi\Delta nL/\lambda$, (b) extinction angles (deg).

ACKNOWLEDGMENTS

This work was supported by National Institute of Aging (2P50 AG 05136), the National Science Foundation (CHE-0092617) and the Petroleum Research Fund of the American Chemical Society (35706-AC6).

REFERENCES

1.   Jin, L.-W. et al. Imaging linear birefringence and dichroism in cerebral amyloid pathologies, *Proc. Natl. Acad. Sci.* 2003, 100, 15294.

2.   Divry, P., Florkin, M., and Firket, J. Sur les propriétés optiques de l'amyloïde, *C. R. Soc. Biol.* 1927, 97, 1808. See also: Wolman, M. and Bubis, J. J. The cause of the green polarization color of amyloid stained with Congo red, *Histochemie*, 1965, 4, 351.

3.   Geday, M. A. et al. Images of absolute retardance *L·Δn*, using the rotating polarizer method, *J. Microsc.* 2000, 198, 1.

4.   http://www.metripol.com

ANALYSIS OF THE INTERACTIONS BETWEEN AMYLOID FIBRILS AND LIPID COMPONENTS

G. P. Gellermann, E. Niekrasz, K. Gührs, S. Shtrasburg[1], M. Pras[1], R. P. Linke[2], M. Fändrich, S. Diekmann, T. R. Appel

Institut für Molekulare Biotechnologie (IMB), Beutenbergstraße 11, D-07745 Jena, Germany; [1]): Heller Institute of Medical Research, Tel Aviv, Israel [2]): Max-Planck-Institut of Biochemistry, Martinsried, Germany. Correspondence should be addressed to G.P.G.; E-mail: gellermann@biocone.com

The formation of amyloid fibrils from soluble polypeptide chains represents a fundamental biochemical process in the initiation and progression of the various amyloid diseases. These fibrils occur in tissues in association with a range of chemically heterogeneous substances that includes glycosaminoglycans, proteins, such as SAP, and lipids [1]. Lipids are probably the most ignored component in this and yet a line of evidences shows that they play a key role in Alzheimer's and Prion disease [2] in addition, they can be co-purified with amyloid fibrils [3]. An improved understanding of lipids and their fatty acid composition in amyloid deposits represents a prerequisite for addressing their possible involvement in amyloid diseases. Therefore we analyzed the molecular species of membrane lipid components of 3 human AA and 6 AL amyloid deposits that had been purified from the surrounding tissues by the method described by Pras [4].

RESULTS AND DISCUSSION

To analyze the fatty acid chains of the polar lipids that are found in the amyloid samples, matrix-assisted laser desorption and ionization time-of-flight mass spectrometry (MALDI-TOF-MS) were applied. The high resolution of the MALDI-MS spectra allows the complete semi-quantitative determination of the length, saturation and distribution of all fatty acids present [5]. However, it is recommended to separated the lipid classes prior to the MS-analysis. Here we adapted existing protocols for normal phase HPLC-methods to a microbore DIOL-column (µHPLC). Baseline separation of the lipid classes present in mammalian brain tissue was achieved by a binary gradient of two solvent mixtures (Fig. 1). Fractions were collected in glass vials. These fractions representing pure lipid classes were analyzed with a Bruker Biflex™ MALDI Mass spectrometer. Spectra of lipids present in *ex vivo* amyloid fibrils were recorded in positive ion mode only when the signal to noise ratio was larger than 4. The masses of the analysed lipid classes have a molar mass range from 700 to 935 Da. Mainly protonated molecular ions [M + 1] are formed and no major peak overlap occurs. Unknown mass peaks were assigned through comparison with known lipid standards. All lipid peaks were identified according to their m/z (Fig. 1). The detection limits for present lipid classes were as follows: galactocerebrosides (GC), 0.05 ng; phosphatidyl-choline (PC), 0.02 ng; phosphatidylethanolamine (PE), 1 ng; phosphatidylinositol (PI), 1 ng; phosphatidylserine (PS), 0.03 ng; sphingomyelin (SM), 0.03; sulfatides (SU), 1.8 ng. The molecular species of lipids present in amyloid deposits are given in table 1. The relative peak intensities were distinguished in 3 categories: ++ denotes high intensity, + moderate and 0 indicates just detectable amounts of lipids.

Table 1. Membrane lipid classes and their molecular species found in ex vivo amyloid fibrils

		AA1	AA2	AA3	AL1	AL2	AL3	AL4	AL5	AL6
GC	18:0				+					
	18:1									
	20:1									
	22:0					+			+	
	22:1				++					
	22:2							+		+
	22:4	++					++			
	24:0								+	
	24:1				++		0	++	++	
	24:2									+
	26:0						++		+	
	26:1	+						+		
	26:2									++
	26:3	0				++	+	++		
	26:4				++				+	
SU	24:0		+					+	+	
	24:3							++	++	
	26:6		+					+	+	
PE	18:0/20:4							++		
	18:0/22:4									+
	18:0/22:6									
	18:0/24:3									+
	(p)18:0/18:2							+		
	(p)18:0/20:4							+		
	PI 16:0/22:4					+				
	18:0/20:4			++		+				
	18:1/20:4			+						
	18:1/22:6			+		++				
PS	16:0/16:1									
	16:1/18:1	++	+		++		+		+	
	16:1/18:2		++							
	16:0/20:2						++		++	
	16:0/22:0									
	18:0/18:1	0								
	18:0/20:4									++
	18:0/22:0		++							
	18:0/22:4					0				
	18:0/22:6					++				
	18:1/18:2									
	(p)20:0/22:1	+			++		+		+	
	(p)20:0/22:2				++					
PC	16:0/16:0						+		+	
	16:0/18:1			−		+	++	+		
	16:0/20:4			+	++	++		+	++	++
	18:0/20:4			++		++		++	+	
	18:1/20:4						0		+	
SM	18:0							+		
	18:1	+	++	++	+	+	+		+	
	22:0									
	24:0			0		+		++		
	24:1							+		++
	24:2	++	++	+	++	++	++	0	++	+

Our analyses revealed that the lipid composition of amyloid fibrils is distinct from healthy tissue in that it is particularly rich in cholesterol and sphingolipids and contain only very rare amounts of those phospholipids, PE and PC, that occur in undiseased tissues in very high amounts (Fig. 1; Table 1). Next to saturated sphingolipids and phospholipids we could detect very high amounts of unsaturated species. This pattern is not obviously related to the chemical class (AA or AL) of the underling amyloid fibril (Fig. 1; Table 1). The most exposed part in lipids are unsaturated fatty acids. Since fatty acid double bonds are highly unstable [6], their loss reflects the extent of degradation. This points out against a strong oxidative or hydrolytic damage of the lipid classes that could have occurred while purifying the amyloid deposits or during lipid analysis.

CONCLUSION:

The method is suitable for analysis of small quantities of lipids present in purified amyloid deposits. The lipids possess a typical pattern that is rich in sphingolipids and cholesterol. The high content of unsaturated molecular species point out against degradation processes that could give raise to the lipid pattern.

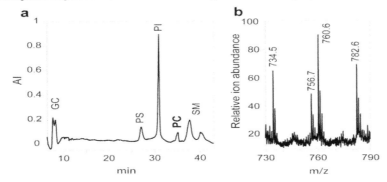

Figure 1. (a) UV-trace (210 nm) of a chromatographic run of amyloid fibril sample of AA. Mass spectrum showing +1 charge states of phosphatidylcholin (b). Assignments and theoretical masses (in brackets) (Schiller et al., 1999): 734,5 ± 0.7: C16:0/C16:0 + H+ (734); 756,7 ± 0.7: C16:0/C16:0 + Na+ (756); 760.6 ± 0.7: 16:0/18:1 + H+ (760); 782.6 ± 0.7: 18:2/18:2 + H+ (782) or 16:0/18:1 + Na+ (782).

METHODS:

Lipid classes were fractionated using a Smart chromatography system (Amersham Biosciences) equipped with a micro bore DIOL-column (1.0 x 100 mm) of 5 µm spheric particles (Macherey-Nagel). Elution was performed in a gradient between solution 1 (n-hexane, 2-propanol, n-butanol, tetrahydrofuran, isooctane, water 64.5:17.5:7:5:5:1 (v/v)) and solution 2 (2-propanol, n-butanol, tetrahydrofuran, isooctane, water 73:7:5:5:10 (v/v)) [7]. Peak fractions were analysed with a Bruker BiflexTM MALDI Mass spectrometer calibrated with 1,2-dipentadecanoyl-sn-glycero-3-phosphocholin (C15:0).

REFERENCES:

1. Kisilevsky R. (2000) amyloidogenesis-unquestioned answers and unanswered questions. *J. Struct. Biol.* **130**: 99-108.
2. Simons K, Ehehalt R. (2002) Cholesterol, lipid rafts, and disease. *J Clin Invest.* 110(5) 597-603.
3. Kim IC, Shirahama T, Cohen AS. (1967) The lipid content of amyloid fibrils purified by a variety of methods. *Am. J. Pathol.* **50**: 869-86.
4. Pras M, Schubert M, Zucker-Franklin D, Rimon A, Franklin EC. (1968) The characterization of soluble amyloid prepared in water. *J. Clin. Invest.* **47**: 924-33.
5. Schiller J, *et al.* (1999) Lipid analysis by matrix-assisted laser desorption and ionization mass spectrometry: A methodological approach. *Anal. Biochem.* **267**: 46-56.
6. Gunstone FD, Harwood, Padley FB (1986) The Lipid Handbook. *Chapman and Hall, New York.*
7. Olsson NU, Harding AJ, Harper C, Salem N Jr. (1996) High-performance liquid chromatography method with light-scattering detection for measurements of lipid class composition: analysis of brains from alcoholics. *J. Chromatogr. B Biomed. Appl.* **681**: 213-8.

DISAGGREGATION EXPERIMENTS AS A TOOL TO DETECT PROTOFIBRILLAR INTERMEDIATES

Martino Calamai', Claudio Canale", Annalisa Relini", Massimo Stefani'", Fabrizio Chiti'" and Christopher M Dobson'

'Department of Chemistry, University of Cambridge, Lensfield Road, Cambridge, CB2 1EW, UK.
"Dipartimento di Fisica, Università di Genova, Via Dodecaneso 33, 16146, Genova, Italy.
'"Dipartimento di Scienze Biochimiche, Università degli Studi di Firenze, Viale Morgagni 50, 50134, Firenze, Italy.

The human muscle acylphosphatase (AcP) has been shown to form amyloid aggregates under mild denaturing conditions, i.e. in presence of 25% trifluoroethanol (TFE) (Figure 1) (1). Our aim was to investigate the changes in the aggregates population after exposure to conditions that favour the native state. We demonstrated that a process reversal to aggregation, the disaggregation process, occurs under conditions where the protein is fully native, i.e. 5% TFE. Our results show that the ability to undergo disaggregation correlates with the size of the aggregates and not just with their structural conformation. Furthermore, the size of the aggregates depends on the time of aggregation. We used a wide variety of complementary techniques to look into the process of disaggregation. Data coming from disaggregation kinetics followed by the recovery of enzimatic activity, intrinsic fluorescence and Thioflavin T (ThT) assay (Table 1), and supported by results from atomic force microscopy, optical microscopy and dynamic light scattering, reveal the presence of aggregates with very distinct sizes during the aggregation process.

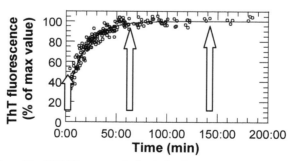

Figure 1. Aggregation followed by ThT. The arrows indicate the times at which the disaggregation process was initiated (10, 70 and 150 min, respectively).

Table 1. Disaggregation rate constants.

Probe and time of aggregation	Fast phase (k_1) (s^{-1})	Slow phase (k_2) (s^{-1})	Native AcP [d]	Non native AcP[e]
ThT [a], 70 min	0.0029±0.0005	0.00034±00010		43±10%
ThT [a], 150 min	Not observed	0.00040±0.00001[f]		53±10%
Act [b], 10 min	0.0025±0.0005	Not observed	89±10%	
Act [b], 70 min	0.0027±0.0005	0.00021±0.00010	80±10%	
Act [b], 150 min	Not observed	0.00038±0.00010	47±10%	
Trp [c], 70 min	0.0018±0.0005	0.00020±0.00005		

The disaggregation of AcP was initiated by diluting five folds with 50 mM acetate buffer, pH 5.5 aliquots of the sample incubated for different times at a concentration of 0.4 mg ml^{-1} in 25% (v/v) TFE, 50 mM acetate buffer, pH 5.5, 25 °C. The data were analysed by fitting to single or double exponential functions.
[a] Disaggregation reaction monitored using ThT fluorescence.
[b] Disaggregation reaction monitored by the recovery of enzymatic activity.
[c] Disaggregation reaction monitored by intrinsic fluorescence.
[d] Extrapolated from the curve showing the recovery of enzymatic activity after completion of the observed exponential phase(s).
[e] Extrapolated from the curve showing the changes in ThT fluorescence after completion of the observed exponential phase(s).
[f] The best estimate and experimental error for $K_2^{ThT, 150}$ are , respectively, the average value and standard error obtained from four independent measurements.

A POSSIBLE MODEL FOR THE AGGREGATION AND DISAGGREGATION PROCESSES OF AcP

After the addition of 25% TFE, the protein partially unfolds (Figure 2a, (1)); globular aggregates (60-200nm), proved to be bundles of protofilaments, develop progressively within 1-2 hours(Figure 2a, (2)); they are able to form clusters of aggregates (400-800nm) (Figure 2a, (3)). It is worth to note that no further conformational changes occur within these structures. The concentration of the globular aggregates increases until it reaches a critical concentration and then larger superstructures (> 5000nm), which could resemble early amyloid plaques, are formed (Figure 2a, (4)). After diluting the aggregated sample to 5% TFE(Figure 2b), the partially unfolded protein is responsible for the observed burst phase (Burst-P^{dis}), the globular aggregates for the fast phase (k_1^{dis}) and clusters of globular aggregates for the slow phase (k_2^{dis}). The larger superstructures cannot be disrupted under the conditions used here.

THERAPEUTIC IMPLICATIONS

Reversion of misfolded proteins to their native state and clearance of amyloid deposits by destabilization of the aggregates have been shown to be winning therapeutic strategies to treat both neurodegenerative and systemic diseases related to protein aggregation (2-3). Our study suggests that the population of aggregates, at least in the case of AcP, is made of species with distinct sizes and thermodynamic properties and could provide a base for the design of appropriate approaches to the treatment of diseases associated with protein aggregation.

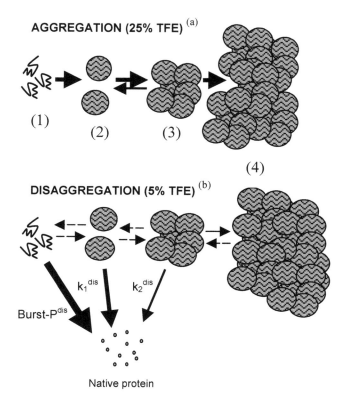

AGGREGATION (25% TFE) [a]

(1)   (2)   (3)

(4)

DISAGGREGATION (5% TFE) [b]

$k_1{}^{dis}$ $k_2{}^{dis}$

Burst-P^{dis}

Native protein

Figure 2. Aggregation and disaggregation model of AcP.

REFERENCES

1. Chiti, F., Webster, P., Taddei, N., Clark, A., Stefani, M., Ramponi, G. & Dobson, C. M. (1999). Designing conditions for in vitro formation of amyloid protofilaments and fibrils. *Proc Natl Acad Sci U S A* 96, 3590-4.

2. Pepys, M. B., Herbert, J., Hutchinson, W. L., Tennent, G. A., Lachmann, H. J., Gallimore, J. R., Lovat, L. B., Bartfai, T., Alanine, A., Hertel, C., Hoffmann, T., Jakob-Roetne, R., Norcross, R. D., Kemp, J. A., Yamamura, K., Suzuki, M., Taylor, G. W., Murray, S., Thompson, D., Purvis, A., Kolstoe, S., Wood, S. P. & Hawkins, P. N. (2002). Targeted pharmacological depletion of serum amyloid P component for treatment of human amyloidosis. *Nature* 417, 254-9.

3. Lachmann, H. J. & Hawkins, P. N. (2003). Novel pharmacological strategies in amyloidosis. *Nephron Clin Pract* 94, c85-8.

CORRELATION BETWEEN STABILITY AND PROPENSITY TO FORM AMYLOID-LIKE FIBRILS

E.Žerovnik, S.Rabzelj, M.Kenig and V.Turk

Department of Biochemistry and Molecular Biology, Jožef Stefan Institute, Jamova 39, 1000 Ljubljana, Slovenia
e-mail-correspondence: eva.zerovnik@ijs.si

In several cases of single point mutations inverse correlation between stability and propensity to form amyloid-like fibrils has been demonstrated (1). Previously, chimeric mutants of human stefins A and B have been studied. Results on their stability and folding rates (Kenig et al., submitted) have shown that there is no correlation between stability and the propensity to fibrillise. The apparent contradiction may be understood if we compare proteins of the same structural class : all stefin B-like and all stefin A-like. Deciding for the propensity to form amyloid-fibrils seems structural factors : the β-sheet structure of both homologues, with stefin's B β-sheet being more hydrophobic and strand-prone already in the denatured state (Žerovnik et al., to be submitted).

In this study, by studying two single point mutations of human stefin B (cystatin B), we further search the correlation between stability and propensity to form amyloid-like fibrils. One iso-form is our recombinant protein (rec-stB), which is based on amino-acid sequence, the other is more commonly observed iso-form as given in the gene database (wt-stB). They differ in only one amino-acid residue at position 31, which is Y31 in rec-stB and is E31 in wt-stB. In addition to these two, we have followed denaturation and amyloid-fibril formation of the mutant G4R observed in some patients with progressive myoclonus epilepsy (EPM1)(2).

To determine stability of the proteins, urea denaturation curves were measured and analysed by a two-state approximation. To see the propensity for amyloid-fibril formation, the proteins were dissolved at slightly acidic conditions (pH 4.8), which promote fibril formation of rec-stB (3,4). Optimal TFE concentration was chosen by titration, where pre-denaturational concentration of the alcohol was taken (which was 9% vol/vol in all three cases).

Our *in vitro* study on stefin B site specific mutant G4R and the two variants, rec-stB and wt-stB shows that order of stability and the propensity to fibrillise differ; there is no correlation. The order of stability is : wt-stB >= G4R mutant >> rec-stB and propensity to fibrillise : wt-stB > rec-stB > G4R mutant. It can be seen that wt-stB forms fibrils after the lowest lag phase of about 100 hours and to the largest extent (judged by final ThT fluorescence). The rec-stB starts fibril growth after a lag phase of 200 hours, even though, this variant is less stable than wt-stB. And, the G4R mutant fibrillises after a lag phase of about 400 hours with the smallest final yield. It should be noted that this mutant is of nearly equal stability than the wild type protein.

Implications for understanding pathology of this type of progressive myoclonus epilepsy would demand more cell-culture and even transgenic animal experiments.

RESULTS

In Figure 1, far UV CD spectra of wild type stefin B (wt-stB) with residue E31, the recombinant stefin B (rec-stB) with residue Y31 and the G4R mutant with an arginine at position 4 are compared. The near UV CD spectra were also recorded (not shown) and reveal that all three proteins are correctly folded.

In Figure 2, are shown urea denaturation curves and in Table 1 the thermodynamic parameters. These were derived by assuming the usual two-state unfolding model.

In Figure 3, maximum in ThT fluorescence is plotted against time. This shows order of amyloid-like fibril growth reaction : lag phase of around 100 hours for wt-stB (E31), 200 hours for rec-stB (Y31) and 400 hours for the EPM1 mutant of stB (R4). Final yield follows the same order : E31 > Y31 > R4. It should be noted here that the protein concentration was kept the same (34 μM) and TFE was added to final 9 % vol/vol in all three cases.

Figure 1: far UV CD spectra   **Figure 2:** urea denaturation curves

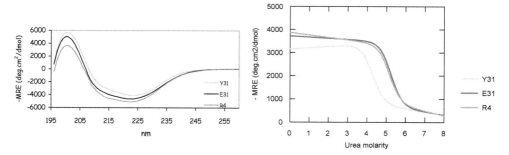

Table 1 : Thermodynamic parameters:

	ΔG°_{N-U} (kJ/mol)	M (kJ/Mmol)	Cm (M)
Y_{31}	- 33.5 ± 6.5	7.9 ± 0.9	4.26 ± 0.02
E_{31}	- 44.4 ± 5.1	8.5 ± 1.0	5.23 ± 0.04
R_4	- 43.4 ± 7.2	8.4 ± 1.4	5.18 ± 0.11

Figure 3: Amyloid-like fibrillization order as measured by ThT fluorescence (pH 4.8, 9% TFE, r.t.)

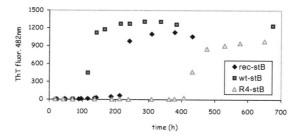

DISCUSSION

The main result of our *in vitro* studies is somewhat surprising. It shows that order of stability has nothing in common with the propensity to fibrillise. Wild-type protein, wt-stB, which is of the highest stability, forms fibrils after the lowest lag phase of about 100 hours. The recombinat variant, rec-stB, which is the least stable, starts fibril growth after a lag phase of about 200 hours, and the EPM1 mutant - G4R, after a lag phase of about 400 hours. The yield of mature fibrils (as judged by intensity of ThT fluorescence) follows the same trend : wt-stB > rec-stB> G4R. Especially striking is the behavior of the EPM1 mutant, G4R, which is of equal stability than the wild-type (see Table 1) yet it has the lowest propensity to fibrillise.

It the literature, amyloid-like fibrils of proteins not involved in any known conformational disease have been shown to exert toxicity (5). Protofibrils and even smaller oligomers were shown to be responsible for toxicity; illustrated by the case of amyloid-beta peptide (6,7). It also was demonstrated that some mutants observed in human inherited disease have an extended lag phase (with oligomers and protofibrils accumulating) (8).

Progressive myoclonus epilepsy of Unverricht-Lundborg type is believed to result in majority of cases from diminished expression and protein activity of stefin B (cystatin B). Our finding that the G4R mutant observed in some EPM1 patients (2) leads to accumulation of prefibrillar aggregates certainly deserves attention. More work on cell-culture and possibly with transgenic animals is warranted to provide evidence if not only loss of function of stefin B as a proteinase inhibitor or loss of some alternative function specific to CNS, but also gain of toxic function may be involved in EPM1 pathology.

To conclude, at least with our series of proteins, direct correlation between stability and the propensity to form amyloid-fibrils does not exist. There are other important factors such as overall charge, sequential and structural traits.

REFERENCES:

1. Hurle, M.R., et al. , 1994. A role for destabilizing amino acid replacements in light-chain amyloidosis. *Proc Natl Acad Sci U S A* **91**, 5446-5450, 1994.

2. Lalioti, M.D., et al., 1997. Identification of mutations in cystatin B, the gene responsible for the Unverricht-Lundborg type of progressive myoclonus epilepsy (EPM1). *Am J Hum Genet*. **60**, 342-351.

3. Zerovnik, E., et al., 2002. Human stefin B readily forms amyloid fibrils in vitro. *Biochim Biophys Acta* **1594**, 1-5.

4. Zerovnik, E., Turk, V., and Waltho, J.P., 2002. Amyloid fibril formation by human stefin B: influence of the initial pH-induced intermediate state. *Biochem Soc Trans*. **30**, 543-547. Review.

5. *Bucciantini, M., et al., 2002. Inherent toxicity of aggregates implies a common mechanism for protein misfolding diseases.* Nature *416, 507-511.*

6. Walsh, D.M., et al., 1999. Amyloid beta-protein fibrillogenesis. Structure and biological activity of protofibrillar intermediates. *J. Biol. Chem*. **274**, 25945-25952.

7. Walsh, D.M., et al., 2002. Naturally secreted oligomers of amyloid beta protein potently inhibit hippocampal long-term potentiation in vivo. *Nature* **416**, 535-539.

8. Conway, K.A., et al., 2000. Acceleration of oligomerization, not fibrillization, is a shared property of both alpha-synuclein mutations linked to early-onset Parkinson's disease: implications for pathogenesis and therapy. *Proc Natl Acad Sci U S A* **97**, 571-576.

LINEARLY ARRANGED OH-GROUPS IN AMYLOID-DEPOSITS AND IN TISSUE ISOLATED AMYLOID FIBRILS

J. Makovitzky

Department of Obstetrics and Gynaecology, University of Rostock, Germany

INTRODUCTION

It has been reported recently that amyloid-deposits and tissue isolated amyloid fibrils are not homogenous, but heterogenous [1,2,3]. Amyloid-deposits contain fibrils and nonfibrillar components. A glycoprotein with a pentagonal structure (P-component) is a constant component of all amyloid-deposits [4].

In this study we have examined the possible sterical orientation of the oligosaccharide-component (P component) and sialic acid in amyloid-deposits and in tissue isolated amyloid fibrils using the "anisotropic" PAS-reaction (Aldehyde-bisulfite toluidine blue-r, ABT-r, [5] and the sialic acid specific reaction [6]. The ABT-r is a useful method: 1. for the selective demonstration of vicinal -OH groups by metachromatic basophilia. 2. for the detection of linear order of the vicinal (glykolytic) -OH groups in polysaccharide chains.

However, on the basis of biochemical studies and the ABT-r it was possible to work out the sialic acid specific topo-optical reaction: in the first step mild periodic acid oxidation produces a Schiff positive C_7-aldehydo neuraminic acid residue, subsequently bisulfite was added and dimethyl methylene blue staining at pH 1.0 with postprecipitation.

MATERIALS AND METHODS

Nine cases of lung amyloidoma, 3 cases of plasmacytoma (including one case of nonsecretory plasmacytoma), 4 cases of breast amyloidoma, 10 cases of laryngeal amyloid, 15 cases of medullary thyroid carcinoma with amyloid, one case of non Hodgkin lymphoma with amyloid and 30 cases with amyloid in the synovial membrane by rheumatoid arthritis. All specimens were fixed in 4% formalin in PBS at pH 7.2. The sections (2-4 μm) were deparrafinized at 80°C for 14-16 h.

Tissue-isolated amyloid fibrils were kindly provided by Dr. R P Linke, Max Planck Institute of Biochemistry, Martinsried, Germany and Dr. M Pras, Heller Institute of Medical Research, Tel Aviv University, Israel). Preparation of microscopic slides is described by Appel and Makovitzky 2003 [7].

Basically the periodic acid solution (1%, 30 min, room temperature) and sodium bisulfite addition (saturated solution, 45 min) is followed by toluidine blue staining at pH1.0 with $K_3[Fe(CN)_6]$/KI postprecipitation. This reaction is a modified anisotropic PAS-reaction.

The sialic acid specific topo-optical reaction is a modified anisotropic mPAS-r (0.01% 10 min at 4°C) and after bisulfite addition is followed by dimethyl methylene blue staining at pH1.0 with postprecipitation.

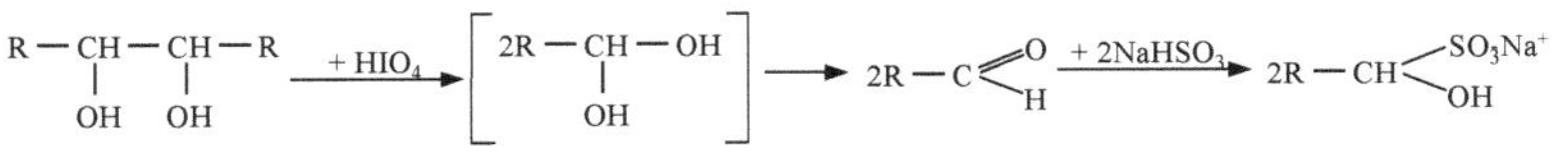

Fig. 1. Chemical reaction of periodic acid bisulfite addition

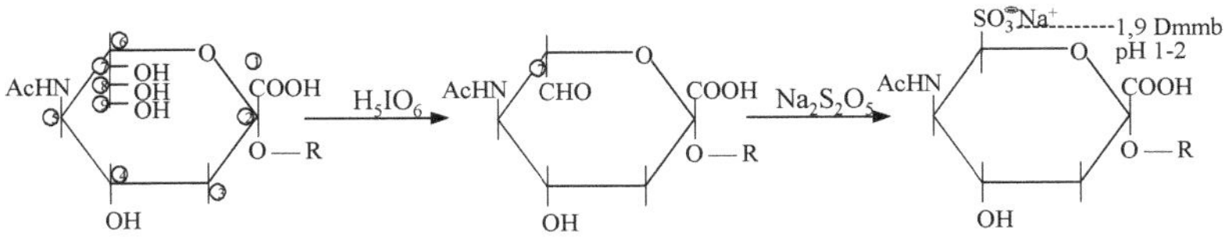

Fig. 2. Scheme of the sialic acid-specific topo-optical staining reaction [6]

RESULTS

The glycoprotein- and sialic acid components were localised in a highly ordered fashion and sterically oriented. Based on the polarisation optical analysis of these reactions acc. Romhányi et al. (1975) the sign of anisotropy was linearly negative (radially positive) Fig. 3-5.

The thiazine dye molecules were bound to the surface of the protein skeleton in a perpendicular orientation. The linearly ordered vicinal OH-groups were oriented perpendicularly to the surface of the protein skeleton.

After sialidase digestion and/or chemical extraction of sialic acid (0,1 N H_2SO_4 for 90 min at 90°C) we have registered similar results: a by far weaker anisotropy after 3-4 hours with green yellow polarisation colour and meta-/ orthochromasia.

In lung amyloidoma we observed interesting differences between (peripheral) parenchyma and tumour-like tissue (central mass). A linear negativity (radially positive) was found at the periphery and linear positivity (radial negative) in the tumour-like region.

We have registered a similar phenomenon in unstained slides, with imbibition analysis, after Congo Red staining and for glycosaminglycan (GAG) components. However, the glycoprotein and sialic acid components are differently oriented in the central and peripheral regions. We showed a similarly phenomenon in breast amyloidoma.

 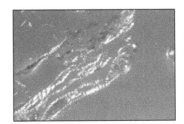

Fig. 3. ABT-r in amyloid-deposits of human menisci, light microscopical picture (80x) and polarization optical picture (80x)

CONCLUSION

1. The amyloid-deposit and tissue isolated amyloid fibrils are heterogenous. The glycoprotein components (P component and sialic acid) and various GAG components: heparan-, keratane

Amyloid fibrils and linearly arranged OH-groups in amyloid

chondroitin sulphate and hyaluronic acid are oriented in a highly ordered fashion. 2. The GAG components are oriented perpendicular to the length of the fibrils. 3. The polysaccharide chains are oriented parallel to the length axis of the amyloid fibrils. The linear ordered vicinal -OH groups are perpendicularly oriented in polysaccharide chains just as sialic acid -OH groups. 4. Our results suggest that amyloid is present in the periphery and in the centre of amyloidomas in different physicochemical phases: The linear arranged vicinal OH-groups in the periphery are oriented perpendicularly, in the centre of the they are oriented parallel to the surface of the protein skeleton. 5. We have found no relevant differences between amyloid-deposits in tissue and in tissue isolated amyloid fibrils (except findings in amyloidomas, see 3). 6. In frozen sections of all amyloid cases, as well as in ex vivo isolated fibrils we have noticed lipid contents.

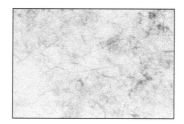 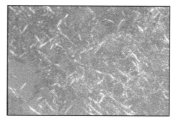

Fig. 4. ABT-reaction on isolated fibrils, light microscopical picture (80x) polarization optical picture (80x)

 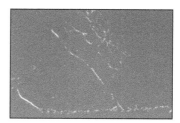

Fig. 5. The sialic acid specific reaction on isolated fibrils, light microscopical picture (80x) and polarization optical picture (80x)

REFERENCES

1.   Cohen, A.S. The constitution and genesis of amyloid. Int Rev Exp Pathol 4:159-243, 1965.

2.   Wolman, M. Amyloid, its nature and molecular structure. Lab Invest 25:104-110, 1971.

3.   Inoue, S. and R. Kisilevsky. A high resolution ultrastructural study of experimental murine AA amyloid. Lab Invest 74:670-683, 1996.

4.   Glenner, G.G. and H.A. Bladen. Purification and reconstution of periodic fibril and unit structure of human amyloid. Science 154:271-272, 1966.

5.   Romhányi, Gy., Deák, Gy. and J. Fischer. Aldehyde-bisulfite-toluidine blue (ABT) staining as a topo-optical reaction for demonstration of linear order of vicinal OH groups in biological structures. Histochemistry 43:333-348, 1975.

6.   Makovitzky, J. Ein topo-optischer Nachweis von C_8-C_9-unsubstituierten Neuraminsäureresten in der Glykokalyx von Erythrocyten. Acta histochem 66:192-196, 1980.

7.   Appel, T.R. and J. Makovitzky. Romhányi´s staining methods applied to tissue-isolated amyloid fibrils. Acta histochem 105:371-372, 2003.

TISSUE DETECTION AND CHARACTERIZATION OF AMYLOID- IS PROTEOMICS THE ANSWER?

Maria M. Picken[1], Vanitha Thulasiraman[3], Roger N. Picken[2], Lee Lomas[3]

[1]*Loyola University Medical Center and* [2]*Hines VA Hospital, IL, USA;* [3]*Ciphergen Biosystems Inc., Fremont, CA, USA.*

Amyloidosis is a diverse group of diseases characterized by extra-cellular tissue deposits of abnormally folded proteins that are generically termed amyloid and that occur in one or many organs. The gold standard of diagnosis is based on detection of deposits in tissues and is based on either a histological approach using affinity to Congo red (which under polarized light produces an apple-green birefringence), or by the detection of fibrillar deposits using electron microscopy (1). However, this diagnostic approach is subject to low sensitivity, and the typing of deposits by immunohistochemical methods is frequently unreliable. Moreover, cumulative research has strongly indicated that, at this stage, the process of amyloidogenesis is relatively advanced and less amenable to treatment.

Surface enhanced laser desorption/ionization-Mass Spectrometry (SELDI-MS) is a technology that combines solid-phase extraction chromatography with direct detection by mass spectrometry. SELDI-MS surfaces can be derivatized in any number of ways to generate uniquely selective ProteinChip® Arrays, and thus be designed to detect specific properties of deposits involved in disease, such as those involved in amyloidosis.

The most frequent type of systemic amyloidosis is amyloid derived from the immunoglobulin light chain; however, other types of amyloid have to be considered in the differential diagnosis (2). By traditional methods amyloid is usually diagnosed at an advanced age (1, 3). The question arises whether technologies such as SELDI-MS could be utilized to allow for a more sensitive and/or earlier diagnosis, and thus increase the probability of successful treatment of the disease. We report here our initial attempts to achieve this goal.

MATERIALS AND METHODS

Affinity reagents targeted against amyloid deposits were prepared by derivatizing a proprietary affinity ligand onto both reactive IDM Affinity Beads (Ciphergen) or RS100 Reactive ProteinChip Arrays (Ciphergen) via a free primary amine (carbonyldiimidizole reactive chemistry). Surfaces were washed extensively overnight in 50mM HEPES buffer (pH 5) containing 0.1% CHAPS. Arrays were used immediately while beads were stored at 4°C until use.

Tissues used in this study were derived from Myocardium, Kidney (cortex and medulla) and Urinary bladder, and consisted of both amyloid negative and amyloid positive material. Tissue was prepared by mechanical disruption in lysis buffer (NP40 1%, glycerol 10%, NaCl 100mM, Tris 20mM pH7.5, EDTA 1mM, DTT 1mM, EDTA free protease inhibitor) using a polytron (Powergen 125) homogenizer. The lysate was spun at 13,000 rpm for 15 min and the supernatant was separated from the pellets, aliquoted and frozen at −80°C until use.

At the time of analysis, supernatants were thawed, and 100ul supernatant incubated with 7ul (packed volume) bead affinity reagent in 100ul of 50mM Tris (pH 7.4).   Reagents were incubated for 90 min with constant agitation at 4°C.   After incubation, affinity beads were washed three times with PBS (20mM, pH 7) containing 0.3% tween-20, twice with PBS (20mM, pH 7) , and once with deionized water.   Bound proteins were extracted with 50% acetonitrile containing 0.3% TFA, then re-concentrated on an H50 ProteinChip® Array.   After a final wash to remove any contaminating salts, 1ul of sinapinic acid (50% saturated solution in 50% acetonitrile and 0.5% trifluoroacetic acid) was applied and the array analyzed using a mass reader (model PBSIIc; Ciphergen).

RESULTS AND DISCUSSION

Positive samples included tissues positive for amyloid by Congo red stain (with apple-green birefringence under polarized light) and deposits that were fibrillar by electron microscopy.   We tested myocardium containing ATTR and AL-lambda and kidney (cortex and medulla) containing ATTR and AL-lambda and AL-lambda amyloidoma of the urinary bladder.   Typing of amyloid deposits in frozen sections was performed against a panel of antibodies and evaluated by immunofluorescence.   ATTR was derived from wild type transthyretin as confirmed by DNA sequencing.   Underlying plasma cell dyscrasia was subsequently confirmed in cases of AL cardiac and renal amyloidosis.   The patient with the urinary bladder AL-lambda amyloidoma does not, thus far, have evidence of an underlying plasma cell dyscrasia.   Corresponding negative tissues were tested in parallel.

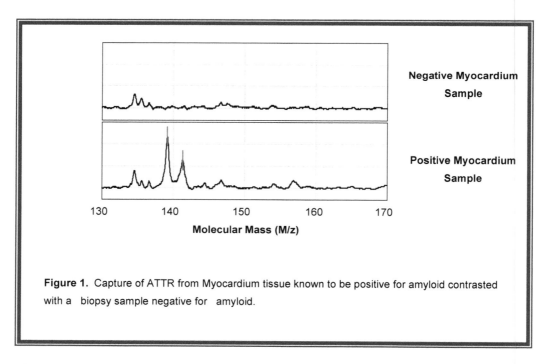

Figure 1. Capture of ATTR from Myocardium tissue known to be positive for amyloid contrasted with a   biopsy sample negative for   amyloid.

Only the known positive samples showed distinct dominant peak(s) with molecular weight(s) corresponding to the monomer (or fragments) of TTR (Fig. 1) or the immunoglobulin light chain (not shown).   Differences in protein profiles were noted between preparations from different organs (myocardium versus kidney).   However, preparations from renal cortex and medulla were similar.   These preliminary studies indicate that this technology can be used to detect and analyse at least two of the most frequent types of amyloid in unfixed tissue samples.   Small sample requirements are particularly desirable for investigations of biopsy material.

Further studies are needed to assess whether this protocol offers a higher sensitivity for the detection of amyloid deposits than currently used methods.

CONCLUSIONS

1. ProteinChip® Arrays derivatized with affinity ligands specific for amyloid deposits may represent a unique opportunity to develop a simple and rapid test for early amyloid detection in tissue biopsies.
2. Preliminary results indicate that such surfaces are able to capture clusters of proteins of the expected masses of amyloid proteins. These clusters were not seen in corresponding negative control tissues.
3. When comparing these protein clusters between Myocardium, Kidney and Bladder tissues, unique profiles were detected. Further characterization of these captured proteins is necessary to confirm their disease process specificity.

ACKNOWLEDGEMENT

This work is supported by a research grant from the Amyloidosis Foundation, Inc.

REFERENCES

1. Picken MM. The changing concepts of amyloid. Arch Pathol Lab Med 2001, 25:25-37.
2. Westermark P., Benson M., Buxbaum JN., Cohen AS., Frangione B., Ikeda S., Masters CL., Merlini G., Saraiva MJ, Sipe JD. Amyloid fibril nomenclature – 2002. Amyloid: J Protein Folding Disord 2002, 10: Suppl. 1, 48-54.
3. Westermark P. , Bergstrom J., Solomon A., Murphy C., Sletten K. Transthyretin-derived senile systemic amyloidosis: clinicopathologic and structural considerations. Amyloid: J Protein Folding Disord 2003, 10: Suppl.1, 48-54

THE CONFORMATION OF AMYLOID β-PEPTIDE IN THE PRESENCE OF FLUORINATED AND ALKYLATED NANOPARTICLES

S. Rocha,[1,2] A. F. Thünemann,[3] M. C. Pereira,[2] M. A. N. Coelho,[2] H. Möhwald,[1] G. Brezesinski[1]

[1] *Max Planck Institute of Colloids and Interfaces, Am Mühlenberg 1, 14476 Golm/Potsdam, Germany*

[2] *Faculty of Engineering, University of Porto, Rua Dr. Roberto Frias, 4200-465 Porto, Portugal*

[3] *Federal Institute for Materials Research and Testing, Richard-Willstätter-Straße 11, 12489 Berlin, Germany*

Amyloid β-peptide is found in cerebrovascular deposits and extracellular neuritic plaques that characterize Alzheimer's disease, a progressive neurodegenerative disorder of the human brain (1). The peptide, with 40 or 42 amino acid residues, is a product of proteolytic cleavage of a transmembranar protein, the amyloid precursor protein. Strong evidences indicate that amyloid β-peptide accumulation represents an early event of Alzheimer's disease (1). The fact that mutations in the amyloid protein precursor gene are associated with Alzheimer's disease is a strong indication of the importance of amyloid in the pathogenesis of the disease. The delineation of the amyloid β-peptide hypothesis predicts therapeutics based on preventing the peptide misfolding and aggregation (2).

It was previously demonstrated that polyelectrolyte-fluorosurfactant complexes are able to dissolve amyloid plaques in sections of animal tissue and to convert β-sheet into α-helix structures (3, 4). These complexes can be, in principle, engineered with biocompatible properties (3). The complexes were in a first approach tested on solid supports. In order to increase the contact area between peptide molecules and complexes, nanoparticles of these complexes were synthesized (5).

NANOPARTICLES CHARACTERIZATION

Polyampholytes with alternating cationic (*N,N'*-diallyl-*N,N'*-dimethylammonium chloride) and anionic charged monomers (maleamic acid) were synthesized with a degree of polymerization in the range of 60 to 80. Complexes of the polyampholytes and dodecanoic and perfluorododecanoic acid, respectively, were prepared. This results, in both cases, in nanoparticles with hydrodynamic diameters of 3 to 5 nm (so-called polyampholyte dressed micelles) (5). Figure 1 shows the particle size distribution of the polyampholyte complexes as determined by analytical ultracentrifugation. The nanoparticles display zeta potentials of $+(25 \pm 10)$ mV when alkylated and $-(47 \pm 5)$ mV when fluorinated and dissolve at concentrations below 0.02 g L^{-1} (i.e., they are not covalently cross-linked). The densities of the nanoparticles were 1.256 g cm^{-3} (alkylated) and 1.654 g cm^{-3} (fluorinated). Due to their small size they have specific surface areas of approximately 1000 m^2 g^{-1}. It can be expected that the high surface area would be useful in providing extensive interactions with the peptide.

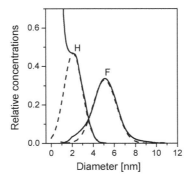

Figure 1. Particle size distributions of the polyampholyte complexes with dodecanoic (H) and perfluorododecanoic acid (F) determined by analytical ultracentrifugation. Concentrations were 1 g L^{-1}.

The nanoparticles consist of a hydrophobic core (formed by the surfactant chains) and a hydrophilic shell. The low molecular weight counter ions (Na$^+$) of the micelle are replaced by the polyampholyte. A simplified structure of the nanoparticles is shown in Figure 2.

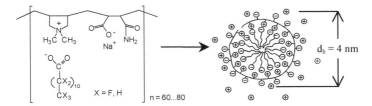

Figure 2. Complexes of poly(N,N'-diallyl-N,N' dimethylammonium-*alt*-maleamic carboxylate) and the sodium salt of dodecanoic acid (X = H) and perfluorododecanoic acid (X = F) (5).

INFLUENCE OF NANOPARTICLES ON AMYLOID β-PEPTIDE STRUCTURE

Circular dichroism (CD) measurements were performed to determine the secondary structure of the amyloid β-peptide [1-40] at pH 7 in presence of different amounts of nanoparticles. Final particle concentrations were 0, 2 and 8 g L^{-1} at a temperature of (20 ± 1) °C. The measured CD spectra are shown in Figure 3. In the absence of particles, amyloid β-peptide [1-40] displays the typical spectrum (curve 1) of a random coil protein (minimum at 198 nm). For a quantitative determination of the secondary structure the content of structure motifs was calculated according to Greenfield (6). This results in a content of 0 % α-helix, 29 % β-sheet and 71 % random coil. The titration of an increasing amount of fluorinated nanoparticles induces a change from random coil to α-helix structure as shown in Figure 3 a (curves 2 and 3). The content of α-helix was 3 % and 18 % after the addition of 2 and 8 g L^{-1} fluorinated nanoparticles, respectively. An isosbestic point can be seen at 203 nm (Figure 3 a), which is a strong indication for an equilibrium between two states.

The alkylated particles, in contrast to the fluorinated nanoparticles, do not induce α-helix rich structures (Figure 3 b). The titration of amyloid β-peptide [1-40] with alkylated nanoparticles resulted in a transition to β-sheet structure (Figure 3 b, curves 2 and 3). The content of β-sheet was about 40 %. The presence of an isosbestic point at 210 nm indicates a two-state transition from random coil to a β-sheet rich structure.

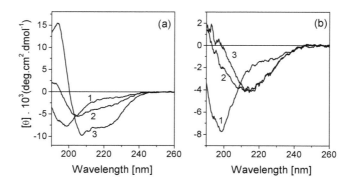

Figure 3. Circular dichroism data of amyloid β-peptide (46 μM) at pH 7 in the presence of fluorinated (a) and alkylated (b) nanoparticles: 0 g L^{-1} (curve 1), 2 g L^{-1} (curve 2), 8 g L^{-1} (curve 3).

Fluorinated nanoparticles induce α-helix rich structures in amyloid β-peptide [1-40] at pH 7 whereas the alkylated ones lead to β-sheet formation. Obviously, both types of nanoparticles interact with the peptide by hydrophobic interactions, which are enhanced in the case of fluorinated nanoparticles due to the presence of CF_3 groups. This might explain why fluorinated particles but not alkylated ones induce α-helix rich structures. Additionally, fluorinated nanoparticles show highly negative zeta potential in contrast to the alkylated ones. This suggests that both electrostatic and hydrophobic interactions might be responsible for stabilizing the helical structures. The strong effect of fluorinated nanoparticles on the secondary structure of the peptide indicates that Figure 2 presents a dynamic structure where a part of the fluorinated chains is in contact with the hydrophilic phase. This suggests that for optimum response proper micelle engineering is required with enough fluorine outside to affect the peptide structure. In conclusion, it has been demonstrated that fluorinated nanoparticles made of a polyampholyte-fluorosurfactant complex induce α-helix rich structures in amyloid β-peptide [1-40], whereas their alkylated analogues do not show this effect. Fluorinated nanoparticles are proposed to be potential candidates for the inhibition and reversion of conformational changes of proteins that lead to amyloid fibril formation.

REFERENCES

1.   Selkoe, D. J., 2001. Alzheimer's disease: genes, proteins and therapy, *Physiol. Rev.* 81: 741-766.

2.   Selkoe, D. J. and Schenk D., 2003. Alzheimer's disease: molecular understanding predicts amyloid-based therapeutics, *Annu. Rev. Pharmacol. Toxicol.* 43: 545-584.

3.   Thünemann, A. F. et al., European Patent application EP1341564.

4.   Vieira, E. P., Hermel, H. Möhwald, H., 2003. Change and stabilization of the amyloid-β (1-40) secondary structure by fluorocompounds, *Biochim. Biophys. Acta* 1645: 6-14.

5.   Thünemann, A. F. et al., 2002. Polyampholyte-dressed micelles of fluorinated and hydrogenated dodecanoic acid, *Langmuir* 18: 5099-5105.

6.   Greenfield, N., Fasman, G. D., 1969. Computed circular dichroism spectra for the evaluation of protein conformation, *Biochemistry* 8: 4108-4116.

SECTION 2

AL AMYLOIDOSIS

SYSTEMIC EQUINE AL AMYLOIDOSIS

Charles L. Murphy,[1] Burnette Crombie,[1] Teresa K. Williams,[1] Sallie D. Macy,[1] Deborah T. Weiss,[1] Alan Solomon,[1] and Yvette S. Nout[2]

[1]Human Immunology & Cancer Program, Department of Medicine, University of Tennessee Graduate School of Medicine, Knoxville, TN, USA; [2]Department of Veterinary Clinical Sciences, College of Veterinary Medicine, Ohio State University, Columbus, OH, USA

I. INTRODUCTION

Heretofore, the few reported cases of equine light-chain-associated amyloidosis had involved animals in which the amyloid was localized exclusively to the nares or skin. Recently, we described the clinical features and postmortem findings in a 16-year-old gelding that succumbed to extensive cardiac amyloidosis, but also had similar deposits in the gastrointestinal tract, spleen, kidney, and liver.[1] We now have determined by protein sequence analysis and immunohistochemistry that the amyloid protein was derived from a horse $V_\lambda 1$ Ig light chain. Notably, this is the first example of systemic AL amyloidosis in this species.

II. MATERIALS AND METHODS

A. AMYLOID EXTRACTION AND CHEMICAL CHARACTERIZATION

Amyloid fibrils were extracted from 4 μm-thick formalin-fixed, paraffin-embedded kidney sections using 6 M/L guanidine HCl. After reduction and carboxymethylation, the protein was purified by reverse-phase HPLC and digested with trypsin under conditions described by Murphy et al.[2]

For sequencing analyses, the amyloid protein extract was dissolved in 0.1% SDS buffer containing 0.1 M/L dithiothreitol and 8 M/L urea and electrophoresed on 10% SDS/PAGE gels. For automated sequence analysis by Edman degradation, the protein band was transferred onto a PVDF membrane and placed in the blot cartridge of an ABI Model 494 Procise gas-phase sequencer.[2]

For tandem mass spectrometric (MS/MS) identification of peptides, samples were separated by reverse phase HPLC and the effluent directed into an ion-trap mass spectrometer.

B. IMMUNOHISTOCHEMISTRY

Tissue was immunostained using the ABC technique. The primary and secondary reagents were, respectively, rabbit polyclonal anti-human $V_\lambda 1$ and C_λ antisera[3] and a biotinylated goat anti-rabbit antibody.

III. RESULTS

Passage of the reduced and alkylated amyloid extract through a reverse-phase HPLC column yielded 2 major peaks. Direct (automated) sequence analyses revealed that one contained hemoglobin, whereas no identifiable residues were found in the second, thus indicating that the N-terminus was blocked, *i.e.*, contained a cyclized form of glutamine. However, after treatment with pyroglutamate amino-peptidase, a sequence of

SLTQPASVSGTLGQT was obtained that was found to be identical to residues 2 through 16 of horse $V_\lambda 1$ light chains[4] and, additionally, was highly homologous to the product of the human $V_\lambda 1b$ germline gene, vl-17.[5] Trypsin digestion of this material was performed and another peptide spanning residues 32 through 47 of horse $V_\lambda 1$ light chains also was identified (Figure 1). These data were confirmed by mass spectrometry (expected and observed masses, 1752.97 and 1752.52, respectively).

```
                        10        20        30        40
Horse Vλ1 gene   QSLTQPASVS GTLGQTVTIS CSGSSSNIGY SYSAVGWYQQ IPGTAPK
     Horse AL    ---------- ------                ---------- V------
```

Figure 1. Comparison of the amino acid sequence of the light chain encoded by the horse $V_\lambda 1$ germline gene with that of the horse light chain-related amyloid fragment. The N-terminal portion (residues 2-16) of the amyloid was determined by direct sequence analysis. Because treatment of the sample with pyroglutamate amino peptidase was required in order to obtain data on the first amino acid, glutamine (Q) was inferred to be the N-terminal residue. The C-terminal sequence (residues 32-47) was obtained by MS/MS analysis. (--), sequence identity.

The AL nature of the green birefringent congophilic deposits was further evidenced by their reactivity with polyclonal antisera specific for human λ1 light chains, as well as the anti-human C_λ reagent.[3] Comparison of sequences encoded by the human $V_\lambda 1b$ and $C_\lambda 2$ germline genes with those derived from their equine $V_\lambda 1$ and $C_\lambda 2$ counterparts revealed a 76 and 69 percent homology, respectively.[4,5]

IV. CONCLUSION

Our findings provide the first conclusive evidence that a horse with systemic amyloid deposits had primary (AL) amyloidosis.

V. ACKNOWLEDGEMENTS

This study was supported, in part, by USPHS Research Grant CA10056 from the National Cancer Institute and the Aslan Foundation. A.S. is an American Cancer Society Clinical Research Professor.

VI. REFERENCES

1. Nout, Y.S., et al., Cardiac amyloidosis in a horse, *J. Vet. Intern. Med.*,17, 588, 2003.
2. Murphy, C.L., et al., Chemical typing of amyloid protein contained in formalin-fixed paraffin-embedded biopsy specimens, *Am. J. Clin. Pathol.*, 116, 135, 2001.
3. Solomon, A. and Weiss, D.T., Serologically defined V region subgroups of human λ light chains, *J. Immunol.*, 139, 824, 1987.
4. Home, W.A., Ford, J.E., and Gibson, D.M., L chain isotype regulation in horse. I. Characterization of Ig lambda genes, *J. Immunol.,* 149, 3927, 1992.
5. Kawasaki, K., et al., One-megabase sequence analysis of the human immunoglobulin λ gene locus, *Genome Res., 7*, 250, 1997.

RADIOIMAGING OF PRIMARY (AL) AMYLOIDOSIS WITH AN AMYLOID-REACTIVE MONOCLONAL ANTIBODY

J.S. Wall, M.J. Paulus[1], S.J. Kennel[2], S. Gleason[1], J. Baba[1], J. Gregor[3], D. Wolfenbarger, D.T. Weiss, and A. Solomon

Human Immunology & Cancer Program, University of Tennessee Graduate School of Medicine, Knoxville, TN; [1]Engineering Science and Technology Division, Oak Ridge National Laboratory, Oak Ridge, TN; [2]Life Sciences Division, Oak Ridge National Laboratory, Oak Ridge, TN; [3]Department of Computer Science, University of Tennessee, Knoxville, TN.

Currently, in the USA, there are no means to visualize amyloid deposits radiographically in patients with primary (AL) amyloidosis. Thus, there exist no non-invasive methods for diagnosing disease or assaying the precise extent of organ involvement ante-mortem. To address this limitation, we have investigated the capability of the amyloid-reactive mAb 11-1F4 to serve as an imaging agent [1]. This antibody, labeled with I-125, was injected i.v. into mice bearing amyloidomas composed of human AL fibrils and the animals were scanned 72 hrs later using micro-single photon emission computerized tomographic (SPECT) and micro-x-ray computerized tomographic (CT) instrumentation. The co-registered SPECT/CT images revealed that the [125]I-11-1F4 mAb localized selectively within the amyloidoma; further documentation that this agent was taken up by the tumors was obtained from analyses of post-mortem derived tissue. Based on the favorable biodistribution data and signal to noise ratio, we plan to determine if the 11-1F4 mAb, labeled with an appropriate isotope, can be used as a diagnostic reagent to ascertain disease extent and document therapeutic response in patients with primary (AL) amyloidosis.

METHODS

Generating the animal model: Human (hu)AL amyloid was extracted and purified from amyloid-laden organs harvested at autopsy using the water floatation method of Pras [2]. Balb/c mice received a sub-cutaneous injection of 50 mg of AL amyloid in a 2 mL volume of sterile phosphate-buffered saline. The amyloidoma, which was readily palpable and visible by microCT imaging, remained resident at the site of injection and became vascularized within 7 d. post-injection (data not shown).

Antibody labeling: The 11-1F4 mAb was radiolabeled with carrier free [125]I (Amersham), using limiting amounts of chloramine T. The mice received a single injection of ^{125}I-labeled 11-1F4 mAb (25 µg protein with a specific activity of 12 µCi/µg) intravenously in the tail vein according to the protocol schematic shown below (Fig 1). As a control, mice received ^{125}I-labeled MOPC-31C, an isotype matched irrelevant mAb.

CT data acquisition: CT data were acquired using a MicroCAT™ II (ImTek Inc., Knoxville, TN) - this circular orbit conebeam system is equipped with a 20-80 kVp microfocus x-ray source. It captures a 90 mm x 60 mm field of view using a 2048 x 3072 CCD array detector, optically coupled to a minR phosphor screen via a fiberoptic bundle [3]. Each CT dataset comprised 360 projections that were acquired in ~ 6 mins. and were reconstructed on 100 μm voxels using a recently developed modified version of the Feldkamp algorithm [4].

SPECT data acquisition: The microSPECT detector, developed by the Detector Group at Jefferson Laboratory (Newport News, VA), consists of a multi-anode photomultiplier tube coupled to a pixelated NaI(Tl) crystal array. The detector has a 55 x 55 mm active area which is discretized into a 64 x 64 array. The resolution was approximately 2mm FWHM when coupled with a parallel hole collimator. The SPECT datasets comprised 60 projections collected over the course of 30 to 60 minutes.

Image Co-registration: MicroSPECT and CT images were co-registered manually. Each mouse image carried fiducial markers supplied by capillaries filled with a solution of ^{125}I which appeared in the SPECT and CT images and could be co-registered in the 3-D reconstructions.

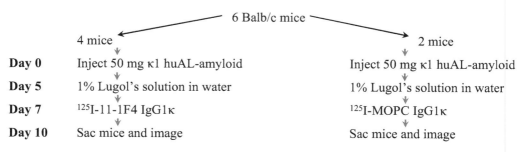

	6 Balb/c mice	
	4 mice	2 mice
Day 0	Inject 50 mg κ1 huAL-amyloid	Inject 50 mg κ1 huAL-amyloid
Day 5	1% Lugol's solution in water	1% Lugol's solution in water
Day 7	[125]I-11-1F4 IgG1κ	^{125}I-MOPC IgG1κ
Day 10	Sac mice and image	Sac mice and image

Figure 1. Protocol for SPECT/CT imaging of huAL amyloidomas in mice with [125]I-11-1F4

RESULTS AND DISCUSSION

Co-registered, CT and SPECT datasets were generated for groups of normal and huAL amyloid-bearing mice and compared (Fig 2). In normal Balb/c mice the images showed a non-specific diffuse distribution of the radiolabeled 11-1F4 mAb consistent with its retention in the blood pool (Fig 2, left).

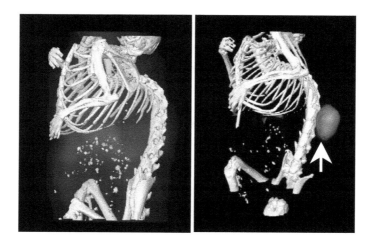

Figure 2. Distribution of [125]I-11-1F4 mAb in a normal mouse (left) and a mouse bearing a huAL κ1 amyloidoma.

The specific activity associated with harvested sections of liver, spleen, heart, kidney and muscle tissue, expressed as % injected dose per gram wet tissue, confirmed that the 11-1F4 mAb was not preferentially sequestered within these organs but rather was located predominantly within the blood pool.

In contrast, when [125]I-11-1F4 was administered to mice bearing huAL amyloidomas the antibody was readily detected associated with the amyloidoma in the SPECT images (Fig 2, arrow in right panel). Indeed, the specific activity associated with the amyloidoma was 5 – 10 times greater than that of the other organs and tissues analyzed (data not shown). At this significantly high signal-to-noise ratio the diffuse blood pool activity could be readily gated from the image and the [125]I-11-1F4-labeled huAL amyloidoma remained unequivocally apparent in the SPECT/CT images (Fig 2, right).

CONCLUSION

The 11-1F4 mAb has been previously shown to accelerate the rate of amyloidolysis in a murine model of AL amyloidosis [1]. The studies presented herein demonstrate that the 11-1F4 mAb co-localizes *in vivo* with huAL amyloid deposits which, in the earlier studies, resulted in their dissolution and also, as shown here, facilitates their visualization using high-resolution micro-imaging techniques. The specific interaction of the 11-1F4 mAb with AL amyloid fibrils presents a unique opportunity for its application as a non-invasive diagnostic and prognostic imaging tool for patients with this invariably fatal and devastating form of amyloid disease.

REFERENCES

1. Hrncic R. *et al.*, Antibody-mediated resolution of light chain-associated amyloid deposits. *Am. J. Pathol.* 157, 1239-1246, 2000.

2. Pras M., Schubert M., Zucker-Franklin D., Ramon A., Franklin E.C: The characterization of soluble amyloid prepared in water. *J Clin Invest.* 47, 924-933, 1968,

3. Paulus MJ, *et al.*, High resolution X-ray computed tomography: an emerging tool for small animal cancer research. *Neoplasia.* 2, 62-70, 2000.

4. Gregor J, Gleason S.S., Paulus, M.J. and Cates J., Fast Feldkamp reconstruction based on focus of attention and distributed computing. *Int. J. Imaging Systems and Technology.* 12, 229-234, 2003.

ACKNOWLEDGEMENTS

AS is an American Cancer Society Professor. This work was supported in part by NCI grant # CA10056 and NIBIB grant # EB00789.

PRIMARY AMYLOIDOSIS: CONFORMATIONAL INTERMEDIATES IN IMMUNOGLOBULIN LIGHT CHAIN PROTEINS UPON INTERACTION WITH CONGO RED UNDER PHYSIOLOGICAL CONDITIONS

Mary T. Walsh[1,3,4], Violet Roskens[2,4], Lawreen Heller Connors[2-4], and Martha Skinner[2,4]

Departments of Physiology and Biophysics[1], Medicine[2], Biochemistry[3], and Amyloid Treatment and Research Program[4], Boston University School of Medicine, 715 Albany Street, Boston, MA, 02118-2526.

Congo red (CR) is a dye used to identify amyloid fibril deposits in tissues (1). CR and its structural analogs have also been suggested as potential therapeutic agents (2). CR has been shown to block amyloid fibril formation and reverse fibril deposition in an animal model of Huntington's disease by unknown mechanisms (3). CR binding to Immunoglobulin Light Chains (IgLCs) may lead to partial unfolding and conformational intermediates. Our goal was to identify and biophysically characterize conformational intermediates in soluble IgLCs upon interaction with CR.

MATERIALS AND METHODS

κ_1 urinary IgLCs of known primary sequence were studied (4). Four were from patients with biopsy-proven primary amyloidosis. One was from an individual with multiple myeloma, and no amyloid disease (Table 1).

Table 1. Characteristics of Amyloidogenic and Non-Amyloidogenic Urinary κ_1 IgLCs used for CD Studies

κ_1 LCs	Disease Severity	Organ Involvement	Comments on LC Sequence
MM-96100	Multiple myeloma; no amyloid	N/A	Cys at Cys-214
AL-96066	Non-aggressive	Extensive soft tissue deposits; synovial membranes, muscle, kidney	Cys at Cys-214
AL-98002	Very aggressive	Heart, kidney, nervous system	Glycosylated at Asn-70 (Asp70Asn) Cys at Cys-214
AL-99067	Rapidly progressive	Extensive soft tissue, kidney	LC dimer
AL-00131	Aggressive	Moderate heart, lung, kidney, GI	Sequencing in progress

CR was added to each LC in phosphate buffered saline, pH 7.4. Immediately after CR addition LC conformation was monitored for 8 hours at 37°C by continuously recording far UV circular dichroic spectra from 250 to 200 nm on an Aviv 62DS CD Spectropolarimeter (Aviv Biomedical, Inc., Lakewood, New Jersey) equipped with a thermoelectric temperature controller (5). Time-dependent changes in β-sheet (217 nm, LC interior), and β-turn/ random coil / loops (206.5 nm, LC surface) were measured for 10 hours at 37°C on separate samples. LC thermal unfolding / stability at 217 nm for β-sheet, and 206.5 nm for β-turn/ random coil / loops were measured from 5-95°C. Thermal stability studies were performed on CR-containing LC samples prepared at 25°C 7 days prior to performing the study, thus ensuring that CR-induced conformational changes were complete. LC concentration = 0.4 mg/ml (6); CR concentration=114 uM; pathlength of cuvette = 0.05 cm. Isothermal titration calorimetry confirmed the binding of CR to each LC.

RESULTS

In the absence of CR all LCs exhibited far UV CD spectra typical of predominantly β-sheet proteins with negative minima at 217 nm. Upon addition of CR, a rapid (within minutes), dramatic change in conformation was observed in which the minimum became more pronounced (more negative), and was shifted to lower wavelengths, typical of a protein that has undergone a conformational change from β-sheet to random coil plus β-sheet. Figure 1 presents CD results for the amyloidogenic LC, AL-98002. In the absence of CR (Left Panel), AL-98002 is predominantly β-sheet, with a negative minimum at 217 nm. Addition of CR alters the spectrum significantly suggesting a change in conformation that increases the amount of random coil. Presented in the inset to the Left Panel is the kinetic response of AL-98002 to CR. CR rapidly alters the surface as shown by the rapid reduction in [θ] at 206.5 nm upon addition of CR. Absence of a kinetic response in [θ] at 217 nm suggests the β-sheet interior is stable toward CR despite its rapid effect on the LC surface. AL-98002 exhibits remarkable thermal stability (Right Panel), perhaps due to the presence of glycosylation. In the absence of CR thermal unfolding is not observed at 217 nm, suggesting a very stable β-sheet interior. In the presence of CR unfolding is observed at high temperature. In the absence of CR cooperative unfolding of the surface (206.5 nm) is observed. In the presence of CR, unfolding cooperativity and T_m are greatly reduced suggesting significant CR-dependent surface destabilization.

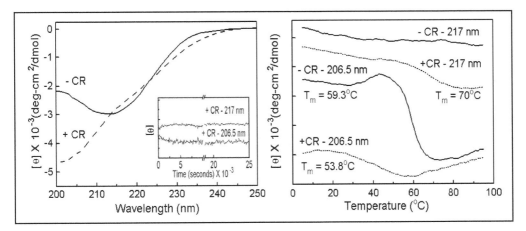

Figure 1. Far UV CD Studies of AL-98002 ± CR.

Table 2 summarizes CD conformation and thermal stability studies for five κ_1 LCs. Molar ellipticity 217 nm (β-sheet) and 206.5 nm (β-turn / random coi l/ loops) values from spectra measured at 37°C, and the mid-point temperatures (T_m) of the thermal unfolding curves at 217 nm and 206.5 nm ± CR are presented. The conformation and stability of these five κ_1 LCs are significantly altered by CR.

Table 2. Summary of LC Conformation and Thermal Stability in PBS, pH 7.4 ± CR.

LC	[θ]$_{217\ nm}$	T$_m$°C, 217 nm	[θ]$_{206.5\ nm}$	T$_m$°C, 206.5 nm
MM-96100				
PBS	-2603	47.4	-987	49.7
+ CR	-2935	72	-3200	37
AL-96066				
PBS	-3153	48.2	-928	~52
+ CR	-4083	37, 67	-4831	Thermally stable
AL-98002				
PBS	-2778	Thermally stable	-2619	59.3
+ CR	-2522	70.1	-4196	53.8
AL-99067				
PBS	-3141	49.4	+250	50.6
+ CR	-3230	51.5	-3453	44.2
AL-00131				
PBS	-1594	52	-77	52
+ CR	-2001	63.4	-1617	41.6, 68.2

CONCLUSIONS

CR binds rapidly to κ$_1$ amyloidogenic and non-amyloidogenic IgLCs. Partial unfolding of LCs is observed by monitoring kinetic and thermal stability of the β-turn / random coil-rich surface of LCs, and destabilization of their β-sheet-rich interiors. The kinetics of the interaction of CR with LCs and subsequent changes in conformation and thermal stability are dependent on the primary sequence of the LC and post-translational modifications. LC structural features critical for preventing partial unfolding, conformational intermediates, and oligomerization must be understood to design therapeutic agents that stabilize the native LC structure, and protect LCs from molecular interactions that promote unstable intermediates and fibril formation.

ACKNOWLEDGEMENTS

This work was supported by the National Institutes of Health, P01-HL26335 (MTW), P01-HL-068705 (MS, LHC, and VR), the Gerry Foundation, and George Burr, Jr. Amyloid Research Fund.

REFERENCES

1. H Puchtler, F Sweat, and M Levine, On the binding of Congo red by amyloid, J. Histochem. Cytochem. , 10:355-364, 1962.

2. MA Findeis, Approaches to discovery and characterization of inhibitors of amyloid β-peptide polymerization, Biochim. Biophys. Acta, 1502: 76-84, 2000.

3. I Sanchez, C Mahlke, and J Yuan, Pivotal role of oligomerization in expanded polyglutamine neurodegenerative disorders, Nature, 421: 373-379, 2003.

4. CM Chung, LH Connors, MD Benson & MT Walsh. Biophysical Analysis of Normal Transthyretin: Implications for Fibril Formation in Senile Systemic Amyloidosis, Amyloid: J. Protein Folding Disorders, 8: 75-83, 2001.

5. A Lim, J Wally, MT Walsh, and CE Costello, Identification and location of a cysteinyl post translational modification in an amyloidogenic Kappa 1 light chain protein by electrospray ionization and matrix-assisted laser desorption/ionization mass spectrometry, Anal. Biochem., 295: 45-56, 2001.

6. OH Lowry, NJ Rosebrough, AL Farr and RJ Randall, Protein measurement with the Folin phenol reagent, J. Biol. Chem., 193: 265-275, 1951.

IMMUNOGLOBULIN LIGHT CHAIN VARIABLE REGION GENES DEMONSTRATE EVIDENCE FOR ANTIGEN SELECTION IN PRIMARY AMYLOIDOSIS

T. Prokaeva, J. E. Ward, E. Fingar, M. Kaut, P. Smith, C. O'Hara, L.H. Connors, M. Skinner, D.C. Seldin

Boston University School of Medicine, Boston MA USA

INTRODUCTION

Primary amyloidosis (AL) is a disorder characterized by overproduction and tissue deposition of monoclonal immunoglobulin (Ig) light chain (LC) fragments as insoluble amyloid fibrils, which leads to progressive organ failure. Small numbers of monoclonal bone marrow plasma cells are responsible for the secretion of amyloidogenic LCs [1]. In several small series of patients with AL amyloidosis, the usage of unusual germline gene repertoire for the rearrangement of amyloidogenic light chain variable region (VL) genes has been shown [2,3]. Only limited information is available about the evaluation of somatic hypermutation process on VL genes in AL amyloidosis [4,5]. However, analysis of somatic mutations in amyloidogenic VL regions can provide insights into the role of antigen (Ag) for the selection and clonal outgrowth of clonal amyloidogenic lymphoid cells. Thus, in the absence of positive or negative pressure on Ig VL regions, a random mutational process would result in an even distribution of replacement (R) and silent (S) mutation throughout the coding sequence [6]. Ag-selected VL genes demonstrate a high frequency of R mutations in complementary-determining regions (CDRs) than in framework regions (FRs), whereas preservation of a functional Ig molecule is associated with a scarcity of R mutations and a higher frequency of S mutations in FRs [6]. Overall, Ag selection analysis can help to trace the developmental stage at which transformation of B-cells has occurred [7]. The purpose of this study was to characterize Ig VL gene usage in AL amyloidosis and to determine if AL plasma cell clones have been subjected to antigen selection.

MATERIALS AND METHODS

The patient population consisted of 150 patients with biopsy-proven AL amyloidosis. Nine AL patients had multiple myeloma. Median age of patients was 58.6 years old (range 32.9-80.3). Sixty-three percent of patients were males and 37% were females. mRNA was isolated from bone marrow samples and subjected to RT-PCR amplification, cloning, and sequencing. The LC isotype, donor germline VL gene segments, and frequency of R vs. S mutations were determined using standard and novel Web-based databases and tools.

RESULTS AND CONCLUSIONS

Of the 150 specimens, 109 expressed λ (72.7%) and 41 expressed κ (27.3%) genes. Figure 1 demonstrates distribution of different LC families in our population of patients with AL amyloidosis. λ1 (24%) and κ1 (23%) accounted for almost half of the VL sequences, followed by λ2 and λ6 (16%), and λ3 (15%). A variety of germline gene donors were used for VL genes rearrangement (Table 1 and Table 2). Two germline genes, namely 1c and 6a, were significantly over-represented and 3h germline gene was significantly under-represented in VLλ monoclonal bone marrow plasma cells. Overall, 43.3% of functional λ and 22.5% of functional κ germline genes were use in amyloidogenic VL gene rearrangement.

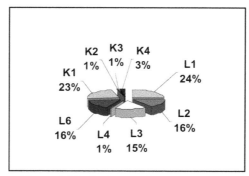

Figure 1. Distribution of different LC families in 150 patients with AL amyloidosis.

Table 1. VLλ germline gene repertoires in AL and in normal polyclonal bone marrow plasma cells

Germline gene	AL group, n=109, (%)	Control group [2], n=264, (%)	P -value
6a	22.0	2	0.0001
1c	14.7	4	0.0008
3h	5.5	15	0.02
2a2	17.4	12	ns
3r	11.9	8	ns
1e	7.3	8	ns
1b	5.5	6	ns
1g	5.5	5	ns
3l	3.7	4	ns
2c	2.8	3	ns
2b2	1.8	5	ns
1a, 4b	0.9	0.8	ns

Table 2. VLκ germline gene repertoire in 41 AL bone marrow plasma cells

Germline gene	Frequency, %
K1: 018/08	34.2
012/02	19.5
LFVK431	12.2
L12, L15	7.3
L5	4.9
K2: A17; K3: A27	2.4
K4: B3	9.8

Somatic mutation rate and estimation of Ag selection pressure was evaluated in 141 VL amyloidogenic sequences. All analyzed sequences were found to have somatic mutations. The percent of mutation varies from as low as 1.0% to as high as 13.3% with mean frequency per sequence of 4.96% (95%CI: 4.62%-5.30%). The assessment of Ag selection on VL genes was performed using multinominal distribution model, as suggested by Lossos et al. [6,7].

Table 3. Examples of R mutations distribution in the CDR and FR regions of VL genes

Selection type	Sequence No	CDR, no. of mutations		P_{CDR}	FR, no. of mutations		P_{FR}
		R_{obs} *	R_{exp} **		R_{obs}	R_{exp}	
Ag selection	00-024	7	3.0	0.02	5	8.4	0.04
No Ag selection	02-107	5	3.9	0.17	7	8.7	0.14

* R_{obs} - number of observed R mutations in cloned VL gene
** R_{exp} - number of expected R mutations that could occur by chance in donor germline gene

Table 3 shows two examples of R mutations distribution in the CDR and FR regions of amyloidogenic VL genes indicating evidence for Ag selective pressure in sequence no. 00-024 and no evidence for Ag selection in sequence no. 02-107. Overall, 58.2% percent of the sequences had either statistically greater than expected frequency of R mutations in the CDRs or statistically lower that expected frequency of R mutations in the FRs or combination of both consistent with antigen selection pressure.

DISCUSSION AND CONCLUSIONS

Ig VL gene usage in our AL patients was markedly different from the normal repertoire, with overrepresentation of a small number of germline genes, specifically 1c and 6a. The spectrum of overrepresented germline genes in this cohort of patients appeared to be slightly different from what has been found in another representative group of 55 AL patients from Italy, where 3r and 6a germline genes were significantly overrepresented [2].

No significant intraclonal variation was observed in the expressed VL regions; homogeneous somatic nucleotide mutations were found in all sequences. However, the mean mutation rate of 4.96% in this series was significantly lower than 6.4% (95%CI: 5.7%-7.2%) (p<0.0001) reported by V.Perfetti et al. [2].

The type and distribution pattern of somatic mutations were used to assess antigen selective pressure on amyloidogenic VL genes. Fifty-eight percent of sequences showed evidence of antigen selection, as reported in smaller series of specimens [4, 5]. It is well known that somatically mutated variable region genes are a hallmark of germinal center B-cells [7]. Evidence of antigen selective pressure on amyloidogenic VL genes as well as homogeneity of somatic mutations in our cohort sequences support the concept that monoclonal transformation of most amyloidogenic plasma cells occurs after B-cell maturation and clonal selection in the lymphoid follicle.

This work was supported by NIH grant: PO1 HL 68705

REFERENCE

1. Perfetti V. et al. AL amyloidosis: characterization of amyloidogenic cells by anti-idiotypic monoclonal antibodies, *Lab Invest.*, 71, 853, 1994

2. Perfetti V. et al. Analysis of Vλ- Jλ expression in plasma cells from primary (AL) amyloidosis and normal bone marrow identifies 3r (λIII) as a new amyloid-associated germline gene segment, *Blood,* 100, 948-953, 2002.

3. Comenzo R.L. et al. Immunoglobulin light chain variable region germline gene use and plasma cell burden in AL amyloidosis contribute to the tropism of organ involvement. *Abst. 42nd Annual Meeting of the American Society of Hematology*, December, 2000

4. Perfetti V. et al. Evidence that amyloidogenic light chains undergo antigen-driven selection, *Blood*, 91, 2948-2954, 1998.

5. Abraham R.S. et al. Immunoglobulin variable region germline genes and mutational hotspots potentially influence protein misfolding in light chain amyloidosis, *Abst. 44nd Annual Meeting of the American Society of Hematology*, December, 2002

6. Lossos I.S. et al. The inference of antigen selection on Ig genes, *J.Immunol.*, 165, 5122-5126, 2000.

7. Lossos I.S. et al. Molecular analysis of immunoglobulin genes in diffuse large B-cell lymphomas, *Blood*, 95, 1797-1803, 2000.

DETECTION AND CHARACTERIZATION OF POST-TRANSLATIONAL MODIFICATIONS IN AL LIGHT CHAIN PROTEINS BY MASS SPECTROMETRY

R Théberge1,2, A Lim1,2, Y Jiang1, LH Connors2, J Eberhard2, M Skinner2, CE Costello1,2

[1]Mass Spectrometry Resource and [2]Amyloid Treatment and Research Program Boston University School of Medicine, Boston, MA 02118 USA

INTRODUCTION

The structural heterogeneity exhibited by immunoglobulin light chains (LC) involved in AL amyloidosis has been a subject of interest in establishing a pathogenesis of the disease. This heterogeneity manifests itself mainly in the variable domain of the protein and as post-translational modifications (PTMs). Analysis of PTMs requires the application of mass spectrometry (MS) methods for detection and characterization. We have implemented such methods for purified LCs isolated from the urine of patients with AL. Electrospray ionization (ESI) and matrix-assisted laser desorption/ionization (MALDI) MS, in combination with enzymatic digestions, were used to detect and characterize PTMs in these proteins. We have identified instances of C-terminal cysteinylation, Kappa chain homodimerization and glycosylation. These results show that mass spectrometry methods are ideally suited for the detection and characterization of PTMs in AL LC proteins.

MATERIALS AND METHODS

Reduction was performed with a 10-fold molar excess of DTT over expected disulfides (100 mM NH_4HCO_3, pH 8, 30 min, 37 °C). Asp-N (enzyme:substrate, 1:300) digestion was carried out in 100 mM NH_4HCO_3 (pH 8, 37 °C, 5 hr). Lys-C (1:100) digestion was performed in 100 mM NH_4HCO_3 (pH 8, 37 °C, 7 hr).

ESI mass spectra of the intact proteins before and after reduction were obtained in the positive-ion mode using a Micromass Quattro II triple quadrupole MS. All enzymatic digests were mixed with 2,5-dihydroxybenzoic acid and analyzed with a Finnigan MAT Vision 2000 MALDI time-of-flight (TOF) MS, using delayed extraction in the linear mode. PNGase F was used to release *N*-linked glycans from the glycosylated light chain. The released glycans were permethylated using standard protocol and analyzed using the Reflex IV MALDI-TOF MS. Peptides of interest were sequenced with an Applied Biosystems/MDS-SCIEX QSTAR™ Pulsar quadrupole/orthogonal acceleration TOF MS.

RESULTS and DISCUSSION

Case 1. ESI mass analysis of the 24-kDa protein indicated its mass to be 23,395 Da, which does not agree with the mass 23,278 Da, calculated from its cDNA-predicted amino acid sequence, allowing for the presence of two disulfide bonds. Reduction of the native protein with DTT produced a mass 23,282 Da (calc. 23,282 Da), corresponding to the amino acid sequence of the gene product containing no disulfide bonds. The −113 Da mass shift is consistent with a loss of a disulfide-linked cysteine and reduction of 2-3 disulfide bonds. MALDI-TOF MS peptide mapping of the Asp-N and Lys-C digests confirmed the sequence of the 24-kDa protein and located the disulfide-linked cysteine at Cys214. Figure 1A shows a MALDI-TOF mass spectrum of a Lys-C digest of the 24-kDa protein. A peak at *m/z* 932.0 was observed, corresponding to a peptide having residues 208-214 containing the cysteinyl post-translational modification disulfide-linked to Cys214. The facile loss of this *S*-cysteinylation due to prompt fragmentation in the MALDI ion source [1] produced the peak observed at *m/z* 812.7. To verify that the peak at *m/z* 932.0 contained disulfide linkage, an aliquot of the Lys-C digest was treated with DTT. The peak at *m/z* 932.0 disappeared, and the relative abundance of the peak at *m/z* 812.9 increased. To verify the position of the *S*-cysteinylation, the Lys-C digest mixture was analyzed using the QSTAR™. The [M+2H]$^{2+}$ ion *m/z* 466.178 of the peptide containing residues 208-214 suspected of having *S*-cysteinylation at Cys214 was isolated and fragmented using collision-induced dissocation to obtain sequence information This MS/MS analysis confirmed the presence of the *S*-cysteinylation at Cys214 (1). Many previous studies have focused on determination of the primary structure of the light chain, and reducing agents have been used routinely in the purification process. Post-translational modifications, such as cysteinylation, of these amyloidogenic LCs may have been altered.

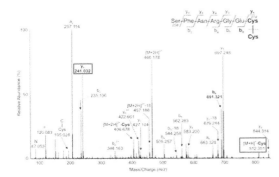

Fig. 1. ESI-CID MS/MS of the [M+2H]2+ (*m/z* 466.178) of the *S*-cysteinylated peptide containing residues 208-214 from the Lys-C digest of the 24-kDa protein.

Case 2. One of the urinary proteins from an AL patient was identified as a covalent homodimer of κ1 light chain. This is demonstrated in **Figure 2** where the effect of DTT on the ESI-MS spectra of the intact protein. Kappa 1 light chains are known to exist primarily as monomers, and λ light chains are found mainly as covalent dimers. Lambda light chains have been shown to be more amyloidogenic than κ light chains. Dimerization of the light chain is hypothesized to be the first step in the polymerization of the protein to amyloid fibrils. Covalent modification of the κ1 light chain may play a role in the protein's amyloidogenicity.

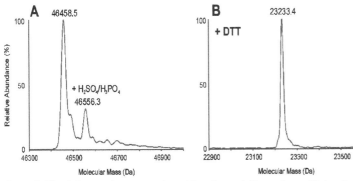

Fig. 2. Deconvoluted ESI-MS spectrum of the κ 1 light chain dimer before (**A**) and after (**B**) treatment with DTT. The effect of DTT treatment indicates that the dimer is disulfide bonded

Case 3. The LC sample was analyzed for glycosylation using a Molecular Probes glycoprotein gel stain kit and tested positive. Glycosylation of both κ and λ LCs has been reported (3, 5). Using MALDI MS, a mass of 22195

Da was obtained for the intact LC. The peak broadness indicated heterogeneity. MALDI MS of the sample after treatment with PNGase F produced a peak at 17624 Da, suggesting that the heterogeneity was due to *N*-linked glycosylation. After PNGase F-released glycans were permethylated using standard protocol (4), MALDI MS of the released and permethylated glycans indicated the presence of bi- and triantennary structures, similar to those found by Omtvedt et al. (5) investigating AL-associated LCs.

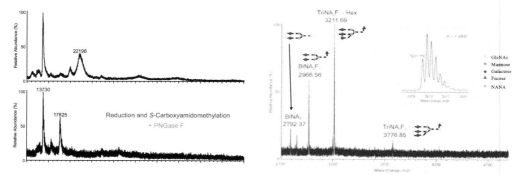

Figure 3. (LEFT) The MALDI-TOF mass spectra of the protein sample (Top) before and (bottom) after treatment with DTT, iodoacetamide, and PNGase F. (RIGHT) MALDI-TOF mass spectrum of the released permethylated glycans from the glycosylated light chain.

CONCLUSION

The underlying factors responsible for initiating the overproduction of the light chain are not fully understood. The mechanism of amyloid fibril formation in AL amyloidosis is unknown. Although *in vitro* studies have shown that amino acid substitutions in the variable region could cause the light chain to undergo a conformational change into amyloid fibrils (6), post-translational modification may also be involved. The use of reduction agents in the purification and/or sample preparation of light chain proteins for structural analysis may destroy relevant post-translational modifications. The mass spectrometry methods presented here enable high-sensitivity determination of post-translational modifications.

REFERENCES

1. Lim A, Wally J, Walsh MT, Skinner M, Costello CE., *Anal. Biochem.*, (2001), **295**, 45-56.

2. Berggård, I. and Peterson, P.A. (1969) *J. Biol. Chem.* **244**, 4299-4307.

3. Milstein, C. P. and Milstein, C. (1971). Glycopeptides from human κ-chains. *Biochem. J.* **121**: 211-215.

4. Ciucanu, I. and Kerek, F. (1984) *Carbohydrate Res.* 131, 209-217

5. Omtvedt, L. A., Bailey, D., Renouf, D. V., Davies, M. J.; Paramonov, N. A., Haavik, S.; Husby, G., Sletten, K. and Hounsell, E. F.   (2000). *Amyloid: Int. J. Exp. Clin. Invest.* **7**: 227-244.

6. Raffen, R., Dieckman, L. J., Szpunar, M., Wunschl C., Pokkuluri, P. R., Dave, P., Stevens, P. W., Cai, X., Schiffer, M. and Stevens, F. J. (1999).. *Protein Sci.* **8**: 509-517.

ACKNOWLEDGEMENTS

This work was supported by NIH grants P01-HL68705 (MS), P41-RR10888 (CEC), S10-RR10493 (CEC), S10-RR15942 (CEC), the Gerry Foundation (MS), and the Young Family Amyloid Research Fund (MS).

GLYCOSYLATION STUDIES OF MONOCLONAL IMMUNOGLOBULIN DEPOSITION DISORDERS.

L. A. Omtvedt[1], L. Royle[3], G. Husby[2], K. Sletten[1], C. Radcliffe[3], D. J. Harvey[3], R. A. Dwek[3], P. M. Rudd[3].

1 Department of Molecular Biosciences, University of Oslo, Norway
2 Department of Rheumatology/ Institute of Immunology, Rikshospitalet, University of Oslo, Norway
3 Glycobiology Institute, Department of Biochemistry, University of Oxford, United Kingdom

INTRODUCTION

Glycosylation of monoclonal immunoglobulins in protein deposition diseases is not well documented. To determine whether specific glycoforms are predisposed to deposit on organs, we have analysed immunoglobulin heavy chains from two patients with overproduction of monoclonal immunoglobulins.

MATERIAL AND METHODS

Patient material. Patient GL was diagnosed with seronegative RA, three years later a monoclonal Ig component (free γ3-heavy chain) was detected in her serum and urine (1). The free heavy chain consisted of the hinge region together with CH2 and CH3 domains, in a dimeric form, which lacked both the variable region and the CH1 domain. This free heavy chain was found deposited in the synovial tissue. The patient was started on a treatment protocol for multiple myeloma, and was very responsive to this treatment and is now in complete remission (1). Patient SKA was diagnosed with systemic amyloidosis confirmed by kidney biopsy. Deposits in tongue, liver, blood vessels, peripheric nerves and bone marrow were also found. The patient excreted a complete IgG2 chain in the urine. The patient has been treated with high dose melphalan supported by autologous bone-marrow transplantation and is in partial remission. The progression of amyloid deposition stopped, with clear regression of hepatic amyloid and no signs in the renal graft. The predominant IgG is still the monoclonal kappa type in both serum and urine. It is not known if the entire IgG is deposited as amyloid fibrils or only the light or heavy chain.

Cabohydrate isolation and analysis. Serum from patient GL and patient SKA was purified using a HiTrap protein G column (Amersham Biosciences). The serum from both patients was also collected after treatment. The glycoproteins was run on SDS-PAGE, the protein bands visualised by coomassie blue and the gel bands cut from the gel (2). The Oligosaccharides were extracted from the gel pieces and released glycans were labelled with 2-AB (2-amino-benzamide) using a LudgerTag 2-AB labelling kit (Ludger Ltd, Oxford, UK) (2,3) . Excess label was removed by ascending chromatography in acetonitrile on Whatmann 3MM paper strips. 2-AB labelled glycans were digested with an array of enzymes as described in Rudd et al. (2). The labelled glycans were analysed by NPHPLC using an external standard of hydrolysed and 2AB-labelled glucose oligomers, by Weak anion exchange (WAX) and by liquid chromatography-electrospray ionisation-mass spectrometry LC-ESI-MS (4-6).

RESULTS

The glycan profile (Figure 1) showed that the monoclonal truncated heavy chain GL (GL-HCDD) contained fully galactosylated biantennary glycans with significantly less fucose but more sialic acid that IgG3 from healthy controls (7). The glycan profile (Figure 1) from serum IgG from the same patient in remission resembled a typical rheumatoid arthritis profile and contained a predominant fucosylated agalactosylated (G0) structure. The glycosylation profile of the heavy chain isolated from urine was the same as that from the serum. The glycan structures from the monoclonal IgG2 molecule isolated from a patient (SKA) with amyloidosis were similar to those of normal IgG in levels of fucosylation, but contained higher than normal levels of G0 type sugars. The carbohydrate profiles (Figure 1) of SKA-treated and SKAs-amyloid revealed no major differences.

Figure 1. Normal Phase HPLC profiles of N-linked glycans from in-gel PNGaseF digestion

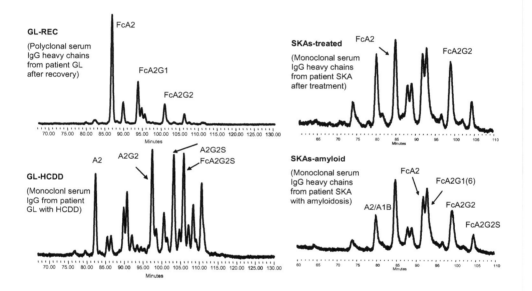

DISCUSSION

The under-fucosylation of the N-glycans of protein GL-HCDD may render the protein more susceptible to deposition by increased flexibility of the Gal 6' antenna and thereby revealing amino acids that normally are hidden by the carbohydrate structure and thus lead to enhanced protein-protein interaction that again may lead to protein deposition (8). Furthermore, IgGs without fucose have an increased binding to FCγRIIIa and this may lead to enhanced cytotoxicity of protein GL (9). In RA, the synovial tissue of affected joints is known to contain activated lymphocytes as well as macrophages, and activated B lymphocytes may be relatively prone to develop clonality (10). This may explain the fact that the patient, diagnosed with seronegativ RA in 1988 prior to be diagnosed with HCDD, developed an overexpression of a monoclonal IgG with a specific oligosaccharide profile. Furthermore, this also indicate that the HCDD may be a consequence of the rheumatoid disease and not the other way around, although this is still elusive (11).The function of the carbohydrate structure in SKA IgG2 remains to be elucidated. However even though the patient continues to produce a monoclonal IgG after the autologous bone marrow transplant, no further progression of amyloid deposit is found. Thus, one may speculate that the pathogenic mechanisms, which govern the progression of this disease, may be other than the glycoprotein alone.

REFERENCES

1. Husby, G., Blichfeldt, P., Brinch, L., Brandtzaeg, P., Mellbye, O. J., Sletten, K., and Stenstad, T., Chronic arthritis and gamma heavy chain disease: coincidence or pathogenic link?, *Scand J Rheumatol* 27 (4), 257-64, 1998.

2. Rudd, P. M., Colominas, C., Royle, L., Murphy, N., Hart, E., Merry, A. H., Hebestreit, H. F., and Dwek, R. A., A high-performance liquid chromatography based strategy for rapid, sensitive sequencing of N-linked oligosaccharide modifications to proteins in sodium dodecyl sulphate polyacrylamide electrophoresis gel bands, *Proteomics* 1 (2), 285-94, 2001.

3. Bigge, J. C., Patel, T. P., Bruce, J. A., Goulding, P. N., Charles, S. M., and Parekh, R. B., Nonselective and efficient fluorescent labeling of glycans using 2-amino benzamide and anthranilic acid, *Anal Biochem* 230 (2), 229-38, 1995.

4. Guile, G. R., Rudd, P. M., Wing, D. R., Prime, S. B., and Dwek, R. A., A rapid high-resolution high-performance liquid chromatographic method for separating glycan mixtures and analyzing oligosaccharide profiles, *Anal Biochem* 240 (2), 210-26, 1996.

5. Zamze, S., Harvey, D. J., Chen, Y. J., Guile, G. R., Dwek, R. A., and Wing, D. R., Sialylated N-glycans in adult rat brain tissue--a widespread distribution of disialylated antennae in complex and hybrid structures, *Eur J Biochem* 258 (1), 243-70, 1998.

6. Royle, L., Mattu, T. S., Hart, E., Langridge, J. I., Merry, A. H., Murphy, N., Harvey, D. J., Dwek, R. A., and Rudd, P. M., An analytical and structural database provides a strategy for sequencing O-glycans from microgram quantities of glycoproteins, *Anal Biochem* 304 (1), 70-90, 2002.

7. Jefferis, R., Lund, J., Mizutani, H., Nakagawa, H., Kawazoe, Y., Arata, Y., and Takahashi, N., A comparative study of the N-linked oligosaccharide structures of human IgG subclass proteins, *Biochem J* 268 (3), 529-37, 1990.

8. Stubbs, H. J., Lih, J. J., Gustafson, T. L., and Rice, K. G., Influence of core fucosylation on the flexibility of a biantennary N-linked oligosaccharide, *Biochemistry* 35 (3), 937-47, 1996.

9. Shinkawa, T., Nakamura, K., Yamane, N., Shoji-Hosaka, E., Kanda, Y., Sakurada, M., Uchida, K., Anazawa, H., Satoh, M., Yamasaki, M., Hanai, N., and Shitara, K., The absence of fucose but not the presence of galactose or bisecting N-acetylglucosamine of human IgG1 complex-type oligosaccharides shows the critical role of enhancing antibody-dependent cellular cytotoxicity, *J Biol Chem* 278 (5), 3466-73, 2003.

10. Husby, G., Amyloidosis, in *Oxford Textbook of Rheumatology*, Madisson IJ, I. D., Woo P, Glass DN Oxford Medical Publications, Oxford, 1998, pp. 1433-1444.

11. Husby, G., Is there a pathogenic link between gamma heavy chain disease and chronic arthritis?, *Curr Opin Rheumatol* 12 (1), 65-70, 2000.

STRUCTURAL LOCALIZATION OF LIGHT CHAIN AMYLOIDOGENICITY

J.S. Wall, C.L. Murphy, A. Tomaszewski, B. Crombie and A. Solomon

Human Immunology & Cancer Program, University of Tennessee Graduate School of Medicine, Knoxville, TN

Although primary (AL) amyloidosis has been preferentially associated with the λ isotype, as well as certain immunoglobulin variable domain (V_L) subgroups, there is only limited information on the molecular factors that cause light chains to form amyloid. We now report the results of our studies on a unique AL protein in which we have identified a fibrillogenic region within the V_L. This component, which was extracted from the thyroid and heart of a patient (DEN), eluted as a single peak by reverse phase-HPLC and was found by LC-MS and amino acid sequencing to consist of a $V_\lambda 1$ heterodimer that, remarkably, lacked the residues located between positions 46 and 76. Disassociation of the intra-chain Cys23-Cys88 disulfide bond by reduction and alkylation yielded 2 discreet peptides: DEN (1-45) and DEN (77-111). As evidenced in a ThT-based microassay, the DEN (1-45) peptide, as well as the native V_L heterodimer, formed fibrils within several hours when incubated at 37°C in 0.15 M NaCl, whereas DEN (77-111) was inert under the same conditions. Based on these results, we posit that the primary structural features responsible for the amyloidogenicity of protein DEN, and perhaps other light chains, reside within the N-terminal 45 residues. Contra wise, those located in the C-terminal portion of the V_L domain are incorporated into the fibril structure and therefore protected from further proteolytic digestion. The localization of amyloidogenic regions of light chains can form the basis for design of therapeutic agents for patients with primary (AL) amyloidosis.

METHODS

Amyloid extraction and chemical characterization

The method used to prepare water-soluble amyloid extracts was essentially that described by Pras et al (1). Briefly, fresh-frozen (-80° C) thyroid obtained post-mortem from patient DEN was in ~300ml of cold saline with a Virtis-Tempest apparatus (Virtis, Gardiner, NY). The homogenate was centrifuged at 6° C for 30 minutes at 17,000 rpm and residual saline-soluble material was removed by repeated homogenization and washing until the resultant supernatant had an OD of <0.10 at A_{280}. The pellet was then repeatedly homogenized, washed with cold, deionized water, centrifuged, and the amyloid-containing supernatants lyophilized. The amount of protein recovered represented ~one-third to one-fifth the weight of the starting material.

The light chain composition and V_L subgroup of the amyloid was determined by amino acid sequencing (Procise Protein Sequencing System, Applied Biosystems, Foster City, CA) and ionizing mass spectroscopy (PE SCIEX API 150 EX, Perkin Elmer, Norwalk, CT) of HPLC-separated peptides obtained by trypsin digestion of reduced and pyridylethylated protein extracted from the water soluble material with $6M$ guanidine HCl (2). The amino acid sequence of the complete V_L domain was determined by translating the nucleotide base sequence of complimentary (c)DNA isolated from the patients plasma cells.

RESULTS AND DISCUSSION

Amino acid sequence analysis of the DEN heterodimer extracted from the fibril confirmed it as a λ1 light chain protein. Gel electrophoresis performed under reducing and non-reducing conditions indicated that the light chain was composed of two disulfide-linked peptides comprising residues 1-45 and 77-111. Residues 46-76 inclusively had been excised, presumably as a result of tryptic-like enzymatic activity. The intervening protein sequence of the V$_L$ DEN domain was established by cDNA sequencing (Figure 1).

```
                1           11          21        27d          35
LOC(λ1):      QSVLTQPPS. ASGTPGQRVT ISCSGSSSNI G....ENSVT WYQHLSGTAP
DEN cDNA:        V          V   A R T A              N A N     Q P K
DEN (amyloid):   V          V   A R T A              N A N     Q P K


                45          55          65        73           83
LOC(λ1):      KLLIYEDNSR ASGVSDRFSA SK..SGTSAS LAISGLQPED ETDYYCAAWD
DEN cDNA:      V    Y DLL  PA          G                   R   S     G
DEN (amyloid): --------- ---------- ---------- ----        S     G


                93          99
LOC(λ1):      DSLDV..AVF GTGTKVTVLG
DEN cDNA:       SA  L    G    L    QPK
DEN (amyloid):  SA  L    G    L    QPK
```

Figure 1. Comparison of the amino acid sequence of λ1 protein LOC with that encoded by cDNA DEN and amyloid protein DEN. The DEN heterodimer was stabilized by an interchain disulfide between Cys23 and Cys88 residues. The cleaved fragment is denoted by the dashed line.

The DEN heterodimer, when incubated in physiological saline solution, pH 7.5 at 37° C, formed fibrils which exhibited tinctorial and structural characteristics of amyloid fibrils, i.e., thioflavin T fluorescence emission at 490 nm (when excited at 450 nm) and aggregates that appeared as twisted unbranching fibrils with a diameter of ~ 9 nm when viewed by electron microscopy (Figure 2). The rate of fibrillogenesis of the DEN heterodimer was notably faster than that of the highly fibrillogenic V$_L$λ6 protein WIL, as manifest in the length of the lag time (3). The reduced, alkylated and purified N-terminal fragment, DEN(1-45), exhibited an even greater propensity to form fibrils under the same conditions. Indeed, no lag time in the reaction was evidenced, perhaps indicative of a downhill polymerization mechanism as is observed during transthyretin polymerization (see e.g., Hurshman et al. this volume). In stark contrast, the C-terminal DEN molecule (77-111) was completely refractory to fibrillogenesis during the course of the experiment. The structural and functional characteristics of the DEN protein and its fragments are summarized in Table 1.

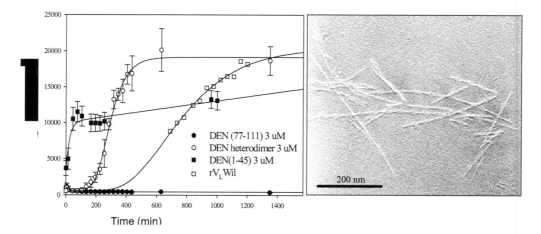

Figure 2. *In vitro* fibrillogenesis of the DEN heterodimer and the isolated N- and C-terminal peptides. Fibrils formed by the DEN heterodimer were viewed by electron microscopy (right).

Table 1. Summary of structural features and fibrillogenic potential of DEN heterodimer and the N- and C-terminal peptides

Feature	Heterodimer	DEN (1-45)	DEN (77-111)
Theoretical pI	4.93	9.79	3.96
MW (expected)	8585.2	4753.3	3951.3
MW (observed)	8586.0	4753.0	3950.0
Residue number	83	46	37
Fibrillogenicity	+	+++	-

Based upon these data we posit that the N-terminal portion of light chains dominates the propensity for fibrillogenesis. Contrawise, the C-terminal portion of the V_L domain does not support self-association into fibrils and when associated with the N-terminal has an inhibitory effect on fibril growth.

REFERENCES

1.   Pras M., Schubert M., Zucker-Franklin D., Ramon A., and Franklin E.C., The characterization of soluble amyloid prepared in water. *J. Clin. Invest.*, 47, 924, 1968

2.   Eulitz M., Ch'ang L.Y., Zirkel C., Schell M., Weiss D.T, and Solomon, A., Distinctive serologic chemical, and molecular properties of human lambda IV light chains. *J. Immunol.*, 154, 3256, 1995

3.   Wall, J.S., Schell, M., Murphy, C., Hrncic, R., Stevens, F., and Solomon, A., Thermodynamic Instability of Human λ6 Light Chains: Correlation with Fibrillogenicity. *Biochemistry,* 38, 14101, 1999.

ACKNOWLEDGEMENTS

We would like to thank Dr. John Dunlap (University of Tennessee, Knoxville, TN) for his work with the electron microscopy and Dr. Fred Stevens (Argonne National Laboratory, Argonne, IL). AS is an American Cancer Society Professor. This work was supported in part by NCI grant # CA10056 and a donation from the Aslan Foundation.

AGGRESSIVE LIGHT CHAIN AMYLOIDOSIS CASES HAVE UNSTABLE IMMUNOGLOBULIN LIGHT CHAINS

L. A. Sikkink, E. M. Baden, S. R. Vance Jr., K. J.-L. Riley, A.T. Moch, and M. Ramirez-Alvarado.

Department of Biochemistry and Molecular Biology, Mayo Clinic College of Medicine. 200 First Street SW, Rochester, Minnesota. 55905, USA. Correspondence should be addressed to:
Ramirezalvarado.marina@mayo.edu

Monoclonal immunoglobulin light chains are responsible for the characteristic pathologic features found in patients with light chain-related diseases that include Multiple Myeloma (MM) and Light Chain amyloidosis (AL). AL is characterized by the deposition of immunoglobulin light chains as amyloid fibrils in vital organs leading to organ failure and death. The organ most frequently affected is the kidney followed by the heart and liver. MM is characterized by bone lesions, hypercalcemia, renal failure, and anemia. We are interested in studying the role of protein folding and stability in amyloid formation by AL proteins. We are particularly interested in the relationship between disease aggressiveness and protein folding and stability parameters for AL proteins. Disease aggressiveness can be defined as the length of time from diagnosis to death of the patient. This study involved different light chains from cardiac amyloid patients with survival ranging from 6 months to 10 years. A kappa MM non-amyloidogenic protein was used as our control. We have followed protein unfolding by thermal and chemical denaturation using Circular Dichroism and Fluorescence spectroscopy. We have found the MM protein to be the most thermodynamically stable protein compared to the AL proteins. The most aggressive AL protein is partially unfolded and with exposed hydrophobic patches shown by binding to 8-Anilino-1-naphthalenesulfonic acid (ANS). There are differences in the thermodynamic stabilities between AL and MM proteins at different concentrations of NaCl, at different pHs, and in the presence or absence of DL-Dithiothreitol (DTT).

INTRODUCTION

Amyloidoses are an important group of protein deposition disorders in which normally soluble protein aggregates to form insoluble extracellular amyloid fibrils (1). Amyloid fibrils and/or their precursors are suggested to lead to cell death and tissue degeneration. These amyloid fibrils are characterized by their ability to bind histological dyes such as Congo Red and Thioflavine T, and they present a cross-β sheet structure by X-ray fibril diffraction.

AL is the only known neoplastic hematological protein deposition disorder. AL is characterized by the deposition of immunoglobulin light chain in multiple vital organs, where the deposits cause organ failure leading to death (2,3). MM is a monoclonal plasma cell proliferative disorder characterized by the presence of monoclonal immunoglobulin light chain. Though the tissue damage of AL is different from MM, they are both plasma proliferative disorders and over expression of monoclonal immunoglobulin light chain is present in both diseases. Immunoglobulin light chains present an attractive system to study the determinants of amyloid formation. As immunoglobulin light chains, this family exhibits wide sequence diversity, making available a large number of sequences associated with these deposition diseases. This characteristic gives an opportunity for comparative studies that might unveil patterns and point to mechanisms for amyloidosis. In addition to sequence diversity, AL presents a rich clinical variability, involving different organs and tissues as well as the aggressiveness with which the disease causes death among individuals affected with this ailment.

Our laboratory is interested in studying the role of protein folding and stability in amyloid formation by AL proteins. We are particularly interested in the relationship between disease aggressiveness and protein folding and stability parameters for AL proteins.

MATERIALS AND METHODS

This study involved different 3 light chains from cardiac amyloid patients (4): **AL-44** is a $\lambda 2$ (VλII λ2B2) light chain variable domain from a male patient that died 6 months after diagnosis at the Mayo Clinic. (AGGRESSIVE CASE). **AL-09** is a $\kappa 1$ (Vκ1 O18/O8) light chain variable domain from a male patient that died one year after diagnosis at the Mayo Clinic. (AGGRESSIVE CASE). **AL-12** is a $\kappa 1$ (Vκ1 O18/O8) light chain variable domain from a female patient that died 10 years after diagnosis at the Mayo Clinic. (MILD CASE). **uMM-01** is a k1 (Vk1 O12/O2) light chain variable domain from a male Multiple Myeloma patient. We use MM-01 as our non-amyloidogenic CONTROL.

cDNAs from patients' bone marrow were used to clone the protein genes in *E. coli* expression vectors except uMM-01 that was purified from a urine sample from the patient. Proteins were purified using size exclusion chromatography. Overall secondary structure was determined by Far UV Circular Dichroism spectroscopy (CD). Protein stability analysis was followed by thermal and chemical denaturation followed by CD or Fluorescence spectroscopy.

RESULTS

We hypothesize that light chains from aggressive AL cases are less stable than mild cases and that the multiple myeloma protein is the most stable of all of them. From our initial secondary structure analysis using Far UV-CD, we found that AL-44 is partially folded. ANS binding experiments showed that AL-44 binds to ANS suggesting that AL-44 has exposed hydrophobic surfaces. The rest of the proteins that we characterize for this study present β-sheet structure and no binding to ANS, suggesting that they are fully folded. Thermal denaturation experiments followed by CD were performed for the rest of the proteins studied. Figure 1 shows that AL-09, AL-12 and uMM-01 unfold following a two-state, cooperative transition at pH 7.4. uMM-01 is the most stable protein followed by AL-12 (MILD), having AL-09 as the least stable protein (AGGRESSIVE) reflected in their melting temperatures (T_m).

Chemical denaturation studies were performed with these proteins. A summary of these results is presented in table 1. The T_m, melting concentration for urea (C_m) and the free energy of folding ($\Delta G_{folding}$) calculated from chemical denaturation plots show that the most stable protein, uMM-01 has the most favorable folding transition. It is followed by AL-12 (MILD), having AL-09 (AGGRESSIVE) as the protein with the lowest T_m, C_m and free energy of folding of this group. These proteins show different stabilities at different pHs as well as different stability tolerance for the changes in pH.

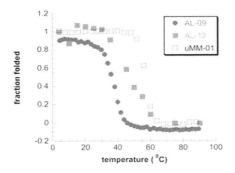

Figure 1. Fraction folded from thermal denaturation experiments followed by CD ellipticity for AL09, AL-12 and uMM-01. All experiments were performed with 10 mM Tris-HCl, pH 7.4

Interestingly, AL-12 is the less tolerant protein to pH changes reflected in its Tm under different experimental conditions even though is the most stable AL protein of our study. These results suggest that we cannot conclude anything about thermodynamic stability under different conditions because AL proteins present different amino acid substitutions that make some of the proteins less tolerant to changes in the environment like pH and salts. We were able to form amyloid fibrils *in vitro* by incubating these proteins at their melting temperature (in the case of AL-09 we incubated it at 37°C). Incubating at the melting temperature maximizes amyloid formation as we have reported previously (5).

Table 1. Summary of thermodynamic stability parameters extracted from thermal and chemical denaturations

Protein	pH	Tm	Cm	ΔG folding
AL-09	7.4	36.16°C	2.5 M	-3.0 kcal/mol
AL-12	7.4	45.6°C	4.4 M	-4.9 kcal/mol
uMM-01	7.4	59.3°C	4.7 M	-5.7 kcal/mol

CONCLUSIONS

Our data indicates that there is correlation between disease aggressiveness and protein stability for this selected group of AL proteins. Protein stability has been previously associated with the ability to form amyloid fibrils. Our current study extends the use of protein stability to try to understand differences in disease aggressiveness.

REFERENCES

1. Buxbaum, J. N. 2003. Diseases of protein conformation: what do in vitro experiments tell us about in vivo diseases?. *Trends in Biochemical Sciences* 28(11): 585-592.
2. Wetzel, R. 1997. Domain stability in immunoglobulin light chain deposition disorders. *Adv Protein Chem* 50: 183-242.
3. Kyle, R.A. and Gertz, M.A.1995. Primary systemic amyloidosis: clinical and laboratory features in 474 cases. *Semin Hematol*. 32 (1):45-59.
4. Abraham R.S. et.al. 2003 Immunoglobulin light chain variable (V) region genes influence clinical presentation and outcome in light chain-associated amyloidosis (AL). *Blood* 101(10):3801-8.
5. Ramírez-Alvarado, M., Merkel, J. S., and Regan, L. 2000. A systematic exploration of the influence of the protein stability on amyloid fibril formation in vitro. *Proc. Natl. Acad. Sci. USA* 97(16):8979-8984.

MOLECULAR CLASSIFICATION OF LIGHT CHAIN (AL) AMYLOIDOSIS BY GENE EXPRESSION PROFILING

R. S. Abraham[1], M. K. Manske[1], K. V. Ballman[2], A. Dispenzieri[1], T. L. Price-Troska[1], M. A. Gertz[1], R. Fonseca[1].

[1]Division of Hematology, [2]Division of Biostatistics, Mayo Clinic Cancer Center, Mayo Clinic, Rochester, MN.

ABSTRACT

Immunoglobulin light chain amyloidosis (AL) is characterized by a clonal expansion of plasma cells within the bone marrow. Gene expression analysis was used to identify a unique molecular profile for AL using enriched plasma cells (CD138+) from the bone marrow of 24 AL and 28 MM patients. Class prediction analysis revealed 29 genes that could distinguish between AL and MM patients with an observed accuracy of 90% and estimated accuracy of 81%. A subset of 12 genes that included TNFRSF7, SDF-1, JUN, PSMA2, DEFA1, NDUFA4, PGK1 and TXN improved the estimated and observed accuracy of classification to 92%. Functional analysis using a novel network mapping software, identified 1051 Focus Genes. Forty-seven genes were further analyzed for their significance to the biology of AL. Notably, these included 8 of the 12 genes as well as CCND1, CDK4, Rb1, BTRC among cell cycle genes, oncogenes-MYC, JUN and p53, XRCC5/XRCC4 and E2A in DNA repair and Ig rearrangement. Importantly, genes deregulated with respect to protein processing and folding included A2M, PSEN1 and 2, APP, UCHL1, SERPINA1 and CASP3. This study provides new insight into the molecular profile of clonal plasma cells and its functional relevance in the pathogenesis of light chain amyloidosis.

MATERIALS AND METHODS

Bone marrow aspirates were collected from 24 AL and 28 MM patients in accordance with IRB and HIPAA regulations. The mononuclear cells (MNCs) from BM were enriched for plasma cells using immunomagnetic bead selection with monoclonal mouse anti-human CD138 antibody. Plasma cell (PC) purity was ascertained by immunocytochemistry analysis for cytoplasmic immunoglobulin light chain (LC). The enriched plasma cells (PCs) were stored in RNAlater® for RNA extraction and preparation for the gene chip experiments. Total RNA was extracted using TRIzol and the RNeasy® Mini kit. The integrity of the RNA was assessed for each sample using an Agilent Bioanalyzer. Double-stranded complementary DNA (cDNA) and labeled complementary RNA (cRNA) were synthesized from the total RNA and hybridized to the Affymetrix Human U133A gene chips. The chips were further processed and scanned according to the manufacturer's protocol. The arrays were scanned with a laser scanner and the data was visualized using the MAS 5.0 Affymetrix software (Affymetrix) to check for obvious failures in the hybridizations. Differentially expressed genes between the AL samples and MM samples were determined using a linear mixed model similar to that described by Chu et al (1). Genes were ranked according to the magnitude (i.e. absolute value) of the grouping variable test statistic, which is a type of two-sample t-statistic. This is equivalent to ranking genes according to their p-value with respect to differential expression between the groups. Genes with a grouping variable test statistic of magnitude 3 or greater

(~ corresponds to p value of <0.01) were considered statistically significantly differentially expressed between the groups. These test statistic values were used for the pathway analysis performed with the Ingenuity software ™. A class prediction approach was used to determine a set of genes that best distinguished among the two groups (AL, and MM). Specifically, the Prediction Analysis of Microarray (PAM) technique was used (2). This method is based on a shrunken centroid approach and uses cross-validation to choose the set of genes with the smallest estimated misclassification error (i.e. the greatest estimated accuracy).

The functional analysis to determine the biological relevance of the data and to identify novel, dysregulated genes was performed using a functional annotation and network-mapping tool, Ingenuity Systems ™ software. A threshold of 3 for the magnitude of the test statistic value was set to identify genes whose expression was significantly differentially regulated. These genes, called Focus Genes, were then used as the starting point for generating biological networks.

RESULTS

The normalized data from all 22,215 genes on the U133A chip were subjected to a class prediction analysis to identify a set of genes that could differentially classify AL and MM. The initial class prediction analysis for the AL and MM groups revealed a set of 29 genes that best differentiated between the groups with an observed accuracy of 90% and an estimated accuracy of 81%. Nineteen of the 24 AL patients were accurately predicted; on the other hand, 23 of the 28 MM patients were correctly classified, with 5 being misclassified in each group.

Functional annotation of the 29 genes revealed that 17 of the 29 genes were associated with the Ig λ light chain locus. Since there is a substantial over-representation of λ light chains in AL it was unclear whether the *a priori* bias of λ light chains artificially skewed the results of the analysis. Therefore, we repeated the centroid analysis with the 12 non-Ig-associated genes (including SDF-1, TNFRS7, JUN, PSMA2 (proteasome alpha 2), DEFA1 (defensin), NDUFA4 (ubiquinone), TXN (thioredoxin), PGK1 (phosphoglycerate kinase 1) from the original set of 29 genes to determine if these could still predict the class of the samples.

The 12 gene subset was able to substantially improve the observed and cross-validated accuracy of classification to 92%. With the revised analysis, 23 of the 24 AL patients were correctly predicted using the set of 12 genes and 25 of the 28 MM patients were also accurately classified (Figure 1). The single "misclassified" AL patient and 2 of the MM patients were represented among the 5 "misclassified" patients in each group in the previous analysis. The 3rd "misclassified" patient was diagnosed with a Stage IIA myeloma. These data suggest that AL patients have a unique pattern of gene expression that distinguishes them from MM patients. A small subset of genes is capable of predicting these disease groups with a high degree of accuracy, substantiating the hypothesis that AL is an entity distinct from MM in a majority of cases.

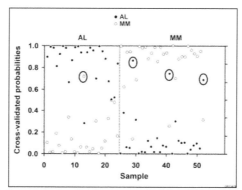

Figure 1. The class prediction analysis after the removal of the 17 Ig λ LC genes, which could artificially skew the accuracy of classification, resulted in an improvement in the estimated and observed accuracy rate to 92%. The class of 23 of the 24 AL and 25 of the 28 MM patients was accurately predicted. The Y-axis shows the threshold value with samples closer to 1 having the highest probability of being an AL or MM sample respectively. The X-axis denotes each of the 24 AL and 28 MM samples.

GENE NETWORK MAPPING AND PATHWAY ANALYSES

We next sought to ascribe biological function and interactions to these genes using the Ingenuity Systems ™ software, which is capable of mapping gene networks and identifying potentially dysregulated pathways in a specific disease.   The use of a stringent test statistic threshold of magnitude 3 resulted in the identification of 1051 Focus Genes from the original 22,215 genes. A total of 200 networks were generated and we analyzed in detail the first 5 highest scoring networks, four each with 35 Focus Genes and a score of 28 (the score is the negative log of the p value) and a fifth network with a score of 10 and 23 Focus Genes, for their biological interactions.   From the analysis of the first five networks of 163 genes, 47 genes were selected for further analysis, 26 genes with significantly higher expression (test statistic value >3) and 21 genes with significantly lower expression (test statistic value < - 3) in AL patients compared to MM. These genes were divided into 11 functional categories. There was an overlap of genes in each category since many of these genes and gene products mediate a number of related biological functions. Interestingly, 8 of the 12 genes (TNFRSF7, SDF-1/CXCL12, PSMA2, JUN, PGK1, thioredoxin, ubiquinone and DEFA1) used in the class prediction analysis were also represented within the selected Focus Genes from the pathway analysis.

 In summary, we have been able to demonstrate that AL can be differentially classified from MM and the use of a novel network and pathway mapping software has permitted us to identify genes that are of potential significance in AL biology.

REFERENCES

1.   Chu TM, Weir B, Wolfinger R. A systematic statistical linear modeling approach to oligonucleotide array experiments. Mathematical Biosciences. 2002; 176:35-51.
2.   Tibshirani R, Hastie T, Narasimhan B, Chu G. Diagnosis of multiple cancer types by shrunken centroids of gene expression. Proceedings of the National Academy of Sciences of the United States of America. 2002; 99: 6567-6572.

ANALYSIS OF B CELL CLONAL EVOLUTION IN BM AND PBL OF AL AND MM PATIENTS USING GRAPHICAL QUANTIFICATION OF LINEAGE TREES

R. S. Abraham[1], M. K. Manske[1], A. Sohni[2], H. Edelman[3], N. Zuckerman[3], G. Shahaf[3], A. Dispenzieri[1], M. Q. Lacy[1], M. A. Gertz[1], R. Mehr[3].

[1]Division of Hematology, Mayo Clinic, Rochester, MN. [2] Eastern Michigan University, Ypsilanti, MI. [3]Bar-Ilan University, Israel.

ABSTRACT

The objective of this study was to examine the clonal relationships between B cells in the periphery and the clonal plasma cells in the bone marrow (BM) in AL, Myeloma (MM) and MM with AL (MM + AL) patients, which use the same light chain germline V gene (VL). The clonal light chain variable (V) gene was identified by PCR, cloning and nucleotide sequencing. Our data show that 1/3rd to 1/2 of the peripheral B cell VL gene repertoire is dominated by the clonal light chain gene in AL patients. This dominance is significantly more pronounced in MM patients (with 85-90% of the B cells using the same germline VL gene as the clonal plasma cell in the BM). The clonal relationship between the PBL and BM populations was further confirmed by Ig VH gene analysis. We used a rigorous computer-aided algorithm for extracting the information contained in the lineage trees, using the terms of mathematical graph theory. Lineage tree analysis has allowed us to obtain a high-resolution analysis of nucleotide sequences from these patients in order to determine the process of clonal selection and evolution in AL and MM. This analysis demonstrates the presence of intra-clonal variations within the BM clones, suggesting that these B cells may be seeded into the BM in "waves". In most patients, the data indicate that the BM clones are derived from the PBL clones, confirming that the clonogenic B cells in the periphery represent a precursor population. The mathematical analysis also reveals that there are significantly more mutations in the VL genes of AL and MM patients compared to normals suggesting either a faster rate of mutation or greater duration of exposure to the mutational process. Additionally, the AL BM sequences show more mutations than the PBL sequences, seeming to suggest lower antigenic selection pressure in the shaping of these trees. In conclusion, lineage tree analysis and the novel mathematical approach of quantifying tree properties provides significant insight into the mechanisms of clonal evolution in the pathogenesis of AL and related plasma cell disorders.

MATERIALS AND METHODS

Mononuclear cells from Bone Marrow (BM) aspirates or peripheral blood (PB) samples were collected from 8 AL (5 λ and 3 κ), 5 MM, 1 MM+AL patients. PB samples (n = 3) from normal individuals served as controls. The clonal light chain V genes from BM and PB samples were identified by PCR and cloning using pGEM ® -T Easy Vector System (Promega). Light Chain nucleotide sequences were analyzed using V BASE and nucleotide alignments were performed with EMBOSS and GCG software. Mathematical analysis was performed using algorithms (Mehr *et al.*) which allow comparison of sequences from clonally related B cells and analysis of B cell diversification between patient groups and normal controls.

RESULTS

The data from our analyses of the germline light chain V gene repertoire in AL patients revealed that between $1/3^{rd}$ to one-half of the B cells in the periphery used the same germline gene donor as the clonal population of plasma cells within the bone marrow. This finding was even more striking in the myeloma (MM) patients where 80-90% of the peripheral repertoire used the same germline gene as the clonal BM population. Therefore, the relevant question was whether there was a clonal relationship between B cells in the periphery using the same germline donor and the clonal plasma cells within the BM. Nucleotide sequence analysis of the light chain V region from the peripheral blood (PB) B cells clones and BM clones showed that indeed these clones were related to each other in the AL, MM and AL+MM patients. The clonal relationship was further substantiated by comparing the Ig heavy chain V region sequences as well. However, it was not clear as to what the nature of the clonal relationship between the PB and BM clones were, in terms of whether the BM clones were derived from a PB precursor or whether both these had a common ancestor elsewhere in the germinal center (GC).

To dissect the clonal relationships between the PB and BM clones and better understand the nature of B cell diversification in these patients, we undertook lineage tree and mathematical analysis of the sequences obtained from both the peripheral B cells as well as the bone marrow plasma cells (1). Using this high-resolution analysis, it was evident that in most of the AL and MM patients, the BM clones appear to be derived from the PB clones suggesting that the latter are precursors to the BM population. Additionally, the lineage tree graphical analysis showed that there are intra-clonal variations within the BM clones, contrary to what has been previously reported in MM. Mathematical analysis quantifying the clonal relationship between the BM and PB B cells in AL and MM patients suggests that in most cases the BM clones are derived from PB precursors (Figures 1-3). However, these precursors may be repeatedly trafficking through the germinal center (GC) where they undergo further mutations that are reflected in subsequent generations of the clonal B cells. Alternatively, there may be a common GC precursor that gives rise to both BM and PB populations.

The trunk and path length measurements (mutations per cell) are significantly different between AL patients and normals as well as MM patients and normals. This suggests that these clonal B cells either have a faster rate of mutation or are exposed to the mutational process longer. Multiple comparison corrections did not support the significance of the data despite the large number of samples and data points, due to the high variability of tree properties between and within individual patients. Nonetheless, these data do demonstrate that there is ongoing diversification in the B cell clones in AL and MM.

These data provide novel and interesting insights into the process of clonal differentiation in neoplastic B cells in patients with monoclonal gammopathies using graphical lineage tree quantification.

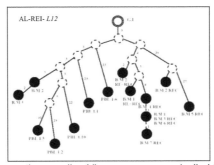

Figure 1. The double circle indicates the germline VL gene sequence (called the root). Each of the dashed circles indicates a deduced intermediate sequence, and black circles denote experimentally obtained sequences. The terminal sequences are called leaves. This AL patient shows 2 main branches of the clonal tree, with the BM sequences originating as a single branch as well as forming a small sub-branch off the PBL branch, establishing the clonal relationship between the 2 populations as well as intraclonal variations within the BM clones. The PBL clones appear to give rise to at least some of the BM clones. The numbers adjacent to each line indicate the # of mutations separating the sequences connected by that line.

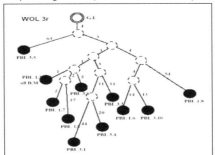

Figure 2. In this lineage tree all the BM clones are derived as offspring from the PBL precursors. The large numbers of mutations separating some leaves from the root suggest that these cells have been exposed to the somatic mutation machinery for a long time or have divided rapidly. The PBL clones represent sub-branches of a main branch.

REFERENCES

1.   Dunn-Walters DK, Belelovsky A, Edelman H, Banerjee M, Mehr R. The dynamics of germinal center selection as measured by graph-theoretical analysis of mutational lineage trees. Developmental Immunology. 2002; 9: 233-243.

IN VITRO AND IN VIVO MOUSE MODELS FOR AL AMYLOIDOSIS

J.E. Ward, D.A. Brenner, J. Eberhard, R. Liao, L.H. Connors, C. O'Hara, D.C. Seldin.

Amyloid Research and Treatment Program, Boston University School of Medicine, Boston, MA USA.

AL amyloidosis is caused by an amyloidogenic immunoglobulin (Ig) light chain secreted by a clonal plasma cell. Mechanisms by which fibrils or their precursors damage tissues are poorly understood, and investigation of this would be facilitated by a reproducible, stable, genetically defined animal model. To develop such a model, we are testing *in vitro* and *in vivo* expression systems for light chains using a variety of vectors and cell lines. Vectors include viral and mammalian promoters for both ubiquitous and lymphoid specific expression and cell lines include COS cells, fibroblasts, and plasmacytoma cells. A full-length human λ6 light chain cloned from a patient with aggressive multi-organ AL amyloidosis has been expressed transiently and stably. The CMV viral promoter confers high level expression in stably transfected SP2/0 Ig$^{-/-}$ mouse plasmacytoma cell lines. The λ6 light chain can be detected by Western blot analysis in the cells and secreted into tissue culture supernatants *in vitro*. These cell lines have been implanted into syngeneic and RAG$^{-/-}$ mice to form tumors intraperitoneally and subcutaneously; RAG$^{-/-}$ mice will not generate mouse anti-human light chain antibody responses. The tumors secrete the human λ6 light chain *in vivo* and it is detected in the circulation by Western blot. Although the duration of these experiments is limited by tumor size, the effect of light chain production is under ongoing investigation. Strategies for longer term expression are being developed, and the optimal constructs will be used to generate transgenic mice.

EXPERIMENTAL DESIGN

The full length rearranged λ6 light chain gene including variable and constant domains was amplified from cDNA prepared from bone marrow RNA from a patient with aggressive multi-organ amyloidosis. This amplicon was cloned into a variety of different expression vectors to test the efficiency of the promoter and enhancer construct by transient and stable *in vitro* transfection. Vectors included the human β-actin promoter and enhancer and the viral CMV promoter for ubiquitous expression and the murine immunoglobulin (Ig) μ heavy chain promoter and enhancer and the proximal lck promoter with the immunoglobulin μ heavy chain enhancer for lymphocyte-specific expression. These constructs were transiently expressed in monkey COS (fibroblast), murine SP2/0 (Ig$^{-/-}$ plasmacytoma) and murine 3T3 (fibroblast) cells. The β-actin promoter construct expressed high levels of λ6 protein assessed by Western blotting in COS cells, but not in SP2/0 or 3T3 cells. The viral CMV promoter construct expressed high levels of λ6 protein in all cell lines tested. Thus the CMV promoter construct was chosen to generate stably transfected cell lines. The immunoglobulin μ heavy chain promoter and enhancer construct failed to express at detectable levels in SP2/0 cells and the lck construct is currently being tested.

Stably transfected λ6 secreting SP2/0 cell lines or control parental SP2/0 cells were injected intraperitoneally or subcutaneously into syngeneic Balb/c and RAG$^{-/-}$ (T and B cell, and thus Ig, deficient) mice. Tumors were allowed to grow from 2-6 weeks before mice were sacrificed due to tumor burden. Echocardiograms were performed prior to sacrifice. Urine, serum, and organs were collected for pathology, immunohistochemistry, and protein analysis.

RESULTS

The λ6 LC was detected in the serum of mice injected with λ6-secreting SP2/0 cells, but not controls which were injected with untransfected parental SP2/0 cells (Figure 1). Albumin and λ6 LC were detected in the urine of the mice receiving λ6 transfected tumors (Figure 2), although there was no renal damage detected upon histologic examination of the kidneys.

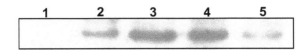

Figure 1. Anti-λ LC Western blot of mouse serum. Lane 1 is a control mouse with untransfected tumor. Lanes 2-5 are mice with λ6 transfected tumors.

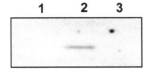

Figure 2. Anti-λ LC Western blot of 5 μl mouse urine. Lane 1, urine from a mouse without tumor; lane 2, urine from a mouse with a λ6 transfected tumor; lane 3, urine from a mouse with an untransfected tumor.

All organs were negative for amyloid deposition by Congo Red staining and anti-λ immunohistochemistry of tissue sections. The λ6 transfected tumors were positive for λ6 by anti-λ immunohistochemistry and Western blot of protein extracts made from homogenized tumor lysates (Figures 3 and 4).

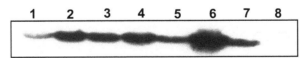

Figure 3. Anti-λ LC Western blot of mouse tumor proteins. Lanes 1-7 are from mice with a λ6 transfected tumor and lane 8 is from a mouse with an untransfected tumor.

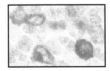

Figure 4. Anti-λ LC immunohistochemistry of a λ6-transfected tumor. The dark intracellular staining is positive for λ LC.

Despite a lack of fibril formation in the short term experimental model, physiologic and biochemical changes were observed in the hearts of the λ6 treated mice. An overall 20% decrease in the left ventricle fractional shortening of the M mode measured by echocardiography was apparent in λ6 treated mice compared to untransfected tumor controls. The fractional shortening, defined by the difference in the measurement of the left ventricle wall in diastole and systole divided by the measurement in diastole, averaged 57% in λ6 treated mice versus 71% in control mice, n=3 per matched group (Figure 5). An upregulation of heme oxygenase-1 protein, a marker of cellular and oxidative stress, was observed in λ6 treated mice by Western blotting (Figure 6).

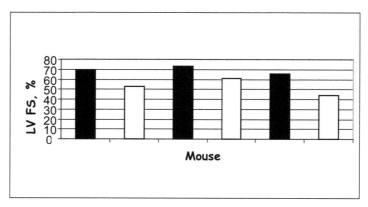

Figure 5. Left ventricular fractional shortening (in percent) for 3 pairs of age and tumor burden matched mice. Mice with λ6 transfected tumors are white bars, black bars are untransfected control tumors.

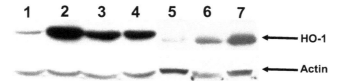

Figure 6. Anti-heme oxygenase-1 Western blot of mouse heart protein. Lane 1 is protein from a wild type mouse without tumor, lane 5 is protein from a mouse with untransfected tumor, and lanes 2-4 and 6-7 are from mice with λ6-transfeced tumors.

DISCUSSION

In this short term mouse model, amyloidogenic λ6 protein is circulating in the blood of mice injected with tumor forming λ6 transfected SP2/0 plasmacytoma cell lines. No fibrils are detected in the mice, however physiologic and biochemical changes are detected in the hearts and kidneys of the mice with circulating λ6 light chain. This provides a model in which different expression constructs can be tested prior to the generation of a transgenic mouse, as well as allowing the observation of early events that occur in the host response to the amyloidogenic light chain. Further study of these tissues for other markers of cellular and oxidative stress, as well as cell death pathways, are ongoing. Ultimately, the production of a transgenic mouse expressing an amyloidogenic λ6 light chain will allow longer term studies of the mechanisms of amyloid formation and the host response, as well as provide a model for screening potential inhibitors and therapeutics.

Supported by NIH P01-HL68705 and the Gerry Foundation. DCS is a scholar of the Leukemia and Lymphoma society

SERUM CARDIAC TROPONINS AND N-TERMINAL PRO-BRAIN NATRIURETIC PEPTIDE: A STAGING SYSTEM FOR PRIMARY SYSTEMIC AMYLOIDOSIS

A. Dispenzieri, R.A. Kyle, M.Q. Lacy, A.S. Jaffe, T.M. Therneau, S.R. Zeldenrust, M.A. Gertz.

Mayo Clinic, 200 First Street S.W., Rochester, MN 55905, USA

I. ABSTRACT

Background: Primary systemic amyloidosis (AL) is a multisystemic disorder resulting from an underlying plasma cell dyscrasia. There is no formal staging system for AL, making comparisons between studies and treatment centers difficult. Our group previously identified elevated serum cardiac troponin T (cTnT) as the most powerful predictor of overall survival. Others have reported that N-terminal pronatriuretic peptide (NT-proBNP) is a valuable prognostic marker. *Objective:* To develop a staging system for patients with AL. *Methods:* 242 patients with newly diagnosed AL seen at the Mayo Clinic between April 1979 and November 2000, who had echocardiograms and stored serum samples at presentation. NT-proBNP measurements were performed on 242 patients in whom cTnT and troponin I (cTnI) had been previously run; overall survival. Two prognostic models were designed using threshold values of NT-proBNP and either cTnT or cTnI (NT-proBNP < 332 ng/L, cTnT <0.035 μg/L and cTnI <0.1 μg/L). Depending on whether NT-proBNP and troponin levels were both low, both high, or only one high, patients were classified as Stage I, III, or II, respectively. *Results:* Using the cTnT/NT-proBNP model 33, 30, and 37% were Stages I, II, and III with median survivals of 26.4, 10.5, and 3.5 months, respectively. The alternate cTnI/NT-proBNP model predicted median survivals of 27.2, 11.1, and 4.1 months, respectively. *Conclusions:* Stratification of AL patients into 3 stages is possible with two readily available and reproducible tests setting the stage for more consistent and reliable cross comparisons of therapeutic outcomes. *Supported in part by CA 62242(R.A.K), CA 91561(A.D) from the National Cancer Institute, and the Robert A. Kyle Hematology Malignancies Fund, Mayo Foundation.*

II. INTRODUCTION

A major drawback in the field of primary systemic amyloidosis (AL) is the absence of a prognostic staging system, making comparisons between different studies and treatment centers difficult to interpret. We have recently reported that in patients receiving standard chemotherapy, circulating cardiac biomarkers are the most powerful independent predictors of survival.[1] Cardiac troponins (cTnT, cTnI) are highly specific markers of myocardial injury.[2] Elevations are common in patients with acute coronary syndromes and in those with acute congestive heart failure.[3] Pro-brain natriuretic peptide is a 108 amino acid propeptide, made by myocytes in response to increased wall stress. It is produced predominantly in the left ventricle and when released, it is cleaved into two fragments, the active brain natriuretic peptide (amino acids 77-108) and a leader sequence known as NT-proBNP (amino acids 1-76). NT-proBNP has been shown to be a sensitive indicator of cardiac abnormalities.[4] Blood levels of N-terminal pro-brain natriuretic peptide (NT-pro-BNP) provide prognostic information in patients with AL treated with conventional chemotherapy.[5] We sought to develop a simple model based on these two cardiac biomarkers.

III. PATIENTS AND METHODS

Between April 1979 and November 2000, there were 261 patients with AL seen at the Mayo Clinic within 30 days of diagnosis who had echocardiograms performed within 30 days of presentation and serum samples stored within 7 days of presentation. cTnT and cTnI measurements were performed on all 261 of these patients.[1] NT-proBNP measurements were done on 242 of these 261. All but 2 patients had follow-up through April 2001. The diagnosis of amyloidosis was made as previously described. Serum samples, stored at -20°C, were thawed and analyzed immediately after thawing. Assays for cTnT and cTnI were performed with sensitive second- and third-generation assays with reagents provided by Roche Diagnostics (Indianapolis, Indiana) and DADE (Newark Delaware). NT-proBNP levels were measured with electrochemiluminescence sandwich immunoassay (ECLIA, Roche) on an Elecsys System 2010. Thresholds for NT-pro-BNP were explored for best fit. The upper reference limit for normal women above the age of 50 years (331.5 ng/L) proved to be the best fit with the highest hazard ratio (2.05, 95% confidence interval 1.56 to 2.71).

IV. RESULTS

The median age of the 242 patients was 64, with 24% under age 55. Sixty-five percent were male. Median values of cTnT, cTnI, and NT-pro-BNP were 0.02 µg/L, 0.11 µg/L, and 580 ng/L. The median left ventricular ejection fraction and septal thickness were 58% and 14 mm. The median number of organs involved was 2. Two hundred and twenty-two patients have died (92%). The median overall survival was 9.2 months with only 25% of patients alive at 30 months.

Patients with a cTnT > 0.035 had a proportional hazard of death of 2.38 (95% CI 1.81 – 3.12) relative to those with lower values. NT-proBNP levels were also valuable univariate predictors of survival. The threshold value of 332 ng/L provided a hazard ratio of 2.05 (95% CI 1.56 – 2.71). Interactions between levels of troponins and NT-proBNP were sought. In multivariable analysis NT-proBNP was an independent--though less powerful--predictor of survival (HR 1.49; 1.09-2.02) than cTnT (1.81; 1.33-2.46); however, NT-proBNP was slightly more powerful (HR 1.71; 1.28-2.28) than cTnI (HR 1.44; 1.09-1.90).

Using the cTnT/NT-proBNP system, 33, 30, and 37% of patients were characterized as Stage I-t, II-t, and III-t, respectively. Median survivals for the 3 groups were 26.4, 10.5, and 3.5 months, p <0.0001, Figure A. Of the Stage II-t patients, 23% have elevated cTnT and 77% elevated NT-proBNP. Because some institutions measure cTnI rather than cTnT, a similar risk model using cTnI instead of cTnT was created for those whom the cTnT assay is not available. The findings were similar. The numbers of Stage I-i, II-i, and III-i patients were 59 (24%), 91 (38%), and 92 (38%) (-i equals cTnI/NT-proBNP model). The respective median survivals for these three groups were 27.2, 11.1, and 4.1 months (p <0.0001), Figure B. Of the Stage II-i patients, 42% have elevated cTnI and 58% elevated NT-proBNP. On multivariate analysis cTnT, NT-proBNP, left ventricular ejection fraction, serum albumin level, and urine monoclonal protein were independent predictors of survival. Age, interventricular wall thickness, and number of organs involved by amyloid were no longer significant. cTnI is not an independent variable once cTnT is in the model.

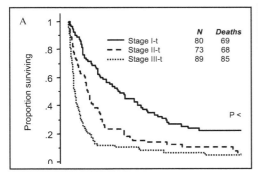 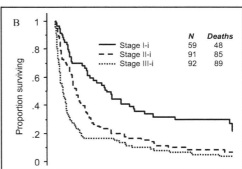

The troponin/NT-proBNP staging systems add information above and beyond the historical cardiac risk factors. Seventeen additional patients who were not suspected of cardiac involvement by either standard echocardiographic or clinical grounds are Stage II-t or III-t. The survival of these 17 patients was inferior to the 33 not suspected of cardiac involvement based on standard predictors but who were Stage I-t (16.9 versus 30.1 months, p = 0.03). The group detected by only biochemical cardiac markers had a better survival as compared the 145 patients who had elevated biochemical markers *and* clinical or echocardiographic suspicion of cardiac involvement (16.9 versus 5.4 months, p = 0.2). In contrast, there were 47 Stage I-t patients who had echocardiographic or clinical features consistent with cardiac involvement. The median survival of this cohort was 26.4 months, which was significantly better than the overall survival (5.4 months) of the 145 patients with both elevated markers (Stage II-t and III-t) and cardiac involvement by historical determinants (p <0.0001). The Stage I-t patients with echocardiographic or clinical evidence of cardiac amyloid did not have a significantly different survival from the 33 Stage I-t patients with normal clinical/echocardiography (26.4 versus 30.1 month, p = 0.7).

V. CONCLUSION

Though our staging systems do not directly supply information about renal, hepatic, or nervous system involvement, they predict survival because cardiac involvement is the primary determinant of prognosis. The risk of death in patients with primary systemic amyloidosis can be accurately stratified with our models using two easily reproducible biochemical measurements, serum troponins (preferably cTnT over cTnI) and NT-proBNP.

VI. REFERENCES

1. Dispenzieri A, et al. Survival in patients with primary systemic amyloidosis and raised serum cardiac troponins. Lancet. 2003;361:1787-1789.
2. Jaffe AS, et al. It's time for a change to a troponin standard. Circulation. 2000;102:1216-1220.
3. Ottani F, et al. Elevated cardiac troponin levels predict the risk of adverse outcome in patients with acute coronary syndromes. Am Heart J. 2000;140:917-927.
4. James SK, et al. N-terminal pro-brain natriuretic peptide and other risk markers for the separate prediction of mortality and subsequent myocardial infarction in patients with unstable coronary artery disease: a Global Utilization of Strategies To Open occluded arteries (GUSTO)-IV substudy. Circulation. 2003;108:275-281.
5. Palladini G, et al. Serum N-terminal pro-brain natriuretic peptide is a sensitive marker of myocardial dysfunction in AL amyloidosis. Circulation. 2003;107:2440-2445.

AMYLOID-INDUCED MYOCARDIAL DYSFUNCTION DEPENDS ON BIOCHEMICAL CHARACTERISTICS OF THE AMYLOIDOGENIC PROTEIN.

Giovanni Palladini[1,2], Maria Joao Saraiva[3], Stefano Perlini[2], Laura Obici[1], Monia Vezzoli[3], Teresa Coelho[4], Hipólito Reis[5], Francesca Lavatelli[1], Giovanbattista Vadacca[6], Remigio Moratti[6], Giampaolo Merlini[1]

Amyloid Center, [1]Biotechnology Research Laboratories, [2]Internal Medicine, [6]Clinical Chemistry Laboratory, "IRCCS – Policlinico San Matteo", and University of Pavia, Italy; [3]Amyloid Unit, Instituto de Biologia Molecular e Celular, IBMC, [4]Unidade Clínica de Paramiloidose and Neurophysiology and Psycophysiology Unit, [5]Unidade Clínica de Paramiloidose and Cardiology Department, Hospital de Santo António, IBMC, Porto, Portugal

Myocardial involvement results in severe dysfunction and is the leading cause of death in patients with amyloidosis. Heart involvement is the main prognostic determinant in AL (1, 2). Patients with familial cardiac amyloidosis caused by variant transthyretin (ATTR) have a significantly better outlook than AL patients with heart involvement (3). Serum N-terminal fragment of natriuretic peptide type B (NT-proBNP) is a sensitive marker of amyloid cardiomyopathy and the most powerful prognostic determinant in AL (4). Several lines of evidence indicate that the degree of myocardial dysfunction does not depend only on the extent of amyloid deposits but also on the biochemical characteristics of the amyloidogenic proteins which dictate the aggregation process (3, 4). We have shown that shutting down the synthesis of the light chain, by chemotherapy, results in prompt reduction of NT-proBNP concentration and recovery of myocardial dysfunction, despite the local persistence of amyloid deposits (4). That amyloidogenic light chains could exert a toxic activity on myocardial tissue is suggested by a) the observation that patients with amyloidosis caused by transthyretin (who do not have pathogenic circulating light-chains) may have important myocardial infiltration but minimal heart failure (3), and b) by the observation that infusion of light chains from patients with cardiac amyloidosis rapidly causes diastolic dysfunction in isolated mouse heart (5). In order to test this hypothesis we have investigated the relationship between amyloid myocardial infiltration and the serum concentration of NT-proBNP in patients with amyloidosis caused by light-chains and by transthyretin.

PATIENTS AND METHODS

NT-proBNP concentration at presentation was measured with an electrochemiluminescence sandwich immunoassay (ECLIA, Roche, Basel, Switzerland) on an Elecsys System 2010 in 17 patients with ATTR amyloidosis, diagnosed at the Pavia amyloid center and at the Porto Amyloid Unit, and in 17 AL patients, referred to the Pavia amyloid centre. Upper reference limits depend on age and gender and are 10.4 pmol/L in men <50 years, 18 pmol/L in women <50 years, 26.4 pmol/L in men ≥50 years and 39.8 pmol/L in women ≥50 years.

All patients had biopsy-proven amyloidosis with cardiac involvement as defined by previously described clinical, electrocardiographic and echocardiographic criteria (4).

RESULTS

Seventeen patients had AL and 17 ATTR. A monoclonal component was detected in all AL patients by high-resolution serum and urine immunofixation electrophoresis (10 λ, 6 κ, 1 biclonal). All ATTR patients had biopsy-proven amyloidosis and carried an amyloidogenic transthyretin gene mutation (6 Glu89Gln, 2 Thr49Ala, 2 Tyr59Lys, 2 Val30Met, 1 Ile68Leu, 1 Phe33Val, 1 Phe64Leu, 1 Tyr78Phe, 1 Val30Ala).

All patients had normal renal function (serum creatinine <1.2 mg/dL). Patients were matched for left ventricular wall thickness, gender and age (Table 1).

Table 1. Patients' characteristics

	ATTR	AL	p-value
gender (male)	13 (76%)	13 (76%)	1.00
age (years)	median 52 (range 31-70)	median 58 (range 46-69)	0.21
IVS thickness (mm)	median 15 (range 11-23)	median 15 (range 11-22)	0.96

Despite comparable echocardiographic parameters, serum NT-proBNP concentration was significantly higher in AL (median: 686.2 pmol/L, range: 295.7-2954.7) than in ATTR patients (median: 84.5 pmol/L, range: 18.8-1088.3), (Figure 1). None of the patients with AL and 2 (12%) of the patients with ATTR had a normal NT-proBNP.

Figure 1

Serum NT-proBNP concentration in patients with AL and ATTR

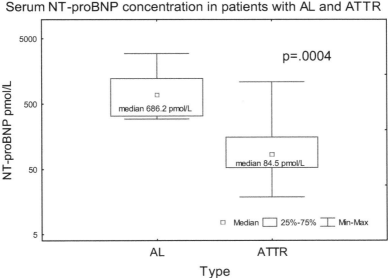

In addition, survival of patients with AL was significantly shorter than that of ATTR patients (Figure 2).

Figure 2

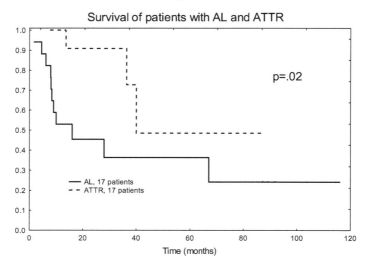

Survival of patients with AL and ATTR

CONCLUSIONS

Our findings confirm the previous observation by Dubrey et al (3) that ATTR cardiomyopathy is less severe than AL cardiac involvement. Amyloidogenic light-chain with cardiac tropism generate more severe myocardial dysfunction than transthyretin amyloidogenic variants despite comparable wall thickening at echocardiography. Thus, the degree of organ dysfunction depends not only on the amount of amyloid deposited (i.e. myocardial wall thickening), but mainly on the biochemical characteristics of the amyloidogenic protein.

REFERENCES

1. Dubrey SW, Cha K, Andersen J, et al (1998). The clinical features of immunoglobulin light-chain (AL) amyloidosis with heart involvement. *Q J Med*; **91:** 141-157.
2. Kyle RA, Gertz MA (1995). Primary systemic amyloidosis: clinical and laboratory features in 474 cases. *Semin Hematol*; **8:** 45-59.
3. Dubrey SW, Cha K, Skinner M, et al (1997). Familial and primary (AL) cardiac amyloidosis: echocardiographically similar diseases with distinctly different clinical outcomes. *Heart*; **78:** 74-82.
4. Palladini G, Campana C, Clersy C, et al (2003). Serum N-terminal pro Brain Natriuretic Peptide is a sensitive marker of myocardial dysfunction in AL amyloidosis. *Circulation*; **107:** 2440-2445.
5. Liao RL, Jain M, Teller P, et al (2001). Infusion of light chains from patients with cardiac amyloidosis causes dyastolic dysfunction in isolated mouse hearts. *Circulation*; **104:** 1594-7.

THE REDUCTION OF THE SERUM CONCENTRATION OF THE AMYLOIDOGENIC LIGHT-CHAIN IN CARDIAC AL RESULTS IN PROMPT IMPROVEMENT OF MYOCARDIAL FUNCTION AND PROLONGED SURVIVAL DESPITE UNALTERD AMOUNT OF MYOCARDIAL AMYLOID DEPOSITS

Giovanni Palladini[1,2], Stefano Perlini[2], Monia Vezzoli[2], Vittorio Perfetti[2], Francesca Lavatelli[1], Ivana Ferrero[2], Laura Obici[1], Riccardo Caccialanza[3], Arthur R Bradwell[4], Giampaolo Merlini[1]

Amyloid Center, [1]Biotechnology Research Laboratories – Department of Biochemistry, [2]Internal Medicine and [3]Department of Applied Health Sciences – Human Nutrition Unit, University Hospital "IRCCS – Policlinico San Matteo", Pavia, and University of Pavia, Italy and [4]The Binding Site Ltd and the Medical School, University of Birmingham, Birmingham, United Kingdom.

Myocardial involvement is common in AL amyloidosis and leads to severe heart dysfunction, which is the main prognostic determinant and the cause of death in the great majority of patients (1, 2).

Together with echocardiography, biochemical markers, such as serum troponins T and I and the N-terminal fragment of natriuretic peptide type B (NT-proBNP), have been recently proposed to evaluate myocardial injury in AL (3, 4). NT-proBNP is a sensitive marker of amyloid cardiomyopathy, the most powerful prognostic determinant and a useful tool in monitoring response to therapy (3). Recently a prognostic index based both on troponins and NT-proBNP has been proposed in AL patients undergoing autologous stem cell transplantation (5).

Current treatment for AL is aimed at reducing the circulating amyloidogenic precursor targeting the bone marrow plasma cell clone responsible for its synthesis. The availability of a quantitative assay for circulating free light chains (FLC) makes it possible to estimate the concentration of the amyloidogenic precursor and its modification after chemotherapy (6).

The molecular mechanisms underlying tissue damage in AL remain undetermined. However, experimental studies suggest that circulating oligomers may be responsible for a substantial proportion of amyloid related toxicity (7). Based on this evidence, we investigated the relation between cardiac function, estimated with NT-proBNP, and hematologic response to chemotherapy in AL patients.

PATIENTS AND METHODS

We determined serum NT-proBNP and FLC concentration at study entry and at the time of evaluation of response to therapy in 21 consecutive AL patients referred to the Amyloid Center in Pavia. All patients had symptomatic disease, with biopsy proven amyloidosis and evidence of a plasma cell dyscrasia. All patients had evidence of cardiac amyloid involvement according to echocardiographic parameters (3). Wall thickness at echocardiography was considered as an estimation of the amount of amyloid deposited in the heart. Patients with renal insufficiency were excluded due to altered NT-proBNP metabolism.

Serum NT-proBNP was measured with an electrochemiluminescence sandwich immunoassay (ECLIA, Roche, Basel, Switzerland) on an Elecsys System 2010. Upper reference limits depend on age and gender and are 10.4 pmol/L in men <50 years, 18 pmol/L in women <50 years, 26.4pmol/L in men ≥50 years and 39.8 pmol/L in women ≥50 years. Circulating FLC concentration was measured using a latex-enhanced immunoassay (The Binding Site, Birmingham, UK) on a Behring BN II nephelometric analyzer (Dade Behring, Deerfield, IL, USA). The reference ranges for polyclonal free κ and free λ light chains were 3.3-19.4 mg/L and 5.7-26.3 mg/L, respectively, with a mean κ to λ ratio of 0.59 (95% CI 0.3-1.2).

Patients were treated with melphalan and high-dose dexamethasone (13, 62%), high-dose dexamethasone alone (3, 14%), thalidomide (3, 14%), melphalan and prednisone (1, 5%) and rituximab (1, 5%).

Response to therapy was defined as a >50% reduction of FLC concentration.

RESULTS

Twenty-one patients were enrolled into the study. Twelve were male and median age was 59 years (range 28-75 years). A monoclonal component was detected in all patients by high resolution agarose gel immunofixation electrophoresis (17 λ, 4 κ). Median serum NT-proBNP concentration in the whole group was 302 pmol/L (40-3762 pmol/L).

Chemotherapy induced a >50% reduction of the initial FLC concentration in 11 (52%) patients. In 9 cases (43%) the FLC concentration and κ/λ ratio normalized. The monoclonal component disappeared at immunofixation (complete remission) in 7 patients (33%).

Response to therapy was accompanied by a decrease of serum NT-proBNP concentration in all patients (median 70%, range: 30-95%), with no evidence of wall thickness reduction at echocardiography. In 6 of the responding patients heart failure resolved. The correlation between FLC and NT-proBNP concentration changes was significant (p<0.02).

CONCLUSIONS

The reduction of circulating FLC induced by chemotherapy was immediately paralleled by a decrease of serum NT-proBNP concentration in all patients and by resolution of heart failure in the majority of them, without concomitant reduction of wall thickness at echocardiography.

These data indicate that reducing the synthesis of the amyloidogenic light-chain promptly translates into an improvement of myocardial dysfunction, despite the amount of amyloid deposited (i.e. wall thickness) remains unaltered. Thus, circulating amyloidogenic light-chains seem to exert a direct toxic effect on cardiomyocytes, highlighting the paramount importance of a rapid reduction of the concentration of the circulating amyloidogenic precursor via chemotherapy in AL patients.

REFERENCES

1. Dubrey SW, Cha K, Andersen J, et al. The clinical features of immunoglobulin light-chain (AL) amyloidosis with heart involvement (1998). *Q J Med*; **91:** 141-157.
2. Kyle RA, Gertz MA (1995). Primary systemic amyloidosis: clinical and laboratory features in 474 cases. *Semin Hematol*; **8:** 45-59.
3. Dispenzieri A, Kyle RA, Gertz MA, et al (2003). Survival in patients with primary systemic amyloidosis and raised serum cardiac troponins. *Lancet*; **361:** 1787-1789.

4. Palladini G, Campana C, Klersy C, et al (2003). Serum N-terminal pro Brain Natriuretic Peptide is a sensitive marker of myocardial dysfunction in AL amyloidosis. *Circulation*; **107**: 2440-2445.

5. Dispenzieri A, Kyle RA, Gertz MA, et al (2004). Prognostication of Survival using Cardiac Troponins and N-terminal pro-Brain Natriuretic Peptide in Patients with Primary Systemic Amyloidosis Undergoing Peripheral Blood Stem Cell Transplant. *Blood* [Epub ahead of print].

6. Bradwell AR, Carr-Smith HD, Mead GP, et al (2001). Highly sensitive, automated immunoassay for immunoglobulin free light chains in serum and urine. *Clin Chem*; **47**: 673-680, 2001.

7. Liao RL, Jain M, Teller P, et al (2001). Infusion of light chains from patients with cardiac amyloidosis causes diastolic dysfunction in isolated mouse hearts. *Circulation*; **104**: 1594-7.

ORGAN SURVIVAL IN AL AMYLOIDOSIS WITH RENAL INVOLVEMENT

Giovanni Palladini[1,2], Vittorio Perfetti[2], Laura Obici[1], Riccardo Caccialanza[3], Francesca Lavatelli[1], Giampaolo Merlini[1]

Amyloid Center, [1]Biotechnology Research Laboratories – Department of Biochemistry, [2]Internal Medicine and [3]Department of Applied Health Sciences – Human Nutrition Unit, University Hospital "IRCCS – Policlinico San Matteo", Pavia, and University of Pavia;

More than 50% of AL patients have renal involvement at diagnosis (1). Patients usually present with nephrotic syndrome or proteinuria. Elevation in serum creatinine is often not prominent at presentation, but end-stage renal failure can occur during the course of the disease and some patients eventually require dialysis (1).

The aim of therapy is to preserve and possibly to restore renal function, but renal impairment itself limits the therapeutic opportunities. Serum creatinine >2 mg/dL represents a contraindication for high-dose chemotherapy and autologous stem cell transplantation (2) and acute renal failure is a frequent complication of this procedure (3). Therefore, it is important to define reversibility criteria of renal damage, in order to design optimal therapeutic strategies.

PATIENTS AND METHODS

We reviewed the clinical records and studied the presentation and outcome of 409 consecutive AL patients who had renal involvement at diagnosis and were referred to the Amyloid Center in Pavia between June 1984 and December 2003. Renal involvement was defined as 24 hour urine protein >0.5 g/24 hours, predominantly albumin.

All patients had a histological diagnosis of amyloidosis and a clinical picture consistent with AL. A monoclonal component was detected by high-resolution agarose gel immunofixation electrophoresis (4) in 398 (97%) patients (288 λ, 79 κ, 31 biclonal). In the remaining 11 patients AL typing relied on immuno-electron microscopy (5).

Hematologic response to chemotherapy was defined as the decrease of the monoclonal component (MC) by at least 50%.

RESULTS

Among 600 patients diagnosed with AL, 409 (68%) presented with renal involvement. Histological diagnosis of amyloidosis was established on renal biopsy in 59 patients (14%) on abdominal fat aspiration in 218 (53%) and on both in 132 (33%). The sensitivity of abdominal fat aspiration was 87%.

Organs other than the kidney were involved in 308 patients (75%) and 224 patients (55%) had cardiac amyloidosis.

Patients' characteristics are reported in Table 1.

Table 1. Patients' characteristics

Median age (years)	62 (range: 28-91)
Gender (men)	226 (55%)
Proteinuria median (g/24h)	5.3 (range: 0.5-56.4)
<1 g/24h	29 (7%)
≥1 and <3.5 g/24h	94 (23%)
≥3.5 g/24h	286 (70%)
Serum creatinine median (mg/dL)	1.16 mg/dL
≤1.2 mg/dL (upper reference limit)	225 (55%)
>1.2 and ≤2 mg/dL	86 (21%)
>2 mg/dL	98 (24%)

Seventy-five patients (18%) progressed to end-stage renal disease and dialysis after a median time of 7.5 months (range: 0.1-97.5 months). Dividing the population into deciles according to the serum creatinine concentration, we observed that the dialysis rate per 100 person years started to increase in patients who presented with a creatinine >1.36 mg /dL (Figure 1).

Figure 1

Estimated dialysis rate in 409 AL patients presenting with renal involvement

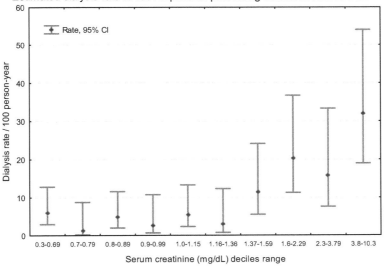

Three-hundred twenty-four patients (79%) started anti-clone therapy before needing dialysis. One-hundred seventy-one patients (53%) received standard melphalan and prednisone, 75 (23%) melphalan plus dexamethasone, 34 (11%) autologous stem cell transplantation, 26 (8%) high-dose dexamethasone, and 18 (5%) other forms of chemotherapy. One-hundred fifty-nine patients (49%) responded to treatment.

The Cox multivariate analysis showed that progression to dialysis was independently affected by younger age (p=.004), proteinuria (p=$7*10^{-5}$), serum creatinine (p=$5*10^{-10}$) and hematologic response to treatment (p=$8*10^{-5}$).

CONCLUSIONS

Nearly one fifth of AL patients presenting with renal involvement eventually progress to dialysis. A relatively low serum creatinine concentration (>1.36 mg/dL) is associated to a dialysis rate greater than 10 per 100 person-year. Thus, in renal AL, a subtle renal dysfunction must not be underestimated.

Higher serum creatinine and proteinuria at diagnosis predict progression to end-stage renal disease requiring dialysis, but hematologic response to chemotherapy can prevent this process. This underlines the importance of rapidly establishing an effective anti-clone therapy.

REFERENCES

1. Palladini G, Adami F, Anesi E, et al. AL amyloidosis: the experience of the Italian Amyloidosis Study Group. In: M. Bély, Á. Apáthy (eds.) Amyloid and Amyloidosis: the proceedings of the 9th International Symposium on Amyloidosis. 2001, pp. 249-251.

2. Comenzo RL, Gertz MA (2002). Autologous stem cell transplantation for primary systemic amyloidosis. *Blood*; **99:** 4276-4282.

3. Fadia A, Casserly LF, Sanchorawala V, et al (2003). Incidence and outcome of acute renal failure complicating autologous stem cell transplantation for AL amyloidosis. *Kidney Int*; **63:** 1868-1873.

4. Merlini G, Marciano S, Gasparro C, et al (2001). The Pavia approach to clinical protein analysis. *Clin Chem Lab Med*; **39:** 1025-1028.

5. Arbustini E, Verga L, Concardi M, et al (2002). Electron and immuno-electron microscopy of abdominal fat identifies and characterizes amyloid fibrils in suspected cardiac amyloidosis. *Amyloid*; **9:** 108-114.

NUTRITIONAL STATUS OF OUTPATIENTS WITH AL SYSTEMIC AMYLOIDOSIS

Riccardo Caccialanza[1], Giovanni Palladini[2], Hellas Cena[1], Christina Vagia[1], Carla Roggi[1], Giampaolo Merlini[2]

From the [1]Department of Applied and Psycho-behavioral Health Sciences, Human Nutrition Unit, Amyloid Center, [2]Biotechnology Research Laboratories, Department of Biochemistry (GP, GM), IRCCS Policlinico San Matteo and University of Pavia University of Pavia, Italy

INTRODUCTION

Although maintenance of good nutritional status is known to be associated with prolonged survival and better quality of life in many chronic progressive diseases (1-3), symptoms correlated to malnutrition, probably also due to the absence of universally standardized diagnostic criteria, still go often unrecognized and untreated (4, 5). Early identification of patients with poor nutritional status or an inadequate nutrient intake is essential in order to identify at risk patients and to target nutritional intervention towards those likely to benefit from it.

To date, the nutritional status of patients with AL has not been exhaustively evaluated, nor has been proposed any specific methodological approach. The aim of the present study was to provide information regarding the nutritional status of a representative sample of AL outpatients and investigate its relation with some clinical features of the disease.

MATERIALS AND METHODS

One-hundred six AL outpatients, 60 (56.6%) males, (median age 60 years; range: 28 – 78 years), evaluated at the Amyloid Center of the University Hospital San Matteo of Pavia, Italy, were consecutively enrolled into the study. All patients had a histological diagnosis of amyloidosis, a clinical history consistent with AL and a monoclonal component (79 λ, 21 κ, 6 biclonal) detected in serum and/or urine at high-resolution agarose gel immunofixation electrophoresis.

The clinical variables recorded were: presence of anorexia; dysphagia; dysgeusia; vomiting, considered as the occurrence of at least two episodes during the week prior to the survey; diarrhea, considered as the occurrence of at least two 24-hour periods with three or more loose or watery stools during the week prior to the survey. Non-edematous usual weight prior to the onset of signs or symptoms of the disease and the onset of unintentional weight loss were referred by the patients. The amount of weight loss and percentage weight loss were calculated. All patients were classified according to the Eastern Cooperative Oncology Group (ECOG) Performance Status scale (6) and those with cardiac involvement were classified according to the New York Heart Association (NYHA) classification (7). Body mass index (BMI) and mid-arm muscle circumference (MAMC) were the two anthropometric parameters recorded. The biochemical markers measured were serum albumin, transthyretin and transferrin.

RESULTS

Sixty-four patients (60%) had echographic signs of cardiac amyloidosis and 34 of them presented with heart failure. Urine protein loss was >1 g/24 h in 69 patients (65.1%) and >3 g/24 h in 52 (49%). Fifty-one patients (48.1%) had a serum creatinine >1.2 mg/L (upper reference limit) and 29 (27.3%) >2 mg/L. Eight patients (7%) were on dialysis. Twenty-five patients (24%) had liver involvement. At physical examination, 58 patients (55%) had peripheral edema and none had ascites. Fourteen patients (13.2%) were on a diet at the moment of the survey.

Thirty-six patients (34%) referred dysgeusia, 32 (30.2%) anorexia, 18 (17%) vomiting, 15 (14.1%) dysphagia, 9 (8.5%) diarrhea. Fifty-eight patients (54.7%) referred unintentional weight loss (median 8 kg, range 2-30 kg, i.e. median 11.3%, range 2.6-34% of usual non-edematous body weight) within the prior 2-72 months (median 12 months; mean weight loss per year 10.9 $\pm$ 1.3 Kg, mean percentage weight loss per year 14.8% $\pm$ 1.5% of usual non-edematous body weight). Mean percentage weight loss was higher among patients with cardiac involvement, heart failure (NYHA class $\geq$II) and ECOG Performance Status $\geq$2. Furthermore, it was distinctly different among NYHA (figure 1) and ECOG Performance Status (figure 2) classes. Median BMI was 24.5 (range 17.5 – 33.7); eight patients (7.5%) had a BMI > 30, 38 (35.8%) a BMI > 25 < 30, 9 (8.5%) a BMI < 20. Twenty-nine patients (27.3%) had a MAMC below the 10th age-sex matched centile. Twenty-six (24.5%) patients had serum transthyretin concentration < 200 mg/L (reference range: 200-400 mg/L). Mean percentage weight loss was higher among subjects with serum transthyretin concentration below the low reference limit in comparison with those who had serum transthyretin concentration within the normal range (10.3 $\pm$ 10.1% vs 5.8 $\pm$ 7.4%; *P* < 0.05). Sixty-seven patients (63.2%) had serum albumin < 35 g/L (reference range: 35-52 g/L). Fifty-six (52.8%) had serum transferrin < 2.0 g/L (reference range: 2.0-3.6 g/L). None of the evaluated anthropometric and biochemical variables was significantly different between patients who were on a diet and those who were not.

Twelve patients (11.3%) died by the end of the study.

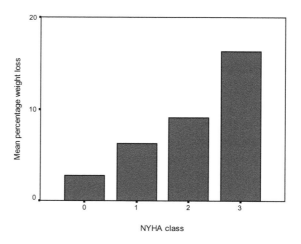

Figure 1. Mean percentage weight loss among NYHA classes

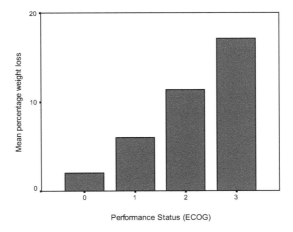

Figure 2. Mean percentage weight loss among ECOG Performance Status classes

CONCLUSIONS

The results of the study show that the nutritional status of patients with systemic AL amyloidosis is impaired with regards to several features. Unintentional weight loss is related to the severity of the disease, while anthropometric measurements are of limited importance in AL. Transthyretin concentration below the lower reference limit is a good indicator of undernutrition.

REFERENCES

1. Chima CS, Barco K, Dewitt MLA, Maeda M, Teran JC, Mullen KD (1997). Relationship of nutritional status to length of stay, hospital costs, and discharge status of patients hospitalized in the medicine service. *J Am Diet Assoc*; **97:** 975-78.

2. Lennard-Jones JE. A Positive Approach to Nutrition as Treatment. King's Fund Report. London: King's Fund Centre, 1992.

3. Gallagher-Allred CR, Coble Voss C, Finn SC, McCamish MA (1996). Malnutrition and clinical outcomes: the case for medical nutrition therapy. *J Am Diet Assoc*; **361:** 369.

4. Wilson MM (2002). Undernutrition in medical outpatients. *Clin Geriatr Med*; **18:** 759-71.

5. Reilly HM (1996). Screening for nutritional risk. *Proc Nutr Soc*; **55:** 841-53.

6. Oken MM, Creech RH, Tormey DC, Horton J, Davis TE, McFadden ET, Carbone PP (1982). Toxicity and response criteria of the Eastern Cooperative Oncology Group. *Am J Clin Oncol* ; **5:** 649-655.

7. The Criteria Committee of the New York Heart Association. Nomenclature and Criteria for Diagnosis of Diseases of the Heart and Great Vessels. 9th ed. Boston, Mass: Little, Brown & Co; 1994: 253-256.

CONGO RED FLUORESCENCE CIRCUMVENTS PITFALLS IN MORPHOMETRIC QUANTITATION OF AMYLOID IN TISSUE SECTIONS

Ugo Donini[1] and Reinhold P. Linke[2]

[1] *CRBA, Policlinico S. Orsola-Malpighi, Bologna/Italy (u.donini@tin.it)*
[2] *Max-Planck-Institut für Biochemie, Martinsried/Germany*

INTRODUCTION

Quantitation of amyloid deposits in tissue sections represents a *direct* method to precisely measure the amount of amyloid in patients and animals. The exact quantifying of amyloid allows evaluating the different therapies used to treat underlying disease in amyloidosis and comparing their efficacy to control amyloidogenic progression. Quantification of amyloid has been performed in brain tissue [1], in renal biopsies [2,3] and in liver biopsies [4] after immunohistochemical identification of amyloid using various chromogens. These results not only demonstrated fairly exact morphometric computer-aided quantification of amyloid [2-4], but also illustrated a remarkable disappearance of amyloid from the liver in a patient with successfully treated juvenile rheumatoid arthritis within several years [4]. In spite of this progress, there remains the question of which of the markers should be used for the identification of amyloid in order to improve precision, a high signal to background value (=*specificity value*), so as to measure the amount of amyloid exactly.

We recently showed that Congo red fluorescence (CRF) can improve the diagnosing of amyloid since it is most sensitive and specific [5] when combined with green birefringence (GB). We, therefore, compared CRF and the three other methods that are usually applied for diagnosing amyloid in tissue sections, that is, Congo red staining on light microscopy (CR), Congo red staining on polarized microscopy (GB) and immunohistochemistry (IHC).

MATERIAL AND METHODS

The formalin-fixed and paraffin-embedded tissue sections were obtained from renal biopsies of 3 patients, 1 with AA and 2 with AL amyloidosis. The 4 μm thick sections were stained with CR [6], and immunohistochemically by means of specific antibodies with the peroxidase-anti-peroxidase system using aminoethylcarbazol (AEC) as chromogen [7]. By video microscopy device applied to a Leitz microscope equipped with fluorescence we have obtained digitized black and white images in 256 grey levels of the glomeruli with amyloid deposits. A software (TOMO), produced in our laboratory, was used which easily distinguishes between the amyloid and background and which, therefore, can automatically recognise the amyloid deposits to be quantitated using a specialized algorithm and *not* a simple grey level-threshold. The objectives of the computer program were the following: 1) to measure the *specificity value* which is the difference between the optical density (OD) of the amyloid deposits and the OD of the background and which indicates the image quality; 2) to measure the percentage of amyloid area on glomerular area; 3) to stain with a new color (homogeneous grey as is shown in Figure 2 or red) the amyloid measured to control the process. The *specificity value* is, therefore, very important in validating a given procedure of computerized morphometry. We have compared the *specificity value* on digitized images obtained in CR, GB, IHC and CRF. With CRF it was possible to obtain by video microscopy the digitized images, which were appropriate for computer quantification that is with a *specificity value* higher than 89, and this technique was used for our study. Ten different glomeruli were measured on the *same* section in each of the renal biopsies examined and the mean value of amyloid percentage with standards deviation was considered as the right measure of amyloid deposits. The method used applied to CRF is illustrated in Figure 1. The glomerulus with amyloid can be processed from TOMO in different way (see inset in Figure 1a). We chose to inscribe the glomerulus in a polygon following the Bowman capsule. The measures are taken inside the polygon as follows: 1) *total area* of the selected polygon expressed in pixels; 2) *total OD* of image examined (mean of the grey levels of the image inside the polygon); 3) *background OD* (mean of the grey levels of the background, of the part of image which was considered automatically from TOMO as not-amyloid deposit); 4) *deposits OD* (mean of grey level of the amyloid deposits, as recognized from TOMO); 5) *specificity value* (absolute difference between deposits OD and background OD); 6) the *amount of amyloid* as percentage of amyloid deposits on glomerulus total area (Figure 1b).

RESULTS

When we obtained the first values of CR and GB, it was clear that they could never match the value of IHC

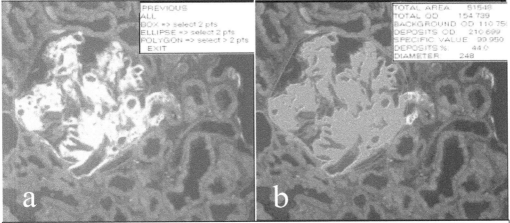

Figure 1. Digitized image of glomerulus on CRF before (a) and after (b) processing with TOMO. The computer-marked amyloid is colored in homogeneous grey and the results are written in the inset (b).

and CRF. The values of CR and in particular GB showed a very low *specificity value*. When, however, values of IHC and CRF were compared, the latter values were the most consistent as reported in Table 1. For this we considered the CRF the valid method for obtaining digitized images for amyloid quantification.

IHC provided moderate contrast and was not equally intensive in all parts of amyloid resulting in a lower *specificity value* while the CRF showed a uniform intensive signal over all parts of the amyloid deposits with very high *specificity value* resulting in a reproducible amyloid measure. We, therefore, continued with only the CRF measurements.

Table 1 Specificity value measured with TOMO applied to digitized images obtained with different methods

Staining and observation method	Specificity value	Standard deviation
Congo red observed on light microscopy: CR	21.2	± 8.35
Congo red observed on polarized light: GB	25.5	± 6.12
Congo red observed on fluorescence microscopy: CRF	118.6	± 20.18
Immunohistochemistry (peroxidase labelling): IHC	49.6	± 6.50

Table 2 shows the results obtained using CRF for quantifying amyloid deposits. The standard deviation

Table 2 Morphometric quantification of amyloid using CRF

Case	Quantity of amyloid (% of glomerular surface)	Standard deviation
N. 1	23.44%	± 3.61
N. 2	11.93 %	± 4.24
N. 3	13.02%	± 3.02

ranges between 3,02 and 3,61, but reflects also the variability in amyloid deposits extension in different glomeruli.

CONCLUSION

A comparison of the four methods CR, GB, CRF and IHC shows that computer amyloid quantification using CRF is the best method since the average values are the most consistent, it is immediate because it doesn't need further staining to be performed and utilizes the standard Congo red staining (6) used for amyloidosis diagnosis. This method of amyloid quantification needing only CRF can, therefore, be recommended to be applied for the exact quantification of amyloid for clinical application.

REFERENCES

1. Gentleman, S.M. et al., A demonstration of the advantages of immunostaining in the quantification of amyloid plaque deposits, *Histochemistry*, 92(4), 355, 1989

2. Donini, U. et al., Computer-aided immunohistochemical quantitation of amyloid in renal biopsies, in *Amyloid and Amyloidosis*, Natvig J.B. et al., Eds., Kluwer Academic Publishers, Dordrecht; 793, 1991

3. Donini, U., Zucchelli, P., Baraldini, M., Linke, R.P., Robot assisted classification of amyloid deposits, in *Amyloid and Amyloidosis1993*, Kisilevsky, R., et al. Eds, Parthenon Publishing, New York-London, 676, 1994

4. Michels, H., Donini, U. and Linke, R.P., Resolution of AA-amyloidosis in three patients with juvenile rheumatoid arthritis (JRA) as shown with computer-aided quantitation of amyloid, in *Amyloid and Amyloidosis 1993*, Kisilevsky, R. et al. Eds, Parthenon Publishing, New York-London, 694, 1994

5. Linke, R.P., Highly sensitive diagnosis of amyloid and various amyloid syndromes using Congo red fluorescence, *Virchows Arch. Path. Anat.*, 436, 439, 2000

6. Puchtler, H., Sweat, F. and Levine, M., On the binding of Congo red by amyloid, *J. Histochem. Cytochem.*, 10, 355, 1962

7. Linke, R.P., Gärtner, V. and Michels, H., High sensitive diagnosis of AA-amyloidosis using Congo red and immunohistochemistry detects missed amyloid deposits, *J. Histochem. Cytochem.*, 43, 863, 1995

CARDIAC TROPONINS AND N-TERMINAL PRO-BRAIN NATRIURETIC PEPTIDE PREDICT SURVIVAL IN PATIENTS WITH PRIMARY SYSTEMIC AMYLOIDOSIS UNDERGOING PERIPHERAL BLOOD STEM CELL TRANSPLANT

A. Dispenzieri, R.A. Kyle, M.Q. Lacy, A.S. Jaffe, T.M. Therneau, M.F. Burrit, and M.A. Gertz.

Mayo Clinic, 200 First Street S.W., Rochester, MN 55905, USA

I. ABSTRACT

Primary systemic amyloidosis (AL) is a disease without cure. Pilot data suggest survival is better in patients undergoing peripheral blood stem cell transplant (PBSCT), but the selection process makes the apparent benefit suspect. We have reported that circulating cardiac biomarkers are the best predictors of survival outside of the transplant setting. We now test whether cardiac troponins (cTnT and cTnI) and N-terminal pro-brain natriuretic peptide (NT-proBNP) are prognostic in transplant patients. Ninety-eight patients with AL undergoing PBSCT had serum cardiac biomarkers measured (cTnT, 98; cTnI, 65; and NT-proBNP, 63 patients). The troponin and NT-proBNP values and scores of these patients were compared to those of our previously reported cohort of 242 patients not undergoing transplant. Elevated levels of cTnT, cTnI, NT-proBNP were present in 14%, 43%, and 48%. Median survival has not been reached for patients with values below the thresholds at 20 months and if above is 26.1, 66.1 and 66.1 months, respectively. Our previously reported risk and staging systems incorporating these markers were also prognostic. Despite the observation that the PBSCT group was a significantly lower risk group than our previously reported non-transplant group, PBSCT was associated with superior survival after correction for risk or stage.

II. INTRODUCTION

A major drawback in the field of primary systemic amyloidosis (AL) is the absence of a prognostic staging system, making comparisons between different studies and treatment centers difficult to interpret. Over the past decade, we and others have introduced peripheral blood stem cell transplant as a therapeutic option for selected patients. Phase II and retrospective analysis suggest that overall survival is better in patients who receive this therapy. However, patients receiving this procedure are highly selected bringing into question the interpretation of the apparent improved survival. We have recently reported that in patients receiving standard chemotherapy, circulating cardiac biomarkers are the most powerful independent predictors of survival.[1] Cardiac troponins (cTnT, cTnI) are highly specific markers of myocardial injury. Pro-brain natriuretic peptide is a 108 amino acid propeptide, made by myocytes in response to increased wall stress. When released from the left ventricle, it is cleaved into two fragments, the active brain natriuretic peptide (amino acids 77-108) and a leader sequence known as NT-proBNP (amino acids 1-76). NT-proBNP has been shown to be a sensitive indicator of cardiac abnormalities. Blood levels of N-terminal pro-brain natriuretic peptide (NT-pro-BNP) provide prognostic information in patients with AL treated with conventional chemotherapy.[2] Moreover, cardiac troponins and NT-pro-BNP are independent predictors of survival in this population of patients, and we have constructed for patients treated with conventional chemotherapy a powerful staging system by combining these two blood tests. We have now used this staging system to determine its prognostic value in patients undergoing PBSCT.

III. PATIENTS AND METHODS

Ninety-nine patients with primary systemic amyloidosis who were transplanted between 3/8/96 and 6/30/03 had serum cardiac troponin levels measured. All samples were obtained within 90 days prior to peripheral blood stem cell infusion. Serum cardiac troponin levels (troponin T [cTnT] and troponin I [cTnI]) were run from stored serum (-20°C) for 65 patients, and an additional 33 had cTnT levels measured prospectively as part of their routine evaluation. NT-proBNP levels were measured from stored serum in 63 of the same patients based on serum availability. Stem cells were mobilized using either cyclophosphamide and GM-CSF or G-CSF alone. Prior to receiving their stem cells, patients received melphalan doses ranging from 100 mg/m^2 to 200 mg/m^2. Patients were given prophylactic antibiotics and supportive care as previously reported.[3]

Four model systems were used: cTnT Idealized Risk Score;[1] cTnI Idealized Risk Score;[1] cTnT/NT-proBNP Staging System; and the cTnI/NT-proBNP Staging System. The cTnT and cTnI Idealized Risk Score are defined as previously described.[1] The two tropoinin/NT-proBNP Staging Systems are defined as follows: Stage I (low risk) when both troponin and NT-proBNP are below the threshold, Stage III (high risk) if both are equal to or above the threshold, and Stage II (intermediate risk) if only one marker is below the threshold. Threshold values are: cTnT <0.035 µg/L; cTnI <0.1 µg/L and NT-proBNP < 332 ng/L. The suffixes [-t or –i] refer to which troponin [cTnT or cTnI] assay was used in the staging system.

IV. RESULTS

For the 98 PBSCT patients, the median time from histologic diagnosis to transplant was 4.2 months and median follow up was 20 months. Twenty-one patients have died. Fourteen percent of patients (14/98) had elevated (≥ 0.035) cTnT values. Forty-four percent of patients tested had a cTnI level greater than or equal to 0.1 ◆g/L. The NT-proBNP level was greater than or equal to 332 ng/L in 48% of patients. The median cTnT and cTnI Idealized Risk Scores were each 1.8. Median overall survival for the entire transplant group has not been reached, and 75% of patients are alive at 26 months. Ten patients (10%) died within 3 months of

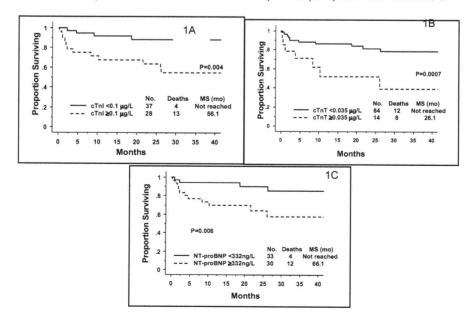

their transplant i.e. potentially treatment related mortality. On univariate logistic regression, only cTnI greater than or equal to 0.1 µg/L, beta-2 microglobulin greater than or equal to 2.7, creatinine greater than or equal to 1.7 mg/dl, and the numbers of organs involved were significant predictors of early death.

  Median overall survival has not yet been reached for the cohort. On univariate analysis all three cardiac biomarkers were predictive for survival as single variables and as part of previously reported prognostic models or staging systems. The Kaplan-Meier survival curves are shown in Figure 1. The respective hazard ratios of death for elevated cTnT, cTnI, and NT-proBNP were: 4.4, 95%CI 1.7-10.9; 4.9, 95CI 1.6-15.1; 3.8 95% CI 1.2-12.0. The models are also of value in the PBSCT population (Figure 2) and they demonstrate that the transplant cohort is a relatively low risk population.

V. DISCUSSION

Our data illustrate several important issues. The first is that serum levels of cardiac troponins and NT-proBNP are valuable predictors of overall survival in AL patients undergoing PBSCT, as they are in patients not undergoing this intensive procedure [1,2]. The second is that elevations of cTnI, but not cTnT and/or NT-proBNP, are predictive for mortality within 90 days of transplant. The third is that patients undergoing PBSCT are lower than average risk because they are highly selected to undergo the rigorous procedure.

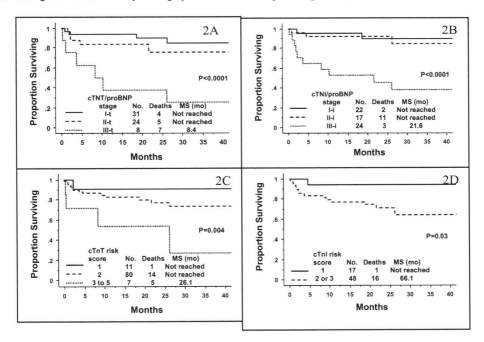

VI. REFERENCES

1.   Dispenzieri A, et al. Survival in patients with primary systemic amyloidosis and raised serum cardiac troponins. Lancet. 2003;361:1787-1789.
2.   Palladini G, et al. Serum N-terminal pro-brain natriuretic peptide is a sensitive marker of myocardial dysfunction in AL amyloidosis. Circulation. 2003;107:2440-2445.
3.   Gertz MA, et al. Stem cell transplantation for the management of primary systemic amyloidosis. Am J Med. 2002;113:549-555.

BNP AS A MARKER OF CARDIAC INVOLVEMENT IN AMYLOIDOSIS

M. Nordlinger, B. J. Magnani, M. Skinner, R. H. Falk,

Boston University Medical Center, Amyloid Research and Treatment Program

INTRODUCTION

Amyloidosis is a multisystem disease commonly affecting the heart. Cardiac amyloidosis (CA), which is usually defined by echocardiographic abnormalities, (increased LV wall thickness) affects the right and left ventricles. Therefore it is generally assumed that heart failure (CHF) in amyloidosis is associated with an elevated JVP. Cardiac involvement in amyloidosis is a marker of poor survival in untreated disease, and a marker of treatment-related morbidity and mortality in patients undergoing high dose chemotherapy (1).

B natriuretic peptide (BNP) has emerged as a sensitive and semi-quantitative tool for the diagnosis of CHF. Levels of BNP fluctuate with the degree of (elevated) LV filling pressures and can be used to monitor patients. Although there is an age related increase in BNP, the upper limit of normal is generally considered to be 100 pg/ml in the absence of renal failure.

In an imunohistochemical study of endomyocardial biopsies from patients with cardiac amyloidosis, Takemura et al demonstrated that BNP expression was augmented compared to controls, and observed that the myocytes adjacent to amyloid deposits tended to show more intense staining for BNP than did those that were not abutting extracellular amyloid (2). They postulated that regional mechanical stress by amyloid deposits in addition to hemodynamic stress induced by diastolic function may therefore be responsible for the expression of BNP in patients with cardiac amyloidosis (2).

We therefore sought to determine whether clinical signs of CHF in cardiac amyloidosis are associated with higher BNP levels than are found in asymptomatic cardiac amyloidosis patients and whether BNP is elevated in cardiac amyloidosis in the absence of clinical signs of CHF when the echocardiogram is abnormal (asymptomatic cardiac amyloid)

METHODS

Without knowledge of the patients' BNP levels, a cardiologist who specializes in cardiac amyloidosis examined all patients and reviewed the patients' studies. Patients with cardiac amyloidosis by echocardiogram were divided into those determined clinically to have CHF (increased JVP and/ or dyspnea with CXR evidence of pulmonary congestion) and those without. Patients were excluded because of dialysis (n=3) and prior heart transplantation (n=2). The concentration of BNP was tested by using a fluorescence immunoassay that does not require radioisotope labeling of the sample. BNP was determined as normal according to the control values set by the manufacturer, Advia Centaur, Bayer Diagnostics, Tarrytown, NY

RESULTS

As anticipated, patients in Group III (CHF) had elevated levels of BNP. However, among Group II patients, (asymptomatic cardiac amyloidosis) 9/10 had a BNP level above 100pg/ ml. There was no statistical difference in mean BNP levels between Groups II and III. (P=NS). However, Group I vs. Group II and Group I vs. Group III were highly significant. *(p< 0.001).*

8/38 patients in Group I had an elevated BNP above 100 pg/ml. Review of the 12-lead ECGs in these patients revealed that 7/8 had abnormal ECGs, four of whom who had low voltage. Of the remaining 31 pts in Group I with a normal BNP, 10 had abnormal ECGs, 3 of those with low voltage. The finding of an abnormal ECG in Group 1 was associated with an elevated BNP (p=0.005 by Fisher's exact test.)

Table 1. Elevation of BNP among Groups I, II, and III

	Group I No Cardiac Amyloidosis	Group II Cardiac Amyloidosis without CHF	Group III Cardiac Amyloidosis with CHF
Mean BNP ($\pm$ SEM)	75 (15.9)	532 (194)	551(84)
BNP >100	8	9	15
BNP $\leq$ 100	30	1	2
TOTAL	38	10	17

CONCLUSION

Among patients without echo evidence of cardiac amyloidosis, BNP elevation above 100 pg/ml occurred 21% of the time. While this may represent undiagnosed non-amyloid heart disease, the finding that a significantly higher proportion of these patients had low voltage on the ECG suggests that subclinical cardiac amyloid infiltrate may produce an elevated BNP despite a normal 2-D echo.

B natriuretic peptide was a very sensitive marker of cardiac involvement based upon the presence of an abnormal echocardiogram. However, mean values of BNP were no greater in Group III patients than in Group II. Although these results may suggest that LV filling pressure is elevated before jugular venous pressure, an alternative hypothesis is that myocyte compression by interstitial amyloid deposits produces BNP elevation even in the absence of CHF To determine which of these hypotheses is correct, a study of hemodynamic, BNP and echo correlation will be needed.

REFERENCES

1. Dubrey, et al, The Clinical Features of Immunoglobulin Light Chain (AL) amyloidosis with heart involvement, QJ Med, 1998, 91:141-157

2. Takemura, et al, Expression of Atrial and Brain Natriuretic Peptides and their Genes in Hearts of Patients with Cardiac Amyloidosis, Journal of the American College of Cardiology, 31:754-765
Supported by grants from NIH P01-HL68705 and the Gerry Foundation.

QUANTITATIVE SERUM FREE LIGHT-CHAIN ASSAY IN THE DIAGNOSTIC EVALUATION OF AL AMYLOIDOSIS

Akar, H.,[1] Seldin D.C,[1,2,3] Magnani, B.,[4] O'Hara, C.,[1,4,5] Berk, J.L.,[1,3] Schoonmaker, C.,[7] Cabral, H.,[7] Dember, L.M.,[1,3] Sanchorawala, V.,[1,2,3] Connors, L.H.[1,5] Falk, R.H.,[1,3] Skinner, M.,[1,3].

From the Amyloid Treatment and Research Program[1] and the Stem Cell Transplant Program[2], Departments of Medicine[3], Laboratory Medicine[4], Pathology[5] and Biochemistry[6] at Boston Medical Center, Boston University School of Medicine and School of Public Health[7], Boston MA 02118.

Supported by grants from the National Institutes of Health (HL 68705), the Gerry Foundation, the Young Family Amyloid Research Fund, and the Amyloid Research Fund at Boston University. H. Akar is a recipient of a fellowship from the Turkish Society of Nephrology. D.C.Seldin is a Scholar of the Leukemia and Lymphoma Society of America.

1.1 INTRODUCTION

Primary systemic amyloidosis (AL) is a plasma cell dyscrasia (PCD) in which clonal plasma cells in the bone marrow produce a monoclonal protein that is amyloidogenic (1). After the histopathological diagnosis of amyloid deposition is made from a biopsy specimen, subsequent typing as AL amyloidosis relies upon the detection of the monoclonal light chain by serum or urine immunofixation electrophoresis (SIFE/UIFE), or bone marrow immunohistochemistry (BM). The serum free light chain (FLC) assay was developed to measure the elevated levels of unbound light chains in the serum and urine of patients with PCD (2, 3, 4). We compared the serum free light-chain immunoassay (FLC assay) for the quantitation of free light chains to the standard assays for plasma cell disease in terms of sensitivity and specificity. FLC levels are elevated in renal insufficiency which is common in AL amyloidosis, thus the utility of the serum FLC κ:λ ratio to detect light chain abnormalities in AL amyloidosis was also examined.

1.2 STUDY DESIGN

The free light chain assay was studied on 169 patients with systemic AL amyloidosis and 20 controls. AL amyloidosis was diagnosed by a positive tissue biopsy and evidence for a PCD by finding a monoclonal light chain by BM, SIFE, and/or UIFE. Control serum samples were obtained from 19 patients with a clinical diagnosis of familial amyloidosis confirmed by genetic testing (17 ATTR, 1 fibrinogen and 1 apolipoprotein A-II amyloidosis) and 1 patient with senile cardiac amyloidosis. Studies were approved by the Institutional Review Board of Boston University Medical Center. Serum samples were obtained from patients and controls at the time of initial evaluation. Serum and urine immunofixation electrophoresis (SIFE/UIFE) were performed using the Sebia HYRYS (Sebia, Norcross, GA) system (serum, 169 cases; urine, 167 cases). SIFE and UIFE were both run on the Sebia HYDRAGEL 4 IF using acid violet staining. Results were interpreted as positive if there was a single restricted band on the gel indicating a monoclonal protein. Serum free light chains were assayed using a latex-enhanced immunoassay on a nephelometric analyzer (Freelite™ the Binding Site, SanDiego, CA), on the Beckman Coulter Immage Immunochemistry System (Beckman Coulter, Fullerton CA).

The FLC was scored as normal or abnormal according to the values set by the manufacturer. The chi-square test was used to compare categorical data and the Wilcoxon rank sum test was used to compare median values.

1.3 RESULTS

In each patient with AL amyloidosis, the clonal light chain type (λ,142; κ,27) was identified by serum or urine immunofixation electrophoresis (SIFE/UIFE), or bone marrow immunohistochemistry (BM-IHC). Table 1 shows the test result of patients and controls. Table 2 and Table 3 show the sensitivity, specificity, positive and negative predictive values (NPV) of traditional tests and the FLC assay in all patients.

1.4 DISCUSSION

As an alternative to traditional tests for AL amyloidosis, a quantitative FLC assay for describing the light chain levels was introduced (2). Elevated levels are interpreted as monoclonal when they occur in only one light chain isotype, resulting in an elevated level or an abnormal ratio. The role of the free light chain assay has been assessed in the diagnostic evaluation of multiple myeloma and AL amyloidosis (4, 5, 6). FLC assay reported to be more sensitive than SIFE or UIFE (5). In our study, an elevated level of κ and λ-FLC among each clonal group of patients (96%, 94%, respectively) was present more often than in any of the traditional tests. However the false positive rate for κ and λ-FLC were each high (30 % and 44 %, respectively). Falsely elevated levels of FLC may be due to age, renal disease, or acute phase response. In one report, the median serum κ and λ FLCs were found to increase with age, with an apparent increase of both in κ and λ FLCs for those > 80 years, while FLC κ:λ ratio remained constant (3). In AL amyloidosis the high incidence of renal disease is a more common cause of falsely elevated FLCs (7, 8) and thus the ratio of κ:λ FLC may be the more useful marker for clonality. Data about the sensitivity of the FLC κ:λ ratio is limited. The FLC κ:λ ratio showed a sensitivity of 89% for the diagnosis of AL among kappa clonal disease and was also higher than that seen with SIFE or UIFE (59% and 67%, respectively). For lambda clonal disease the sensitivity of the FLC κ:λ ratio (73 %) was lower than either SIFE or UIFE (79% and 92 %, respectively). The abnormal κ:λ ratio appears to be both sensitive and specific for the detection of clonality and it may be more useful to combine both κ and λ FLC and FLC κ:λ ratio in the diagnostic evaluation of AL amyloidosis.

The positive predictive value (PPV) was better for immunofixation results among kappa clonal disease, but was comparable for all tests in patients with lambda clonal disease. NPV was comparable for all tests among patients with kappa and lambda clonal disease.

The FLC assay may be useful screening test in patients who have symptoms suggestive of AL amyloidosis and supplement other diagnostic testing. Our data are given for patients with known clonal disease. A drawback of the FLC assay that it is not a test for monoclonal proteins, and for each patient this must be defined by one of the traditional tests. The false positive elevations that occur in renal disease and acute phase responses are of less concern when the κ:λ ratio is normal.

The quantitative nature of the FLC assay are likely to have value in monitoring patients after treatment. At present, diagnosis of AL rests upon a panel of hematologic tests, and in some cases, genetic testing to exclude other forms of amyloidosis.

1.5 REFERENCES

1. Glenner GG, Terry W, Harada M, Isersky C, Page D. Amyloid fibril proteins: proof of homology with immunoglobulin light chains by sequence analyses. Science. 1971;172:1150-1151.
2. Bradwell AR, Carr-Smith HD, Mead GP, Tang LX, Showell PJ, Drayson MT, Drew R. Highly sensitive, automated immunoasay for immunoglobulin free light chains inserum and urine. Clin Chem 2001;47:673-680.
3. Katzman JA, Clark RJ, Abraham RS, Bryant S, Lymp JF, Bradwell AR, Kyle RA. Serum refence intervals and diagnostic ranges for free kappa and lambda immunoglobulin light chains : relative sensitivity for detection of monoclonal light chains. Clin Chem 2002;48:1437-1444.
4. Abraham RS, Clark RJ, Bryant SC, Lymp JF, Larson T, Kyle RA, Katzmann JA. Correlation of serum immunoglobulin free light chain quantification with urinary Bence Jones protein in light chain myeloma. Clin Chem 2002;48:655-7.
5. Abraham RS, Katzmann JA, Clark RJ, Bradwell AR, Kyle RA, Gertz MA. Quantitative analysis of serum free light chains. A new marker for the diagnostic evaluation of primary systemic amyloidosis. Am J Clin Pathol 2003;119:274-8.
6. Lachmann HJ, Gallimore R, Gillmore JD, Carr-Smith HD, Bradwell AR, Pepys MB, Hawkins PN. Outcome in systemic AL amyloidosis in relation to changes in concentration of circulating free immunoglobulin light chainsfollowing chemotherapy. Br J Haematol 2003;122:78-84.

7.   Fadia A, Casserly LF, Sanchorawala V, Seldin DC, Wright DG, Skinner M, Dember LM. Incidence and outcome of acute renal failure complicating autologous stem cell transplantation for AL amyloidosis. Kidney Int 2003;63:1868-1873.
8.   Casserly LF, Fadia A, Sanchorawala V, Seldin DC, Wright DG, Skinner M, Dember LM. High-dose intravenous melphalan with autologous stem cell transplantation in AL amyloidosis-associated end-stage renal disease. Kidney Int 2003;63:1051-1057.

Table 1. Test Result of Patients and Controls

	AL (N = 169)	AL Kappa clonal (N = 27)	AL Lambda clonal (N = 142)	Control (N = 20)	Comparing Kappa to Lambda
	N (%)	N (%)	N (%)	N (%)	p-value
Abnormal Bone Marrow Clonality	150 (89%)	25 (93%)	125 (88%)	ND	0.49
Abnormal SIFE	128 (76%)	16 (59%)	112 (79%)	0*	0.03
Abnormal UIFE	148 (88%)	18 (67%)	130 (92%)	0*	0.0003
Abnormal κ-FLC	69 (41%)	26 (96%)	43 (30%)	8 (40%)	NA
Abnormal λ-FLC	145 (86%)	12 (44%)	133 (94%)	1 (5%)	NA
Abnormal κ:λ ratio	127 (75%)	24 (89%)	103 (73%)	2 (10%)	0.07
Serum Creatinine <= 1.5 mg/dl	128 (76%)	18 (67%)	110 (77%)	19 (95%)	0.23

*UIFE and SIFE were each not determined for one different control patient.
SIFE/UIFE: monoclonal light chain by serum/urine immunofixation electrophoresis
Abnormal κ-FLC if quantitative κ-FLC was greater than 19.4, abnormal λ-FLC if quantitative λ-FLC was greater than 26.3, abnormal FLC k/l ratio if the ratio of free κ:λ was less than 0.26 or greater than 1.65.
Chi-Square test is used to compare categorical data among kappa/lambda patients.
ND: Not determined
NA: Not applicable

Table 2. Kappa Clonal Disease: Validity of Tests for the Diagnosis of AL Amyloidosis

	Sensitivity (95% CI)	Specificity(95% CI)	PPV(95% CI)	NPV(95% CI)
Bone Marrow Clonality	93%(83% - 100%)	100%(100% - 100%)	100%(100% - 100%)	91%(79% - 100%)
Abnormal SIFE	59%(41% - 78%)	100%(100% - 100%)	100%(100% - 100%)	65%(48% - 81%)
Abnormal UIFE	67%(49% - 84%)	100%(100% - 100%)	100%(100% - 100%)	69%(52% - 86%)
Abnormal SIFE or UIFE	81%(67% - 96%)	100%(100% - 100%)	100%(100% - 100%)	80%(64% - 96%)
Abnormal SIFE andUIFE	44%(26% - 63%)	100%(100% - 100%)	100%(100% - 100%)	57%(41% - 74%)
Abnormal κ-FLC	96% (89% - 100%)	60% (39% - 81%)	76%(62% - 91%)	92%(78% - 100%)
Abnormal κ:λ ratio	89%(77% - 100%)	90%(77% - 100%)	92%(82% - 100%)	86%(71% - 100%)

*Traditional Tests: One of the following positive: BM, SIFE or UIFE

Table 3. Lambda Clonal Disease: Validity of Tests for the Diagnosis of AL Amyloidosis

	Sensitivity (95% CI)	Specificity(95% CI)	PPV(95% CI)	NPV(95% CI)
Bone Marrow Clonality	88%(83% - 93%)	100% (100% - 100%)	100%(100% - 100%)	54%(83% - 93%)
Abnormal SIFE	79%(72% - 86%)	100% (100% - 100%)	100%(100% - 100%)	40%(26% - 54%)
Abnormal UIFE	92%(87% - 96%)	100% (100% - 100%)	100%(100% - 100%)	63%(46% - 79%)
Abnormal SIFE or UIFE	99%(98% - 100%)	100% (100% - 100%)	100%(100% - 100%)	95%(86% - 100%)
Abnormal SIFE and UIFE	71%(64% - 79%)	100% (100% - 100%)	100%(100% - 100%)	33%(21% - 45%)
Abnormal λ-FLC	94%(90% - 98%)	95%(85% - 100%)	99%(98% - 100%)	68%(51% - 85%)
Abnormal κ:λ ratio	73%(65% - 80%)	90%(77% - 100%)	98%(95% - 100%)	32%(20% - 44%)

*Traditional tests: One of the following positive: BM, SIFE or UIFE

AL AMYLOIDOSIS-ASSOCIATED HYPOTHYROIDISM: PREVALENCE AND RELATIONSHIP TO NEPHROTIC SYNDROME

J.S. Bhatia, V. Sanchorawala, D.C. Seldin, L.F. Casserly, J.L. Berk, R. Falk, M. Skinner, L.M. Dember

Renal Section and Amyloid Treatment and Research Program, Boston University School of Medicine, Boston, Massachusetts, USA

INTRODUCTION

The thyroid gland is a common site for amyloid deposition in AL amyloidosis [1]. However, it is not known how often there is associated hypothyroidism. Renal involvement is also common in AL amyloidosis and usually manifests as persistent nephrotic syndrome. Nephrotic range proteinuria results in urinary loss of thyroid hormones and their binding proteins and has been reported in small series to be associated with thyroid function abnormalities [2-5]. The objectives of this study were to determine the prevalence of hypothyroidism in AL amyloidosis and to evaluate the impact of nephrotic syndrome on its occurrence. In addition, we explored the effect of treatment with high dose intravenous melphalan and autologous stem cell transplantation on AL amyloidosis-associated hypothyroidism.

METHODS

Records of all patients with AL amyloidosis evaluated at Boston University between January 1, 1996 and December 31, 2001 were reviewed. Serum thyroid stimulating hormone (TSH) and 24-hour urinary protein excretion were measured at initial evaluation and all follow-up visits.

Hypothyroidism was defined as TSH >5.5µIU/ml or treatment with thyroid replacement hormone. Nephrotic-range proteinuria was defined as urinary protein excretion >3 grams per 24 hours. Patients who were dialysis-dependent or taking medications that can affect thyroid function (amiodarone, methimazole, propylthiouracil, lithium carbonate, interferon alpha, cholestyramine, phenytoin, and carbamazepine) were excluded from the analysis. In addition, patients with a history of thyroidectomy, radioiodine therapy, or external neck irradiation were excluded.

RESULTS

Four hundred fifty-five patients with AL amyloidosis were evaluated at Boston University during the five-year period. Of these patients, 379 patients were included in the analysis. Forty-eight patients did not have TSH values, 14 patients were on dialysis, 13 patients were on medications that affect thyroid function, and 1 patient had a prior thyroidectomy.

Hypothyroidism was present in 100 of the 379 patients (26%). This compares with a 9.5% prevalence of hypothyroidism reported for the general adult population [6]. As shown in Figure 1, the proportion of AL amyloidosis patients with hypothyroidism was similar among those with and those without nephrotic syndrome

(28% vs 24%, p=0.45). Urinary protein excretion (median, range) was not different between the hypothyroid (4.4 g/day, 0.06-27.5) and euthyroid patients (3.6 g/day, 0.47-24.6) (p=0.63). In addition, among patients with nephrotic syndrome, the degree of urinary protein excretion did not differ between those who were euthyroid (7.4 g/day, 3.2-24.6) and those who were hypothyroid (7.6 g/day, 3.3-27.5) (p=0.8) (see Figure 2).

Figure 1. Hypothyroidism in Nephrotic versus Non-nephrotic Patients

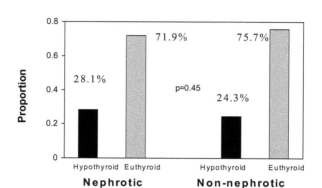

Figure 2. Proteinuria in Hypothyroid versus Euthyroid Nephrotic Patients

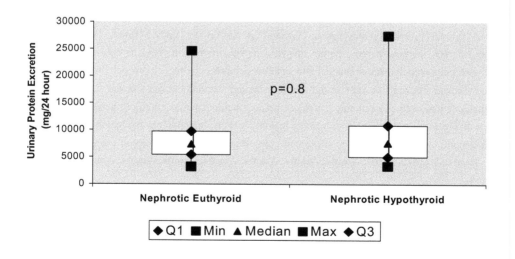

Thirty-five patients with hypothyroidism underwent treatment for AL amyloidosis with high dose intravenous melphalan and autologous stem cell transplantation. Eleven of these patients (31%) had a complete hematologic response evident 12 months after treatment. Two of the 11 patients (18%) with a complete hematologic response and 4 of 24 patients (17%) with persistence of the plasma cell dyscrasia had resolution of the hypothyroidism defined as normalization of TSH concentration in the absence of thyroid replacement hormone therapy.

CONCLUSIONS

The prevalence of hypothyroidism in AL amyloidosis is substantially greater than that in the general adult population. There appears to be no relationship between the either the presence or the quantity of proteinuria and hypothyroidism. The absence of a relationship between thyroid function and proteinuria suggests that thyroid infiltration rather than urinary losses of thyroid hormone and/or binding proteins underlies the hypothyroidism in AL amyloidosis. AL amyloidosis-associated hypothyroidism appears to be reversible following high dose chemotherapy in a small proportion of patients.

REFERENCES

1. Arean VM, Klein, Robert E: Amyloid Goiter. Am J Clin Pathol 36:341-355, 1961
2. Gavin LA MF, Castle JN, Cavalieri RR: Alterations in Serum Thyroid Hormones and Thyroxine-binding Globulin in Patients with Nephrosis. J Clin Endocrinol Metab 46:125-130, 1977
3. Afrasiabi MA VN, Gwinup G, Mays D, Barton CH, Ness RL, Valenta LJ: Thyroid Function studies in the Nephrotic Syndrome. Annals of Internal Medicine 90:335-338, 1979
4. McLean RH, Kennedy TL, Rosoulpour M, Ratzan SK, Siegel NJ, Kauschansky A, Genel M: Hypothyroidism in the Congenital Nephrotic Syndrome. J Pediatr 101:72-75, 1982
5. Rich MW: Hypothyroidism in Association with Systemic Amyloidosis. Head Neck 17:343-345, 1995
6. Canaris, GJ et al: Colorado Thyroid Disease Prevalence Study. Arch Int Med 160: 526-534, 2000

Supported by grants from NIH P01-HL68705 and the Gerry Foundation.

RENAL INVOLVEMENT IN SYSTEMIC AMYLOIDOSIS: REPORT OF AN ITALIAN RETROSPECTIVE STUDY

Bergesio F, Ciciani A.M (1)., Palladini G (2)., Brugnano R (3), Bizzarri D (4), Pasquali S (5), Salvadori M on behalf of the Immunopathology Group of the Italian Society of Nephrology

Nephrology Dept of Careggi University Hospital and (1) Renal Unit of Torregalli,Hospital, Florence. (2) Amyloid Center and Internal Medicine University Hospital S.Matteo, Pavia. (3) Nephrology Unit Perugia Hospital. (4) Nephrology Unit Arezzo Hospital. (5) NephrologyDept University Hospital of Bologna.

INTRODUCTION

Renal involvement is quite common in systemic amyloidosis and often occurs as the presenting clinical feature of the disease. The major clinical signs are characterized by proteinuria, often in the nephrotic range, with or without renal failure (1,2). When present, renal failure strongly influences the natural history of the disease and its prognosis. Dialysis and kidney transplantation have been increasingly employed in the past years to treat renal insufficiency and improve the quality of life of these patients (3). However, If diagnosis has now became much easier than in the past, the same cannot be said for the definition of the type of deposits and consequently of the type of amyloidosis. At the moment, no data are known about demographic characteristics , overall and renal survival concerning a large cohort of patients with different types of renal amyloidosis. In addition, many different approaches are actually employed both for diagnosis and treatment by different nephrological Centers.

STUDY DESIGN

In order to determine the impact of renal involvement in systemic amyloidosis in Italy and to analyze the demographic characteristics and survival of people affected by the disease we set out a retrospective survey of all cases of renal amyloidosis diagnosed in Italy between January 1995 and december 2000.
Informations were collected by a data form at diagnosis and at follow-up including an investigation on diagnostic and therapeutic attitudes of each Center.
44 nephrological Units and 3 Internal Medicine Depts throught the country participated to the study and contributed to the collection of a database at diagnosis and at Follow-up.

RESULTS

Table 1. Clinical Characteristics of patients according to the type of amyloidosis at diagnosis

	AL	AA	MYELOMA	AF
Number of patients	202 (53%)	123 (32%)	36 (9%)	6 (2%)
Gender (M/F)	113/89	49/74	22/14	3/3
Age (years)	65 (34-87)	63 (19-86)	63 (41-88)	62
Serum Creatinine (mg/dl)	1,2 0,6-10,3)	1,7 (0,5-12,4)	1.0 (0,5-9,7)	1,9
Urinary Protein Excretion (g/day)	5.0 (0-21,4)	5.0 (0,4-29,8)	3,8 (0,5-20)	3,2
MC (serum and/or urine)	170/190 (89%)	18/117 (15%)	34/36 894%)	-
κ/λ ratio	1/4	2/1	1/2	-
Heart Involvement	59/157 837%)	16/94 (17%)	18/35 (51%)	-

367 patients were finally selected for the study. Clinical Characteristics of patients are reported in Tab.1 In 14 patients (4%) the type of deposits remained unknown. Medians of Bone marrow plasmacells were 8% in AL and 29% in Myeloma. Rheumatoid arthritis was the most common disease observed in AA (37%) followed by Crohn disease (9%) and Tuberculosis (5%). Heredofamilial forms were made up of 4 cases of Apo A1 (Leu75Pro), 1 case of Fibrinogen (Glu526Val) and 1 case remained undetermined. Patients without proteinuria and/or renal failure who were missing a renal biopsy were not included in the study. In 5 patients renal biopsy showed the presence of another glomerular nephropathy associated to amyloidosis.

DIAGNOSIS

Renal biopsy was the principal means for diagnosis in 328 (85%) cases. Fat tissue aspirate (FTA) and other tissues' biopsies were employed respectively in 175 (45%) and 46 (12%) cases. However they were "diagnostic" only in 15% of patients who did not perform renal biopsy. FTA was employed by 35 out of 47 centers, mainly as a needle aspirate (20), but also as a needle biopsy or even surgically in 11. Usually the material was evaluated by the pathologist (24) and only in 8 cases by the same operator who did FTA. Coincidence between FTA and renal biopsy was only observed in 66% of cases.

Amyloid typing was mainly done by using immunoistochemistry (IIC) (usually κ and λ, rarely AA antisera) on renal tissue (39 Centers). In 4 cases elctron microscopic IIC was employed wheras 3 Centers used IIC on bone marrow. Only 2 centers declared to use genetic investigation as a routine procedure.

Follow-up

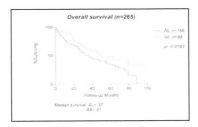

Fig.1. Overall Survival in AL and AA Amyloidosis

Overall survival was significantly longer in patients with AA than in patients with AL (51 vs 37 months) as shown in Fig.1. In both groups of patients survival was markedly influenced, as expected, by heart involvement. On the contrary, the presence of serum creatinine > 1.3 mg/dl only affected survival in AA group. Renal transplantation was performed only in one patient affected by AL. Immunosuppressive therapy was employed in 117 cases of AL and 63 of AA. In AL the most common drug regimens included melphalan + prednisone, melphalan + high-dose dexamethasone, Thalidomide + intermediate-dose dexamethasone and autologous stem cell transplantation (ASCT) performed in 18 patients. Provided it was carried out for at least 6 months , no matter the drug regimen, immunosuppression significantly improved overall survival of AL patients (Tab.2). All patients were offered supportive therapy.

Table 2. Clinical Characteristics of patients at Follow-up

	AL	AA
Number of patients with available data	202	123
Number of patients without available data	32	22
Median Time of follow-up (months)	24 (1-88)	30 (1-99)*
Dead	88 (43%)	44 (36%)
Dialysis	58 (35%)	45 (44%)
Median Survival according to heart involvement (Yes/No)	26/55**	21/79**
Median Survival according to sCr</>1,3 mg/dl	37/33	/39**
Median Survival according to therapy </> 6 months	30/69***	-

Median values and range are reported in months. *p<0.01 vs AL ; **p<0.002 ; ***p=0.005

Renal outcome (evaluated as the median time needed to reach dialysis) was similar in patients with AL and AA (45 vs 33 months) and appeared significantly influenced in both groups by creatinine levels >1,3 mg/dl at diagnosis (Tab.2). Creatinine levels were significantly higher in the AA group who ended in dialysis.

58 patients with AL and 44 with AA entered a regular dyalisis programme, 13 of whom were put on peritoneal dialysis. Urinary protein excretion (UPE) > 3g/day significantly influenced the renal outcome in patients with AA. In addition a significantly higher UPE was observed in AL group who entered dialysis. Median time from diagnosis to dialysis was similar in the two groups (12 months in AL vs 10 months in AA). Median time on dialysis was 5 months for AL and 13 for AA. Immunosuppressive therapy carried out for .≥ 6 months markedly improved renal survival in AL group.

CONCLUSIONS

 Patients with AA renal Amyloidosis showed a significantly longer survival than patients with AL, likely due to the less frequent cardiac involvement found in these patients. In agreement with previous reports (1,2), cardiac involvement is by far the major factor affecting survival in both groups (Tab2), followed by serum creatinine at diagnosis in patients with AA amyloidosis. Renal outcome is not different between AA and AL and is influenced by both serum creatinine and urinary protein excretion at diagnosis. Renal biopsy (RB) is still the commonest way to diagnose renal involvement in amyloidosis in nephrology units. The low coincidence observed between FTA and RB likely depends on a non correct application of the FTA procedure by nephrologists. Amyloid typing is not an homogeneous diagnostic process and mainly is based on κ/λ IIC on renal tissue. The use of DNA analysis to rule out hereditary forms is still very uncommon. To date, amyloid typing cannot yet be considered a relyable diagnosis until diagnostic guidelines will not be provided and widely accepted by most clinicians in the country. ASCT and kidney transplantation still remain rare therapeutic options for renal patients with systemic amyloidosis.

REFERENCES

1. Gertz MA, Lacy MQ and Dispenzieri A : Immunoglobulin light chain amyloidsis and the kidney.Kidney Int 61:1-9, 2002
2. Joss N, McLaughlin K, Simpson K and Boulton-Jones JM: Presentation, survival and prognostic markers in AA Amyloidosis. Q J Med 93:535-542, 2000
3. Moroni G,Banfi G, Montoli A et all: Chronic dialysis in patients with systemic amyloidosis: the experience in Northern Italy. Clinical Nephrology 38:81-85,1992

BIOCHEMICAL CHARACTERIZATION OF AN IMMUNOGLOBULIN LAMBDA I LIGHT CHAIN AMYLOID PROTEIN ISOLATED FROM FORMALIN-FIXED PARAFFIN-EMBEDDED TISSUE SECTIONS

M. Yazaki(1), J.J. Liepnieks(1), M.D. Benson(1,2)

(1) Department of Pathology and Laboratory Medicine, Indiana University School of Medicine (2) Richard L. Roudebush Veterans Affairs Medical Center, Indianapolis, Indiana USA

INTRODUCTION

Over 20 types of amyloidosis, systemic and localized, have been identified by the subunit protein in amyloid deposits. Since prognosis and treatment depends on the type of amyloid, it is essential to correctly identify the amyloid protein. Specific identification of amyloid type requires the isolation of the amyloid protein and its biochemical characterization. Recently, a procedure for isolation and chemical typing of amyloid protein in formalin-fixed paraffin-embedded tissue sections using Edman degradation analysis and mass spectroscopy analysis was reported (1). We have used this procedure to identify almost the entire amino acid sequence of an amyloid immunoglobulin light chain from formalin-fixed paraffin-embedded heart tissue sections of an autopsied patient with primary (AL) amyloidosis.

METHODS

Ten $4\mu m$ thick approximately $2cm^2$ sections of formalin-fixed cardiac tissue on glass slides were deparaffinized in xylene, rehydrated sequentially in 100%, 95%, 80% ethanol and distilled water, and air dried. Tissue was scraped from the slides and 1ml of 6M guanidine hydrochloride, 0.5M Tris pH8.2 containing 10mg dithiothreitol/ml and 1mg EDTA/ml was added. After incubation at 37°C for five days, the sample was alkylated with 24.1mg iodoacetic acid and filtered through a $0.2\mu m$ filter. $100\mu l$ was applied to a ProSorb PVDF cartridge for sequence analysis. The remainder was exhaustively dialyzed in Spectra/Por 6 dialysis tubing against water and lyophilized. The dried sample was suspended in 1ml of 10% acetic acid. Five percent was analyzed by SDS-PAGE and silver staining. Sixty percent was digested with trypsin, fractionated by reverse phase HPLC on a synchropak RP-8 column (100 x 4.6mm), and eluted peaks analyzed by Edman degradation. Twenty-five percent was digested with pyroglutamate aminopeptidase, applied to a ProSorb cartridge, and analyzed by Edman degradation.

RESULTS

Congo red stained sections of the cardiac tissue revealed massive amyloid deposition in the myocardium with green birefringence under polarized light. SDS-PAGE analysis of the guanidine hydrochloride solubilized material showed 2 bands at approximately 16kDa and 18kDa. Edman degradation analysis of the extract before and after pyroglutamate aminopeptidase digestion indicated the protein contained an N-terminal

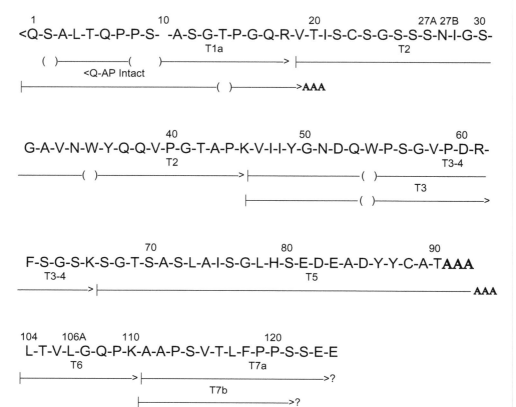

Figure 1. Amino acid sequence of amyloid protein LYN. The arrows indicate the tryptic peptides used to determine the sequence. The arrow labeled <Q-AP Intact indicates the sequence from the intact protein after pyroglutamate aminopeptidase digestion. Parentheses in arrows denote residues not completely verified. Dots at end of arrows indicate that the peptide continued but was not sequenced further. <Q denotes pyroglutamic acid. Whether the C-terminus of peptide T7a was E123 or E124 and of peptide T7b was S121 or S122 could not be determined and is indicated by ? at the ends of T7a and T7b. Residues are numbered according to Kabat *et al* (2).

pyroglutamate residue and was derived from a lambda light chain. Analysis of tryptic peptides identified residues 1-90 of the variable region and residues 104-123 or 124 (Figure 1).

DISCUSSION

The variable region sequence of the amyloid protein is most homologous to the lambda I light chain subgroup. While previous studies fractionated the guanidine hydrochloride extract from formalin-fixed paraffin embedded tissue sections by reverse phase HPLC before identification of the amyloid protein, no further purification after guanidine hydrochloride extraction was necessary in this case before biochemical identification of the amyloid protein. The extract contained only minor amounts of contaminating proteins, and contained sufficient material to determine almost the entire sequence of the amyloid lambda light chain except for approximately 15 residues in the carboxyl end of the variable region. Biopsy and archived formalin-fixed tissues will become a valuable source for isolation and extensive biochemical characterization of amyloid proteins, as these tissues are now a valuable source for isolation of DNA for genetic analysis. This is further demonstrated by a recent study in which the amyloid protein in calcifying epithelial adontogenic tumors or Pindborg tumors was isolated from paraffin-embedded formalin-fixed sections of the tumor and characterized as the N-terminal approximately 46 residues of a hypothetical 153 residue protein encoded by the FLJ20513 gene cloned from the human KATO III cell line (3).

REFERENCES

1.   Murphy CL, Eulitz M, Hrncic R, Sletten K, Westermark P, Williams T, Macy SD, Wooliver C, Wall J, Weiss DT and Solomon A (2001).   Chemical typing of amyloid protein contained in formalin-fixed paraffin embedded biopsy specimens. *Am J Clin Pathol* **116**: 135-142.

2.   Kabat EA, Wu TT, Perry HM, Gottesman KS and Foeller C (1991).   *Sequences of Proteins of Immunological Interest, 5th Edition*, Vol 1; pp. 131-134, U.S. Department of Health and Human Services, Public Health Service, National Institutes of Health, Bethesda, MD.

3.   Solomon A, Murphy CL, Weaver K, Weiss DT, Hrncic R, Eullitz M, Donnell RL, Sletten K, Westermark G and Westermark P (2003).   Calcifying epithelial odontogenic (Pindborg) tumor-associated amyloid consists of a novel human protein. *J Lab Clin Med* **142**: 348-355.

CASE REPORT: UNUSUAL CLINICAL PRESENTATION OF PRIMARY AL AMYLOIDOSIS

Maria Grazia Chiappini[1], Marco Di Girolamo[2], Giovanni Selvaggi[1], Konstantinos Giannakakis[3]

1.   Nephrology and Dialysis Unit, Fatebenefratelli Hospital, AFaR CRCCS, Isola Tiberina, Rome, Italy.
2.   Department of Internal Medicine, Fatebenefratelli Hospital, AFaR CRCCS, Isola Tiberina, Rome, Italy.
3.   Department of Experimental Medicine and Pathology "La Sapienza" University, Rome, Italy.

The clinical hallmark of amyloidosis AL is the presence of circulating monoclonal immunoglobulin light chain detectable in the serum and/or in the urine in above of 80% of patients. Any organ, excluding the brain, can be a target of amyloid deposition (kidney 49% of cases, heart 28%, liver 10%, peripheral nervous system 6%, gastrointestinal tract 4%, other organs 3%.[1]

Renal involvement is present from one-third to one-half of patients at diagnosis. The clinical manifestation of the renal disease is related to the predominant site of involvement. [2, 3] Most patients (about 75%) have predominant glomerular deposition and present with proteinuria that is usually in the nephrotic range.[4]

We report a case of primary AL amyloidosis presenting as progressive renal failure without evidence of proteinuria.

CASE REPORT

A 64-year-old man was referred to the Nephrology Unit because of a progressive increase in serum creatinine, from 1.84 mg/dl, 6 months before, to 3.7 mg/dl. Family history was positive for end-stage renal disease (mother) and vascular disease (father and brother). Two years prior to admission erectile dysfunction occurred and, six months before, oral drowsiness and metallic taste. Patient did not show signs of cardiac involvement, no episodes of arrythmias or bradycardia, no postural hypotension, no fatigue or weight loss. Physical examination revealed a well being man. Blood pressure was 130/85 mmHg. Pulse was regular 76/min. No periorbital nor peripheral edema or other clinical abnormalities were found. On admission, laboratory studies revealed: ESR 40 mm/hr, WBC 6,300 per mm^3, Hb 14,2 g/dl, Ht 40.8 % glucose 91 mg/dl, creatinine 3.7 mg/dl, urea nitrogen 134 mg/dl, uric acid 9.9 mg/dl, sodium 136 mmol/l, potassium 5.2 mmol/l, chloride 97 mmol/l, bicarbonate 21 mmol/l, Total protein 77 g/l, Albumin 48 g/l, serum β_2Microglobulin 6.2 ng/ml, AST and ALT respectively 18 U/l and 23 U/l, total bilirubin 0.8 mg/dL, alkalin phosphatase 180 U/l, total cholesterol 188 mg/dl, LDL cholesterol 115 mg/dl, IgG 7.1 g/l, IgA 0.8 g/l, IgM 0.4 g/l, serum light chains κ 1.4 g/l and λ 1.3 g/l. Urinalysis was negative for protein, no red cells or casts were found. Immunofixation of serum and urine was negative for monoclonal immunoglobulin light chain. Serology for HBV, HCV, HIV and immunological diseases was negative too. A further worsening of renal function was observed during the hospitalization with increasing of serum creatinine up to 5.2 mg/dl.

The patient underwent a kidney biopsy that revealed 15 glomeruli. An intense Congo red positivity was detected in renal arterial and arteriolar vessels (+++) with less involvement in glomerular (++) and tubular (+)

structures   (Fig. 1).   Immunofluorescent   microscopy   showed   high   positivity   (+++)   for κ light chains. Immunohistochemistry for AA amyloidosis was negative. Subsequently labial mucose and periumbelical subcutaneous fat biopsy was performed resulting in a high positivity for Congo red.

The other diagnostic procedures (EKG and 24-hours EKG monitor, abdominal and cardiac ultrasonographic examination, skeletal radiography, electromyography) did not detect any abnormalities. Bone marrow specimen showed 12% plasma cells. A further more accurate serum electrophoresis test showed a light increase in β_1 zone identified as a small amount of IgGλ in the central γ-catodo zone. Urine immunofixation detected the physiological ladder phenomenon of κ light chains without free λ light chains.

The patient began chemotherapy (pulse high-dose Dexametasone plus oral Melphalan) and serum creatinine progressively decreased to a stable value of 1.8-2.0 mg/dl, after 9 cycles.

DISCUSSION

The most frequently affected organ in AL Amyloidosis is the kidney and the clinical manifestations of the renal disease vary with the site of involvement.[5] Most patients have predominant glomerular deposition presenting with proteinuria, that is usually in the nephrotic range with peripheral edema. About 75 percent of patients present proteinuria. Proteinuria and nephrotic syndrome, as well as renal insufficiency, are also common in secondary AA amyloidosis. [2,3]

The urine sediment is typically benign, due to the lack of inflammation. In non-diabetic adults with nephrotic syndrome, amyloidosis accounts for 12% of renal biopsies. End-stage renal disease develops in approximately 20% of subjects with nephrotic syndrome. Fewer than 5% show urinary proteins loss less than 150 mg/day. These occasional patients have different renal presentation because of different site of amyloid deposition, primarily limited to the vessels, leading to narrowing of the vascular lumens. [6,7] Even less common is heavy tubular deposition, potentially leading to signs of tubular dysfunction, such type 1 (distal) renal tubular acidosis or polyuria due to nephrogenic diabetes insipidus. Extremely rare is crescentic glomerulonephritis superimposed upon renal amyloidosis. In these cases acute renal failure is associated with an active urine sediment.[8]

Recent research efforts have been directed mainly towards the characterization of the structural and biochemical features of amyloidogenic light chains. κ-light chain derived amyloidosis has been found to be less frequent and less nephrotoxic that λ-light chain derived amyloidosis.[1,9,10]

Our case is characterized by an unusual clinical presentation of AL amyloidosis (k light chain) in absence of all common symptoms, signs, laboratory and instrumental data suggesting this disease, except an unexplained progressive renal failure. We obtained a correct diagnosis of AL amyloidosis only by kidney biopsy. Heritable amyloidosis, suspected from the familial history, was excluded. Lack of proteinuria could be explained by the the higher vessel involvement in respect of glomerular and tubular one. Serum and urine immunofixation, revealing a detectable light chain in the majority of patients with AL renal amyloidosis (near 90%), was considered initially negative in our case. It is to note that laboratory tests currently employed could lack of sensitivity when the monoclonal peak is too small to be detected.   Recent, new immunodiagnostic assay systems permit to identifie small amounts of monoclonal light chains. [11] The reason for this unusual picture of renal amyloidosis remains to be defined. It should be better explain trough the comparison with other cases characterized by the same clinical feature.[12]

REFERENCES

1.   Merlini, G. and Bellotti, V., Mechanism of disease: molecular mechanisms of amyloidosis, *N. Engl. J. Med.*, 349, 583, 2003.

2.   Gertz, M.A., Lacy, M.Q. and Dispenzieri, A., Immunoglobulin light chain amyloidosis and the kidney, *Kidney International*, 61,1, 2002.

3.   Schena, F.P., Pannarale, G. and Carbonara, M.C., Clinical and therapeutic aspects of renale amyloidosis, *Nephrol. Dial. Transpant.*, 11 (Suppl 9), 63, 1996

4.   Sezer, O. et al., Diagnosis and treatment of AL amyloidosis, *Clin Nephrol.*, 53, 417, 2000.

5.   Paueksakon, P. et al., S. Monoclonal gammopathy: Significance and possible causality in renal disease, *Am. J. Kidney Dis., 42, 87,* 2003.

6.   Falck, H.M., Tornroth, T. and Wegelius, O., Predominantly vascular amyloid deposition in the kidney in patients with minimal or no proteinuria, *Clin. Nephrol., 19, 137,* 1983.

7.   Westermark, G.T. et al., AA-amyloidosis. Tissue component-specific associated of various protein AA subspecies and evidence of a fourth SAA gene product, *Am. J. Pathol., 137, 377,* 1990.

8.   Moroni, G. et al., Extracapillary glomerulonephritis and renal amyloidosis, *Am. J. Kidney Dis.*, 28, 695, 1996.

9.   Odani, S., Komori, Y. and Gejyo, F., Structural analysis of amyloidogenic λ Bence Jones Protein (FUR), *Amyloid: Int. J. Exp. Clin. Invest.*, 6, 77, 1999.

10. Rochet, J.C. and Lansbury, P.T. Jr, Amyloid fibrillogenesis: themes and variations, *Curr. Opin. Struct. Biol.*, 10, 60, 2000,

11. Roshini, S.A. et al., A new marker for the diagnostic evaluation of primary systemic amyloidosis, *Am. J. Pathol., 119, 274,* 2003.

12. Pasquali, S. et al., Vascular renal amyloidosis, presented at X[th] Int. Symp. on Amyloid and Amyloidosis, Tours, April 18-22, 2004.

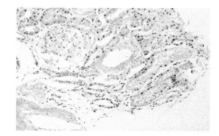 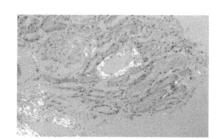

A   B

Figure 1: Histological findings of renal tissue (PAS staining) with marked deposition of amyloid in to the wall vessel (A). Congo-red stain viewed under polarized light green berefringence of vessel amyloid deposits. The glomerulus appears to be spared by amyloidosis (B).

SERUM LEVELS OF FREE KAPPA AND LAMBDA LIGHT CHAINS IN PATIENTS WITH SYSTEMIC AL, AA, AND ATTR AMYLOIDOSIS

J. van Steijn [1], J. Bijzet [1], H. de Wit [2], B.P.C. Hazenberg [1], I.I. van Gameren [1], K. van de Belt [2] and E.Vellenga[3]

Departments of Rheumatology (1), Pathology and Laboratory Medicine (2), and Haematology (3), University Hospital, Groningen, The Netherlands

INTRODUCTION

Immunoglobulin light chains are precursors of AL amyloid fibrils. Free light chains can be quantified with a recently described immunoassay and appear to be useful in AL amyloidosis (1, 2).

OBJECTIVE

To study the value of serum free light chains for the diagnosis of systemic AL amyloidosis and for monitoring effects of therapy.

PATIENTS AND METHODS

Twenty-two patients with AL-kappa and 53 with AL-lambda amyloidosis were studied at diagnosis. Controls were 40 healthy blood donors, 9 patients with arthritis, 10 patients with AA amyloidosis, and 10 patients with ATTR amyloidosis. Clonal responses after chemotherapy were evaluated in 20 patients with AL amyloidosis and follow-up of one year or longer.

 Free kappa and lambda light chains in stored serum samples were quantified with a nephelometric immunoassay (FreeLite™) with specific antibodies raised against hidden epitopes. The 99% confidence limits of controls were calculated for kappa, lambda, and the ratio of both light chains.

RESULTS

Control lower and upper 99% confidence limits were for kappa 1.50 and 77.5 mg/l, for lambda 2.54 and 144 mg/l, and for the K/L ratio 0.29 and 1.09 (see figure 1A). Twenty-one (95%) AL-kappa and 50 (94%) AL-lambda patients were outside the K/L ratio reference range (see figure 1B).

 During follow-up of 20 AL patients the K/L ratio did not normalise in 6 (30%) AL patients, although > 50% response was seen in 4 (20%) of them. The ratio normalised transiently in 5 (25%), and normalised until the end of follow-up in 9 (45%) patients (see figure 2). The clonal response was reflected by the K/L ratio in most patients.

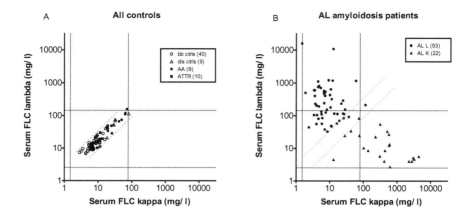

Figure 1. A. Control lower and upper 99% confidence limits (dotted lines) for free light kappa, lambda, and kappa/lambda ratio in serum. B. Serum values of free light kappa, lambda, and kappa/lambda ratio of 53 AL-lambda and 22 AL-kappa patients.

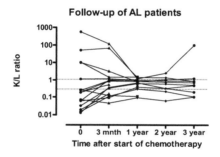

Figure 2. Free kappa/lambda ratio in serum during follow-up after start of chemotherapy in 20 patients with AL amyloidosis.

CONCLUSIONS

Serum free kappa and lambda light chains, and especially the ratio of both, are valuable tools to diagnose AL amyloidosis and to monitor the clonal effect of therapy.

REFERENCES

1. Bradwell AR, Carr-Smith HD, Mead GP, et al. Highly sensitive, automated immunoassay for immunoglobulin free light chains in serum and urine. Clin Chem 2001; 47:673-80.
2. Lachmann HJ, Gallimore R, Gillmore J, et al. Outcome in systemic AL amyloidosis in relation to changes in concentration of circulating free immunoglobulin light chains following chemotherapy. Br J Haematol 2003; 122:78-84.

DIAGNOSTIC VALUE OF FREE KAPPA AND LAMBDA LIGHT CHAINS IN FAT TISSUE OF PATIENTS WITH SYSTEMIC AL AMYLOIDOSIS

B.P.C. Hazenberg [1], J. Bijzet [1], H. de Wit [2], J. van Steijn [1], E. Vellenga [3] and M.H. van Rijswijk [1]

Departments of Rheumatology (1), Pathology and Laboratory Medicine (2), and Haematology (3), University Hospital, Groningen, The Netherlands

INTRODUCTION

Congo red stained fat tissue is used to diagnose systemic amyloidosis. A recently described immunoassay quantifies free immunoglobulin light chains in serum and urine (1).

OBJECTIVE

To study the value of free light chain concentrations in fat tissue for diagnosing systemic AL amyloidosis.

PATIENTS AND METHODS

Subcutaneous abdominal fat tissue specimens of 56 patients with AL amyloidosis (14 AL-K and 42 AL-L) were studied. Control specimens were from 35 patients without amyloid, 20 with AA amyloidosis, and 15 with ATTR amyloidosis.

The washed fat tissue was extracted in 6 M guanidine hydrochloride (2). Free kappa and lambda light chains in fat tissue supernatants were quantified with a nephelometric immunoassay (FreeLite™) with specific antibodies against hidden epitopes (1). The 95% confidence limits of controls were calculated for kappa, lambda, and the ratio of both light chains. Congo red stained slides were scored semi-quantitatively ranging from 0 to 3+ (2).

RESULTS

Control lower and upper 95% confidence limits were for kappa 4.73 and 402 ng/mg fat, for lambda 4.06 and 286 ng/mg fat, and for the K/L ratio 0.24 and 6.89 (see figure 1A).

Seven (50%) of the kappa values of AL-L and 20 (48%) of the lambda values of AL-K patients were above the normal ranges. Eight (57%) AL-K and 30 (71%) AL-L patients were outside the normal K/L ratio range (see figure 1B). Thirty-six (88%) of 41 AL specimens with Congo red score 2+ or 3+ were outside the normal K/L ratio range.

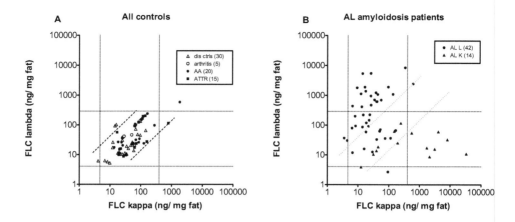

Figure 1. A. Control lower and upper 95% confidence limits for free light kappa, lambda, and kappa/lambda ratio in fat tissue. B. Fat tissue values of free light kappa, lambda, and kappa/lambda ratio of 42 AL-lambda and 14 AL-kappa patients.

CONCLUSIONS

Free kappa and lambda light chains in fat tissue, and especially the ratio of both, are valuable to diagnose AL type of amyloid in 50-70% of cases, increasing to 88% of cases with 2+ or 3+ amyloid present in the Congo red stain.

REFERENCES

1. Bradwell AR, Carr-Smith HD, Mead GP, et al. Highly sensitive, automated immunoassay for immunoglobulin free light chains in serum and urine. Clin Chem 2001; 47:673-80.
2. Hazenberg BPC, Limburg PC, Bijzet J, van Rijswijk MH. A quantitative method for detecting deposits of amyloid A protein in aspirated fat tissue of patients with arthritis. Ann Rheum Dis 1999; 58:96-102.

EXPERIENCE WITH A PATIENT-ADAPTED APPROACH TO AUTOLOGOUS PERIPHERAL BLOOD STEM CELL TRANSPLANTATION IN AL AMYLOIDOSIS

V. Perfetti, G. Palladini, [2]S. Siena, [2]M. Bregni, [2]M. Di Nicola, V. Navazza, [2]M. Magni, [2]A.M.Gianni, [1]G. Merlini

Internal Medicine and Medical Oncology and [1]Dept. of Biochemistry, Laboratories of Biotechnology, Univ. of Pavia-IRCSS Policlinico S. Matteo, Pavia, and [2]Division of Medical Oncology C, NCI, Milan, ITALY

Primary systemic amyloidosis is an aggressive plasma cell dyscrasia whose prognosis is largely dependent on the extent of heart involvement and the number of clinically involved organs. Present therapy is aimed to the annihilation of the marrow plasma cell clone. Several groups experimented high-dose melphalan (MEL) with autologous peripheral blood transplantation (HDCT) with high response rates and apparently prolonged survival. However, HDCT was characterized by severe toxicity and high transplant related death rates (up to 40%)(1). The number and extent of organ involvement (principally the heart) appear to be the most important parameters predicting toxicity, and should be carefully weighted in patient selection and MEL dose determination. We report our experience with a patient-adapted approach to HDCT, with an option for a second transplant in selected cases.

PATIENTS AND METHODS

Twenty-two patients were recruited and treated. Selection criteria are reported in Table 1.

Table 1. Selection criteria for HDCT and relative MEL dose.

Low risk (all criteria must be satisfied)	Intermediate risk	Not eligible
• <60 yrs	• 60-65 yrs	• >65 yrs
• organs involved ≤2	• s. creatinine ≥1.5 mg/dL	• organs involved ≥3
• ejection fraction >50%	• compensated heart involvement (NYHA II)	• NYHA III or IV
• s. creatinine ≤1.4 mg/dL	• performance status 2	• s. bilirubin >2 mg/dL
• CO lung diffusion >50% predicted		• systolic BP <90 mmHg
• asymptomatic heart involvement*		• performance status 3
• performance status ≤1		
MEL dose	**MEL dose**	
200 mg/m^2	140 mg/m^2	
	100 mg/m^2	

*with evidence of echocardiographic heart involvement

  Hematologic response was assessed by immunofixation of serum and urine using high-resolution agarose gel electrophoresis and by bone marrow morphological examination. Complete response (CR) was defined by negative immunofixation with plasma cell infiltration <5%. Partial response (PR) was characterized by more than 50% reduction in monoclonal component concentrations in serum and urine. Mobilization and harvesting of peripheral blood CD34+ cells was obtained mainly by rHG-CSF alone (10µg/kg/day, divided in two doses). Patients were stratified according to the main organ involved and by number of organs. Organ involvement was assessed clinically and by biochemistry, and by ecographic/endoscopic investigations when needed. Organ response was defined as follows: a) heart, ≥2 mm reduction of the interventricular septum thickness, or amelioration of one NYHA class; b) kidney, ≥ 50% reduction of daily proteinuria in the absence of progressive renal insufficiency; c) liver, ≥2 cm reduction of the liver size with concomitant ≥50% reduction of alkaline phosphatase; d) neuropathy, resolution of orthostatic signs or other neurologic abnormalities.

RESULTS

Twenty-two patients were treated. Eleven had dominant kidney (50%), 5 had major heart involvement (23%). A single clinically involved organ was present in 54% of cases.

Hematologic response. As three patients died in the peri-transplant period, 19 patients were evaluable at +3 months (86%). Hematologic response comprised 53% PR and 11% CR. Six patients (32%) were unresponsive. At +12 months, CR was observed in 6 additional cases that were already in PR at +3 months, for an overall intention to treat CR of 36% (8 of 22 treated patients). PR was observed in 5, for an overall response (CR+PR) of 59%. Only 1 of the 6 unresponsive patients obtained PR at +12 months.

Survival. Four of 22 (18%) patients died within the first 12 months, three in the peritransplant and one for amyloid progression (despite PR). With a median follow-up of 36 months (range 12-73) of surviving patients at +12 months, 72% are still alive. Median overall survival is 68 months (Kaplan-Meyer). Some patients in CR experienced prolonged disease-free survivals and are alive at +70 (kidney), +73 (heart) and +38 months (heart). Survival curves (Kaplan-Meyer) of the transplanted population discriminated according to response at +3 months (CR+PR vs NR) showed that remission of the plasma cell dyscrasia was associated with significantly prolonged survival (CR+PR, median not reached, more than 70 months; NR, median 13 months) (*P=.033*, log rank test).

Organ response. 1) Heart. Of the 10 patients with heart involvement (predominant in 5), 7 survived at least 12 months. Four of the seven survivors obtained hematological CR and amelioration of heart involvement. 2) Kidney. Fourteen patients with kidney involvement (predominant in 11) were transplanted, and 12 (86%, 9 with primary kidney involvement) were alive at the +12 months follow-up. Kidney function was improved in 42% of cases, stable in 25%, worsen in 33%. In responsive patients, median 24hour pre- and post-treatment proteinuria were significantly different (*P=0.027*, Wilcoxon's sum test). Overall organ response was observed in 10 of 13 patients achieving hematologic remission (77%).

Toxicity. 1) Peritransplant mortality. Mortality in the first 90 days following HDCT was 14%. One patient with primary heart involvement died of sudden death, one for VOD in hepatic amyloidosis, and one for CMV infection (primary kidney). 2) Kidney. Despite the personalized approach to MEL dosing, significant toxicity was still observed. Two of the 11 kidney patients (creatinine 1.4 and 1.6 mg/dL, MEL 200 and MEL 140, respectively) experienced grade 2 toxicity and progressed to renal insufficiency and dialysis. Grade 3 toxicity was observed in a patient with a pretransplant creatinine of 2.6 mg/dL (MEL 140), who did not respond to treatment and is now in dialysis. One patient with kidney amyloidosis and normal renal function performed a second transplantation while in PR after the first transplant. Serum creatinine was 1.4 mg/dL at one year and slowly progressed to dialysis and is now in CR at 6 years, waiting for a kidney transplant. 3) Heart. Two of 10 patients with heart involvement deceased in the peritransplant period (sudden death, VOD). Pre-transplantation serum NT-proBNP concentrations, a new sensitive marker of cardiac dysfunction (2), were subsequently found above the cut-off in 6 of 7 cases, a very unfavorable prognostic factor. Toxicity was significant, particularly infectious complications, which were clustered in this group.

DISCUSSION

Overall hematologic response at +12 months was 59% on an intention to treat analysis (36% CR, 23% PR). These results compare very favorably with conventional dose chemotherapy, either with melphalan and prednisone (overall response 20-30%, 5% CR) or high-dose dexamethasone (overall response 35%, 9% CR)(3). This superior response rate appears to confer a significant survival benefit, with a median survival of responsive patients not yet reached (more than 70 months) at a median follow-up of 36 months. However, this population is highly selected and the impact of HDCT in amyloidosis will have to wait for randomized studies vs conventional chemotherapy.

Despite the fact that MEL conditioning was tailored on patient disease extension and fragility, toxicity was still high (TRM 14%) and with unusual complications. The amyloid patient manifests its fragility under stressful conditions, with possible irreversible organ damage. A clinically relevant finding of our study is the importance of the hematologic response at +3 months. Only one of the patients who failed to show response at +3 months obtained PR at +12 months, suggesting that these patients should soon be treated with alternative therapy. On the contrary, the presence of a response at +3 months, even partial, is a favorable factor, with a substantial proportion of patients getting a CR at +12 months.

Patients with heart involvement did not tolerate well HDCT and complications were more frequent in this group. The introduction of serum NT-proBNP determination may be useful in this setting, helping in patient stratification and selection.

In conclusion, this study describes our approach to HDCT in amyloidosis, an approach aimed at reducing toxicity while preserving efficacy. We observed remission rates that are superior to conventional therapy with reduced treatment-related mortality and toxicity compared to early transplantation reports. However, the risks of death and transplant morbidity are still substantially higher than in other hematological malignances, stressing the importance of accurate selection of candidates and the need for less toxic therapies (4). With these limitations, HDCT represents a therapeutic option able to induce very prolonged remissions.

REFERENCES

1. Comenzo, R.L., and Gertz, M.A. (2002). Autologous stem cell transplantation for primary systemic amyloidosis. *Blood* **99**, 4276-4282
2. Palladini, G., Campana, C., Klersy, C., Balduini, A., Vadacca, G., Perfetti, V., Perlini, S., Obici, L., Ascari, E., d'Eril, G. M., Moratti, R., and Merlini, G. (2003). Serum N-terminal pro-brain natriuretic peptide is a sensitive marker of myocardial dysfunction in AL amyloidosis. *Circulation* **107**, 2440-2445
3. Palladini, G., Anesi, E., Perfetti, V., Obici, L., Invernizzi, R., Balduini, C., Ascari, E., and Merlini, G. (2001). A modified high-dose dexamethasone regimen for primary systemic (AL) amyloidosis. *Br J Haematol* **113**, 1044-1046
4. Palladini, G., Perfetti, V., Obici, L., Caccialanza, R., Semino, A., Adami, F., Cavallero, G., Rustichelli, R., Virga, G., and Merlini, G. (2004). Association of melphalan and high-dose dexamethasone is effective and well tolerated in patients with AL (primary) amyloidosis who are ineligible for stem cell transplantation. *Blood* **103**, 2936-2938

RISK-ADAPTED INTRAVENOUS MELPHALAN WITH ADJUVANT THALIDOMIDE AND DEXAMETHASONE FOR NEWLY DIAGNOSED UNTREATED PATIENTS WITH SYSTEMIC AL AMYLOIDOSIS: INTERIM REPORT OF A PHASE II TRIAL

RL Comenzo, P Zhou, L Reich, S Costello, A Quinn, S Fircanis, L Drake, C Hedvat, J Teruya-Feldstein, DA Filippa, M Fleisher

Memorial Sloan-Kettering Cancer Center, New York, New York, USA

From previous clinical trials employing intravenous (iv) melphalan and mobilized stem cell support for patients with systemic AL amyloidosis we have learned that survival depends on the number of major viscera involved with amyloid, on the degree of cardiac involvement, and on the response of the plasma cell disease to melphalan (1-3). We have also learned that adjusting the dose of melphalan modulates toxicities and that prompt treatment with intravenous melphalan is appropriate, as opposed to several cycles of oral therapy as a prelude to transplant (4, 5). Therefore, this clinical trial builds on those results and seeks to answer two questions: does risk-adapted melphalan dosing reduce treatment-related mortality and does adjuvant therapy for persistent disease after SCT enhance the response of the plasma cell disease?

Eligible patients have systemic AL with symptomatic visceral involvement, are within 12 months of diagnosis and untreated. Transthyretin and fibrinogen A-α genes are sequenced to rule out hereditary amyloid (6, 7). Based on risk assignment, patients receive 100, 140 or 200mg/m2 with stem cell support (low-risk) or two cycles of 40mg/m^2 (high-risk) (8). Patients with persistent plasma cell disease at 3 months receive adjuvant therapy with either thalidomide and dexamethasone or dexamethasone alone. Primary endpoints are survival, and organ system and hematologic responses. We report interim results.

PATIENTS AND METHODS

The target population of MSKCC #02-031 is patients with primary systemic amyloidosis (AL) and a clonal plasma cell disease within 12 months of diagnosis and not previously treated. Organ involvement, plasma cell disease and responses are defined as previously described (5). DNA-based polymerase chain reaction assays with cloning and sequencing of amplicons for exons 2, 3 and 4 of transthyretin (TTR) and for fibrinogen A-α were employed to further exclude patients with hereditary amyloid variants and incidental monoclonal gammopathies. In this phase II trial, the efficacy of iv melphalan is tested as initial therapy and, in patients with persistent clonal plasma cell disease at 3 months after initial therapy, the efficacy of oral thalidomide and dexamethasone is tested as adjuvant therapy. Melphalan dosing is risk-adapted based on patient age and extent of organ involvement (8). Patients with neuropathy, deep venous thrombosis and excessive edema receive dexamethasone alone without thalidomide. The primary endpoint is two-year overall and progression-free survival. Secondary endpoints are the response of the plasma cell disease at 3, 12 and 24 months after initial therapy, and the response of the amyloid-related disease at the 12 and 24 months. Additional secondary endpoints include the prognostic significance of Ig V_L gene use, and prospective evaluations of the serum free light chain assay (FLC), troponin I and BNP levels.

RESULTS

From 9/02 until 4/04 32 patients were enrolled on this trial, 21 men and 11 women with a median age of 57 years (range, 34-72). The time from diagnosis to enrollment was a median of 2 months. Dominant organ involvement was cardiac (n=10), renal (n=12) and other (n=10). All had monoclonal gammopathies (λ=24, κ=8). Eighty-eight percent (28/32) had elevated pathologic FLC with abnormal κ::λ ratio. Twenty-nine of the 32 patients have been or are being treated, 3 in the high-risk and 26 in the low-risk group. One patient in the low-risk group died on day 2 of stem cell mobilization with G-CSF, the only treatment-related death thus far. In the low-risk group, patients received melphalan doses of 100 (n=6), 140 (n=10) and 200 mg/m^2 of melphalan (n=9). There have been six deaths *in toto*, five of them due to progression of disease (POD, low-risk=2, high-risk=3) a median of 5 months after initial melphalan (4-10).

Twenty-four patients have been evaluated for hematologic response at 3 months after initial therapy. Sixteen responded (4 CR, 12 PR), while 7 have stable (SD) and 1 progressive disease (PD). Of 13 patients with either CR or PR and baseline abnormal FLC, the κ::λ ratio has normalized in 7 while the pathologic light chain level has been reduced by $\geq$ 50% in 11. Conversely, in 8 patients with SD or PD the pathologic free light chain level has been reduced by $\geq$ 50% in 2 of them.

Of 20 patients eligible to receive adjuvant therapy, 3 refused or were too ill, 9 received thalidomide and dexamethasone and 8 received dexamethasone alone (neuropathy=4, prior DVT=3, excessive edema=1). Thalidomide dosing is from 50mg to 200mg nightly, while dexamethasone dosing can range from 20 to 40mg daily in 4 day pulses from one to three times monthly. Two patients on thalidomide and dexamethasone completed 9 months of therapy and both showed responses, including 1 CR, while 2 had POD despite adjuvant therapy and in 3 cases adjuvant therapy was discontinued (DVT=1, RSV pneumonia=1, intolerance=1); 2 continue on both drugs within the first year of therapy. Two patients on dexamethasone alone completed 9 months of therapy and showed responses, including 1 CR, while 4 had POD, and 2 continue on dexamethasone alone within the first year of therapy.

Seven patients have been evaluated for hematologic and amyloid-disease responses at 12 months. All have hematologic responses, including 3 CR (two of whom had been on adjuvant therapy) and 4 PR. All 7 experienced objective improvement in amyloid-related disease.

DISCUSSION

The concepts of risk-adapted melphalan dosing and adjuvant therapy after initial melphalan therapy are being explored in this phase II clinical trial. There have been no deaths within 100 days of stem cell transplant in 25 low-risk patients although there has been one death in mobilization, a rare but previously described adverse event in AL patients (5). The risk-adapted approach, therefore, appears to significantly reduce treatment-related mortality. Furthermore, adjuvant therapy is feasible and appears to benefit about 30% of patients (4/13) with improvement in hematologic response, including the achievement of CR. It is important to note how tailored the use of adjuvant therapy is in this trial; the emphasis is on tolerance of these drugs, a problem area particularly with thalidomide in AL patients (10).

The use of the serum FLC in conjunction with standard measures of hematologic response is likely to provide valuable information on the utility of FLC testing. Of 13 patients with either CR or PR and baseline abnormal FLC, the κ::λ ratio normalized in 7 while the pathologic light chain level was reduced by $\geq$ 50% in 11, demonstrating that reduction in pathologic FLC may be more frequently associated with response at 3 months than normalization of the ratio. Assuming that the pathologic free light chain is the amyloid-precursor protein in AL, one could conclude that a minority of "responding" patients may remain at risk for progression of amyloid. Alternately, in 2 of 8 patients with SD or PD the pathologic free light chain level has been reduced by > 50%, suggesting that a minority of "non-responding" patients may derive a benefit nonetheless with respect to progression of amyloid. These interim results add further evidence in support of a role for FLC testing in patient management during treatment of systemic AL amyloidosis (9).

ACKNOWLEDGEMENTS

We thank the Food and Drug Administration Orphan Products Development Program and Celgene for their support.

REFERENCES

1. Kyle RA. High-dose therapy in multiple myeloma and primary amyloidosis: an overview. Semin Oncol 1999;26:74-83.
2. Comenzo RL. Hematopoietic cell transplantation for primary systemic amyloidosis: what have we learned. Leuk Lymphoma 2000;37:245-58.
3. Skinner M, Sanchorawala V, Seldin DC, Dember LM, Falk RH, Berk JL, Anderson JJ, O'Hara C, Finn KT, Libbey CA, Wiesman J, Quillen K, Swan N, Wright DG. High-dose melphalan and autologous stem-cell transplantation in patients with AL amyloidosis: an 8-year study. Ann Intern Med 2004;140:85-93.
4. Comenzo RL, Sanchorawala V, Fisher C, Akpek G, Farhat M, Cerda S, Berk JL, Dember LM, Falk R, Finn K, Skinner M, Vosburgh E. Intermediate-dose intravenous melphalan and blood stem cells mobilized with sequential GM+G-CSF or G-CSF alone to treat AL (amyloid light chain) amyloidosis. Br J Haematol 1999;104:553-9.
5. Sanchorawala V, Wright DG, Seldin DC, Falk RH, Finn KT, Dember LM, Berk JL, Quillen K, Anderson JJ, Comenzo RL, Skinner M. High-dose intravenous melphalan and autologous stem cell transplantation as initial therapy or following two cycles of oral chemotherapy for the treatment of AL amyloidosis: results of a prospective randomized trial. Bone Marrow Transplant 2004;33:381-8.
6. Lachmann HJ, Booth DR, Booth SE, Bybee A, Gilbertson JA, Gillmore JD, Pepys MB, Hawkins PN. Misdiagnosis of hereditary amyloidosis as AL (primary) amyloidosis. N Engl J Med 2002;346:1786-91.
7. Jacobson DR, Pastore RD, Yaghoubian R, Kane I, Gallo G, Buck FS, Buxbaum JN. Variant-sequence transthyretin (isoleucine 122) in late-onset cardiac amyloidosis in black Americans. N Engl J Med 1997;336:466-73.
8. Comenzo RL, Gertz MA. Autologous stem cell transplantation for primary systemic amyloidosis. Blood. 2002;99:4276-82.
9. Lachmann HJ, Gallimore R, Gillmore JD, Carr-Smith HD, Bradwell AR, Pepys MB, Hawkins PN. Outcome in systemic AL amyloidosis in relation to changes in concentration of circulating free immunoglobulin light chains following chemotherapy. Br J Haematol 2003;122:78-84.
10. Seldin DC, Choufani EB, Dember LM, Wiesman JF, Berk JL, Falk RH, O'Hara C, Fennessey S, Finn KT, Wright DG, Skinner M, Sanchorawala V. Tolerability and efficacy of thalidomide for the treatment of patients with light chain-associated (AL) amyloidosis. Clin Lymphoma 2003;3:241-6.

THALIDOMIDE PLUS INTERMEDIATE-DOSE DEXAMETHASONE IN REFRACTORY/RELAPSED AL PATIENTS

Giovanni Palladini[1,2], Vittorio Perfetti[2], Stefano Perlini[2], Laura Obici[1], Francesca Lavatelli[1], Rosangela Invernizzi[2], Franco Bergesio[3], Benedetto Comotti[1], Salvatore Iannaccone[4], Giampaolo Merlini[1]

Amyloid Center, [1]Biotechnology Research Laboratories – Department of Biochemistry and [2]Internal Medicine, University Hospital "IRCCS – Policlinico San Matteo", Pavia, and University of Pavia; [3]Nephrology Unit, Careggi Hospital, Firenze; [4]Nephrology Unit, "G. Moscati" Hospital, Avellino, Italy.

Current treatments for AL are based on reducing the supply of the circulating amyloidogenic light chain by targeting the plasma cell clone responsible for its synthesis. Although the monoclonal plasma cell population in AL is usually small, chemotherapy regimens are based on those used in multiple myeloma. High-dose melphalan followed by autologous stem cell transplantation (ASCT) induces complete remissions in a considerable proportion of eligible patients (1). Current effective treatments in patients not eligible for ASCT include oral melphalan associated with prednisone (2) or dexamethasone (3), or high-dose dexamethasone alone (4). Still, a significant proportion of AL patients do not respond to first-line therapy. Based on the observation that thalidomide is effective in refractory and relapsed multiple myeloma (5, 6) and on the reported synergy of thalidomide and dexamethasone on myeloma plasma cells in vitro (7), we evaluated the combination of oral thalidomide and intermediate-dose dexamethasone (T-Dex) in AL patients who did not respond to or relapsed after first-line therapy.

PATIENTS AND METHODS

Thirty-one refractory/relapsed AL were treated with oral dexamethasone 20 mg on days 1-4, every 21 days, and thalidomide given continuously (100 mg daily, with 100 mg increments every 2 weeks if well tolerated, up to 400 mg).

Hematologic response was defined as the decrease of the monoclonal component (MC) by at least 50%. Complete remission consisted in the disappearance of the MC from both serum and urine documented by high-resolution agarose-gel immunofixation electrophoresis (8). Organ response was assessed according to the criteria proposed by the Mayo Clinic Group (2). The response was evaluated 1 month after Thalidomide 400 mg/day was reached and subsequently every three months. The patients who did not tolerate the 400 mg/day dosage were also evaluated every 3 months and continued treatment at the maximum tolerated dose. Toxicity and adverse events were recorded according to the National Cancer Institute Common Terminology Criteria for Adverse Events version 3.0 (CTCAE). To detect possible adverse reactions to thalidomide (i.e. peripheral neuropathy, arrhythmia), neurological evaluations (electromyography when indicated) and 24 h Holter electrocardiography were repeated monthly in all patients.

RESULTS

Thirty-one (18 men) patients were enrolled into the study. Patients' characteristics are reported in Table 1.

Table 1. Patients' characteristics.

Median age (years)	62 (range: 34-71)
>2 organs involved	19 (61%)
Median creatinine (mg/dL)	1.1 (range 0.6-6.8)
Serum creatinine >2 mg/dL	7 (23%)
Median proteinuria (g/24h)	4 (range 0-29)
Proteinuria ≥3 g/24 h	18 (58%)
Heart involvement	12 (38%)
Median interventricular septum thickness (mm)	14 (range 8-19)
Interventricular septum thickness >15 mm	5 (16%)
Median EF (%)	55 (range 35-72)
Orthostatic hypotension	7 (23%)
Relapsed / refractory after previous treatment	21 (68%) / 10 (32%)

Eleven patients (35%) tolerated 400 mg/day dosage and received thalidomide for a median time of 5.7 months (range: 4-14 months). The remaining 20 patients (65%) did not reach the target dosage and received thalidomide (median dosage 200 mg/day, range: 100-300 mg/day) for a median time of 3 months (range: 0.5-13 months). Response to treatment is reported in Table 2. Median time to response was 3.6 months (range: 2.5-8.0 months). Hematologic response to treatment resulted in a survival benefit (p=.01)

Table 2. Response to T-Dex

Thalidomide dose tolerated	Hematologic response	Organ response
400 mg/day (11 patients)	8 (73%), of whom CR 3 (27%)	4 (36%)
<400 mg/day (20 patients)	7 (35%), of whom CR 3 (15%)	4 (20%)
Total (31 patients)	15 (48%), of whom CR 6 (19%)	8 (26%)*

CR: complete remission

* >50% reduction in proteinuria in 6 patients, >50% decrease in alkaline phosphatase concentration in 1 and resolution of orthostatic hypotension in 1.

Overall, 20 patients experienced severe (CTCAE grade ≥3) treatment related toxicity. Adverse reactions included symptomatic bradycardia (8 patients, 26%), sedation/fatigue (4, 13%), constipation (2, 7%), acute dyspnea (2, 7%), deep venous thrombosis, skin lesions, epilepsy and renal failure (1 patient, 3%, each). There was no treatment-related mortality. Median follow-up of living patients is 24 months (range 9-34 months). Seven patients died after a median follow-up of 8 months (range: 3-28 months) due to heart failure (6), sudden death (1) and renal failure (1).

DISCUSSION

The association of dexamethasone and thalidomide provided a 35% hematologic response rate and functional improvement of the organs involved in 20% of patients. These figures rise up to 73% and 36%, respectively, in patients who tolerate the 400 mg/day dose of thalidomide. These results compare favorably with that reported in previous study of thalidomide alone in AL (9, 10). It is possible that the association of dexamethasone improves the efficacy of thalidomide as observed in multiple myeloma (11). As previously reported (9, 10) we observed a high-incidence (65%) of severe thalidomide related toxicity in AL. Notably, symptomatic bradycardia emerges as a common adverse reaction to thalidomide in AL, as recently observed in multiple myeloma (12), thus underlying the need of strictly monitoring for bradyarryhthmia patients receiving thalidomide. In conclusion, the low tolerability of thalidomide discourages its introduction in first line regimens. However, given the extremely poor prognosis of AL patients who do not respond to treatment, a combination of thalidomide and dexamethasone may represent a valid option for refractory and relapsed patients.

REFERENCES

1. Comenzo RL, Gertz MA (2002). Autologous stem cell transplantation for primary systemic amyloidosis. *Blood*; **99:** 4276-4282.
2. Kyle RA, Gertz MA, Greipp PR, et al (1997). A trial of three regimens for primary amyloidosis: colchicine alone, melphalan and prednisone and melphalan, prednisone and colchicine. *N Engl J Med*; **336:** 1202-1207
3. Palladini G, Perfetti V, Obici L, et al (2004). The association of melphalan and high-dose dexamethasone is effective and well tolerated in patients with AL (primary) amyloidosis ineligible for stem cell transplantation. *Blood* **103:** 2936-2938
4. Palladini G, Anesi E, Perfetti V, et al (2001). A modified high-dose dexamethasone regimen for primary systemic amyloidosis. *Br J Haematol*; **113:** 1044-1046.
5. Singhal S, Mehta J, Desikan R, et al (1999). Antitumor activity of thalidomide in refractory multiple myeloma. *N Engl J Med*; **341:** 1565-1571.
6. Barlogie B, Desikan R, Eddlemon P, et al (2001). Extended survival in advanced and refractory multiple myeloma after single-agent thalidomide: identification of prognostic factors in a phase 2 study of 169 patients. *Blood*; **98:** 492-494
7. Hideshima T, Chauhan D, Shima Y, et al (2000). Thalidomide and its analogs overcome drug resistance of human multiple myeloma cells to conventional therapy. *Blood*; **96:** 2943-2950.
8. Merlini G, Marciano S, Gasparro C, et al (2001). The Pavia approach to clinical protein analysis. *Clin Chem Lab Med*; **39:** 1025-1028.
9. Seldin DC, Choufani EB, Dember LM, et al (2003). Tolerability and efficacy of thalidomide for the treatment of patients with light chain-associated (AL) amyloidosis. *Clin Lymphoma*; **3:** 241-246.
10. Dispenzieri A, Lacy MQ, Rajkumar SV, et al (2003). Poor tolerance to high doses of thalidomide in patients with primary systemic amyloidosis. *Amyloid*; **10:** 257-261.
11. Alexanian R, Weber D, Anagnostopoulos A (2003). Thalidomide with or without dexamethasone for refractory or relapsing multiple myeloma. *Semin Hematol*; **40:** 3-7.
12. Fahdi IE, Gaddam V, Saucedo JF, et al (2004). Bradycardia during therapy for multiple myeloma with thalidomide. *Am J Cardiol*; **93:** 1052-1055.

VAD CHEMOTHERAPY FOLLOWED BY AUTOLOGOUS STEM CELL TRANSPLANTATION (ASCT) FOR THE TREATMENT OF NEPHROTIC SYNDROME IN LIGHT CHAIN (AL) AMYLOIDOSIS

J.B. Perz[1], S.O. Schonland[1], R.P. Linke[2], M. Zeier[1], M. Hundemer[1], A.D. Ho[1], H. Goldschmidt[1]

[1] Department of Internal Medicine, University of Heidelberg, Germany, [2] Max-Planck Institute of Biochemistry, Munich, Germany

1. INTRUDUCTION

AL (amyloid light chain) amyloidosis is a rare disease with an incidence of 5 - 13 patients / million / year. The disease is the result of a clonal plasma cell expansion, in which monoclonal light chains transformed to amyloid can be deposited in various tissues leading to organ dysfunction and organ failure. The median survival without therapy is 10-14 months from diagnosis (1). Renal disease is the most common organ involvement and it can be seen in up to 65% of patients with AL amyloidosis. With high-dose melphalan (HDM) and autologous stem cell transplantation (ASCT) haematological and clinical remission rates in approximately 50% of treated patients could be reported in several phase II studies (2). HDM followed by ASCT appears to prolong survival in patients, in whom haematological remission can be reached (3). The treatment toxicity of HDM is on the other hand known to be severe with mortality rates of 15 - 40%.

We evaluated a therapy proposal with cyclic VAD chemotherapy followed by stem cell mobilisation prior to HDM and ASCT (4) in a prospective single centre phase II study with the main objective to assess its feasibility, toxicity and efficacy in patients with newly diagnosed systemic AL amyloidosis and nephrotic syndrome. VAD may suppress the clonal plasma cell disease and stabilise the clinical organ impairment prior to HDM with ASCT, thus it can be advantageous with respect to improved haematological response and survival rate.

2. PATIENTS

Between 1998 and 2003 16 patients with AL amyloidosis and nephrotic syndrome were treated in our unit with the proposed regimen. Kidney biopsy was performed in all cases and it revealed light chain (AL) amyloid deposits. All patients had nephrotic syndrome with an urine protein excretion of > 3 000 mg/day. Patients should have met the following additional inclusion criteria: age between 18 and 65 years, performance status 0-2 according to WHO criteria, ventricular ejection fraction of more than 40% and haemodynamic stability. Patients on haemodialysis were not excluded, if other eligibility criteria were met. The median age at diagnosis was 54 years. There were 8 women and 8 men in our collective.

3. METHODS

Enrolled patients received 2-5 cycles of VAD chemotherapy in a standard dosing schedule as described by Samson (5). Stem cells were mobilised after chemotherapy by G-SCF in a dose of 30 ug/kg twice a day from day 6 until leucapheresis were completed. Non-selected stem cells were stored in DMSO by –70 degree.

Patients received intravenous HDM at a dose of 200 mg/m² or at a reduced dose of 140 mg/m² in case of renal function impairment, haemodialysis or heart function impairment. Autologous stem cells were infused 48 hours after HDM infusion. See Figure 1.

4. RESULTS

4.1 Therapy

The median number of administered VAD cycles were 3. The median interval from diagnosis to HDM and ASCT was 9 months and ranged from 4 to 16 months. 2 out of 16 patients (12,5%) succumbed within 10 days after the first VAD cycle. The causes of death were: sepsis in severe neutropenia with multi-organ failure and sudden cardiac death after fluid retention. Both patients suffered from extensive amyloidosis with more than two involved organs. In one patient the disease progressed despite VAD chemotherapy and the patient was not eligible for HDM. HDM and ASCT could be therefore performed in 13 out of 16 initially treated patients (81%). Two patients died during the first 50 days after HDM and ASCT. Causes of death were cerebral bleeding on day +10 and profound intestinal bleedings refractory to surgery on day +48. One of these two patients was 65 years old. The main toxicity of HDM was mucositis including prolonged nausea and anorexia.

Figure 1. Treatment schedule.

1. VAD therapy: 2-4 cycles
   (1,6 mg vincristine, 36 mg/m² adriamycin, dexamethasone)
2. stem cells mobilisation (one of the three possible regimens):
   ifosfamide 12 g/m²CAD (cyclophosphamide 1g/m²,
   adriamycin, dexamethasone)   cyclophosphamide 4g/m² G-
   CSF 600 µg/day
3. high dose melphalan (140 - 200 mg/m²)
4. autologous stem cell transplantation (ASCT) >2,5 10⁶ CD34+/kg

4.2 Survival

The median follow up after HDM and ASCT is until now 31 months with a range from 0,3 to 70 months. The 1-year survival from diagnosis is 75% and the 3-years survival 71% (see Figure 2). 11 out of 13 patients (85%) survived at least 3 months after HDM and ASCT. The overall therapy related mortality is 25% (4 out of 16 patients), whereas the HDM related mortality is 15% (2 out of 13 patients). All 4 patients, who died from treatment complications had a high mortality risk at therapy start: extensive amyloidosis with more than two involved organs or age of more than 60 years.

Figure 2..
Survival from time of diagnosis

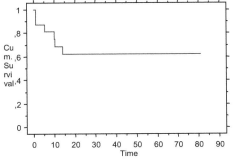

4.3 Haematological Response

After HDM with ASCT a haematological remission rate of 50% was observed: 6 complete remissions and 2 partial remissions were seen out of 16 patients initially treated. In all patients with response of plasma cell disease an improvement or stabilisation of renal function was seen. In four patients (25%) the plasma cell disease progressed after therapy: two of them died due to progressive disease, two are still alive and have a stable clinical status.

4.4 Clinical Response

Out of 16 patients with nephrotic syndrome nine had a normal initial renal function with serum creatinine lower than 1,5 mg/dl and seven had a serum creatinine higher than 1,5 mg/dl at treatment start. Two patients were initially on haemodialysis. In five out of the six patients with complete haematological remission (31% out of all 16 patients) a clinical remission of nephrotic syndrome with a decline of urine protein excretion to lower than 3 000 mg/day was observed. All patients with haematological and clinical remissions had initially a normal renal function with a serum creatinine lower than 1,5 mg/dl. In the group of patients with high serum creatinine at time of treatment start three became haemodialysis dependent over therapy time and three died due to treatment complications or disease progress. Patients with initially impaired renal function are therefore at high risk of severe treatment complications after VAD and HDM with ASCT. On the other hand one patient with haemodialysis at treatment start achieved stable disease and benefited from treatment in terms of improved performance status.

5. CONCLUSIONS

Nephrotic syndrome is a frequent manifestation of AL amyloidosis. Untreated amyloidosis is in almost all cases a progressive and fatal disease. High dose melphalan coupled with stem cell support is used widely after it has been reported to be feasible in the treatment of patients with AL amyloidosis (2,3). The use of additional induction chemotherapy prior to high dose melphalan has been evaluated in only few studies until today (4). We worked out that HDM with ASCT after VAD induction is feasible in patients with light chain amyloidosis and nephrotic syndrome. 50% of patients in our collective reached a haematological remission of the plasma cell disorder and 31% of initially treated patients a clinical remission of the renal disease. The achievement of clinical remission of the organ manifestation is therefore possible in patients with AL amyloidosis. The therapy toxicity is high in patients with kidney amyloidosis: 12,5% treatment associated deaths were seen after VAD, 15% after HDM with ASCT. Due to this high toxicity rate prophylactic use of antibiotics and antimycotics seems to be essential for the treatment of patients with AL amyloidosis with chemotherapy (6). Patients with initial serum-creatinine < 1,5 mg/dl have a high probability of complete haematological remission and remission of nephrotic syndrome. In contrast patients with initial serum-creatinine > 1,5 mg/dl may become haemodialysis dependant during chemotherapy. ASCT should thus be especially recommended to patients with initial serum creatinine < 1,5 mg/dl. Randomised phase III studies are necessary to elucidate the role of induction chemotherapy prior to HDM and ASCT and new treatment strategies have to be sought to improve the results of nephrotic syndrome treatment in AL amyloidosis.

REFERENCES

1. Kyle, R.A. et al., Long-term survival (10 years or more) in 30 patients with primary amyloidosis, Blood, 93, 1062-6, 1999.
2. Comezo, R.L. et al., Autologous stem cell transplantation for primary systemic amyloidosis, Blood, 99(12), 4276-82, 2002.
3. Skinner, M. et al., High-dose melphalan and autologous stem-cell transplantation in patients with AL amyloidosis: an 8-year study, Ann Intern Med, 140(2), 85-93, 2004.
4. Gono, T. et al., Nephrotic syndrome due to primary AL amyloidosis, successfully treated with VAD and subsequent high-dose melphalan followed by autologous peripheral blood stem cell transplantation, Intern Med 2003, 42(1), 72-7, 2003
5. Samson, D. et al., Infusion of vincristine and doxorubicin with oral dexamethasone as first-line therapy for multiple myeloma, Lancet, 14,2(8668), 882-5, 1989.
6. Palladini, G., et al., A modified high-dose dexamethasone regimen for primary systemic (AL) amyloidosis, Br J Haematol 2001, 113(4), 1044-6, 2001.

SUPERIOR SURVIVAL IN PRIMARY SYSTEMIC AMYLOIDOSIS PATIENTS UNDERGOING PERIPHERAL BLOOD STEM CELL TRANSPLANT : A CASE CONTROL STUDY

A. Dispenzieri, R.A. Kyle, M.Q. Lacy, T.M. Therneau, and M.A. Gertz.
Mayo Clinic, 200 First Street S.W., Rochester, MN 55905, USA

I. ABSTRACT

Primary systemic amyloidosis (AL) is a plasma cell dyscrasia resulting in multisystem failure and death. High dose chemotherapy with peripheral blood stem cell transplant (PBSCT) has been associated with higher response rates and seemingly higher overall survival than standard chemotherapy. Selection bias, however, confounds interpretation of these results. We performed a case match control study comparing overall survival of 63 transplanted AL patients to 63 non-transplanted patients. Matching criteria included age, gender, time to presentation, left ventricular ejection fraction, serum creatinine, septal thickness, nerve involvement, 24 hour urine protein, and serum alkaline phosphatase. According to design, there was no difference between the groups with respect to gender (57% males), age (median 53 years), left ventricular ejection fraction (65%), number of patients with peripheral nerve involvement (17%), cardiac interventricular septal wall thickness (12 mm), serum creatinine (1.1 mg/dL), and bone marrow plasmacytosis (8%). Sixty-six patients have died (16 cases and 50 controls). For PBSCT and control groups respectively, the 1, 2, and 4 year overall survival rates are: 89 and 71%; 81 and 55%; and 71 and 41%. Outside a randomized clinical trial, these results present the strongest data supporting the role of PBSCT in selected patients with AL. *Supported in part by CA 62242(R.A.K), CA 91561(A.D) from the National Cancer Institute, and the Robert A. Kyle Hematology Malignancies Fund, Mayo Foundation.*

II. INTRODUCTION

Peripheral blood stem cell transplant was introduced as a therapeutic option for primary systemic amyloidosis (AL) in the late 1990s, but transplant related mortality was high [1,2]. With better patient selection, transplant related mortality rates dropped to approximately 15% [3,4] to 25% [5], but concern over whether the improved outcome was an artifact of selection grew. We previously demonstrated that if strict transplant eligibility criteria were applied to patients with amyloid, only 16% of patients were candidates for transplant and that the highly selected patients had a 24 month survival comparable to that reported in the literature for patients undergoing PBSCT.[6] With longer follow-up, however, there is a suggestion that the 48 month survival is better in transplanted patients (60%) [4] than in selected, but conventionally treated patients (43%) [6]. Intuitively, the increased hematologic and organ response rates observed with PBSCT should result in superior survival, but this has not yet been demonstrated in a randomized controlled trial. We therefore sought to address the question in a different fashion—that is, a case control study matching those patients who have received high dose chemotherapy with peripheral blood stem cell transplant (cases) to those who have comparable clinical features, but who have not been transplanted (controls).

III. PATIENTS AND METHODS

Seventy-one consecutively transplanted (7/26/96 through 5/29/01) AL patients were initially selected as the case group. Seven were excluded because they received a cadaveric heart transplant, and 1 other was excluded because of the lack of echocardiogram data, leaving 63 cases. In order to minimize the effect of potentially confounding variables, the transplant patients were matched one-to-one to amyloid patients who did not receive transplants. For each transplant patient, the set of potential controls was restricted to those of the same gender, such that their time from initial diagnosis to being seen at our institution was comparable to that of the case (to control for biases in referral patterns), and who were alive and under our care for at least as long as the time to transplant of the associated case (deaths very soon after referral would never have been eligible for a transplant). Within this group of potential controls, that subject was chosen who best matched the case in terms of clinical risk variables including level of nerve involvement, serum creatinine, septal thickness, ejection fraction, alkaline phosphatase, and urine protein. Matching can never be exact on such a long list of factors; the relative importance of each of these was based on the coefficients from a survival model that was fit to the combined set of cases and potential controls. The analysis focused on comparing PBSCT cases versus matched controls that did not receive PBSCT. All statistical tests were two-sided and p-values less than 0.05 were considered significant.

IV. RESULTS

As shown in Table 1, there was overlap between time periods of diagnosis for the cases (June 1992 to November 2000) and the controls (March 1983 to August 2000). Despite the fact that our first AL stem cell transplant was performed March 1996, less than a third of the 63 cases were transplanted before 1999. Only fourteen percent of the controls were diagnosed in 1998 or later. Reasons for not transplanting the controls during the overlap period included the stricter transplant eligibility criteria used during the middle to late 1990s, patient preference, and financial restrictions.

PATIENT CHARACTERISTICS	Case (n=63)	Control (n=63)	All (n=126)
		No., median (range)	
Diagnosis to Mayo, months	1.5 (0, 71.6)	1.4 (0, 11.0)	1.4 (0, 71.6)
Diagnosis to transplant/treatment, mo	4.4 (1.3, 74.6)*	1.4 (0, 13.2)	2.7 (0, 74.6)
Males, n (%)	36 (57%)	36 (57%)	72 (57%)
Age, years	53 (30, 69)	53 (32, 69)	53 (30, 69)
Nerve involvement, n (%)	11 (17%)	11 (17%)	22 (17%)
Left ventricular ejection fraction, %	65 (30, 79)	65 (30, 80)	65 (7, 80)
≤ 50%, n (%)	4 (6%)*	12 (19%)	16 (13%)
Interventricular septum, mm	12 (7, 23)	12 (7, 25)	12 (7, 25)
> 15 mm, n (%)	11 (17%)	10 (16%)	21 (17%)
Serum alkaline phosphatase, IU/L	196 (57, 1195)	196 (76, 1926)	196 (57, 1926)
Serum creatinine, mg/dl	1.1 (0.5, 2.1)	1.0 (0.6, 10.3)	1.0 (0.5, 10.3)
Serum M-protein, g/dl	0.5 (0.0, 2.9)	0.0 (0.0, 2.8)	0.0 (0.0, 2.9)
Serum albumin, g/dl	2.7 (1.4, 3.8)	2.7 (0.9, 4.1)	2.7 (0.9, 4.1)
24 hour urine protein, g/24 hours	4.6 (0.0, 20.7)	5.3 (0.1, 24.7)	4.8 (0.0, 24.7)
24 hour urine M-protein	0.1 (0.0, 3.1)	0.2 (0.0, 4.0)	0.2 (0.0, 4.0)
Bone marrow plasma cells, %	10 (1, 91)	8 (1, 66)	8 (1.0, 91)
> 30%, n (%)	8 (16%)	3 (5%)	11 (10%)

* $p < 0.05$

Twenty-seven of the transplanted patients received some therapy prior to transplant conditioning. Peripheral blood stem cells were mobilized with cyclophosphamide and GM-CSF (n=33) or GCSF alone (n=30). Seventeen received melphalan (140 mg/m2) and total body radiation (1200 cGy), 37 melphalan 200 mg/m2, 9 melphalan 100-140 mg/m2, followed by the usual supportive care.[3] Transplant related mortality—death within 100 days of transplant—was 13% (8/63). Initial therapies in the control group included alkylator based oral chemotherapy regimens (n=52), high-dose steroid based therapy (n=4), colchicine (n=5), or experimental regimens (n=2; IDOX [1], vitamin E [1]). The median follow-up from diagnosis was 3.8 years for the cases and 8.8 years for the controls. Sixteen patients have died in the PBSCT group whereas 50 have died in the control group. For PBSCT and control groups respectively, the 1, 2, and 4 year overall survival rates from diagnosis are: 89 and 71%; 81 and 55%; and 71 and 41%, p < 0.001 (Figure). The 1, 2, and 4 year survivals calculated from date of transplant (cases) or initiation of therapy (controls) were as follows: 82 and 68%; 81 and 53%, and 70 and 40%, p<0.001.

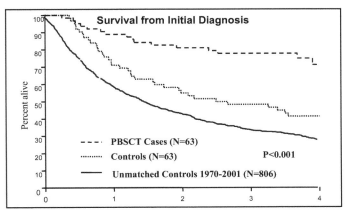

V. DISCUSSION

Our study of 126 AL patients demonstrates that patients receiving PBSCT enjoy better overall survival than their matched counterparts not undergoing this procedure. Though our data do not suggest that PBSCT is the best therapy for *all* amyloid patients, they do illustrate that patients deemed candidates for the procedure at an Amyloid Treatment Facility do better with the procedure than without. One must recall that the initial treatment related mortality figures for AL patients undergoing transplant approached the sobering value of 50% [1], but with more stringent selection criteria are now approximately 15% [3,4] at specialty Amyloid Transplant Centers and 25% at other centers.[5] Though this study cannot substitute for a randomized controlled study, it does lend additional evidence to the belief that in selected patients PBSCT results in superior overall survival as compared to standard chemotherapy.

VI. REFERENCES

1. Moreau P, et al. High-dose melphalan and autologous bone marrow transplantation for systemic AL amyloidosis with cardiac involvement. Blood. 1996;87:3063-3064.

2. Comenzo RL, et al. Intermediate-dose intravenous melphalan and blood stem cells mobilized with sequential GM+G-CSF or G-CSF alone to treat AL amyloidosis. Br J Haematol. 1999;104:553-559.

3. Gertz MA, et al. Stem cell transplantation for the management of primary systemic amyloidosis. Am J Med. 2002;113:549-555.

4. Sanchorawala V, et al. An overview of the use of high-dose melphalan with autologous stem cell transplantation for the treatment of AL amyloidosis. Bone Marrow Transplant. 2001;28:637-642.

5. Vesole DH, et al. High Dose Therapy with Autologous HSCT for Patients with Primary Systemic Amyloidosis (AL): Results from the ABMTR. Blood. 2003;102:Abstr 402.

6. Dispenzieri A, et al. Eligibility for hematopoietic stem-cell transplantation for primary systemic amyloidosis is a favorable prognostic factor for survival. J Clin Oncol. 2001;19:3350-3356.

TREATMENT OF AL AMYLOIDOSIS WITH TANDEM CYCLES OF HIGH DOSE MELPHALAN AND AUTOLOGOUS STEM CELL TRANSPLANTATION

V. Sanchorawala, D. G. Wright, K. Quillen, R. Kunz, K.T. Finn, K. Malek, R. H. Falk, J. Berk, L.M. Dember, M. Skinner, D. C. Seldin

Boston University School of Medicine, Boston Medical Center, Boston, MA, USA

INTRODUCTION

AL or primary systemic amyloidosis is characterized by widespread deposition of amyloid fibrils, derived from monoclonal immunoglobulin light chains as a result of a clonal plasma cell dyscrasia. In the last decade, high-dose melphalan and autologous stem cell transplantation (HDM/SCT) has been shown to induce hematologic and clinical remissions and prolong survival in patients with AL amyloidosis (1). Moreover, the treatment outcomes following HDM/SCT are related to whether or not a complete hematologic response, defined as disappearance of the underlying monoclonal gammopathy from serum and urine by immunofixation electrophoresis and of clonal plasma cell dysrasia by bone marrow biopsy, is achieved. The median survival of patients who achieve a hematologic complete response (CR) is greater than 8 years compared to 5.2 years for those who do not achieve a CR. Furthermore, clinical improvements in affected organ systems occurs in 66% of patients achieving a hematologic CR compared to only 30% of those who do not. Similar differences with respect to improvements in performance status were observed in association with hematologic CR.

Because hematologic CR is such a critical determinant of treatment outcome following HDM/SCT in patients with AL amyloidosis, we questioned whether tandem cycles of treatment might increase the proportion of patients who ultimately achieve a hematologic CR. Experience with tandem cycles of high-dose chemotherapy in multiple myeloma have shown that hematologic CR increases following tandem cycles of treatment from 24% to 43% (2). Moreover, event-free and overall survival has been reported to be superior for tandem cycles of treatment (3).

These considerations prompted us to conduct a prospective trial of tandem cycles of HDM/SCT in AL amyloidosis. The specific objectives of this trial have been to determine whether a second cycle of HDM/SCT improves the hematologic CR rate and to determine the feasibility and tolerability of tandem cycles of this aggressive treatment in AL amyloidosis.

PATIENTS AND METHODS

Eligibility criteria for this trial, begun in 2000, have required a tissue diagnosis of amyloidosis with evidence of an underlying clonal plasma cell dyscrasia, age 65 or younger, and less than 300 mg of prior oral melphalan therapy. Other eligibility criteria have included minimum measures of performance status (SWOG 0-2) and of cardiac (LVEF by ECHO > 45%), pulmonary (DLCO > 50% of predicted) and hemodynamic (supine systolic BP > 90mm Hg) function. Patients in renal failure on dialysis have not been excluded if other eligibility criteria were met.

Peripheral blood stem cells were collected by leukopheresis following G-CSF mobilization and a minimum yield of 7.5 x 10^6 CD 34+ cells/kg was required for patients on the trial. All patients received a first cycle of high dose intravenous melphalan (200 mg/m^2) divided over 2 successive days. Stem cells were infused into patients 48 hours following completion of chemotherapy.

Following the first cycle of treatment, patients were evaluated for hematologic response at 6 months. Patients with evidence of persistent plasma cell dyscrasia, received a second cycle of high-dose melphalan (140 mg/m^2) followed by autologous transplant of their cryopreserved stem cells, within 6 –12 months of the first cycle.

All patients were followed for hematologic and clinical responses at 3, 6, 12 months and annually thereafter.

RESULTS

From 11/2000 to 3/2003, 38 patients were enrolled in this trial. The median age is 55 years (range, 32-65) with a male to female ratio of 1.7 to 1.0. Twenty nine % (11/38) of patients had isolated renal involvement, a good prognosis group, but 32% (12/38) had cardiac involvement and 63% (24/38) had greater than 2 organ systems involved, features indicative of poor prognosis.

Of the 38 patients enrolled, 8 (21%) were removed from the protocol because of either inadequate stem cell collection, 6 patients, or complications during stem cell mobilization and collection, 1 patient. 30 patients received a first cycle of high-dose melphalan at 200 mg/m2 and stem cell transplant under the protocol, and there was 1 death within 90 days of this treatment.

6 months after this initial cycle of treatment, 15 or 50% patients were found to have achieved a hematologic CR. Of the 14 patients without a hematologic CR, 11 patients received a second cycle of HDM/SCT (140 mg/m^2), while 3 patients did not either because of patient choice or because of excessive toxicities during the first cycle of treatment.

Morbidity and mortality were not excessive in this trial. There were no deaths during the stem cell mobilization and collection phase of treatment and the overall early death rate within 90 days of SCT was 6% (2/30). One of 30 (3%) patients died after the first cycle of HDM/SCT with a diffuse interstitial pneumonitis, and 1 of 11 (9%) died after the second cycle due to parainfluenza 3 viral pneumonia. Grade IV non-hematologic toxicities were somewhat more frequent (27%) after the second cycle of treatment, but overall such toxicities were comparable in frequency to those we have observed with single cycle of HDM/SCT treatment.

After the second cycle of treatment 3 of 10 (30%) of patients achieved a hematologic complete response. This was in addition to the 15 or 50% of patients who had achieved a hematologic CR after the first cycle of treatment. Thus, the ultimate hematologic CR rate was 60% (18/30). Free light chain measurements were available for 23 patients enrolled on the study and 83% (19/23) of the patients, who completed the protocol, achieved a greater than 50% reduction in the free light chain measurements.

SURVIVAL

The median survival for all 38 patients enrolled in the trial, has not yet been reached with a median follow up of 31 months, range 14 – 40 months. Estimated survival at 1 year is 95% and at 2 years is 84%. This survival compares favorably with a matched cohort of 162 patients whom we treated with a single cycle of 200 mg/m^2 of HDM/SCT on prior trials.

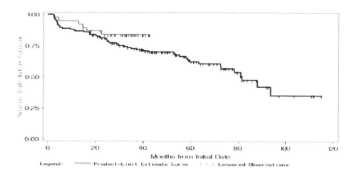

Top curve - Tandem HDM/SCT 200 mg/m2 (current trial) n=38
Bottom curve - Single HDM/SCT 200 mg/m2 (matched historical cohort treated on prior protocols) n=162

CONCLUSIONS

In summary, tandem cycles of HDM/SCT are feasible and tolerable for selected patients with AL amyloidosis. Moreover, tandem cycles of HDM/SCT can increase the proportion of patients who ultimately achieve a hematologic complete response. Further follow-up will be required to determine whether this is associated with improved survival.

REFERENCES

1. Skinner M, Sanchorawala V, Sedlin D.C. et al. High-dose melphalan and autologous stem cell transplantation in patients with AL amyloidosis: An 8-year study. Annals of Internal Medicine 2004; 140: 85-93.

2. Barlogie B, Jagannath S, Vesole D et al. Superiority of tandem autologous transplantation over standard therapy for previously untreated multiple myeloma. Blood 1997; 89: 789-793.

3. Attal M, Harousseau J, Fancon et al. Double autologous transplantation improves survival in multiple myeloma patients: Final analysis of a prospective randomized study of the "Integroupe Francophone du Myelome (IFM 94). Blood 2002; 151a.

THE USE OF SEQUENTIAL LIVING DONOR KIDNEY AND AUTOLOGOUS STEM CELL TRANSPLANTATION FOR PATIENTS WITH PRIMARY (AL) AMYLOIDOSIS AND ADVANCED RENAL FAILURE

N. Leung[1], M. D. Griffin[1], A. Dispenzieri[2], J. M. Gloor[1], T. R. Schwab[1], S. C. Textor[1], M. Q. Lacy[2], T. S. Larson[1], M. D. Stegall[3], and M. A. Gertz[2].

Mayo Clinic, 200 First Street SW, Rochester, Minnesota, 55905, USA. Division of Nephrology[1] Division of Hematology[2] Division of Transplant Surgery[3]

I. Abstract

Patients with primary (AL) amyloidosis and renal failure have limited life expectancy on dialysis and face higher risk during autologous stem cell transplantation (ASCT). We have pursued a novel strategy of living donor kidney transplant (KTx) prior to ASCT for these patients. Clinical and laboratory outcomes of all 7 patients accepted for this protocol at Mayo Clinic between Dec 1999 and Feb 2003 were reviewed. Five of the 7 were on dialysis prior to KTx. All 7 patients underwent living donor KTx and experienced good initial graft function. Immunosuppression was tacrolimus (n = 4), cyclosporine (n = 2), or sirolimus (n = 1) in combination with mycophenolate mofetil and prednisone. Five patients subsequently underwent ASCT, 4 with good outcome. For these 4 patients, stem cell harvest was carried out prior to KTx for 3 and after KTx for 1. Average times to neutrophil and platelet engraftment were 15.5 and 23.5 days respectively. The fifth developed progressive hepatic amyloidosis after KTx and was advised against ASCT, died following ASCT elsewhere. Two did not undergo ASCT. One died of a demyelinating disease 4 weeks after KTx. The other has thus far elected not to undergo ASCT and has minimal recurrent disease after 3 years. Living donor KTx followed by ASCT is a feasible approach for AL patients with advanced renal impairment. Prudent patient selection is crucial as KTx will delay ASCT. Stem cell engraftment was not impaired by triple-drug immunosuppression.

II. Introduction

Primary systemic (AL) amyloidosis is a disease characterized by soft tissue fibrillar deposition of light chain fragments. Survival is poor despite standard therapy with melphalan and prednisone.[1] Better response has been achieved with myeloablative therapy with high dose intravenous melphalan followed by autologous stem cell transplantation.[2] This treatment however is not recommended for patients with severe renal insufficieny due to the high treatment mortality associated with posttransplant renal failure.[3] In order to overcome this, we have pursued a strategy of performing living kidney transplantation prior to ASCT in patients who were otherwise not candidate for ASCT secondary to severe renal insufficiency. We propose that improvement of renal function by kidney transplantation can reduce the risk of ASCT to baseline level.

III. Patients and Methods

Patients were recruited from the Dysproteinemia and Nephrology Clinic. Sequential transplantations were offered to patients who were ineligible for ASCT because of poor renal function. Patients with multiple myeloma were excluded. All patients underwent a kidney biopsy for diagnostic purpose. AL was confirmed by the presence of light chain deposits on the kidney biopsy by kappa/lambda immunofluorescence, a circulating monoclonal light chain or a clonal expansion of plasma cells on bone marrow biopsy.

Standard screening process for KTx was performed on all patients. Living donor KTx was performed in the usual manner with donor surgery performed by hand-assisted laparoscopic nephrectomy. Basiliximab (Simulect®) or anti-thymocyte globulin (Thymoglobulin ®) was used for induction. Oral immunosuppression regimens consisted of either tacrolimus, cyclosporin or sirolimus in combination with mycophenolate mofetil and prednisone. Immunosuppression was changed or adjusted according to existing protocol and were continued throughout ASCT. Glomerular filtration rate (GFR) was measured pre and posttransplant by non-radioactive labeled iothalamate clearance. Patients underwent protocol renal allograft biopsy at 3 months and 1 year posttransplant.

Stem cells mobilization was achieved by growth factor alone. Harvesting continued until 5×10^6 CD 34^+ cells were collected. Patients proceeded to ASCT if renal allograft function was determined to be stable and had no signs of rejection at the 3 month posttransplant follow-up. Conditioning consisted of melphalan (L-PAM) given over 2 consecutive days. Standard antibiotic prophylaxis was given. Engraftment is defined as absolute neutrophil count (ANC) is $> 0.5 \times 10^9$/L or platelets is $> 50 \times 10^9$/L.

IV. Results

Between December of 1999 to February of 2003, 5 males and 2 females underwent kidney transplantation in anticipation of ASCT. Characteristics of these patients are described in Table 1. Heart and liver/gastrointestinal tract involvement occurred in 2 patients each. All Ktx were successfully performed with median follow up of 581 days. Induction agents and immunosuppressants documented in Table 2 reflected our usual practice at the time of KTx. Sirolimus was converted to tacrolimus during ASCT. One patient had borderline rejection prior to ASCT and another developed an acute cellular rejection after ASCT, both of which were successfully treated with corticosteroid.

Table 1. Patient characteristics

Patient	Age	Gender	Renal Function Pretransplant	LC	ASCT	Organ Involvement
1	61.9	M	dialysis	λ	Yes	2
2	52.6	M	dialysis	λ	No	2
3	57.0	M	dialysis	λ	No	1
4	64.9	M	dialysis	κ	Yes	1
5	60.5	F	dialysis	λ	Yes	2
6	64.4	F	Cr = 6.3	λ	Yes	1
7	50.8	M	Cr = 4.6	κ	Yes	2

Table 2. Induction and Immunosuppression for Kidney Transplantation

Patient	Induction Agent	Primary Oral Immunosuppressant	SCr after KTx	GFR after KTx	Follow up (days)
1	Basiliximab	cyclosporin	1.4		295
2	none	tacrolimus	1.5		1340
3	Basiliximab	cyclosporin	0.9		25
4	ATG	sirolimus	1.3	52	1078
5	ATG	tacrolimus	1.2	44	740
6	ATG	tacrolimus	1.1	45	581
7	ATG	tacrolimus	1.9	57	417

To date, 5 patients received ASCT after kidney transplantation with median follow-up of 467 days (range 272-697) after ASCT. Stem cell collection and conditioning are listed in Table 3. The patient on sirolimus required a second mobilization with GM-CSF in order to collect sufficient stem cells. No data was available for Patient #1 who died at another institution during the peri-ASCT period. Another patient died 4 weeks after kidney transplantation with a demyelinating condition. Both patients were known to have died with a functioning renal allograft. A third renal allograft recipient has elected not to proceed with ASCT. He has minimal histological evidence of recurrent disease in the allograft but renal allograft function remains stable. All patients engrafted. All 4 patients developed neutropenic fever, 2 with positive cultures but no one required vasopressor therapy or ventilatory support. At last visit, none of the 4 patients had evidence of recurrence. ASCT was performed at our institution at a mean of 221 ($\pm$ 114) days after KTx and 420 ($\pm$ 140) days after initial consultation.

Table 3 Stem Cell Collection and Conditioning

Patient	Timing of PBSC Collection	Growth Factor	# of apheresis	Yield of CD 34$^+$ cells (x 10^6)	L-PAM Dose (mg/m^2)
4	Post KTx	GCSF, GMCSF	3	0.72/ 4.26	140
5	Pre KTx	GCSF	2	10.09	200
6	Pre KTx	GCSF	1	5.25	140
7	Pre and Post	GCSF	6	5.93/ 5.05	200

V. Discussion

Our experience showed that sequential living donor kidney and autologous stem cell transplant is a safe and effective treatment for AL patients with severe renal impairment. Our limited data also suggest engraftment of stem cells was not altered by triple drug immunosuppression. However, the success of this treatment depends on the careful candidate selection. First, the patient must pass the usual screening criteria for kidney transplantation. Those with significant cardiac or hepatic disease are unlikely to survive the surgery. Candidates should also have slowly progressive disease with a life expectancy of at least 12 months. Experience from multiple myeloma patients showed that cytotoxic therapy after kidney transplantation often resulted in death from complications of immunosuppression. Those with shorter life expectancy are unlikely to live long enough to go through the sequential transplants as the shortest time interval between the initial Nephrology consultation and ASCT was 215 days. The availability of a living donor is crucial since it is extremely unlikely patients will receive a deceased donor kidney in time. Finally, we recommend that mobilization and collection should be completed prior to initiating immunosuppression. While it is based on very limited data, it seems that patients collected after kidney transplantation required more apheresis sessions and growth factors. This will need to be confirmed by a larger study. With the proper selection, sequential transplantation can be an useful option in patients who are otherwise poor candidates for ASCT.

VI. References

1.   Kyle RA, Gertz MA, Greipp PR, et al: A trial of three regimens for primary amyloidosis: colchicine alone, melphalan and prednisone, and melphalan, prednisone, and colchicine. *New Engl J Med* 336:1202-1207, 1997

2.   Skinner M, Sanchorawala V, Seldin DC, et al: High-dose melphalan and autologous stem-cell transplantation in patients with AL amyloidosis: an 8-year study. *Ann Intern Med* 140:85-93, 2004

3.   Fadia A, Casserly LF, Sanchorawala V, et al: Incidence and outcome of acute renal failure complicating autologous stem cell transplantation for AL amyloidosis. *Kidney Intl* 63:1868-1873, 2003

DELAYED RESOLUTION OF NEPHROTIC SYNDROME FOLLOWING STEM CELL TRANSPLANT FOR AMYLOIDOSIS (AL)

Morie Gertz, M. D., Nelson Leung, M. D., Martha Lacy, M. D., and Angela Dispenzieri, M. D.

Divisions of Hematology and Nephrology, Department of Medicine, Mayo Clinic College of Medicine, Rochester, MN

INTRODUCTION

In most malignancies, response following stem cell transplantation is assessed by the maximal reduction in tumor mass. In multiple myeloma, response is assessed using the surrogate of 50% reduction in M protein level (1). In AL, the goal of high-dose chemotherapy is eradication of the amyloidogenic protein production. The final endpoint is improvement in organ function for those organs involved with amyloid deposits.

PATIENTS AND METHODS

We retrospectively reviewed the changes in 24-hour urinary protein excretion among 39 amyloid patients who were serially monitored for a change in their urinary protein level. Data were abstracted among those patients who responded to treatment to better understand the kinetics of the response.

RESULTS

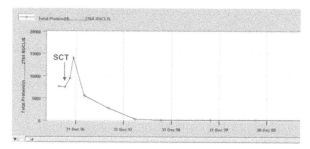

Figure 1. Example of a patient whose urinary protein rose immediately following stem cell transplant and took well over one year to decline into the normal range.

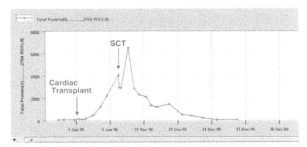

Figure 2. A patient who had cardiac transplant for amyloid and developed renal amyloid nephrotic syndrome. Urinary protein rose and did not reach its lowest level until two years following transplant.

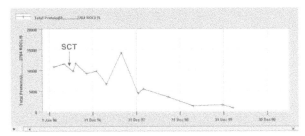

Figure 3. Demonstrates another patient with delayed resolution of the nephrotic syndrome where the maximal urinary protein reduction took three years to be achieved.

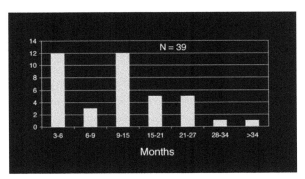

Figure 4. Time to a 50% reduction in urinary protein loss for 24 hours.

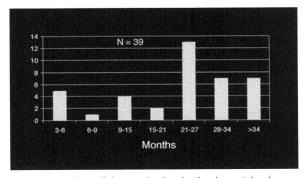

Figure 5. Time in months until the maximal reduction in proteinuria was observed.

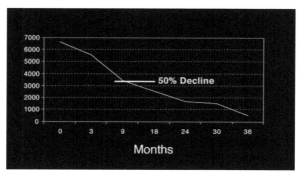

Figure 6. Median urinary protein loss for all 39 responders at three-month intervals. Although a 50% reduction in urinary protein loss was seen at nine months, further declines occur up to three years.

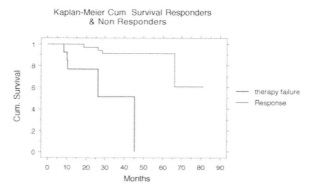

Figure 7. The impact of a response to transplant on survival in patients with renal AL.

Twelve of our 39 patients (30%) had a post-transplant rise in their urinary protein loss. Thirty-five of the 39 patients remain alive. The four patients who died did so at 66 and 26 months due to progressive amyloidosis, and two died post-transplantation of unexplained bacteremia and aspiration pneumonia.

DISCUSSION

Patients with renal amyloid commonly demonstrate a delay in reduction of urinary protein loss following stem cell transplant. Assessment of organ response in patients with renal amyloidosis should be delayed until one year post-transplant. It is premature to consider patients therapeutic failures and initiate salvage therapies before one year if any downward trend is seen in urinary protein loss. It is difficult to distinguish therapeutic failures from delayed responders. The median time to a 50% reduction in urinary protein is approximately nine months, and the median time to maximal response in urinary protein loss is approximately 27 months.

REFERENCE

1.   Blade J, Samson D, Reece D, Apperley J, Bjorkstrand B, Gahrton G, Gertz M, Giralt S, Jagannath S, Vesole D: Criteria for evaluating disease response and progression in patients with multiple myeloma treated by high-dose therapy and haemopoietic stem cell transplantation. Myeloma Subcommittee of the EBMT. European Group for Blood and Marrow Transplant. Br J Haematol 102 (5):1115-1123, 1998.

AMYLOID VERSUS NON-AMYLOID IMMUNOGLOBULIN DEPOSITS

G. Merlini

Amyloidosis Centre, Biotechnology Research Laboratories, University Hospital IRCCS Policlinico San Matteo, Department of Biochemistry, University of Pavia, Italy

Monoclonal immunoglobulin deposits can be organized as fibrils, microtubules, crystals or casts, or be unorganized, as granular deposits. The two main prototypic conditions caused by highly organized fibrillar deposits or amorphous aggregates are light chain amyloidosis (AL) and light chain deposition disease (LCDD), respectively.

MOLECULAR BASIS OF THE DIFFERENT ORGANIZATION OF LIGHT CHAIN DEPOSITS

In AL, the fibril formation is triggered by a combination of factors: a) the amino acid sequence of the light chain (LC) variable region leading to a relatively less stable native state; b) local tissue concentration of the amyloidogenic light chain; c) mildly destabilizing environmental conditions (pH, ions, solutes); d) post-translational modifications including proteolysis (the major constituent of AL deposits are LC fragments of approximately 16 kDa comprising the V region and variable portions of the C region sequence) and glycosylation (this seems particularly relevant for the κI LC variability subgroup). That the amino acid sequence is the main determinant of LC aggregation is supported by several findings. There is a strong preponderance of λ over κ light chains, and certain LC germline genes (*Vλ6a* and *Vλ3r*) (1) or gene families (VκI and VκIV) are significantly associated with amyloid formation. Analysis of the primary structures of more than 100 κI LCs, including 37 amyloidogenic ones, revealed structural features that may contribute to amyloidosis and "flaws" in the VκIV germline gene that could cumulatively decrease the stability of the variable region (2). That the loss of LC variable domain stability is the major driving force in fibril formation is supported by data from several authors who performed site-directed mutagenesis studies on recombinant VL regions. Although these studies are informative, we must remember that they were performed in the absence of a constant region or a part of a constant region which may affect the folding properties, so their translation to the *in vivo* process needs caution.

As expected, the primary structure of the LC is the main determinant of their deposition in tissues also in LCDD, as suggested by data obtained in animal models. Analysis of the primary structures of LCDD-related LC showed unusual hydrophobic residues exposed to the solvent in CDR1 or CDR3 regions. Sequence alignment of VκI and VκIV revealed the presence of hydrophobic residues, leucine and isoleucine, or tyrosine at positions 27 and/or 31 in all known cases of LCDD (3). In fact, in the LCDD-associated κ LC (SCI) sequenced by our group, tyrosine was present in position 31 in addition to two isoleucines in positions 53 and 77: both residues are exposed to the solvent with potential perturbation of protein folding. Radiolabeled κ SCI injected into mice was rapidly (within an hour) accumulated, without a lag phase, in the kidney at concentrations 3 to 5 times higher than in the liver or spleen.

Both fibrillar and non-fibrillar monoclonal LC deposits may coexist in the same patient, and the identity of the amino acid sequence of the deposited protein has been reported. The molecular mechanisms underlying the possibility that the same protein can adopt dramatically different structural organizations have been investigated. The data obtained from the analysis of a recombinant amyloidogenic VκIV (SMA) indicate that a given protein might have more than one critical intermediate conformation during the aggregation pathway, and that such different conformations may lead to different types of deposits (4). This same VκIV (SMA) was analyzed in a series of *in vitro* assays: the protein assembled into amyloid-like fibrils when agitated and could also form granular aggregates upon exposure to copper. Two distinct molecular surfaces of the LC underlie each type of aggregate (5). In conclusion, the experimental data indicate that: a) sequence-specific intrinsic properties of the protein determine whether it will remain soluble or aggregate; b) fibrillar and granular aggregates form along different pathways which are governed by the environmental conditions; c) distinct molecular surfaces of the LC underlie each type of aggregate; d) multiple alternative molecular hydrophobic surfaces would favour formation of a granular aggregate, while a single surface would lead to docking of subunits, as is required for polymerization into amyloid fibrils.

CLINICAL FEATURES

AL and LCDD have similarities and also marked differences. While LCDD is frequently associated with multiple myeloma (in 65% of the cases in a recently reported large LCDD population) (6), only one tenth to one fifth of AL are associated with clinically overt myeloma. The kidney is the organ most frequently involved in both conditions: in practically all patients (96%) with LCDD and in approximately three quarters of AL patients. Renal function is more severely and rapidly compromised in LCDD than in AL, with almost all LCDD cases having renal insufficiency at presentation, while urinary protein excretion is important in both conditions, exceeding 3.5 g/day in 53% and 40% of AL and LCDD patients, respectively. The mesangial cells play a pivotal role in the pathogenesis of renal damage in both conditions, and the peculiar interaction of λVI LC with mesangial cells is probably at the basis of the documented association of this LC subclass with amyloid kidney involvement (7). Extrarenal symptomatic deposition in AL patients involved the heart (56%), liver (28%), peripheral nervous system (21%), soft tissues (15%) and gastrointestinal tract (8%), while in LCDD patients the heart was much less frequently involved (21%), as was the liver (19%), and peripheral nervous system (8%).

The diagnosis of AL is biopsy-based. It is mandatory to demonstrate the presence of a clonal plasma cell disorder. The quantification of serum free light chain may complement immunofixation. The latter can detect a monoclonal protein in serum or urine of 96% of patients, with a κ/λ ratio of 29/71. However, the possibility of being misled by the presence of a monoclonal component and the potential risk of typing a hereditary amyloidosis as AL must always be kept in mind. At our Center, the diagnostic gold standard is immunoelectron microscopy which allows unequivocal typing of amyloid deposits (8).

LCDD is usually diagnosed by renal biopsy which shows: a) by immunofluorescence, linear LC deposits within glomerular and tubular basement membranes (97% of cases); b) by electron microscopy, the prototypic granular electron-dense deposits (77% of cases); and c) by light microscopy, nodular sclerosing glomerulopathy (51% of cases) (6). Myeloma cast nephropathy can be associated in 16%-32% of the cases and AL amyloidosis in 3%-33% of patients. Monoclonal protein is detected by immunofixation in serum or urine in 94% of LCDD patients with a κ/λ ratio of 68/32. The deposits may be formed by heavy chain (HC) instead of LC constituting the so-called heavy chain deposition disease (HCDD). Although these HC have a deletion of the CH1 domain, this deletion seems necessary but not sufficient for deposition, since VH probably also contributes to tissue deposition. In ~10% of LCDD monotypic HC is associated with monotypic LC in deposits (light and heavy chain deposition disease, LHCDD).

The overall survival of patients with AL and LCDD is similar, the median value being 45.3 (Pavia AL population) and 49.2 months (6), respectively: 24% of AL and 31% of LCDD patients are alive at 8 years. Death in AL patients is most commonly due to heart involvement (75% of patients), while cachexia and heart involvement accounted for 44% and 11% of deaths, respectively, in LCDD. The prognostic factors in AL were heart involvement and response to therapy (protective), while in LCDD survival was correlated to age, the presence of multiple myeloma, and extrarenal LC deposition (6). In AL patients in whom the value of serum N-terminal-pro-natriuretic peptide type B (NT-proBNP) was recorded, this marker was by far the most significant prognostic indicator (9). A significant correlation was observed between the post-chemotherapy changes of the concentration of serum free light chains and those of NT-proBNP, indicating that the amyloidogenic circulating LC may directly contribute to the progression of amyloid cardiomyopathy independently of the amount of amyloid deposited.

REFERENCES

1   Perfetti, V., Casarini, S., Palladini, G., Vignarelli, M. C., Klersy, C., Diegoli, M., Ascari, E., and Merlini, G. (2002). Analysis of V lambda-J lambda expression in plasma cells from primary (AL) amyloidosis and normal bone marrow identifies 3r (lambda III) as a new amyloid-associated germline gene segment. *Blood* **100**, 948-953

2   Stevens, F. J. (2000). Four structural risk factors identify most fibril-forming kappa light chains. *Amyloid* **7**, 200-211

3   Deret, S., Chomilier, J., Huang, D. B., Preud'homme, J. L., Stevens, F. J., and Aucouturier, P. (1997). Molecular modeling of immunoglobulin light chains implicates hydrophobic residues in non-amyloid light chain deposition disease. *Protein Eng* **10**, 1191-1197

4   Khurana, R., Gillespie, J. R., Talapatra, A., Minert, L. J., Ionescu-Zanetti, C., Millett, I., and Fink, A. L. (2001). Partially folded intermediates as critical precursors of light chain amyloid fibrils and amorphous aggregates. *Biochemistry* **40**, 3525-3535

5   Davis, D. P., Gallo, G., Vogen, S. M., Dul, J. L., Sciarretta, K. L., Kumar, A., Raffen, R., Stevens, F. J., and Argon, Y. (2001). Both the environment and somatic mutations govern the aggregation pathway of pathogenic immunoglobulin light chain. *J Mol Biol* **313**, 1021-1034

6   Pozzi, C., D'Amico, M., Fogazzi, G. B., Curioni, S., Ferrario, F., Pasquali, S., Quattrocchio, G., Rollino, C., Segagni, S., and Locatelli, F. (2003). Light chain deposition disease with renal involvement: Clinical characteristics and prognostic factors. *Am J Kidney Dis* **42**, 1154-1163

7   Comenzo, R. L., Zhang, Y., Martinez, C., Osman, K., and Herrera, G. A. (2001). The tropism of organ involvement in primary systemic amyloidosis: contributions of Ig V-L germ line gene use and clonal plasma cell burden. *Blood* **98**, 714-720

8   Arbustini, E., Morbini, P., Verga, L., Concardi, M., Porcu, E., Pilotto, A., Zorzoli, I., Garini, P., Anesi, E., and Merlini, G. (1997). Light and electron microscopy immunohistochemical characterization of amyloid deposits. *Amyloid* **4**, 157-170

9   Palladini, G., Campana, C., Klersy, C., Balduini, A., Vadacca, G., Perfetti, V., Perlini, S., Obici, L., Ascari, E., d'Eril, G. M., Moratti, R., and Merlini, G. (2003). Serum N-terminal pro-brain natriuretic peptide is a sensitive marker of myocardial dysfunction in AL amyloidosis. *Circulation* **107**, 2440-2445

THE ROLE OF MATRIX METALLOPROTEINASES (MMPs) AND TISSUE INHIBITORS OF METALLOPROTEINASES (TIMPs) IN MESANGIAL MATRIX REPLACEMENT IN AL-AMYLOIDOSIS: AN IN-VIVO AND IN-VITRO CORRELATIVE STUDY

J. Keeling[1], S. Dempsey[1], L. Joseph[2], E.A. Turbat-Herrera[1,3,4] and G.A. Herrera[1,5]

[1]Department of Pathology, [3]Department of Obstetrics and Gynecology, [4]Radiology and [5]Cellular Biology and Anatomy, Louisiana State University Health Sciences Center, Shreveport, Louisiana, USA
[2]Department of Pathology, University of Arkansas for Medical Sciences, Little Rock, Arkansas, USA

INTRODUCTION

Renal amyloid deposition begins in the mesangium. As the disease progresses, the mesangial matrix is replaced by amyloid fibrils. Mechanisms involved in the replacement of the mesangial extracellular matrix (ECM) are not well understood. The present study aimed at understanding the roles of MMPs and TIMPs in this process. Involved glomeruli from AL-amyloidosis (AL-Am) patient specimens and human mesangial cells (HMCs) in culture incubated with amyloidogenic light chains (LCs) purified from the urine of patients with biopsy proven glomerular AL-Am were used for the study.

MMPs are a group of ubiquitous, zinc dependent endopeptidases (1). More than 20 MMPs have been characterized. They are divided into 8 classes, five of which are secreted as inactive forms while the remaining three are membrane-bound (MT-MMPs). These MMPs are involved in multiple activities including ECM remodeling. All MMPs share key domains and some possess specific additional domains which result in the cleavage of different substrates and their ability to be activated by different mechanisms. The proteolytic activities of MMPs are inhibited by the action of natural inhibitors-tissue inhibitor of MMPs (TIMPs). TIMPs bind to MMPs in a 1:1 ratio and prevent the activation of the bound MMP until the TIMP molecule is cleaved off. Five MMPs (MMPs 1,2,3,7 and 9) and TIMPs 1 and 2 were studied using immunohistochemistry and ELISA techniques for protein measurements. These MMPs belong to the following three types of MMPs: Minimal domain MMPs (MMP 7), simple hemopexin-domain-containing MMP (MMPs 1 and 3) and gelatin-binding MMPs (MMPs 2 and 9). These MMPs and TIMPs were used because of the availability of well-characterized antibodies allowing proper evaluation.

Microarray technology has become a powerful technique used in the investigation of gene expression. Many variations on the Affymetrix method have been developed designed to narrow the amount of genes to be probed to permit a more focused evaluation. In the current study, a selective MMP/TIMP microarray (SuperArray Biosciences Corporation) was utilized to analyze the role of gene expression in glomerular AL-amyloidogenesis in-vitro and in-vivo. Analysis of the data was performed using appropriate software.

MATERIALS AND METHODS

Twenty-five renal biopsies and autopsy specimens from patients with AL-Am and HMCs incubated with LCs purified from the urine of patients with renal biopsy proven AL-amyloidosis were used in the experiments. Light chains were not pooled and three amyloidogenic LCs were used for the experiments, all lambda. Normal renal tissue obtained from renal parenchyma away from renal cell carcinomas was used as negative control and cases of thrombotic microangiopathy (TMA), a condition characterized by thrombotic occlusion of the microvasculature in the kidney and secondary mesangiolysis where MMP activation is striking, were used as positive controls. Cases of MCN and LCDD provided additional specimens for comparison. LCs were extracted from the urine of patients with AL-amyloidosis and purified using ammonium sulfate precipitation and ion exchange chromatography. Second passage HMCs were used for the experiments. The HMCs were incubated with the various LCs and without LCs and with albumin for up to 96 hours.

Immunohistochemical stains for MMPs 1,2,3, 7 and 9 and TIMPs 1 and 2 were performed in the tissue specimens (surgical and autopsy sections) and HMCs incubated with LCs, albumin and without LCs. The sections were graded manually 0-no staining and 1 to 3+ by two independent viewers in a blind fashion and with the aid of the Automatic Cellular Imaging System (ACIS). Colocalization of the various MMPs and TIMPs was also performed using different markers (dyes)- diaminobenzidine and aminoethyl carbazole so that areas of colocalization would appear yellow. The cell lysates and supernatants from the HMCs incubated with and without LCs were also used for Western analysis for the various MMPs at various time frames up to 96 hours after incubation under the various conditions. Also performed were ELISA protein determinations for the same MMPs. Finally, mRNA for MMPs/TIMPs using microarrays was quantitated from HMCs treated under the various conditions and glomeruli microdissected from 12 biopsy and autopsy specimens from patients with AL-Am, TMA, LCDD, and MCN using laser capture microdissection (LCM). An average of 120 glomeruli were microdissected from each case. Total RNA was isolated from cells for this study and stored until use. The array membranes contained the genes of 19 MMPs, 4 TIMPs and 1 negative control (PUC18) blotted in duplicates. Critical analysis, however, was restricted to the 5 MMPs (MMPs 1,2,3,7 and 9) previously selected and TIMPs 1 and 2 to be able to establish correlations with the other data.

RESULTS

Overall MMPs expression was markedly increased in renal tissues from patients with AL-amyloidosis when compared with negative controls (normal kidneys) and MCN and LCDD. Analysis of glomeruli in renal biopsy and autopsy specimens revealed a marked increase in MMPs expression in AL-amyloidosis cases compared with normal kidneys. There was a statistically significant increase in all MMPs when compared with controls (from an eight fold increase in MMPs 1 and 7 to a 2.3 fold increase in MMP 2) (figure 1). There were no statistical differences in TIMPs 1 and 2 expression in glomeruli from patients with AL-Am as compared with normal tissue or other conditions noted above. Manual quantitation results of immunoreactivity for the various MMPs/TIMPs were confirmed with the ACIS. Colocalization of various MMPs and TIMPs could also be demonstrated.

Staining of HMCs incubated with no LCs, MCN, LCDD and AL-Am-LCs revealed only expression of MMP 7. No staining for MMPs 1, 2, 3 or 9 was detected in the HMCs incubated under the various conditions. This information was corroborated and no MMPs could be detected in Westerns of cell lysates, except for weak expression of MMP-7 in all conditions. Therefore, emphasis was placed on examination of the supernatants where all MMPs were found. MMP-3 was statistically significantly increased at 72 hours and MMP-9 at 96 hours when compared with control (no LCs) and LCDD-LC. There were cyclical oscillations in MMP expression over time, reflecting the dynamic nature of the system.

MMPs and TIMPs mRNA microarrays performed on microdissected renal glomeruli from patients with AL-Am, compared with glomeruli with patients with MCN, LCDD and TMA revealed significant, statistically significant increases in MMPs, especially 9 and 1(9>1) when compared with MCN and LCDD cases (figure 2). In the HMCs there were significant increases in MMPs 1 and 7 (1>7) mRNA at 72 hours with AL-Am-LCs and in MMPs 2,3 and 1(2>3>1) at 96 hours post-incubation when compared with HMCs incubated with MCN and LCDD-LCs and without LCs. mRNA levels for both MMPs and TIMPs showed modulations in their expression over time in the in-vitro model. There were no significant differences in TIMPs 1 and 2 mRNA levels detected when glomeruli from patients with AL-Am were compared with normal glomeruli or those from patients with MCN, LCDD or TMA or in HMCs exposed to the different experimental conditions.

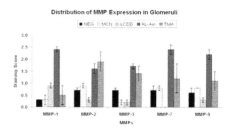
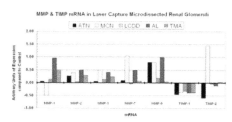

Figure 1   **Figure 2**

DISCUSSION AND CONCLUSIONS

There are relatively few papers addressing the role of MMPs and TIMPs in mesangial homeostasis. Most of the early work was centered on gelatinases A and B (MMPs-2 and 9). MMP-2 has been reported to play a key role in the remodeling of the glomerular basement membranes and mesangial matrix, as well as in the regulation of mesangial cell proliferation and differentiation (2). MMPs can be induced by the proteins present in the ECM (3); therefore, changes in the constituents of the mesangial matrix can significantly alter the expression and activation of MMPs. MMPs are produced by mesangial cells (4). Injurious agents to the mesangial cells may alter the MMP/TIMP mesangial equilibrium and shift it to either degradation or an increase of ECM.

The experimental data shown here clearly demonstrated statistically significant increases in MMP expression and mRNA levels in patients with AL-Am and in HMCs incubated with LCs obtained form the urine of these patients. Overall, the in-vitro and in-vivo data obtained correlated well.

The translational study reported herein shows the importance of MMPs in the pathogenesis of glomerular Am. It is important to realize that the study of tissue specimens reflects a one-time evaluation of MMP/TIMP expression, usually after the damage to the glomerulus is well underway or already accomplished. In the in vitro system the dynamic alterations in the various MMPs and TIMPs were documented with significant fluctuations noted due to the fact that the HMCs reacted in a dynamic fashion. Also as the HMCs laid down matrix, the newly formed matrix further influenced mesangial matrix turnover.

This study has enhanced our understanding of how mesangial matrix destruction occurs in AL-Am. New therapeutic approaches can be designed to decrease, avert or abolish the amyloid-induced mesangial alterations.

REFERENCES

1.   Nagase H, Woessner JF Jr: Matrix metalloproteinases. J Biol Chem 274:21491-21494, 1999.
2.   Turck J, Pollock AS, Lee LK, Marti HP, Lovett DH: Matrix metalloproteinase 2 (gelatinase A) regulates glomerular mesangial cell proliferation and differentiation. J Biol Chem 271:15074-15083, 1996.
3.   Martin J, Eynstone L, Davies M, Steadman R: Induction of metalloproteinases by glomerular mesangial cells stimulated by proteins of the extracellular matrix. J Am Soc Nephrol 12:88-96, 2001.
4.   Knowlden J, Martin J, Davies M, Williams JD: Metalloproteinase generation by human glomerular epithelial cells. Kidney Int 47:1682-1689, 1995.

PATHOGENESIS OF RENAL AL-AMYLOIDOGENESIS (AL-AM): AN IN-VITRO APPROACH TO UNDERSTANDING MECHANISMS INVOLVED

G.A. Herrera[1,2], J. Keeling[1], J. Teng[1]

[1]*Department of Pathology and* [2]*Cellular Biology and Anatomy, Louisiana State University Health Sciences Center,*
Shreveport, Louisiana USA

INTRODUCTION

The pathogenesis of renal AL-amyloidosis (AL-Am) has remained elusive until recently when information has become available to begin understanding mechanisms involved. The in-vitro approach has served the purpose of elucidating how glomerulopathic light chains (LCs) exert their pathogenetic effects. Renal AL-Am begins in the mesangium. Interactions between amyloidogenic LCs and mesangial cells (MCs) are crucial in the pathogenesis of renal AL-Am. Monoclonal LC processing by MCs is crucial in leading to the formation of amyloid. As renal AL-Am progresses, the mesangial matrix is replaced by amyloid fibrils. By incubating amyloidogenic LCs purified from the urine of patients with biopsy proven renal AL-Am with MCs in culture, the steps that take place can be dissected. There are a number of events involved in this complex process beginning with amyloid-forming light chain-mesangial cell interactions, followed by intracellular LC trafficking in MCs, intralysosomal LC processing, fibril formation and subsequent replacement of the normal mesangial matrix.

MATERIALS AND METHODS

Human mesangial cells (HMCs) were obtained from nephrectomy specimens away from the tumor mass following standard procedures. These HMCs were maintained in Petri dishes in a chamber under incubator-like condition and a fluorescence microscope was attached to the system to monitor LC-MC interaction (figure 1). LysoSensor and LysoTracker fluorescent acidotropic probes freely permeable through cell membranes were used to label acidic organelles such as lysosomes in HMCs. These HMCs were incubated with amyloidogenic LCs from patients with biopsy proven renal AL-Am. The LCs were not pooled and were purified from the urine of 6 different patients to perform the experiments noted below. Five LCs were lambda and one kappa. All LCs were fluorescent labeled with Texas red. The sequence of events that occurs as the amyloidogenic LCs interact with HMCs were sequentially studied using a variety of techniques including light and fluorescence microscopy, immunohistochemistry, electron microscopy, ultrastructural immunogold labeling, and molecular biology techniques. Control samples included HMCs incubated with no LCs, albumin and LCs from patients with myeloma cast nephropathy (MCN) light chain deposition disease (LCDD).

RESULTS

Texas red labeled amyloidogenic LCs interacted with the HMCs resulting in a phenotypic conversion of HMCs from a smooth muscle phenotype to a macrophage phenotype(1) and avid internalization through a clathrin–mediated process, observed by colocalization of clathrin and internalized LCs. Lysosomal compartment expansion was noted in the transformed MCs with the LysoSensor probes. Maximum surface interaction occurred at 15-30 minutes with some variability form LC to LC. This contrasted with no interaction of MCs with MCN-LCs and significant interactions with LCDD-LC(2,3), but minimal internalization and transformation to myofibroblastic phenotype. The AL-LCs trafficked in the MCs and were delivered to the mature (late) lysosomal system approximately 60-75 minutes after the LCs were incubated with the HMCs(4), as evidenced by colocalization of the LCs (red labeled) with mature lysosomes (fluorescein labeled), resulting in a yellow signal. Processing of the amyloidogenic LCs occurred in the late lysosomes. Next time that LCs were identifiable in the system, they were in the form of amyloid fibrils. Our laboratory is in the process of studying the details of how LCs are processed in the lysosomes, how amyloid fibrils are formed and eventually extruded to the extracellular matrix (ECM).

The interactions in AL-Am LCs and HMCs are mediated through a receptor, which is not yet fully characterized (4). This receptor is not the cubilin/megalin receptor demonstrated to be present in proximal tubular cells mediating internalization of LCs in proximal tubules (5,6). Our studies have also shown that once amyloid is extruded from the cells, MMP activity is enhanced resulting in destruction of the mature mesangial ECM and replacement by amyloid. The activity of transforming growth factor beta (TGF-β), normally responsible for restoring mesangial matrix is also inhibited, precluding matrix rebuilding(3). The entire process can be broken down in 4 steps (figure 2), and the details of step 3 remain unclear.

Crosslinking studies of amyloidogenic LCs with HMC membranes were utilized to characterize the receptor complex by mass spectroscopy. Shifted bands were shown to be present at approximately 325 and 480 Kd presumably representing the LC-HMC membrane complex. Mass spectroscopy revealed components of caveolin 1, a structural component in the HMC membrane, which is involved in HMC-LC interactions (4).

Figure 1

Figure 2

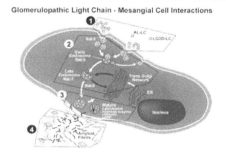

Glomerulopathic Light Chain - Mesangial Cell Interactions

DISCUSSION

Approximately 30% of patients with myeloma are associated with glomerulopathies (7). Only a minority of patients with myeloma develop AL-Am. LCs being of low molecular weight are usually freely filtered through the glomerulus and delivered to the proximal tubules. Therefore, there must be a pathologic reason for affinity of certain LCs to the mesangium where pathological changes ensue. The current research has demonstrated the presence of a receptor for glomerulopathic LCs on the HMCs. This receptor, not yet fully characterized, is different from the one identified in proximal tubular cells.

The amyloidogenic LCs trafficking in MCs is unique, in that they are the only LCs delivered to the mature lysosomal system for processing. Other internalized LCs are catabolized in the endosomal compartment (4). Crucial steps in the process of amyloid formation in the mesangium were clearly delineated in the in-vitro system. There are a number of potential targets amenable to therapeutic intervention. The data that has been obtained is of paramount importance in addressing amelioration, prevention and/or reversal of the renal damage that takes place in renal AL-Am.

The physicochemical and conformational characteristics of the glomerulopathic LCs (those attracted to the mesangium, actively engaging in interactions with HMCs and producing glomerulopathies such as AL-Am and LCDD) are responsible for centering their pathogenic effects on the mesangium. These intrinsic characteristics of the variable portion of the LC molecule also govern their amyloidogenic potential. There are structural features that have been shown to predispose certain LCs to engage in amyloidogenesis(8-10). The propensity of a given protein to aggregate and form amyloid fibrils under certain conditions varies with its composition and sequence(9). Presumably there may also be host factors that affect the capability of cells to engage in amyloid formation. However, these host factors are not entirely known at the present time(10). The system described here could be used to test the amyloidogenic potential of a given LC isolated from the urine of a patient with plasma cell dyscrasia. The system can also be of value in testing therapeutic maneuvers to treat patients with renal AL-Am before irreversible renal failure occurs.

REFERENCES

1. Comenzo RL, Zhang Y, Martinez C, Osman K, Herrera GA: The tropism of organ involvement in primary systemic amyloidosis: Contributions of Ig V_L germline gene use and clonal plasma cell burden. The Proceedings of the IXth International Symposium on Amyloidosis, (M. Bely, A. Apathy Eds.), 2001, pp. 273-275.
2. Russell WJ, Cardelli J, Harris E, Baier RJ, Herrera GA: Monoclonal light chain-mesangial cell interactions: Early signaling events and subsequent pathologic effects. Lab Invest 81:689-703, 2001: Monoclonal light chain-mesangial cell interactions: Early signaling events and subsequent pathologic effects. Lab Invest 81:689-703, 2001.
3. Herrera GA, Russell WJ, Isaac J, Turbat-Herrera EA, Tagouri Y, Sanders PW, Dempsey S: Glomerulopathic light chain - mesangial cell interactions modulate in vitro extracellular matrix remodeling and reproduce mesangiopathic findings documented in vivo. Ultrastruct Pathol. 23:107-126, 1999.
4. Teng J, Russell WJ, Gu X, Cardelli J, Jones ML, Herrera GA: Different types of glomerulopathic light chains interact with mesangial cells using a common receptor but exhibit different intracellular trafficking patterns. Lab Invest 84:440-451, 2004.
5. Batuman V, Dreisbach AW, Cyran J: Light-chain binding sites on renal brush-border membranes. J Physiol 258:F1259-1265, 1990.
6   Batuman V, Verroust PJ, Navar GL, Kaysen JH, Goda FO, Campbell WC, Simon E, Pontillon F, Lyles M, Bruno J, Hammond TG: Myeloma light chains are ligands for cubilin (gp280). Am J Physiol 275:F246-254, 1998.
7. Herrera GA: Renal manifestations in plasma cell dyscrasias: An appraisal from the patients' bedside to the research laboratory. Annals Diagn Pathol 4:174-200, 2000.
8. Stevens FJ, Myatt EA, Chang CH, Westholm FA, Eulitz M, Weiss DT, Murphy C, Solomon A, Schiffer M: A molecular   model for self-assembly of amyloid fibrils: Immunoglobulin light chains. Biochemistry 34:10697-10702, 1995.
9. Chiti F, Stefani M, Taddei N, Ramponi G, Dobson CM: Rationalization of the effects of mutations on peptide and protein aggregation rates. Nature 424:805-808, 2003.
10. Solomon A, Weiss DT: Protein and host factors implicated in the pathogenesis of light chain amyloidosis (AL amyloidosis). Amyloid. Int J Exp Clin Invest 2:269-279, 1995.

INTERNALIZATION OF SYSTEMIC AMYLOID LIGHT CHAINS AND HEPARAN SULFATE PROTEOGLYCAN DEPOSITION

V. Trinkaus-Randall, S. Steeves, G. Monis, M.T.Walsh, J.D. Chiu, J. Eberhard, L.H. Connors, M. Skinner

Departments of Biochemistry, Physiology and Biophysics, Pathology and Laboratory Medicine, Medicine Boston University School of Medicine Boston, MA

Proteoglycans are associated with amyloid fibrils in patients with systemic amyloidosis. There is extensive data implicating heparan sulfate proteoglycans (HSPGs) in the genesis of amyloid (1). Our goal was to evaluate the response of primary cardiac fibroblasts to exogenous amyloidogenic light chains (LCs) and to correlate the localization of LCs with expression and localization of heparan sulfate proteoglycans. The response of cells to 11 urinary IgG LCs of kappa1, lambda 6 and 3 subtypes was evaluated and compared to a urinary LC from a patient with multiple myeloma and no amyloid.

METHODS

Primary cardiac fibroblasts were cultured. 11 amyloid LCs were added cells and the response was assessed. To monitor the localization of the LCs, they were conjugated to Oregon Green 488 and imaged. Cells were fixed and localization of HSPG was assessed. Sulfation of the secreted proteoglycans was determined after elution over Q1-columns with dimethylmethylene blue. The sulfated fractions were treated with polysaccharidases and glycosaminoglycans (GAG) identified. The conformation and stability of selected LCs in association with the HS was evaluated.

RESULTS AND DISCUSSION

LOCALIZATION OF AMYLOIDOGENIC LIGHT CHAINS ADDED TO CARDIAC FIBROBLASTS

To determine the response of cells to amyloidogenic LCs, 11 urinary Ig LCs of $\kappa 1$ and $\lambda 1$, 3 and 6 subtypes were conjugated to Oregon Green 488 and added to primary cardiac fibroblasts. Live cell confocal images were collected. All of the LCs entered the cells; (over 7 independent experiments with individual conjugations). LCs were detected in a punctate pattern around the nucleus and remained for over 24 hours. Unconjugated Oregon Green was added to cell cultures and did not enter the cells. To determine the localization of LCs within the cell, a z-series through cells was taken for each LC (Figure 1).

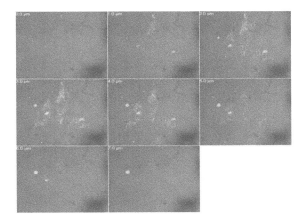

Figure 1. AL-k1 LC present in a punctate pattern Simultaneous DIC and fluorescent images were collected..

As the length of exposure to LC increased there was a transition from a punctate pattern to a filamentous appearance. The extensiveness of the filaments within the cells varied with the specific LCs and did not co-localize with F-actin. Cells displayed a transition from a fibroblast to myofibroblast morphology with extensive actin filaments. This change has been noted in other fibroblast cells in response to injury.

LIGHT CHAIN MEDIATED THE TRANSLOCATION OF HEPARAN SULFATE PROTEOGLYCAN TO THE NUCLEUS

HSPGs translocated to the nucleus in response to short term incubation with AL LC (K1) but not with the non-amyloid LC. When the epitope was not exposed there was no detectable HSPG. When cells were incubated with heparitinase prior to the end of the incubation, fixed and stained, HSPGs were only localized at the cell border.

LONG TERM CHANGES IN GLYCOSAMINOGLYCANS IN RESPONSE TO LIGHT CHAINS

Experiments were performed to evaluate the distribution of the HSPG for amyloidogenic LCs. AL -LCs resulted in HSPGs to be localized throughout the cell with prominent staining at the interstices where actin fibers coalesced. Cells were treated with FGF-2, LC or medium (2,3). Media and cell lysates were subjected to anion exchange column chromatography. Fractions were collected and analyzed by DMB after NaCl was increased to 1.5M. Sulfated GAGs were found in the cell and medium, but not in the matrix, however both glypican and syndecan were present in all 3 fractions according to western blot analysis. All LCs induced an increase in sulfated GAG over control in the medium. In fact the sulfation was greater than that induced by FGF-2, which mediates sulfation. There was substantially less sulfated GAG in the cellular fraction and LC's showed a 20% decrease in CS from control and 50% from FGF-2. However, the UV monitor and WBA demonstrated the presence of core protein. The data indicate that GAGs that are detected throughout the cell are not highly sulfated.

ASSOCIATION OF LIGHT CHAINS WITH GLYCOSAMINOGLYCANS

Far UV CD spectroscopy was used to monitor the binding of HS to LCs by measuring the secondary structure at 25°C and the thermal stability or folding-unfolding of the β-sheets at 217 nm with HS (4,5). Wavelength spectra of AL-k1 LCs (Figure 2, top panels) (+/- HS), showed a change in secondary structure with HS (more negative molar ellipticity). Folding-unfolding studies (Figure 4, lower panels) showed that the midpoint temp of unfolding of the β-sheets of the LCs was reduced by 1.4 and 1.8° with HS, respectively, suggesting that HS interaction destabilized the LCs .

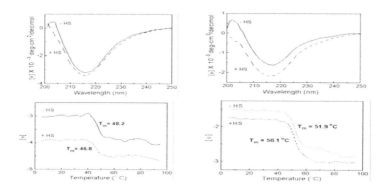

Figure 2. Changes in secondary structure mediated by 2 k1 LCs (96066 left half, 200131 right half). Upper panels show Far UV spectra of AL-k1 LCs with and without HS. Lower panels show thermal unfolding of same LCs with and without HS at 217 nm.

Internalization of the light chains mediates the expression and localization of HSPGs. Translocation of heparan sulfate proteoglycans to the nucleus in response to FGF-2 has been demonstrated in primary fibroblasts (2,3). As FGF-2 is a growth factor that requires nuclear localization to induce biological responses, HSPGs may be a vehicle for heparin binding growth factors. This is interesting as other amyloid proteins have been demonstrated to have HS binding sequences (6).

ACKNOWLEDGEMENTS
We acknowledge assistance with cells by D. Brenner in the laboratory of Dr. Liao and assistance with confocal imaging from Christopher Schultz. NIH NHLBI P01-HL-068705 and P01-HL-26335T32-HL-07291. Gerry Amyloid Research Fund

REFERENCES

1. Ohishi H, Skinner M, Sato-Araki N, Okuyama T, Gejyo F, Kimura A, Cohen AS, Schmid K. Glycosaminoglycans of the hemodialysis-associated carpal synovial amyloid and of amyloid-rich tissues and fibrils of heart, liver, and spleen. Clin Chem. 1990 36:88-91.
2. Richardson TP, Trinkaus-Randall V, Nugent MA. Regulation of heparan sulfate proteoglycan nuclear localization by fibronectin. J Cell Sci. 114:1613-23, 2001
3.   Hsia E, Richardson TP, Nugent MA. Nuclear localization of basic fibroblast growth factor is mediated by heparan sulfate proteoglycans through protein kinase C signaling. J Cell Biochem. 88:1214-25,2003
4. N Greenfield & GD.Fasman, Computed circular dichroism spectra for the evaluation of protein conformation, Biochemistry, 8: 4108-4116, 1969
5. CM Chung, LH Connors, MD Benson & MT Walsh. Biophysical Analysis of Normal Transthyretin: Implications for Fibril Formation in Senile Systemic Amyloidosis, Amyloid: J. Protein Folding Disorders, 8: 75-83, 2001
6. Stevens FJ, Kisilevsky R. Immunoglobulin light chains, glycosaminoglycans, and amyloid. Cell Mol Life Sci.57:441-449, 2000

HUMAN AMYLOIDOGENIC LIGHT CHAINS DIRECTLY INVOKE OXIDANT STRESS AND RESULT IN IMPAIRED CARDIOMYOCYTE FUNCTION

Daniel A Brenner, Mohit Jain, David R Pimentel, Bo Wang, Lawreen H Connors,Martha Skinner, Carl S Apstein, Ronglih Liao

Amyloid Treatment and Research Program Boston University School of Medicine, Boston, MA

INTRODUCTION

Primary (AL) amyloidosis is a plasma cell dyscrasia resulting in the clonal production of immunoglobulin light chain proteins (LC) and subsequent multi-organ dysfunction.[1] Congestive heart failure remains the greatest cause of death in AL amyloidosis, due to the development of a rapidly progressive amyloid cardiomyopathy. Patients with amyloid cardiomyopathy are largely unresponsive to current heart failure therapies,[2,3] and have a median survival of less than six months and a five-year survival of less than 10%.[3,4] The mechanisms underlying this disorder, however, are yet to be determined. Prior theories have suggested that interstitial fibril deposition of AL proteins are the main cause of contractile dysfunction and cardiomyopathy.[2] This, however, is inconsistent with clinical observations, which have detailed a lack of correlation between myocardial fibril deposition and cardiac dysfunction in AL patients.[5]

In this report, we demonstrate that physiologic levels of human amyloid LC proteins, isolated from patients with amyloid cardiomyopathy, specifically alter cellular redox state in cardiomyocytes, marked by an increase in intracellular reactive oxygen species (ROS), upregulation of the redox-sensitive protein, heme oxygenase-1 (HO-1), and oxidative modification of intracellular proteins. Oxidant stress imposed by cardiac-LC proteins resulted in *direct* impairment of cardiomyocyte contractility, independent of neuro-hormonal stimulants, vascular factors, or extracellular fibril deposition, and was prevented through antioxidant treatment.

MATERIALS AND METHODS

Human LC proteins were isolated from patients with non-amyloidogenic myeloma, non-cardiac involved AL amyloidosis or cardiac-involved AL amyloidosis. Adult rat ventricular cardiomyocytes were isolated and cultured as previously described.[6] Cells were treated with ultrapurified water (vehicle), myeloma LC and non-cardiac involved amyloid LC (Con-LC, n=2), or cardiac-involved amyloid LC (Cardiac-LC, n=2) proteins at physiologic concentrations (20 mg/L) for 24 hours, unless specified. For antioxidant experiments, cardiomyocytes were co-treated with the superoxide dismutase/catalse mimetic Mn(III)terakis(1-methyl-4-pyridyl)porphyrin pentachloride (MnTMPyP) at 50 μmol/L. Intracellular ROS levels were determined using the cell-permeable, redox-sensitive flurophore, dichlorofluorescein-diacetate (DCF). Protein expression of HO-1 was determined through Western blot. Cardiomyocyte contractility was determined in cells paced at 5 Hz at 37°C, using video-edge detection, as previously described.

RESULTS

To determine whether physiologic levels of LC proteins directly alter redox status in cardiomyocytes, intracellular ROS levels were determined following exposure to amyloidogenic proteins. Human Con-LC proteins did not influence ROS levels relative to vehicle exposure. In contrast, Cardiac-LC proteins directly induced cellular oxidant stress, as demonstrated by enhanced DCF fluorescence. Elevated ROS levels in Cardiac-LC cells were prevented through antioxidant treatment with MnTMPyP.

Redox-sensitive protein expression was also determined following incubation with LC proteins, using the cellular stress marker, HO-1.[7] Cardiac-LC resulted in an upregulation of HO-1 expression relative to vehicle and Con-LC. Importantly, induction of HO-1 was prevented through antioxidant treatment with the superoxide dismutase / catalase mimetic, MnTMPyP and adenoviral catalase, confirming redox-sensitive expression.

Importantly, LC proteins did not form either LC aggregates or amyloid fibrils, as determined by spectrophotometric analysis or Congo red staining (data not shown), suggesting that Cardiac-LC *directly* altered redox state in cardiomyocytes, independent of fibril formation.

Finally, Cardiac-LC directly reduced cardiomyocyte contractile function, whereas Con-LC did not alter cellular shortening. The effects of Cardiac-LC proteins on cardiomyocyte function were reversed with antioxidant treatment, suggesting that altered cellular shortening with Cardiac-LC proteins was secondary to increased oxidant stress.

SUMMARY AND CONCLUSION

We demonstrate that human amyloidogenic cardiac-LC proteins specifically alter cellular redox state in isolated cardiomyocytes, resulting in direct impairment of cardiomyocyte contractility. Cardiomyocyte dysfunction induced by cardiac-LC proteins was prevented through antioxidant treatment.

These results may explain clinical observations detailing a discrepancy between the degree of myocardial amyloid fibril deposition and cardiac dysfunction.[5] Interestingly, only LC proteins associated with cardiomyopathy, rather than non-cardiac associated LC proteins, resulted in increased ROS and cardiomyocyte dysfunction in isolated cells, suggesting that LC primary sequence and/or post-translation modifications, rather than osmolar stress or non-specific protein-receptor interaction, dictate end-organ targeting and dysfunction. Moreover, significant cellular oxidant stress and cardiomyocyte dysfunction were observed with physiologic concentrations of LC proteins, and with proteins isolated from multiple patients, thereby eliminating artifact associated with supra-normal protein concentration or single protein variants. These results are consistent with prior findings documenting direct impairment of cellular function, independent of fibril deposition, in non-cardiac tissue by other amyloidogenic proteins.[8,9]

The mechanisms by which amyloid cardiac-LC proteins increase cellular ROS alter redox state may include activation of cellular oxidase complexes,[10] mitochondrial dysfunction,[11] or metal ion reduction.[12]

This study suggests for the first time that cardiac dysfunction in amyloid cardiomyopathy is directly mediated by LC protein-induced cardiomyocyte oxidant stress and alterations in cellular redox status, independent of fibril deposition. Antioxidant therapies or treatment strategies aimed at eliminating circulating LC proteins may therefore be beneficial in the treatment of this fatal disease.

ACKNOWLEDGEMENTS

This study was supported by grants from the Evans Medical Foundation, Gerry Foundation, Young Family Amyloid Research Fund, and National Institutes of Health HL-68705 (M.S.), HL-67297 and HL-73756 (R.L.). The authors thank Ms. Violet Roskens and Mr. Jeremy Eberhard for assistance with protein purification.

REFERENCES

1.   Merlini G, Bellotti V. Molecular mechanisms of amyloidosis. *N Engl J Med*. 2003;349:583-96.

2.   Braunwald E. *Heart Disease*. 5 ed. Philadelphia: W.B. Saunders Company; 1997.

3.   Falk RH, Skinner M. The systemic amyloidoses: an overview. *Adv Intern Med*. 2000;45:107-37.

4.   Skinner M, Anderson J, Simms R, Falk R, Wang M, Libbey C, Jones LA, Cohen AS. Treatment of 100 patients with primary amyloidosis: a randomized trial of melphalan, prednisone, and colchicine versus colchicine only. *Am J Med*. 1996;100:290-8.

5.   Dubrey SW, Cha K, Skinner M, LaValley M, Falk RH. Familial and primary (AL) cardiac amyloidosis: echocardiographically similar diseases with distinctly different clinical outcomes. *Heart*. 1997;78:74-82.

6.   Jain M, Brenner DA, Cui L, Lim CC, Wang B, Pimentel DR, Koh S, Sawyer DB, Leopold JA, Handy DE, Loscalzo J, Apstein CS, Liao R. Glucose-6-phosphate dehydrogenase modulates cytosolic redox status and contractile phenotype in adult cardiomyocytes. *Circ Res*. 2003;93:e9-16.

7.   Yet SF, Tian R, Layne MD, Wang ZY, Maemura K, Solovyeva M, Ith B, Melo LG, Zhang L, Ingwall JS, Dzau VJ, Lee ME, Perrella MA. Cardiac-specific expression of heme oxygenase-1 protects against ischemia and reperfusion injury in transgenic mice. *Circ Res*. 2001;89:168-73.

8.   Andersson K, Olofsson A, Nielsen EH, Svehag SE, Lundgren E. Only amyloidogenic intermediates of transthyretin induce apoptosis. *Biochem Biophys Res Commun*. 2002;294:309-14.

9.   Miranda S, Opazo C, Larrondo LF, Munoz FJ, Ruiz F, Leighton F, Inestrosa NC. The role of oxidative stress in the toxicity induced by amyloid beta- peptide in Alzheimer's disease. *Prog Neurobiol*. 2000;62:633-48.

10.   Shimohama S, Tanino H, Kawakami N, Okamura N, Kodama H, Yamaguchi T, Hayakawa T, Nunomura A, Chiba S, Perry G, Smith MA, Fujimoto S. Activation of NADPH oxidase in Alzheimer's disease brains. *Biochem Biophys Res Commun*. 2000;273:5-9.

11.   Eckert A, Keil U, Marques CA, Bonert A, Frey C, Schussel K, Muller WE. Mitochondrial dysfunction, apoptotic cell death, and Alzheimer's disease. *Biochem Pharmacol*. 2003;66:1627-34.

12.   Huang X, Cuajungco MP, Atwood CS, Hartshorn MA, Tyndall JD, Hanson GR, Stokes KC, Leopold M, Multhaup G, Goldstein LE, Scarpa RC, Saunders AJ, Lim J, Moir RD, Glabe C, Bowden EF, Masters CL, Fairlie DP, Tanzi RE, Bush AI. Cu(II) potentiation of alzheimer abeta neurotoxicity. Correlation with cell-free hydrogen peroxide production and metal reduction. *J Biol Chem*. 1999;274:37111-6.

MANAGEMENT OF PRIMARY AMYLOIDOSIS (AL)

Morie Gertz, M. D.

Division of Hematology, Department of Medicine, Mayo Clinic and Mayo Clinic College of Medicine, Rochester, MN

INTRODUCTION

Most therapies for primary amyloidosis are directed at the plasma cell clone responsible for the synthesis of the insoluble immunoglobulin light chain. These light chain fragments assemble in a beta-pleated sheet conformation rather than the native alpha helical state and become resistant to proteolysis by physiologic enzymes. At this time, no fibril-degrading agents exist. The rationale for therapy is to interrupt precursor protein synthesis by reducing the plasma cell burden in the bone marrow from which the light chain originates. It is then hoped that the normal turnover of amyloid fibril protein will favor catabolism over synthesis resulting in resolution of light chain deposits in tissues.

STANDARD CHEMOTHERAPY FOR AMYLOIDOSIS

Two prospective randomized studies compared melphalan-containing regimens to colchicine. The melphalan was administered orally from four to seven days, with cycles repeated every four to six weeks based on the degree of myelosuppression produced by the melphalan. Both studies demonstrated a significant survival benefit for melphalan-containing regimens with the melphalan-containing arms showing a median survival of 17 months and the colchicine-alone arm showing a median survival of only 12 months. Although statistically significant, p = 0.001, a survival improvement of only five months is far from ideal, but it does demonstrate the principle that cytotoxic chemotherapy can improve the outcome for patients with AL. Recently, a report on 137 patients treated with either high-dose therapy and stem cell replacement or one of two intermediate-dose regimens demonstrated that the best outcomes were in those patients whose free light chain level fell by 50% as a surrogate measure of a reduced plasma cell burden in the bone marrow (1). In this report, survivals were similar in the responding group whether they received high-dose therapy with stem cells or intermediate-dose chemotherapy. The question is whether the proportion of patients responding is similar between the two groups. The reported median survival for those patients who serum free light chain fell by more than 50% exceeded 96 months. The median survival of those patients who failed to achieve this level of suppression was just over 24 months.

High-dose dexamethasone as a single agent has been reported to be an effective salvage therapy in patients with amyloidosis, particularly those who do not have cardiac involvement. Durable responses have been reported, although no head-to-head comparisons with melphalan/prednisone chemotherapy have been conducted. Dexamethasone therapy orally should be considered a viable regimen for the management of AL.

Recently, a regimen of oral melphalan and oral dexamethasone has been reported for patients who are not suited, by virtue of age or performance status, for stem cell transplant. Melphalan was given 0.22 mg/kg per day for four days with dexamethasone 40 mg for four consecutive days. Cycles were repeated every 28 days to a maximum of nine cycles. Patients received omeprazole, ciprofloxacin, itraconazole, and amiodarone prophylaxis. Thirteen of 46 patients responded—15 achieving a complete hematologic response defined by eradication of light chain in the serum and urine. There was a 48% organ response and no treatment-related mortality. The median survival for responders was significantly superior to the nonresponders.

Thalidomide has been proposed for the treatment of amyloidosis. In one study, 16 patients failing prior therapy were treated. Median tolerated dose of thalidomide was 300 mg. However, this produced 50% grade 3 and 4 toxicity, and one-quarter of the patients had to discontinue therapy due to toxicity. There were no complete responses. Four showed a reduction in light chain proteinuria, but the urine total protein fell only 9-20%. In a second study of 12 patients, the median duration of thalidomide therapy was 72 days. Treatment was stopped in 6 of the 12 due to toxicity, 4 progressions, and 2 deaths. Cognitive difficulties, edema, and constipation were seen in 75%; dyspnea, dizziness, and rash in 50%. The conclusion was that patients with AL tolerate thalidomide poorly and cannot be treated with significant doses (2).

STEM CELL TRANSPLANT

We have transplanted, to date, 174 patients; 79% were male. The median age was 55. The median serum albumin was 2.8. Nine percent of our patients had a serum creatinine level greater than 1.9, and 15% had an alkaline phosphatase greater than 1 ½ times the institutional norm. A serum monoclonal protein was found in 125 and a urine monoclonal protein in 154; 55% were nephrotic, and two-thirds had greater than 1 g of protein in the urine. The median percent bone marrow plasma cells were 7 with 17% of the patients having over 20% plasma cells in the bone marrow. One, two and greater than two-organ involvement was seen in 47, 36, and 17%, of patients respectively. Stem cell mobilization was accomplished with growth factor alone with a median number of collections of two. Our minimum for CD34 yield was two million. The median achieved was six million.

The day-100 mortality for stem cell transplant was 12% but has fallen to 6% in the last calendar year. The median time to 500 neutrophils was 14 days. The median time to 20,000 platelets was 17 days. Hospitalization was a median of nine days; 16% of patients were never hospitalized. The group transplanted included cardiac, renal, liver, and neuropathy patients in 49, 68, 19, and 16%, respectively. The median survival for all patients has not been reached but is projected to be greater than five years. Patients who had greater than two-organ involvement have a median survival of 22 months, and the median survival has not been reached in patients with one- or two-organ involvement. Hematologic response has a profound effect on outcome with responders having a survival greater than five years, and nonresponders having a median survival of approximately 25 months. Important predictive factors for outcome included the absolute lymphocyte count on day 15 post-transplant with those patients having greater than 500/mcL, having a superior survival to those who do not achieve this level. The dose of chemotherapy administered is also important in predicting survival. Patients receiving melphalan 200 mg/m^2 or its equivalent had a superior survival to those patients who received lower-dose conditioning. In a multivariate analysis, the predictors of survival after transplant include the lymphocyte count on day 15, the serum beta 2 microglobulin, the number of organs involved, and the troponin T level (3).

Organ response following the therapy of amyloidosis is often delayed beyond one year because the resolution of amyloid deposits is a time-dependent process after precursor protein production is interrupted. In our group of patients who responded, a 50% reduction of urinary protein was not seen until a median of nine months post transplant, and the maximum response did not occur until 48 months post transplant. We now have a risk-adapted approach to the treatment of amyloidosis where patients who are a higher risk by virtue of the number of organs involved, the extent of cardiac involvement, their renal function and age are conditioned with lower doses of melphalan between 100 and 140 mg/m2.

Patients who are eligible for a stem cell transplant do inherently better than similar patients who are not eligible for transplant, a reflection of the selection that occurs in accepting patients for transplant. However, in a case-controlled study where 63 patients having a stem cell transplant were matched with 63 controls, there was a significant benefit for those patients who were transplanted. Attempts to improve the outcome of transplant by giving patients two cycles of melphalan, prednisone chemotherapy prior to transplant resulted in no difference in hematologic, organ response, or survival.

ORGAN TRANSPLANTATION

Heart transplantation has been accomplished for amyloidosis, and ten-year survivors have been reported. In our experience, the five-year survival in patients with advanced cardiac amyloidosis who have a heart transplant is 44%. This is inferior to those patients transplanted for primary cardiomyopathy. Cardiac transplant is a legitimate option for patients with advanced disease, but prevention of recurrent amyloid in the transplanted organ remains a major impediment to the wider application of the technique. Investigational new therapies that are promising include an anti-amyloid fibril antibody. This has been demonstrated to cause regression in a mouse model of amyloidoma.

Agents have been developed that inhibit P component, a glycoprotein that is bound to all forms of amyloid. The hope is that by the inhibition of P component binding that the fibrillar structure will be destabilized and more prone to denature into subunit light chains. The anthracycline iodo-doxorubicin, has been shown to dissolve soft tissue deposits of amyloid, and ongoing studies of this agent are underway.

In conclusion, the optimal therapy for light chain amyloidosis remains unknown. There is an ongoing phase-3 study in Europe of stem cell transplant compared to melphalan/dexamethasone. The current policy at Mayo Clinic is to transplant those patients who would not suffer excess morbidity or mortality risk until final results of a phase-3 study become available.

REFERENCES

1.   Lachmann HJ, Gallimore R, Gillmore JD, Carr-Smith HD, Bradwell AR, Pepys MB, Hawkins PN. Outcome in systemic AL amyloidosis in relation to changes in concentration of circulating free immunoglobulin light chains following chemotherapy. Br J Haematol 2003; 122 (1):78-84.

2.   Dispenzieri A, Lacy MQ, Rajkumar SV, Geyer SM, Witzig TE, Fonseca R, Lust JA, Greipp PR, Kyle RA, Gertz MA. Poor tolerance to high doses of thalidomide in patients with primary systemic amyloidosis. Amyloid 2003; 10 (4):257-261.

3.   Gertz MA, Lacy MQ, Dispenzieri A, Gastineau DA, Chen MG, Ansell SM, Inwards DJ, Micallef IN, Tefferi A, Litzow MR. Stem cell transplantation for the management of primary systemic amyloidosis. Am J Med 2002; 113 (7):549-555.

DEFINITION OF ORGAN INVOLVEMENT AND TREATMENT RESPONSE IN PRIMARY SYSTEMIC AMYLOIDOSIS (AL): A CONSENSUS OPINION FROM THE 10TH INTERNATIONAL SYMPOSIUM ON AMYLOID AND AMYLOIDOSIS

Morie A. Gertz, MD, Ray Comenzo, MD, Rodney H. Falk, MD, Jean Paul Fermand, MD, Bouke P. Hazenberg, MD, Philip N. Hawkins, MD, Giampaolo Merlini, MD, Philippe Moreau, MD, Pierre Ronco, MD, Vaishali Sanchorawala, MD, Orhan Sezer, MD, Alan Solomon, MD, Giles Grateau, MD

From the Dysproteinemia Clinic (M.A.G.), Mayo Clinic, Rochester, Minnesota; the Hematology Service, Department of Medicine (R.C.), Memorial Sloan-Kettering Cancer Center, New York, New York; the Amyloid Treatment and Research Program (R.H.F., V.S.), Boston University, School of Medicine, Boston, Massachusetts; the Hôpital Saint-Louis (J.P.F.), Paris, France; the Department of Rheumatology (B.P.H.), University Hospital, Groningen, The Netherlands; The National Amyloidosis Center, Department of Medicine (P.N.H.), Royal Free Hospital, London, England; the Amyloid Center, Biotechnology Research Laboratory (G.M.), University Hospital, IRCCS Policlinico, San Mateo, Pavia, Italy; the Department of Hematology (P.M.), University Hospital, Nantes, France; the Inserm 489 Nephrology Service (P.R.), Hospital Tenon, Paris, France; the Myeloma and Lymphoma Research Unit (O.S.), Hospital Charite, Department of Hematology and Oncology, Humboldt University, Berlin, Germany; the Human Immunology and Cancer Program (A.S.), University of Tennessee, Graduate School of Medicine, Knoxville, Tennessee; and the Public Assistance Hospital (G.G.), Hôtel-Dieu, Paris, France.

INTRODUCTION

Thirty years ago, the treatment of AL was primarily supportive. Specific criteria for recognizing organ involvement and defining response were of little benefit and minimal utility. During the last ten years, however, new therapies directed at the plasma cell, the source of the light chain, and specific therapies designed to destabilize the amyloid fibril have been developed. Deposits of amyloid have been shown by SAP scanning to be in dynamic equilibrium. Dissolution of amyloid deposits is possible if the production of the precursor light chain can be reduced. Systemic chemotherapies—high dose requiring stem cell support, intermediate-dose, and conventional-dose chemotherapies—are now being used regularly to treat patients. Each institution has its own specific criteria for evaluating the number of organs involved with amyloid and defining what constitutes a response. Often, the criteria differed from institution to institution making it difficult to compare directly outcomes whose endpoint in large part was determined by response. Several centers have also reported that outcomes are related directly to the number of organs involved with amyloid. Therefore, counting the number of organs becomes more than an academic exercise (1).

Thirteen leaders in the field were invited to submit institutional criteria for organ involvement and response, from which guidelines were developed. With the adoption of these guidelines it is hoped that uniform reporting criteria will be used in the diagnosis, assessment of organ involvement, and evaluation of response in AL.

RESULTS

Table 1. Organ Involvement: Biopsy of Affected Organ or Biopsy at an Alternate Site [*]

Kidney	24-hour urine protein >0.5 g/d, predominantly albumin
Heart	Echo: mean wall thickness >12 mm, no other cardiac cause
Liver	Total liver span >15 cm in the absence of heart failure or alkaline phosphatase >1.5 times institutional upper limit of normal
Nerve	Peripheral: clinical; symmetric lower extremity sensorimotor peripheral neuropathy Autonomic: gastric-emptying disorder, pseudo-obstruction, voiding dysfunction not related to direct organ infiltration
Gastrointestinal tract	Direct biopsy verification
Lung	Direct biopsy verification Interstitial radiographic pattern
Soft tissue	Tongue enlargement, clinical Arthropathy Claudication, presumed vascular amyloid Skin Myopathy by biopsy or pseudohypertrophy Lymph node (may be localized) Carpal tunnel syndrome

[*] Alternate sites available to confirm the histologic diagnosis of amyloidosis: fine-needle abdominal fat aspirate, biopsy of the minor salivary glands, rectum, gingiva.

Table 2. Organ Response

Heart	Mean interventricular septal thickness decreased by 2 mm, 20% improvement in ejection fraction, improvement by 2 New York Heart Association classes without an increase in diuretic use or improvement by 1 New York Heart Association class associated with a 50% reduction in diuretic requirements and no increase in wall thickness
Kidney	50% decrease (at least 0.5 g/d) of 24-hour urine protein (urine protein must be >0.5 g/d pretreatment). Creatinine and creatinine clearance must not worsen by 25% over baseline.
Liver	50% decrease in abnormal alkaline phosphatase value Decrease in liver size radiographically at least 2 cm
Nerve	Improvement in electromyogram or nerve conduction velocity (rare)

Table 3. Organ Disease Progression

Heart	Interventricular septal thickness increased by 2 mm compared with baseline An increase in New York Heart Association class by 1 grade with a decreasing ejection fraction of $\geq 10\%$
Kidney	50% increase (at least 1 g/d) of urine protein to greater than 1 g/d or 25% worsening of serum creatinine or creatinine clearance
Liver	50% increase of alkaline phosphatase above the lowest value
Nerve	Progressive neuropathy by electromyography or nerve conduction velocity

Table 4. Hematologic (Immunochemical) Response Criteria

Complete response	Serum and urine negative for a monoclonal protein by immunofixation
	Free light chain ratio normal
	Marrow <5% plasma cells
Partial response	If serum M component >0.5 g/dL, a 50% reduction
	If light chain in the urine with a visible peak and >100 mg/d a 50% reduction;
	If free light chain >10 mg/dL a 50% reduction
Progression	From CR, any detectable monoclonal protein or abnormal free light chain ratio (light chain must double)
	From PR or stable response, 50% increase in serum M protein to >0.5 g/dL or 50% increase in urine M protein to >200 mg/d; a visible peak must be present
	Free light chain increase of 50% to >10 mg/dL
Stable	No CR, no PR, no progression

CR, complete response; PR, partial response.

CONCLUSION

Defining organ involvement and response criteria for amyloidosis has always been challenging. The mission of the 13 members of the consensus panel was to define criteria that could be used worldwide by physicians who treat patients with this disease and to permit uniform reporting criteria of treatment-related outcomes. It is certain that these criteria will undergo further revision at the 11th International Symposium on Amyloidosis to be held at Woods Hole, Massachusetts, in 2006, as new imaging techniques and new biomarkers such as troponin and brain naturetic peptide result in a better understanding of organ involvement and response in amyloidosis.

REFERENCE

1. Sezer O, Niemoller K, Jakob C, Langelotz C, Eucker J, Possinger K: Novel approaches to the treatment of primary amyloidosis. Expert Opin Investig Drugs 9 (10):2343-2350, 2000.

MEASUREMENT OF SERUM FREE LIGHT CHAINS IN AL AMYLOIDOSIS

H.D.Carr-Smith,[1] R. Abraham,[2] G.P. Mead,[1] H. Goodman,[3] P. Hawkins,[3] A.R. Bradwell[4]

[1]The Binding Site Ltd., PO Box 1172, Birmingham B14 4ZB, UK, [2]The Mayo Clinic, Rochester, USA., [3]National Amyloidosis Centre, Royal Free & University College, London,[4]Division of Immunity & Infection, University of Birmingham, B15 2TT, UK

SERUM FREE LIGHT CHAIN ASSAYS IN THE DIAGNOSIS OF AL AMYLOIDOSIS

The role of free light chain assays has been assessed for diagnostic utility in a large series of patients with AL amyloidosis at the National Amyloidosis Centre in London. In a retrospective analysis of stored serum, 98% of 262 patients had abnormal free light chain concentrations at presentation [1]. In contrast, only 3% of patients had sufficient serum concentrations of monoclonal free light chains to be quantitated by electrophoresis. Many patients had elevated free light chains in the urine but serum measurements are preferable. Comparison of the results with electrophoresis tests is shown in Table 1 and Figure 1.

Table 1: Comparison of electrophoresis tests and serum free light chain immunoassays in 262 patients with AL amyloidosis studied at the UK National Amyloidosis Centre.

Serum and urine protein immunofixation and electrophoresis	Number of patients in each group
Quantifiable intact monoclonal immunoglobulin	140(53%)
Quantifiable serum monoclonal light chains	8(3%)
Monoclonal light chains detected by immunofixation alone	67 (26%)
No detectable monoclonal protein in serum or urine	55 (21%)
Serum free light chains abnormal	**257 (98%): κ 79(30%): λ 178 (68%)**

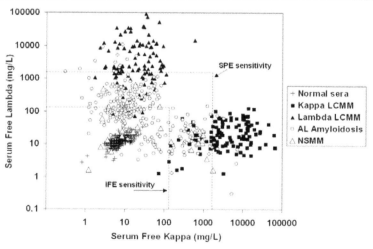

Figure 1: Serum free light chain concentrations in 262 patients with AL amyloidosis at the time of diagnosis. Comparison is made with 282 normal individuals, 224 patients with light chain multiple myeloma (LCMM) and 28 patients with nonsecretory multiple myeloma (NSMM). SPE = serum protein electrophoresis: IFE = immunofixation electrophoresis.

Concentrations of free light chains in the AL amyloidosis patients were similar to those observed in nonsecretory myeloma but lower than those found in light chain myeloma. Classification into kappa or lambda types by free light chain assays was corroborated by immunofixation or bone marrow phenotyping. In most patients, the concentrations of monoclonal free light chains were within the range of 30-500 mg/L. There was no correlation with the concentrations of intact monoclonal immunoglobulins.

The diagnostic accuracy of traditional serum and urine tests has been compared with serum free light chain assays in a study from The Mayo Clinic [2]. Samples from 95 patients with AL were selected, based upon whether serum or urine tested positive or negative for monoclonal proteins by immunofixation electrophoresis and bone marrow immunohistochemistry. For samples that were serum and urine positive by immunofixation electrophoresis, the sensitivity of serum free light chain immunoassays was marginally lower. In patients whose serum was immunofixation negative but urinr positive, the serum free light chain results showed a sensitivity of 95% (kappa) and 100% (lambda) respectively. In patients who were negative by serum and urine immunofixation electrophoresis (but confirmed by bone marrow tests), serum free light chains had a sensitivity of 86%.

The study above was based on samples from highly selected diagnostic groups. A further study was undertaken to evaluate free light chain assays in 34 unselected patients with AL who had undergone stem cell transplantation [2]. Of the 34 patients, 26 were abnormal by serum immunofixation, 28 by urine immunofixation and 24 by both serum and urine immunofixation. However, only 19 patients had a monoclonal serum and 17 a monoclonal urine protein that could be quantified by electrophoresis. In contrast, changes in serum free light chains could be used for assessing all 34 patients although, in four, the concentrations were within the normal range.

The above studies indicate that it is only rare patients that have normal free light chain concentrations but have serum or urine monoclonal proteins identified by electrophoresis. The reasons are unclear but it is not related to light chain genotypes.

SERUM FREE LIGHT CHAINS FOR MONITORING PATIENTS WITH AL AMYLOIDOSIS

The potential utility of serum free light chain measurements for assessing response to chemotherapy in AL amyloidosis was studied in a retrospective analysis of 164 patients [1]. Patients received high-, intermediate- or low-dose treatments, and all were monitored on a six-monthly basis. Stored blood samples were analysed for serum free light chains in the 137 patients who survived at least six months. Results showed that a reduction in the amyloidogenic free light chains by 50% or more, following chemotherapy, was associated with a 10-fold survival advantage. There was an 88% probability of survival for five years if the free light chains had fallen by more than 50%, compared with 39% if the free light chains had reduced by less than 50% (P < 0.001) (*Figure 2*). Median survival was only 15 months in the 27 patients whose free light chains showed no response (P < 0.0001). A greater than 50% fall in serum free light chains was more significantly related to good outcome than any other clinical or biochemical measure. In future, chemotherapy might be altered to take into account favourable or unfavourable responses in free light chain concentrations.

While complete suppression of the clonal plasma cells is desirable, reduction in the amyloidogenic serum free light chain concentrations by 50-70% is often sufficient to lead to stabilisation or regression of amyloid deposits. Continuation of toxic chemotherapy when light chain concentrations have responded may be unnecessary or even harmful. Several recommendations on the use of serum free light chain measurements are in the new UK guidelines [3].

The use of serum free light chain assays for monitoring patients following stem cell transplantation was recently reported from the Mayo Clinic [2]. Serum free light chains were either the only marker for measurement, or decreased before the other markers in 21 of the 34 patients. Overall, changes in free light chain concentrations showed a better correlation with changes in organ function than changes in electrophoresis tests.

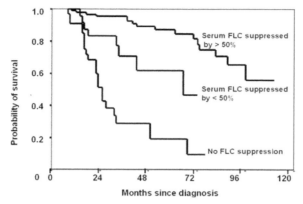

Figure 2: Kaplan-Meier estimate of survival probability in 137 patients with AL showing that a reduction of free light chain concentrations by greater than 50% following chemotherapy was associated with increased survival.

An important link between cardiac dysfunction in AL amyloidosis and serum free light chain concentrations was observed by Palladini et al [4]. 21 AL amyloidosis patients with symptomatic myocardial involvement were given chemotherapy and monitored for serum free light chains and N-terminal pro-natriuretic peptide type B (NT-proBNP), a sensitive marker of myocardial dysfunction. During treatment, 11 of the patients had a reduction of serum free light chains by more than 50%. In all of the 11 patients there was a corresponding reduction of NT-proBNP levels and in six, the heart failure resolved. These observations suggest that the amyloidogenic precursors equate to the circulating monoclonal free light chains and are directly toxic to cardiac muscle cells. It also emphasises the utility of the free light chain assays and the importance of rapidly reducing their concentrations by treatment in order to help control heart failure.

SUMMARY: In patients with AL amyloidosis, serum free light chain concentrations:-Are elevated in over 95% of patients at disease presentation. Provide a quantitative assessment of circulating fibril precursors. Correlate with changes in amyloid load during treatment. Are helpful in measuring early responses to treatment and identifying disease relapses because of their short serum half-life. Are useful for monitoring patients undergoing stem-cell transplantation.

REFERENCES

1. Lachmann HJ, Gallimore R, Gillmore JD, Carr-Smith HD, Bradwell AR, Pepys MB, Hawkins PN. 2003 Outcome in systemic AL amyloidosis in relation to changes in concentration of circulating free immunoglobulin light chains following chemotherapy. *British Journal of Haematology*; 122: 78-84.
2. Abraham RS, Katzmann JA, Clark RC, Bradwell AR, Kyle RA, Gertz MA. 2003. Quantitative analysis of serum free light chains. A new marker for the diagnostic evaluation of primary systemic amyloidosis. *American Journal of Clinical Pathology* ,119: 274-278.
3. Bird JM, Samson D, Mehta A, Cavenagh J, Hawkins P, Lachmann H. 2004. Guidelines on the diagnosis and management of AL amyloidosis. *British Journal of Haematology. In press.*
4. Palladini G, Bosoni, Lavatelli F, D'Eril GM, Moratti R, Merlini G. Circulating free light chain concentration correlates with degree of cardiac dysfunction in AL amyloidosis. *In press.*

MELPHALAN PLUS HIGH-DOSE DEXAMETHASONE IN AL PATIENTS WHO ARE NOT CANDIDATES FOR STEM CELL TRANSPLANTATION

Giovanni Palladini[1,2], Vittorio Perfetti[2], Laura Obici[1], Francesca Lavatelli[1], Riccardo Caccialanza[3], Fausto Adami[4], Giobatta Cavallero[5], Roberto Rustichelli[6], Giovambattista Virga[7], Giampaolo Merlini[1]

Amyloid Center, [1]Biotechnology Research Laboratories – Department of Biochemistry, [2]Internal Medicine and [3]Department of Applied Health Sciences – Human Nutrition Unit, "IRCCS – Policlinico San Matteo", Pavia, and University of Pavia; [4]Department of Clinical and Experimental Medicine – Padova University, Padova, [5]Internal Medicine – Hospital "Santa Croce e Carle", Cuneo, [6]Nephrology Unit – "Arcispedale Santa Maria Nuova", Reggio Emilia, [7]Nephrology Unit – Provincial Hospital, Camposampiero (Padova).

The prognosis of patients with systemic AL is poor, with a median survival of 12-18 months and the main prognostic determinant is amyloid heart involvement (2). At present, therapy of AL is aimed at suppressing the plasma cell clone producing the amyloidogenic light-chain. The most effective treatment so far is autologous stem cell transplantation (ASCT), which can induce complete remissions in more than 50% of patients who can undergo this procedure (2, 3). However, a considerable proportion of AL patients is not eligible for ASCT due to advanced disease (3). The patients in whom ASCT is contraindicated are those in greatest need of prompt care. They are usually treated with oral melphalan plus prednisone (MP). However, response rate to this regimen is unsatisfactory and time to response can be unaffordably long. Our previous experience indicated that MP can provide a 37% response rate with a long time to response (median 9 months) (4). Kyle et al. reported that 28% of AL patients respond to MP and 30% of them require to be treated for more than 1 year (5). Time to response to high-dose dexamethasone (HD-Dex) is shorter (median 4 months), but response rate to HD-Dex is still only 35% (6). In an attempt of synergizing the therapeutic actions of melphalan and HD-Dex, we treated with this association 46 patients who were not eligible to ASCT.

PATIENTS AND METHODS

We evaluated the combination of melphalan (0.22 mg/Kg) and HD-Dex (40 mg) given orally on days 1-4 every 28 days (M-Dex) in 46 AL patients who did not meet the eligibility criteria for ASCT. All patients had symptomatic disease, with histological diagnosis of amyloidosis and a monoclonal component was detected in serum and/or urine of all patients at high-resolution immunofixation. ASCT was contraindicated because of heart failure in 32 patients (70%), involvement of >2 organs in 24 (52%), age >65 years in 17 (37%), creatinine >2 mg/dL in 2 (4%) and abnormal respiratory function tests in 1 (2%). Twenty-four patients (52%) were not eligible for ASCT for ≥2 criteria. All patients received prophylactic omeprazole, itraconazole and ciprofloxacin. Since, in our preliminary experience with standard HD-Dex, we observed a significant proportion of treatment-related fatal arrhythmias (6), amiodarone was administered to patients who presented repetitive ventricular arrhythmias at 24 h Holter electrocardiogram. This Holter feature is associated to an increased risk of sudden death in AL (7). Hematologic response was defined as the decrease of the monoclonal component (MC) by at least 50%. Complete remission consisted in the disappearance of the MC from both serum and urine documented by high-resolution agarose-gel immunofixation electrophoresis (8). Organ response was assessed according to the criteria proposed by the Mayo Clinic Group (5). Treatment was continued for up to 9 courses in patients who achieved the hematological response and discontinued if a complete hematological remission was obtained, if the monoclonal component increased and in the case of treatment-related toxicity. The response was evaluated every 3 months and established at the nadir of serum and urine monoclonal protein.

RESULTS

Forty-six patients (33 men) were enrolled into the study. Patients' characteristics are reported in Table 1.

Table 1. Patients' characteristics.

Median age (years)	62 (range: 34-79)
>2 organs involved	35 (76%)
Median creatinine (mg/dL)	1 (range 0.5-3.6)
Serum creatinine >2 mg/dL	2 (4%)
Median proteinuria (g/24h)	4 (range 0-24)
Proteinuria $\geq$3 g/24 h	27 (59%)
Heart failure	32 (70%)
Median interventricular septum thickness (mm)	14 (range 8-23)
Interventricular septum thickness >15 mm	15 (33%)
Median EF (%)	50 (range 34-75)
Orthostatic hypotension	16 (35%)

Patients were treated for a median of 4 courses and 2 died before completing the third cycle. In 31 patients (67%), M-Dex caused a >50% reduction of MC and in 15 (33%) the MC disappeared at high-resolution immunofixation. Median time to response was 4.5 months (range: 2-10 months). Complete remission is maintained in 14 of 15 patients after a median follow-up of 16 months (range: 3-34 months). Organ response was observed in 22 (48%) patients, all of whom also achieved hematologic response.

Table 2. Organ response.

>50% reduction of proteinuria	14
resolution of heart failure	6
improvement of respiratory function tests in lung involvement	1
normalization of alkaline phosphatase in liver involvement	1

Complete hematologic response was accompanied by organ response in 13 out of 15 cases (87%). Organ and hematologic responses were simultaneous in 16 patients (72%), in the remaining cases, organ function improvement followed hematologic response after a median time of 5 months (range: 3-18 months). The median follow-up of living patients is 20 months (range: 6-43 months). Hematological response resulted in a significant survival benefit (p=.009). Nine patients (20%) died after a median follow-up of 5 months (range: 1-27 months). Eight had severe heart involvement and died of heart failure (6 patients) and sudden death (2 patients), one died due to unrelated causes. No patient died because of treatment.

Table 3. Severe toxicity.

Respiratory infections	3* (6%)
Cytopenia	1 (2%)
Myelodysplastic syndrome (after 2 cycles)	1 (2%)
Total	5 (10%)

*2 of whom were the only patients who did not follow the antibiotic prophylaxis

CONCLUSIONS

Despite the fact that we selected patients with advanced disease, M-Dex proved effective and very well tolerated in AL patients ineligible for ASCT. There was no treatment-related mortality. Time to response was short (4.5 months). The 67% response rate compares favorably with that achievable in unselected patients with standard melphalan and prednisone (4, 5) and with HD-Dex alone (6). Response to M-Dex translates into a significant survival advantage. Given the relatively high rate of organ response (48%) a considerable proportion of patients who do not reach complete remission with M-Dex, but who do obtain organ response, might be re-evaluated as candidates to ASCT, although the exposure to the alkylating agent may jeopardize stem cell harvesting.

REFERENCES

1. Kyle RA, Gertz MA (1995). Primary systemic amyloidosis: clinical and laboratory features in 474 cases. *Semin Hematol*; **8**: 45-59.

2. Sanchorawala V, Wright DG, Seldin DC, et al (2001). An overview of the use of high-dose melphalan with autologous stem cell transplantation for the treatment of AL amyloidosis. *Bone Marrow Transplant*; **28**: 637-642.

3. Comenzo RL, Gertz MA (2000). Autologous stem cell transplantation for primary systemic amyloidosis. *Blood*; **99**: 4276-4282.

4. Palladini G, Adami F, Anesi E, et al. AL amyloidosis: the experience of the Italian Amyloidosis Study Group. In: M. Bély, Á. Apáthy (eds.) Amyloid and Amyloidosis: the proceedings of the 9th International Symposium on Amyloidosis. 2001, pp. 249-251.

5. Kyle RA, Gertz MA, Greipp PR, et a (1997)l. A trial of three regimens for primary amyloidosis: colchicine alone, melphalan and prednisone and melphalan, prednisone and colchicine. *N Engl J Med*; **336**: 1202-1207.

6. Palladini G, Anesi E, Perfetti V, et al (2001). A modified high-dose dexamethasone regimen for primary systemic amyloidosis. *Br J Haematol*; **113**: 1044-1046.

7. Palladini G, Malamani G, Cò F, et al (2001). Holter monitoring in AL amyloidosis: prognostic implications. *Pacing Clin Electrphysiol*; **24**: 1228-1233.

8. Merlini G, Marciano S, Gasparro C, et al (2001). The Pavia approach to clinical protein analysis. *Clin Chem Lab Med*; **39**: 1025-1028.

HIGH DOSE MELPHALAN AND AUTOLOGOUS STEM CELL TRANSPLANTATION (HDM/SCT) FOR TREATMENT OF AL AMYLOIDOSIS: TEN YEAR SINGLE INSTITUTIONAL EXPERIENCE

D. C. Seldin, V. Sanchorawala, K. Malek, D. G. Wright, K. Quillen, K. T. Finn, R. Falk, J. Berk, L. Dember, J. Wiesman, J. J. Anderson, M. Skinner

Amyloid Treatment and Research Program and Stem Cell Transplant Program, Section of Hematology-Oncology, Department of Medicine, Boston University School of Medicine, Boston MA 02118.

In an effort to increase treatment responses for AL amyloidosis patients and improve organ function, quality of life, and survival, we and other centers have explored the safety and efficacy of HDM/SCT over the last 10 yrs. More than 700 pts have been evaluated for treatment with HDM/SCT on institutionally approved clinical trials at Boston Medical Center since 1994. 56% of evaluated patients met eligibility criteria for transplantation, and 363 patients have intitiated treatment. 330 transplants have been performed with an overall treatment related mortality of 13% and a complete hematologic response (CR) rate of 40%. With a median followup of >4 years, 54% of treated pts remain alive, and median survival exceeds 4 years. Survival and hematologic response is associated with improvement in organ function and in quality of life (QOL).

BACKGROUND

AL amyloidosis is a plasma cell dyscrasia in which clonal immunoglobulin light chains form fibrillar deposits in tissues, leading to organ failure and death. Following upon the recognition of the pathogenesis of the disease and its similarity to multiple myeloma, treatment with standard oral melphalan and prednisone chemotherapy was begun. This treatment is modestly effective, with about a third of patients having improvements in bone marrow plasmacytosis, serum or urine paraprotein production, and end organ function. However, hematologic CRs are rare and the survival benefit is small [1,2]. In the early 1990's, case reports describing responses following the use of HDM/SCT appeared. We began such treatment protocols in 1994 and published a small case series in 1996 [3] and followup on 25 patients in 1998 [4]. In 2003, we carried out an analysis of results of screening of 701 patients and treatment of 312 [5]. For the meeting in Tours, these data were updated, with nearly 10 years experience and followup information on almost all patients.

RESULTS AND DISCUSSION

While the mortality during the peri-transplant period is much higher for AL amyloidosis patients than patients with most other hematologic disorders due to complications related to compromised organ function, those patients who are carefully selected and managed during transplant enjoy an excellent hematologic response rate (40%) and median survival of 4.75 years, much longer than that seen with other treatments (Figure 1).

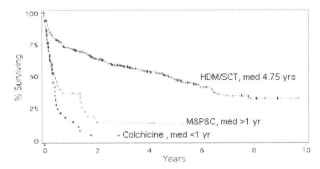

Figure 1. Overall survival of patients treated with HDM/SCT compared with historical control groups treated with oral melphalan, prednisone, and colchicine (M&P&C) or colchicine alone [1].

The benefits of having a plasma cell disorder that is chemosensitive are seen in Figure 2, in which survival of those patients who achieved a hematologic CR (defined as normalization of the bone marrow immunohistochemistry and disappearance of serum and urine monoclonal proteins on immunofixation electrophoresis at one year after treatment) is compared to that of those who do not achieve CR. The median survival of the CR group has not been reached, but is expected to exceed 8 years. The median survival of the non-CR group is 5.1 years. Note that these curves do not include those patients who died within the first year and could not be evaluated for CR, which is why the medians are greater than in Figure 1.

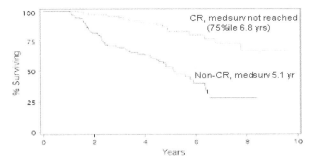

Figure 2. Survival depends upon hematologic response.

Survival is also dependent upon the presence or absence of cardiac disease (Figure 3). Patients with cardiac disease had to have well-compensated symptoms or asymptomatic cardiac disease; patients with more severe cardiac symptoms or depressed ejection fraction by echocardiogram were excluded from treatment. There was, however, no exclusion based upon wall thickness alone. It is important to note that cardiac disease primarily led to early mortality, and some patients did have long survival, suggesting that this is still a beneficial treatment for some patients with cardiac disease, particularly if early complications can be reduced.

In addition to the hematologic responses, organ responses (reported in reference 5), and excellent survival, patients who undergo treatment with HDM/SCT actually have an improvement in quantitative measures of quality of life, based upon self-reporting using the the Medical Outcomes Study 36-item Short Form General

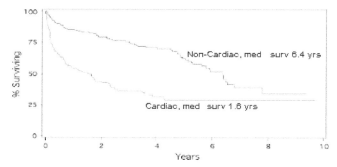

Figure 3. Survival depends upon presence or absence of cardiac disease, but some patients with cardiac disease that are able to under treatment have long survival.

Health Survey (SF-36) questionnaire. In figure 4, we see that the baseline QOL measures in even the best-risk transplant eligible patients are reduced, compared with age-matched U.S. controls (Norm). However, at one and two years after treatment, scores improve markedly.

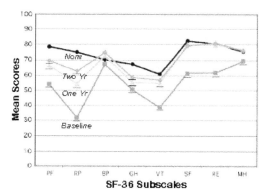

Figure 4. Annual improvement in mean QOL scores after treatment with HDM/SCT.

REFERENCES

1. Skinner, M., et al., Treatment of 100 patients with primary amyloidosis: a randomized trial of melphalan, prednisone, and colchicine versus colchicine only. Am J Med, 1996. 100(3): p. 290-8.
2. Kyle, R.A., et al., A trial of three regimens for primary amyloidosis: colchicine alone, melphalan and prednisone, and melphalan, prednisone, and colchicine. N Engl J Med, 1997. 336(17): p. 1202-7.
3. Comenzo, R.L., et al., Dose-intensive melphalan with blood stem cell support for the treatment of AL amyloidosis: one-year follow-up in five patients. Blood, 1996. 88(7): p. 2801-6.
4. Comenzo, R.L., et al., Dose-intensive melphalan with blood stem-cell support for the treatment of AL (amyloid light-chain) amyloidosis: survival and responses in 25 patients. Blood, 1998. 91(10): p. 3662-70.
5. Skinner, M., et al., Survival and Clinical Response to Treatment with High-Dose Melphalan and Autologous Stem-Cell Transplantation in Patients with AL Amyloidosis: An 8-Year Study. Ann Intern Med, 2004.

Supported by grants from NIH P01-HL68705 and the Gerry Foundation. DCS is a scholar of the Leukemia and Lymphoma society

THALIDOMIDE TREATMENT IN 80 PATIENTS WITH AL AMYLOIDOSIS: TOLERABILITY, CLONAL DISEASE RESPONSE AND CLINICAL OUTCOME

H.J.B. Goodman, H.J. Lachmann, R. Gallimore and P.N. Hawkins

Centre for Amyloidosis and Acute Phase Proteins, Royal Free and University College Medical School, London, United Kingdom

Myeloma responds to thalidomide in 30% of patients who have already undergone chemotherapy, and in 60% of untreated patients when combined with dexamethasone. We report the use of thalidomide in 80 patients with AL in whom cytotoxic therapy had either been ineffective or deemed too toxic to pursue. Median age was 63 yrs (range 32-83), and the patients had had a median of one type of chemotherapy (0-5). Thalidomide was taken for a median of six months (0.4-34), at a median dose 100mg/day (50-600). Thalidomide was used alone in 51 patients, with dexamethasone in eight, cyclophosphamide in three, cyclophosphamide and dexamethasone in thirteen, melphalan and dexamethasone in two, and in other combinations in two cases. Median follow-up was 9 months. Recognised adverse events of somnolence, constipation and/or neuropathy occurred in 50 patients. Three had venous thromboses and one had a major arterial clot. Thalidomide was discontinued due to lack of effect in eight cases and adverse effects in 25. Clonal disease responses of >50% were recorded in 39 out of 64 patients who were followed by serum free light chain measurements, and in 16 out of 49 patients who had paraproteins that were quantified conventionally. Organ function improved or remained stable in 26% and 40% of evaluable cases respectively, and SAP scintigraphy showed regression of amyloid in 18%. 17 patients on thalidomide died of progressive disease. Thalidomide alone or in combination with other agents is probably as effective in AL amyloidosis as it is myeloma, but has frequent adverse effects.

HIGH DOSE THERAPY WITH AUTOLOGOUS HEMATOPOIETIC STEM CELL TRANSPLANTATION (HSCT) FOR PATIENTS WITH PRIMARY SYSTEMIC AMYLOIDOSIS (AL): AN ABMTR STUDY

DH Vesole, W Pérez, D Reece, M Akasheh, M Horowitz and C Bredeson

For the Autologous Blood and Marrow Transplant Registry (ABMTR) Multiple Myeloma Working Committee, Milwaukee, WI USA

INTRODUCTION

Primary systemic amyloidosis (AL) is a rare plasma cell dyscrasia with approximately 2000 cases diagnosed yearly in North America. In this disease, monoclonal immunoglobulin light chains are produced by malignant plasma cells resulting in amyloid fibril deposition in vital organs leading to progressive, fatal multi-system failure. The median survival for treated patients is approximately 18 months compared to 12 months for untreated patients; the median survival for patients with symptomatic cardiac involvement is 6 months (1).

With the promising outcomes observed with high dose chemotherapy and autologous hematopoietic stem cell transplantation in the more common plasma cell dyscrasia, multiple myeloma, pilot trials to test this approach in AL were pursued. These pilot trials, which almost uniformly utilized high dose melphalan, again based upon the experience in multiple myeloma, showed encouraging results: hematologic response rates were appreciated in approximately 60% of the patients; organ response rates were also observed but at a lower rate. Unfortunately, these early pilot trials were associated with high transplant-related mortality: 15-25%, and significantly higher, approaching 65% in patients with AL cardiomyopathy. More recent clinical trials have been conducted with more stringent patient selection criteria: transplant-related mortality has decreased to 10-15% (2).

Since AL is a rare disease, there are only a few academic programs with an in-depth focus on the treatment of this disease: in the United States the primary research centers are the Mayo Clinic in Rochester, MN and Boston University in Boston, MA. However, the majority of AL patients are treated in other research centers. It is important to determine the outcomes of AL patients who undergo high dose therapy with hematopoietic stem cell transplant in the general academic community. To this end, in 2001, the Autologous Blood and Marrow Transplant Registry initiated a study of transplant outcomes in patients with primary systemic amyloidosis.

PATIENT CHARACTERISTICS

From 1995 through 2001, 107 recipients of autologous transplants for AL were reported from 48 transplans were reported to the ABMTR. The median age was 55 years (range 31-71) with the male to female ratio of 60:40. The paraprotein isotype was IgG (35%), IgA (10%), light chain (45%), other (2%) and none (9%).

Table 1. No. and Type of Organs Involved Pre-HSCT	
No. Organs	**%**
0	20
1	58
2	20
3	2
Cardiac	22
Liver (elevated alkaline phosphatase)	47
Renal (nephrotic syndrome)	70

Table 2. Prior Therapy	
Regimen	**%**
No prior therapy	35
Melphalan +/- other	26
Multiagent	23
Others	16
Lines of Prior Therapy	**%**
0	35
1	40
2	12
3+	13

Approximately one-third of patients had received no prior therapy. Melphalan-based therapies were the most commonly used (Table 2). Three fourths of the patients were minimally treated: 35% with no prior therapy and 40% with one prior treatment regimen. The time from first treatment to transplant was within 6 months in 74% of patients: 35 patents with no prior therapy; 38 with less than 6 months. Seventeen percent of patients had 6 to 12 months of prior therapy whereas only 9% had more than 1 year of prior treatment.

TRANSPLANT CHARACTERISTICS

Most transplant centers reported only a single case (25 centers) whereas only 3 centers reported 5or more transplants.

As has been observed in other AL transplant studies, the majority utilize single agent melphalan as the preparative regimen: in this analysis, 81% received melphalan alone, 8% TBI-containing regimens and 11% other regimens. The majority of patients received $\geq$ melphalan 100 mg/m^2.

RESPONSE AND SURVIVAL

There is no universally accepted response criteria for amyloidosis. Due to limitations of obtaining registry data, the response criteria for this study were more loosely defined than those followed by individual centers or cooperative groups. The response criteria for this study were defined as follows: hematologic response required a $\geq$ 50% decrease in paraprotein-urine and/or serum was considered a partial response. Complete remission required a negative urine and/or serum protein electrophoresis. An organ response was recorded as either improved or not improved for three organ systems: renal, hepatic and cardiac.

Since AL patients may improve over a long course of time, response criteria were evaluated at 1 year post-transplant (Table 4). Thirty four percent of patients had an objective response to transplant; 68% had stable disease or response.

Table 4. Response at 1-Year Post-Transplant	
Best Response	**Prob % (95% CI)**
Complete remission	17 (16)
Partial response	17 (16)
No response/Stable disease	33 (31)
Progressive disease	11 (10)
Transplant-related mortality	29 (27)

Table 5. Causes of Death: N = 45 (42%)		
Etiology	**N**	**% Total**
Primary disease	23	51
Cardiac failure	8	18
Pulmonary	6	13
Infection	3	7
Multiple organ failure	2	4
Renal failure	1	2
Other cause	2	4

Forty two percent of patients have expired: the transplant-related mortality at 1 year was 29%. The major causes of death were disease progression (51%), cardiac (18%), pulmonary (13%) and infection (7%) as shown in Table 5 (above). Overall survival from transplant is shown in Figure 1 below. The 1-year survival was 66% (56-75; 95% CI) and the 3 year survival was 56% (45-56; 95% CI) for the 107 patients.

Figure 1. Autotransplants for amyloidosis

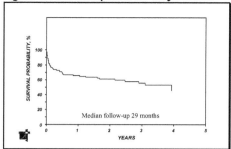

PROGNOSTIC FACTORS

An extensive numbers of variables were analyzed in a multivariate analysis which included patient-related (age, sex, Karnofsky score); transplant-related (donor type, time from diagnosis to transplant) or disease-related (organ involvement pre-HSCT, cardiac involvement, number of prior therapies, time from diagnosis to transplant). No variable had predictive value for survival post-transplant including cardiac involvement or number of organs involved.

CONCLUSION

Select patients with primary AL benefit from high dose therapy with HSCT: a case-match control study has demonstrated the superiority of HSCT to conventional therapy (3). Although the ABMTR transplant-related mortality was quite high (29%), the 3-year survival (56%) was comparable to results reported by single institutions. With a better understanding of the recently documented prognostic factors (e.g. pro-brain natuiretic peptide, cardiac troponin (cTNT), beta-2 microglolbuin, and the number of organs involved) more stringent patient selection will undoubtedly result in lower transplant-related mortality and improved survival (4,5). In addition, primary AL patients have a more difficult post-transplant course due to multi-organ disease involvement and familiarity with these types of toxicities should also reduce the transplant-related mortality. To test this hypothesis, it would be of future interest for the ABMTR to compare transplant outcomes in primary AL before 2003 with those after 2003.

REFERENCES

1. Kyle RA, Gertz MA, Greipp PR, et al. A trial of three regimens for primary amyloidosis: colchicines alone, melphalan and prednisone, and melphalan, prednisone, and colchicines, N Engl J Med 1997; 366:1202-7.
2. Skinner M, Sanchorawala V, Seldin D et al High-dose melphalan and autologous stem-cell transplantation in patients with AL amyloidosis: An 8-year study. Ann Int Med 2004: 140:85-93.
3. Dispenzieri A, Kyle RA, Lacy MQ, et al. Superior survival in primary systemic amyloidosis patients undergoing peripheral blood stem cell transplantation: a case control study. Blood 2004; 103:3960-3963.
4. Dispenzieri A, Lacy MG, Kyle RA et al. Eligibility for hematopoietic stem-cell transplantation for primary systemic amyloidoisis is a favorable prognostic factor for survival. J Clin Oncol 2001; 19:3350-6.
5. Comenzo RL, Gertz MA. Autologous stem cell transplantation for primary systemic amyloidosis. Blood. 2002; 99:4276-4282.

PROSPECTIVE EVALUATION OF THE UTILITY OF THE SERUM FREE LIGHT CHAIN ASSAY (FLC), CLONAL Ig V_L GENE IDENTIFICATION AND TROPONIN I LEVELS IN A PHASE II TRIAL OF RISK-ADAPTED INTRAVENOUS MELPHALAN WITH ADJUVANT THALIDOMIDE AND DEXAMETHASONE FOR NEWLY DIAGNOSED UNTREATED PATIENTS WITH SYSTEMIC AL AMYLOIDOSIS

RL Comenzo, P Zhou, L Reich, S Costello, A Quinn, S Fircanis, L Drake, C Hedvat, J Teruya-Feldstein, DA Filippa, M Fleisher

Memorial Sloan-Kettering Cancer Center, New York, New York, USA

The availability of the serum free light chain assay and of soluble markers of myocardial strain and injury (troponin, brain natriuretic peptide (BNP)) will likely improve the evaluation and treatment of patients with AL amyloidosis (1-4). In addition, we and others have noted correlations between immunoglobulin light chain variable region germline gene (Ig V_L) utilization by AL clones and the pattern of AL organ involvement (5-7). Indeed, one study suggests that Ig V_L gene use contains prognostic value (5). The applicability of new tests for patient management, however, requires prospective evaluation. In an on-going phase II clinical trial for newly diagnosed untreated patients with systemic AL (MSKCC #02-031), we are prospectively evaluating the utility of the serum free light chain assay (FLC), clonal Ig V_L gene identification, troponin I and BNP levels. We report an interim analysis of our findings.

PATIENTS AND METHODS

The target population of MSKCC #02-031 is patients with primary systemic amyloidosis (AL) and a clonal plasma cell disease who are within 12 months of diagnosis and not previously treated. Organ involvement and plasma cell disease were defined as previously described (8). DNA-based polymerase chain reaction assays with cloning and sequencing of amplicons for exons 2, 3 and 4 of transthyretin (TTR) and for fibrinogen A-α were employed to further exclude patients with hereditary amyloid variants and incidental monoclonal gammopathies (9, 10).

In this phase II trial, the efficacy of intravenous (iv) melphalan is tested as initial therapy and, in patients with persistent clonal plasma cell disease at 3 months after initial therapy, the efficacy of oral thalidomide and dexamethasone is being tested as adjuvant therapy. Melphalan dosing is risk-adapted as previously described, based on patient age and extent of organ involvement (11).

The primary endpoint of the trial is the two-year overall and progression-free survival of recently diagnosed untreated patients with AL treated in this manner. Secondary endpoints are the response of the plasma cell disease at 3, 12 and 24 months after initial therapy, and the response of the amyloid-related disease at the 12- and 24-month time-points. Additional secondary endpoints include the prognostic significance of Ig V_L gene use, and prospective evaluations of the serum free light chain assay (FLC), troponin I and BNP levels.

Evaluation of Ig V_L gene use was as previously described. Serum free kappa and lambda light chain levels were measured with the FreeLite assay after a period of validation with institutional normal controls (The Binding Site, LTD; Birmingham, UK). Troponin I levels were measured by immunoassay according to the manufacturer's instructions (Tosoh Bioscience, Inc; San Francisco, CA). Statistical analyses were performed with PRISM (GraphPad, San Diego, CA).

RESULTS

From 9/02 until 4/04 32 patients were enrolled on this trial. The time from diagnosis to enrollment was a median of 2 months. Dominant organ involvement was cardiac (n=10), renal (n=12) and other (n=10). All had monoclonal gammopathies (λ=24, κ=8). Eighty-eight percent (28/32) had elevated pathologic FLC with abnormal κ::λ ratio. Twenty-nine of the 32 patients have been or are being treated, 3 in the poor risk and 26 in the good risk group. One patient in the good risk group died on day 2 of stem cell mobilization with G-CSF, the only treatment-related death thus far. In the good risk group, patients received iv melphalan doses of 100 (n=6), 140 (n=10) and 200 mg/m^2 of melphalan (n=9). There have been six deaths *in toto*, five of them due to progression of disease.

The results of FLC testing at baseline, 3 months and 12 months are shown in Figure 1. The axes are logarithmic and in units of mg/dl. Each data-point symbol represents a patient's kappa/lambda values. The closed triangles are cases in which the kappa:lambda ratio is normal while the open circles are data-points with skewed ratios. The levels of pathologic kappa and lambda light chains were compared for 7 patients with kappa disease and 21 with lambda disease at baseline, and there was no significant difference (median (range); kappa = 30.5 (5.89-290) and lambda = 14.6 (2.1-168) mg/dl, p =0.37 by Mann-Whitney).

Figure 1. The FLC values for patients at baseline, 3 months and 12 months after therapy (left to right)

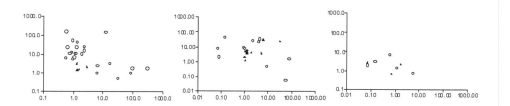

We asked whether or not changes at 3 months post-therapy in the kappa:lambda ratios or in the level of the pathologic FLC correlated with patient survival and found a clear-cut trend toward significance in both instances as shown in Figure 2. In addition, we assessed the relationship between changes in the ratios and pathologic levels and the achievement of a response of the plasma cell disease (> 50% reduction in M protein). Eleven of 13 patients who responded had > 50% reductions in the pathologic FLC, while only 2 of 8 patients with stable or progressive disease had such reductions (p=0.02).

Figure 2. Correlation of changes in the kappa:lambda ratio (κ:λ) and in decreases in pathologic light chain levels (Δ-) with patient survival

K:λ at 3 months	Dead	Alive
Abnormal	4	7
Normal	1	9

X^2=2.00, df=1, p=0.08, RR=3.64 with 95% CI 0.48-27.34

Δ- at 3 months	Dead	Alive
< 50%	3	4
$\geq$ 50%	2	12

X^2=2.10, df=1, p=0.07, RR=3.0 with 95% CI 0.64-14.03

Twenty-one patients have been evaluated at 3 months after initial therapy. Five have died of progression of disease (POD) a median of 5 months after initial melphalan (4-10). Sixteen are alive at a median of 8 months after initial melphalan (4-15). Seven of 21 had negative serum and urine immunofixation at 3 months; of these, 2 had persistent pFLC with abnormal ratios, 1 with microscopic clonal plasma cells on marrow biopsy and 1 without. The median fractional change in pFLC was significantly lower in the survivors compared to those who died of POD (p<0.01 by two-tailed Mann-Whitney). Seven patients have been evaluated at 12 months. All responded and had improved amyloid-related organ disease. Of 24 clonal Ig V_L genes identified, the most frequent are 3r (5, including 2 cardiac), 2a2 (5, including 3 cardiac), and 6a (3, including 2 renal). Median baseline troponin levels were significantly higher in cardiac patients, 0.09 (0-0.61) versus 0 (0-0.12) in others (p<0.05), as was the case for BNP levels also, 918 (234-3450) versus 74.5 (37.5-413) (p<<0.01).

DISCUSSION

Interim results of this phase II clinical trial thus far support the utility of the FLC assay and troponin I and BNP levels. Normalization of FLC levels and ratio at 3 months after initial therapy will likely correlate with survival within the first 12 months of therapy. Troponin I and BNP levels will likely be of prognostic significance. Ig V_L gene identification is of limited utility, although in 2 instances it did provide useful confirmatory data. Two African-American patients with wildtype TTR sequences had monoclonal gammopathies but normal FLC levels; both had typical AL Ig V_L gene utilization (6a, DPK21). Whether these assays provide independent prognostic information awaits the completion of the trial and a planned multivariate analysis.

An important question exists with respect to the definition of hematologic responses. How useful are the FLC levels in follow-up in comparison to the standard measures of M protein such as quantitative immunoglobulins, immunofixation and urine electrophoresis? Clearly it is possible to have a conventionally defined complete remission (CR) and persistently abnormal FLC with elevated pathologic light chains. Alternately it is possible to have stable disease and normalized FLC with a normal ratio. It is likely that the former group of patients will not do as well as CR patients with normal FLC, and the latter group will likely do better than those with persistent unchanged M protein levels. Whether a small phase II trial of the sort we are conducting will be able to identify trends or relative risks in this regard is not clear.

ACKNOWLEDGEMENTS

We thank the Food and Drug Administration Orphan Products Development Program and Celgene for their support.

REFERENCES

1. Abraham RS, Katzmann JA, Clark RJ, Bradwell AR, Kyle RA, Gertz MA. Quantitative analysis of serum free light chains. A new marker for the diagnostic evaluation of primary systemic amyloidosis. Am J Clin Pathol 2003;119:274-8.
2. Lachmann HJ, Gallimore R, Gillmore JD, Carr-Smith HD, Bradwell AR, Pepys MB, Hawkins PN. Outcome in systemic AL amyloidosis in relation to changes in concentration of circulating free immunoglobulin light chains following chemotherapy. Br J Haematol 2003;122:78-84.
3. Dispenzieri A, Kyle RA, Gertz MA, Therneau TM, Miller WL, Chandrasekaran K, McConnell JP, Burritt MF, Jaffe AS.Survival in patients with primary systemic amyloidosis and raised serum cardiac troponins. Lancet 2003;361:1787-9.
4. Palladini G, Campana C, Klersy C, Balduini A, Vadacca G, Perfetti V, Perlini S, Obici L, Ascari E, d'Eril GM, Moratti R, Merlini G. Serum N-terminal pro-brain natriuretic peptide is a sensitive marker of myocardial dysfunction in AL amyloidosis. Circulation 2003;107:2440-5.
5. Comenzo RL, Zhang Y, Martinez C, Osman K, Herrera GA. The tropism of organ involvement in primary systemic amyloidosis: contributions of Ig V(L) germ line gene use and clonal plasma cell burden. Blood 2001;98:714-20.
6. Perfetti V, Casarini S, Palladini G, Vignarelli MC, Klersy C, Diegoli M, Ascari E, Merlini G. Analysis of VI-JI expression in plasma cells from primary (AL) amyloidosis and normal bone marrow identifies 3r (IIII) as a new amyloid-associated germline gene segment. Blood 2002;100:948-53.
7. Abraham RS, Geyer SM, Price-Troska TL, Allmer C, Kyle RA, Gertz MA, Fonseca R. Immunoglobulin light chain variable (V) region genes influence clinical presentation and outcome in light chain-associated amyloidosis (AL). Blood 2003;101:3801-8.
8. Comenzo RL. Primary systemic amyloidosis. Curr Treat Options Oncol 2000;1:83-9.
9. Lachmann HJ, Booth DR, Booth SE, Bybee A, Gilbertson JA, Gillmore JD, Pepys MB, Hawkins PN. Misdiagnosis of hereditary amyloidosis as AL (primary) amyloidosis. N Engl J Med 2002;346:1786-91.
10. Jacobson DR, Pastore RD, Yaghoubian R, Kane I, Gallo G, Buck FS, Buxbaum JN. Variant-sequence transthyretin (isoleucine 122) in late-onset cardiac amyloidosis in black Americans. N Engl J Med 1997;336:466-73.
11. Comenzo RL, Gertz MA. Autologous stem cell transplantation for primary systemic amyloidosis. Blood. 2002;99:4276-82.

SECTION 3

AA AMYLOIDOSIS

CHARACTERISTIC AND CLINICAL OUTCOME OF 340 PATIENTS WITH SYSTEMIC AA AMYLOIDOSIS

H.J.Lachmann, H.J.B.Goodman, J.Gallimore, J.A.Gilbertson, J.Joshi, M.B.Pepys, and P.N.Hawkins

Centre for Amyloidosis and Acute Phase Proteins, Royal Free and University College Medical School, London, United Kingdom

We report 340 patients with AA amyloidosis evaluated in our centre over 15 years. 55% were men and the most frequent underlying conditions were inflammatory arthritides in 219 patients (64%), chronic sepsis in 48 (14%), hereditary periodic fever syndromes in 25 (7%) and Crohn's disease in 13 (4%). The underlying disorder remained covert in 20 (6%) cases.

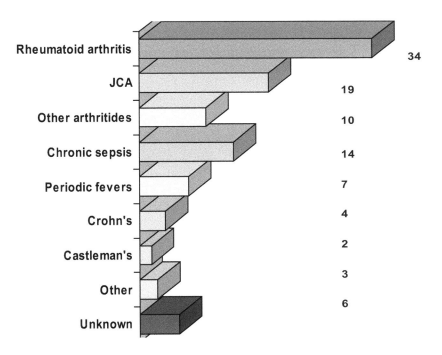

Figure 1. Underlying causes of chronic inflammation in 340 patients with AA amyloidosis
Median age at presentation was 50 years (range 11 to 87) and the median latency was 17 years.

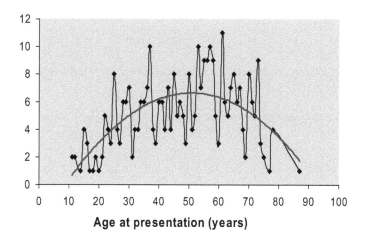

Figure 2. Age at presentation with AA amyloidosis

At presentation, 12% were on dialysis and among the remainder 97% had significant proteinuria, 58% were nephrotic, and the serum creatinine exceeded 250 µmol/L in one-fifth.   Echocardiography in 222 cases identified numerous abnormalities ranging from LVH, to dilated cardiomyopathy, HOCM, ischaemic damage and valvular lesions, but was characteristic of amyloidosis in only two cases.

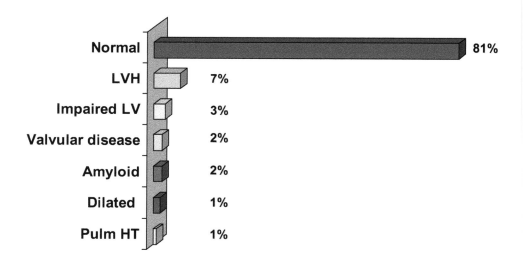

Figure 3. Predominant echocardiographic findings in 222 patients with AA amyloidosis

106 patients have needed dialysis and 24 have had renal transplants. Graft and patient survival at one year were 89% and 92% respectively, comparable with other causes of ESRF. Only one graft was lost due to recurrent amyloid. SAP scintigraphy was diagnostic at presentation in 98% of cases and was repeated annually, and treatment was focussed on reducing production of SAA guided by monthly serum measurements. Median survival by Kaplan-Meier analysis was 10 years for the whole group and 6.2 years for patients whose SAP scans showed progression. In contrast, 45/67 patients whose amyloid regressed are alive.

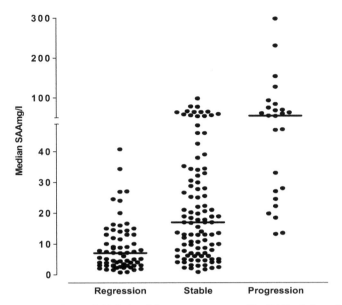

Change in amyloid during follow up as assessed by SAP scintigraphy

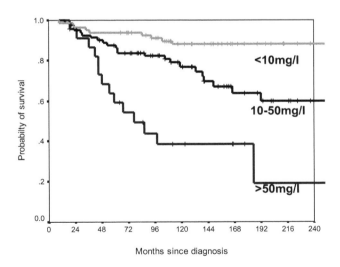

Months since diagnosis

Figure 4. Outcome in 195 patients with AA amyloidosis in relation to their underlying inflammatory actively as assessed by median SAA concentration.

The clinical picture in AA amyloidosis is determined predominantly by renal involvement, and long term outcome is excellent among patients whose SAA is suppressed sufficiently to facilitate regression of their amyloid deposits.

SAA1 GENE ANALYSIS IN FINNISH PATIENTS WITH AA AMYLOIDOSIS

C. Terai, H. Kaneko, M. Moriguchi, Y. Koseki, H. Kajiyama, N. Kamatani, S. Tiitinen, K. Kaarela, M. Hurme and C. P. J. Maury

Institute of Rheumatology,Tokyo Women´s Medical University,Tokyo,Japan and Rheumatism Foundation Hospital,Heinola, Department of Imunology,Tampere University,and Department of Medicine, University of Helsinki,Helsinki,Finland

Linkage between *SAA1* exon 3 polymorphisms and susceptibility to reactive amyloidosis in patients with rheumatoid arthritis (RA) has been the subject of several previous studies. The *SAA1.3* allele, a haplotype determined by the two SNPs 2995C and 3010C in the exon 3, has been shown to be associated with reactive amyloidosis in Japanese patients with RA (1),while the *SAA1.1* allele which has 2995T and 3010C, has shown a negative association with AA-amyloidosis risk in Japanese patients. A positive association between the *SAA1.1 allele* and AA-amyloidosis risk was reported in English patients (2). Discrepancy in the associations of the exon 3 polymorphisms with AA-amyloidosis risks among different ethnic groups suggests that the exon 3 polymorphisms are merely markers linked to another polymorphism more closely associated with AA-amyloidosis.

In the search of polymorphisms in the 5'-flanking region of *SAA1 gene*, we found 3 common SNPs, -61C/G, -13T/C, and –2G/A and demonstrated that –13T allele is more closely associated with AA-amyloidosis risk than *SAA1.3 allele* (3). Recently, -13T allele has shown to be associated with AA-amyloidosis in American Caucasian patients also (4).

The aim of this study is to examine whether –13T and/or *SAA1.3* haplotype of the *SAA1* gene is associated with AA amyloidosis risk in Finnish patients with RA.

PATIENTS AND METHODS

We studied 54 adult Finnish patients with seropositive RA and biopsy-proven amyloidosis (RA+A), 54 matched Finnish RA patients without amyloidosis (RA), and 100 Finnish blood donors (Co). For comparison of the haplotype estimation, 59 Japanese RA patients with amyloidosis and 58 Japanese controls were studied.

SNPs in the 5´flanking region (-61C/G, -13T/C, -2G/A) and exon 3 (2995T/C, 3010C/T) of the *SAA1* gene were determined by PCR-RFLP as described previously (3). Two sets of PCR primers were used to obtain specific amplification of the 5'-flanking region and a part of exon 1 of SAA1. PCR products amplified by second primer set were digested with either *Mnl* I, *Aci* I, or *Cac8* I restriction enzymes. Statistical analyses of genotype and allele frequency comparisons of the various SNPs were performed using the chi-square test. Haplotype frequencies of the 5 SNPs were estimated using LD support software program.

RESULTS

The frequencies of 5 SNPs of the *SAA1* gene in Finnish population are shown in Table1. The frequency of the – 13T allele was 0.02 in RA+A, 0.12 in RA, and 0.11 in Co. The frequency of the –61C allele was 0.88 in RA+A, 0.46 in RA, and 0.72 in Co. The two SNPs in the exon 3 showed difference between RA+A and RA or Co. Thus, the frequencies of 2995T and 3010Cin RA+A were significantly higher than those in RA and Co. The level of association between the genotypes involving the 5 SNPs sites and AA-amyloidosis was assessed using contingency tables (Table 1).   Among the 5 SNPs, 2995T SNP showed most strong association with the disease risk. While –13T showed significant negative association with amyloidosis risk, -61C showed more tight association with the disease risk than –13C.

The frequency of the *SAA1.3* allele in Finnish population was 0.03 in RA+A, 0.17 in RA and 0.11 in Co. The frequency of the *SAA1.1* allele was 0.87 in RA+A, 0.52 in RA and 0.64 in Co (P<0.0001 RA+A *vs.* RA and Co). The association with AA-amyloidosis of the *SAA1.1* allele that is determined by 2995T and 3010C is more significant than the 2995T alone. The Odds ratio and chi-square of SAA1.1 allele 6.24 and 31.51 in RA+A vs. RA and 3.86 and 19.17 in RA+A vs. Co, respectively.

Table 1. Comparison of the allele frequencies of 5 SNPs among amyloidosis, RA, and controls.

SNPs	RA+A n = 54	RA n = 54	Co n = 100	RA+A vs. RA Odds ratio	chi-square	RA+A vs. Co Odds ratio	chi-square
-61 at the SAA1							
C	0.88	0.65	0.72	3.97	16.04	2.84	10.28
G	0.12	0.35	0.28				
-13 at the SAA1							
C	0.98	0.88	0.895	7.25	8.67	6.22	7.59
T	0.02	0.12	0.105				
-2 at the SAA1							
G	0.96	0.91	0.925	2.65	2.75	2.11	1.75
A	0.04	0.09	0.075				
2995 at the SAA1							
T	0.87	0.54	0.68	5.79	28.78	3.16	13.44
C	0.13	0.46	0.32				
3010 at the SAA1							
C	0.90	0.69	0.735	3.88	13.81	3.18	11.34
T	0.10	0.31	0.265				

Table 2. Comparison of haplotype frequencies of 5 SNPs between Finnish and Japanese populations

SNP sites in the SAA1 gene						Haplotype frequency			
5'-flanking			Exon 3			Finnish		Japanese	
-61	-13	-2	2995	3010	(exon 3 haplotype)	AA+RA	Co	AA+R	Co
C	C	G	T	C	*SAA1.1*	0.852	0.605	0.102	0.297
G	C	G	C	T	*SAA1.5*	0.055	0.175	0.205	0.203
C	T	G	C	C	*SAA1.3*	0.009	0.105	0.568	0.381
G	C	A	C	T	*SAA1.5*	0.037	0.075	0.011	0.042
G	C	G	T	C	*SAA1.1*	0.019	0.025	0.000	0.000
C	C	G	C	T	*SAA1.5*	0.000	0.015	0.000	0.000

The EH software estimates haplotype frequencies of the 5 SNPs of the *SAA1* gene. Table 2 shows haplotype frequencies at the SAA1 locus between Finnish and Japanese. The haplotype CCGTC was the most common in RA+A (0.85) and Co (0.61), while that was the second common in Japanese. Among 6 haplotypes of the *SAA1* gene found in Finnish populations, the haplotype CCGTC alone associated with AA-amyloidosis risk and other haplotypes including the haplotype GCGTC having *SAA1.1* polymorphism did not associated with the disease risk.

DISCUSSION

AA-amyloidosis is one of the most serious complications of RA and is associated with severe and sustained RA inflammation. The incidence of AA-amyloidosis secondary to RA varies among different ethnic groups. In Japan, AA-amyloidosis was found in more than 20% of autopsied patients with RA and AA-amyloid deposition was confirmed in 13.3% of RA patients by gastrointestinal biopsy. In Finland, amyloidosis was detected in 5.8% of cases in a population-based series of RA subjects who had died (5). In the United States, on the other hand, it is estimated that less than 1-2% of RA patients develop AA-amyloidosis. It remains unknown whether these differences are genetic or results of environmental influences.

In this study, The *SAA1.1* and *SAA1*-61C alleles are associated with AA-amyloidosis in Finnish RA patients. Although the haplotype *SAA1.1* that defines protein sequence showed stronger association with the disease risk than the 2995T allele, this does not mean SAA 1.1 protein is critical for amyloidgenecity in Finnish population since the frequency of the haplotype GCGTC whose exon 3 polymorphism is *SAA1.1* was decreased in AA-amyloidosis.

Finnish AA-amyloidosis did not show association with −13T allele of the *SAA1* gene that showed significant association with Japanese and American Caucasian AA-amyloidosis. This indicates that −13T and −61C are a marker SNP that is in linkage disequillibrium with the true disease associated polymorphism. Another possible explanation for this is the genetic specialty in Finnish population who differs in many genetic aspects from other European populations. A distinct population structure and isolated history in Finns could produce a special polymorphism associated with AA amyloidosis there. Further study on *SAA1* gene is needed to obtain true association of this gene with AA-amyloidosis.

REFERENCES

1. Moriguchi, M. et al. Influence of genotypes at SAA1 and SAA2 loci on the development and the length of latent period of secondary AA-amyloidosis in patients with rheumatoid arthritis, *Human Genetics*, 105, 360, 1999.
2. Booth,D.R. et al. SAA1 alleles as risk factors in reactive systemic AA amyloidosis, *Amyloid*, 5, 262, 1998.
3. Moriguchi, M. et al. A novel single-nucleotide polymorphism at the 5'-flanking region of *SAA1* associated with risk of type AA amyloidosis secondary to rheumatoid arthritis, *Arthtitis Rheum.*, 44, 1266, 2001.
4. Yamada, T. et al. An allele of serum amyloid A1 associated with amyloidosis in both Japanese and Caucasians, *Amyloid*, 10, 7, 2003.
5. Myllykangasl-Luosujärvi, R. et al. Amyloidosis in a nationwide series of 1966 subjects with rheumatoid arthritis who died during 1989 in Finland, Rheumatology, 38, 499, 1999.

A PROSPECTIVE ANALYSIS OF DEMOGRAPHY, ETIOLOGY, AND CLINCIAL FINDINGS OF AA AMYLOIDOSIS PATIENTS ENROLLED IN THE INTERNATIONAL CLINICAL PHASE II/III FIBRILLEX ™ STUDY

W. Hauck[†], L.M. Dember[‡], P.N. Hawkins[§], B.P.C. Hazenberg[#], M. Skinner[‡], M.-C. Bouwmeester[†], R. Briand[†], E. Chicoine[†], C. Gurbindo[†], L. Hughes[†], D. Garceau[†], on behalf of the Fibrillex[TM] Amyloidosis Secondary Trial (FAST) Group[*].

[†]Neurochem Inc., Laval, Quebec, Canada; [‡]Boston University School of Medicine, Boston, U.S.A.; [§]Royal Free Hospital, London, U.K.; [#]University Hospital, Groningen, The Netherlands

AA amyloidosis is a multisystem disorder caused by the deposition of serum amyloid A protein as fibrillar aggregates that interfere with the structural integrity and function of targeted organs or tissues[1,2]. It occurs in association with chronic inflammatory disorders that give rise to an acute phase response, notably including rheumatoid arthritis, ankylosing spondylitis, inflammatory bowel disease, familial Mediterranean fever (FMF), and chronic infections (e.g. tuberculosis, osteomyelitis). No specific treatment is currently available for this life-threatening disease and its management has traditionally focused on control of the underlying chronic inflammatory disease. Fibrillex™ is a novel therapeutic agent under investigation for the treatment of AA amyloidosis. It is thought to act by interfering with the binding of glycosaminoglycans (GAGs) to the GAG binding site on amyloid A protein, thus preventing polymerization into aggregates and subsequent deposition of AA amyloid fibrils in organs[3]. This novel class of therapeutic agents has been named GAG mimetics.

PATIENTS AND METHODS

The Fibrillex™ Amyloidosis Secondary Trial (FAST) is a Phase II/III study that is currently being conducted in 27 sites in 13 countries to evaluate the efficacy and safety of Fibrillex™. Two hundred and sixty patients have been screened and 183 of them were randomized to treatment with placebo or Fibrillex™. The dose regimen was dependent on creatinine clearance (CrCl): CrCl > 80 mL/min (1200 mg bid), CrCl = 30-80 mL/min (800 mg bid), 20 ≤ CrCl < 30 mL/min (400 mg bid). Following the 24-month double-blind study, patients are offered enrollment in an open-label extension study in which all patients receive Fibrillex™ for an additional 2 years.

All participants are 18 years of age or over. The diagnosis of AA amyloidosis in patients was confirmed prior to their starting in the study by positive Congo red staining of biopsy material and immunohistochemistry or immunoelectron microscopy by a qualified pathologist. Participants were eligible for the study if they had either proteinuria ≥ 1 g/day or creatinine clearance ≤ 60 mL/min. Participants whose creatinine clearance was ≥ 20 mL/min were eligible only if they also had serum creatinine ≤ 3 mg/dL. In addition, participants taking cytotoxic agents, colchicine therapy or ACE inhibitors were required to be on stable treatment (no new initiation or dose changes) for 3 months prior to screening.

The primary outcome measure is a composite index of clinical improvement or worsening of both renal and GI functions. Renal function deterioration will be evaluated using a composite endpoint which includes one of the following: doubling of serum creatinine, 50% reduction in creatinine clearance, progression to nephrotic syndrome, progression to dialysis or end-stage renal disease (ESRD). GI function deterioration will be evaluated on the basis of progression or new-onset chronic diarrhea or a ≥ 10 % loss in body weight.

As of April 5, 2004, out of the total of 183 patients enrolled, all patients had completed the first 12 months of follow-up, 24 patients had completed 24 months and were enrolled in the on-going open-label extension study. Eighteen had progressed in their disease to end-stage renal failure and dialysis; 11had discontinued due to serious adverse events; 11 had discontinued for personal reasons. There were a total of 95 serious adverse events reported in 63 patients, but only 3 have been reported as related to study medication (placebo or Fibrillex™). Four data safety monitoring board meetings have reviewed blinded, interim data, and since there were no safety issues each time the board concluded with a unanimous recommendation to proceed with the study.

BASELINE PATIENT CHARACTERISTICS AND CLINICAL PROFILES

At baseline, 42% of patients were male. The median age was 51 years (range 21-77). The average weight was 65 kg (range 34-142), the average height was 163 cm (range 94-190) and the average body mass index was 24.

Overall, rheumatic diseases were the most common form of underlying illness, occurring in 64% of patients. FMF and infectious diseases accounted for 19% and 8% of cases, respectively (Table 1). This distribution was uniform by geographic region, except for Israel, Tunisia, and Turkey, where FMF was the most common underlying disease, accounting for 70% of cases (Table 2). The time between diagnosis of the underlying disease and enrollment in the study (biopsy confirmed AA amyloidosis) appeared to be shorter for patients with FMF and non-identifiable underlying disease. The significance of this difference has not yet been explored statistically (Table 1).

Table 1. Underlying disease among FAST study patients

Underlying disease	% of patients (n=183)	Time between diagnosis of underlying disease and study start (years)
Rheumatic diseases (rheumatoid arthritis, ankylosing spondylitis, juvenile chronic arthritis, psoriatic arthritis, lupus arthritis)	64%	14.3
Familial Mediterranean Fever	19%	8.5
Infectious diseases (osteomyelitis, pulmonary tuberculosis, bronchiectasis, recurrent pulmonary infection)	8%	17.4
Inflammatory bowel diseases (Crohn's disease)	2%	13.1
Other (goiter, non-identifiable)	7%	4.7

Source: Data on file, Neurochem Inc.

Table 2. Underlying disease among FAST study patients by region

	U.S.	Western E.U.	Eastern E.U.	Israel/Tunisia/ Turkey	Total
Underlying disease (# patients)	18	50	71	44	183
Rheumatic disease	9 (50%)	39 (78%)	62 (87%)	7 (16%)	117 (64%)
Familial Mediterranean Fever	1 (5.5%)	3 (6%)	0	31 (70%)	35 (19%)
Infectious diseases	1 (5.5%)	4 (8%)	7 (10%)	2 (5%)	14 (8%)
Inflammatory bowel diseases	3 (17%)	1 (2%)	0	0	4 (2%)
Other	4 (22%)	3 (6%)	2 (3%)	4 (9%)	13 (7%)

Source: Data on file, Neurochem Inc.

Patients' clinical profiles at baseline are summarized in Table 3. Importantly, 67% of patients had proteinuria ≥ 1 g/day, 43% had creatinine clearance ≤ 60 mL/min and 32% had nephrotic syndrome (defined as proteinuria >

3 d/day, albuminemia < 3.4 g/dL, and peripheral edema or the use of diuretics). Baseline serum amyloid A (SAA) levels were normal (<10 mg/L) in 33% of patients, elevated (10-50 mg/L) in 46% of patients, and high (50 mg/L) in 21% of patients.

Table 3. Clinical profile of FAST patients at baseline

	% of patients
Edema	57%
Proteinuria ≥1 g/day	67%
Creatinine clearance ≤60 mL/min	43%
Serum creatinine >2 mg/dL	14%
Nephrotic syndrome	32%
Symptomatic gastrointestinal AA amyloidosis	2%
Mean BP	130/78 mmHg
Antihypertensive therapy	46%
Orthostatic hypotension	7%
Mean C-reactive protein	1.81 ± 2.01 mg/L

SAA: <10 mg/L
            10-50 mg/L
            >50 mg/L
	33% of patients
	46% of patients
	21% of patients

Source: Data on file, Neurochem Inc.

CONCLUSIONS

Fibrillex™ is the first member of a new therapeutic class (GAG mimetics) being tested in the largest ever study in patients with AA amyloidosis. This international, multicenter trial provides valuable data that allows us to examine and compare the natural history of AA amyloidosis with respect to character and extent of organ system involvement in many countries. All patients had amyloid A fibril deposits affecting at least 2 organ functions despite having received treatment for their underlying inflammatory conditions. The majority of patients recruited (64% overall) had rheumatic diseases as the underlying disease, whereas those recruited in Israel, Tunisia, and Turkey (70%) had FMF as the major underlying chronic inflammatory condition. The clinical profiles of these confirmed AA amyloidosis patients is as follows: 67% of all patients presented with proteinuria ≥ 1 g/day, 43% of all patients had creatinine clearance ≤ 60 mL/min, and 32% of all patients already had developed nephrotic syndrome.

This Phase II/III trial is ongoing and remains blinded. Results of the study are expected to be available in 2005.

REFERENCES

1. Falk RH, Comenzo RL, Skinner M. (1997) The systemic amyloidoses. *N Engl J Med,* 337(13), 898-909
2. Cunnane G. (2001) Amyloid proteins in pathogenesis of AA amyloidosis. *Lancet,* 358, 24-25
3. Garceau D, Gurbindo C, Laurin J. (2001) Safety, tolerability and pharmacokinetic profile of Fibrillex™ (anti-AA amyloid agent) in healthy and renal impaired subjects. In: Bely M and Apathy A. Amyloid and Amyloidosis, Proceedings of the IXth International Symposium on Amyloidosis, July 15-21, 2001, pp.116-118.

ACKNOWLEDGEMENTS

FAST Group: Finland: Dr. K. Kaarela (Heinola); France: Dr. G. Grateau (Paris), Dr. E. Hachulla (Lille), Dr. X. Puéchal (Le Mans); Israel: Dr. A. Livneh (Tel-Hashomer), Dr. I. Rosner (Haifa); Italy: Dr. G. Merlini (Pavia); Lithuania: Dr. I. Butrimiene (Vilnius); Netherlands: Dr. B.P.C. Hazenberg (Groningen); Poland: Dr. A. Filipowicz-Sosnowska (Warsaw), Dr. P. Wiland (Wroclaw); Russia: Dr. O. Lesnyak (Yekaterinburg), Dr. E.I. Nasonov (Moscow); Spain: Dr. J.A. Jover (Madrid), Dr. J. Munoz Gomez (Barcelona), Dr. X. Tena Marsa (Badalona), Dr. J. Valverde Garcia (Llobregat); Tunisia: Dr. H. Ben Maïz (Tunis); Turkey: Dr. H. Direskeneli (Istanbul), Dr. A. Gul (Istanbul), Dr. H. Ozdogan (Istanbul); U.K.: Dr. P.N. Hawkins (London), Dr. J. A. Hunter (Glasgow); U.S.: Dr. M. D. Benson (Indianapolis), Dr. L. M. Dember (Boston), Dr. A. Dispenzieri (Rochester), Dr. P. D. Gorevic (NY)

This study was supported in part by FDA Orphan Products Development (OPD) grant #2007-01.

TNFRSF1A MUTATIONS IN ITALIAN PATIENTS AFFECTED BY APPARENTLY SPORADIC PERIODIC FEVER SYNDROME

L. Obici[1], S. Marciano[1], G. Palladini[1], F. Lavatelli[1], S. Donadei[1], A. Cigni [2], A.E. Satta[2], L. Praderio[3], M. Tresoldi[3], G. Merlini[1].

Biotechnology Research Laboratory[1], IRCCS Policlinico San Matteo, Pavia; Institute of Medical Pathology[2], University of Sassari, Sassari; Division of Internal Medicine[3], IRCCS Ospedale San Raffaele, Milano, Italy.

Hereditary periodic fevers are autoinflammatory disorders characterized by recurrent febrile episodes with serositis, synovitis and cutaneous inflammation, potentially leading to development of AA amyloidosis (1).

Genetic testing has greatly improved their diagnosis and treatment, particularly in patients without typical ethnic background. Familial Mediterranean fever (FMF), caused by mutations in *MEFV*, is the more prevalent and best-characterised hereditary periodic fever in Italy. However, occurrence of other hereditary fever syndromes has not been assessed. The Tumor Necrosis Factor (TNF) receptor–associated periodic syndrome (TRAPS) is a genetically distinct disorder mainly affecting people of Northern European origin. This dominantly inherited syndrome is caused by mutations in the *TNFRSF1A* gene, which codifies for the p55 subunit of the TNF receptor. Typical features include localized myalgia, periorbital edema, conjunctivitis and long duration of attacks, usually lasting more than one week. To date, 24 different mutations have been identified, all clustering in the first and second extracellular cysteine rich domains (CDRs) of this receptor (2).   We searched for mutations in *MEFV* and *TNFRSF1A* in 48 Italian patients with unexplained, recurrent fever not fulfilling the diagnostic criteria for familial Mediterranean fever. None of these patients had a significant family history for the disease.

PATIENTS AND METHODS

We studied 48 patients (age 5-61 years) referred to our Centre over a 3 year-period for unexplained, periodic fever consistent with TRAPS and/or AA amyloidosis. All patients had no significant family history for the disease. Clinical features at presentation included long-lasting attacks (>1 week) (96%), serositis (75%), arthralgias (42%), lymphadenopathy (25%), presence of skin rash (21%), prolonged myalgia (21%), conjunctivitis (8%) and AA amyloidosis (6%).

Genomic DNA was extracted from peripheral blood according to standard procedures. Mutation analysis of *TNFRSF1A* and *MEFV* was performed by direct sequencing. Restriction endonuclease assays for the Cys30Thr, Cys43Arg and the Arg92Gln mutations were developed in order to screen patient's relatives and 200 control chromosomes. Paternity testing was performed in three patients carrying a *de novo* mutation.

Soluble plasma levels of TNFRSF1A (p55) were measured by solid-phase ELISA (R&D Systems).

RESULTS

A *TNFRSF1A* mutation was identified in eight patients (17%) affected by apparently sporadic periodic fever syndrome. Four different variants were found, namely Cys30Tyr, Cys43Arg, Thr50Met and Arg92Gln, the first two being novel (Figure 1). These new substitutions abrogate a disulfide bond in the first extracellular cysteine-rich subdomain of *TNFRSF1A*. Screening of 200 Italian control chromosomes did not reveal any carrier for these variants, further supporting their pathogenic role.

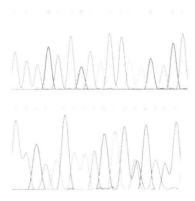

Figure 1. Sequence analysis of exon 3 of the *TNFRSF1A* gene showing a G→A transition at nucleotide 176 resulting in a Cys to Tyr substitution at codon 30 (upper case) and a T→C transition at nucleotide 214, resulting in a Cys to Arg substitution at codon 43 of the polypeptide chain (lower case).

Three patients presented a *de novo* mutation, also confirmed by paternity testing. Five patients carried the known low-penetrance Arg92Gln mutation. Detection of the Arg92Gln substitution in their asymptomatic relatives confirmed that this is a low-penetrance mutation Three of these patients were also heterozygous for a *MEFV* variant (Table 1). The Arg92Gln variant was found in 5/200 control chromosomes. This frequency is higher than that previously reported in other populations. Blood levels of soluble TNFRSF1A were below 1 ng/mL in patients carrying the Cys30Thr, Cys43Arg and Thr50Met variants but were normal in individuals with the Arg92Gln mutation.

Table 1. Genotype-phenotype relationship in five patients carrying the Arg92Gln mutation.

Clinical features	#1	#2	#3	#4	#5
Genotype	Arg92Gln *MEFV* Glu148Gln	Arg92Gln *MEFV* Val726Ala	Arg92Gln	Arg92Gln	Arg92Gln *MEFV* Glu148Gln
Age at onset	6 months	12 years	50 years	23 years	40 years
Fever	1-2 weeks	7-10 days	> 2 weeks	> 3 weeks	absent
Inflammatory attacks	abdominal pain	abdominal pain arthralgias	arthralgias	abdominal pain, pleurisy	absent
Skin rash	present	absent	absent	absent	absent
Lymphadenopathy	present	absent	absent	absent	absent
sTNFRSF1A (pg/mL) (range 1094-1972)	1250	1510	1200	1460	1170
Amyloidosis	absent	absent	absent	absent	present

DISCUSSION

Diagnosis and treatment of hereditary periodic fever syndromes presently rely on genetic testing. Analysis of *TNFRSF1A* gene allowed to establish a proper diagnosis in 8 Italian patients with sporadic periodic fever and inflammatory attacks without significant family history for the disease.

 Occurrence of *de novo* mutations and recurrence of low-penetrance variants may account for apparently sporadic hereditary periodic fever syndromes in rarely affected populations. Moreover, the high frequency of the Arg92Gln variant in the Italian population suggests that this substitution might variably modulate the disease phenotype when associated with one *MEFV* mutation. We recommend genetic testing even in patients without a typical family history and/or ethnic background to avoid misdiagnosis and to prevent amyloid-related renal complications.

REFERENCES

1.   Drenth JPH, van der Meer JWM. Hereditary periodic fever. N Engl J Med 2001; 345:1748-1757

2.   Aganna E, Hammond L, Hawkins P, et al. Heterogeneity among patients with Tumor Necrosis Factor receptor-associated periodic syndrome phenotypes. Arthritis and Rheumatism 2003; 48:2632-2644.

AA-AMYLOIDOGENESIS REPLICATED IN CELL CULTURE: A SERUM AMYLOID A PEPTIDE INHIBITOR PROVIDES DIRECT EVIDENCE FOR THE ROLE OF HEPARAN SULFATE

E. Elimova[1], R. Kisilevsky[1, 2], W. A. Szarek[3] and J. B. Ancsin[2]

Departments of Biochemistry[1], Pathology and Molecular Medicine[2] and Chemistry[3], Queen's University, Kingston, Ontario Canada K7L 3N6

INTRODUCTION

Investigations into the mechanism of amyloidogenesis have been performed either by using *in vitro* systems to study fibril assembly of isolated and synthetic amyloidogenic polypeptides, or by examination of different amyloids *in situ* (1,2). *In vitro* studies frequently employ non-physiological conditions for fibrillogenesis and fail to study biological factors promoting amyloidogenesis, while analysis of *in vivo* systems are hampered by a lack of control over a large number of potential variables. A cell culture model would provide a bridge between these two systems and in this report we describe a monocytic cell culture system that appears to recapitulate all the main features of AA-amyloidogenesis. It has also allowed us to demonstrate that a synthetic peptide (27-mer), corresponding to the heparan sulfate (HS) binding site of Serum amyloid A 1.1 (SAA1.1) (3) is a potent inhibitor of amyloidogenesis providing direct evidence for the role of HS in the disease process. The anti-amyloid activity of this 27-mer suggests that anti-amyloid strategies based on the HS binding sequences for other amyloid polypeptides such as Aβ or IAPP may be highly successful.

MATERIALS AND METHODS

The murine monocytic cell line J774A.1 (American Type Culture Collection, Manassas, VA) was cultured in RPMI (Sigma) media which contained 25 mM HEPES, 15% fetal bovine serum (FBS) and 50 µg/ml penicillin-streptomycin, at 37°C, 5% CO_2, essentially as described in (4). The cell stocks were passaged every four days, and the media replaced every other day. Cells were seeded at a minimal density in 8-well chamber slides (Lab-Tek®, Nalge Nunc International, Naperville, IL) in 350µl medium/well and allowed to reach about 2.2×10^5 cells per well. To initiate AA-amyloidogenesis cells were treated with 10µg of AEF in the culture media. After 24h the media was removed and replaced with fresh media containing either 0.3 mg/ml HDL-SAA, HDL, 0.05 mg/ml SAA1.1 or SAA2.1, which was replenished every two days for 7 days. The effect of low molecular weight-heparin (LMW-hep), polyvinyl sulfonate (PVS) and SAA1.1 peptide, residues 77-103 (27-mer) on amyloidogenesis was evaluated in cell culture by co-incubation at different concentrations with HDL-SAA. At the end of the treatment period the amyloid was assayed by thioflavin-T (Th-T) fluorescence (5). Plasma HDL-SAA concentrations were experimentally elevated in CD1 mice (Charles Rivers, Montreal, Quebec, Canada) by a subcutaneous injection of 0.5 ml of 2% (w/v) $AgNO_3$ producing a sterile abscess and isolated by sequential density flotation as described previously (3). SAA1.1 and 2.1 were isolated from HDL-SAA denatured with 6 M guanidine-HCl and purified by reversed phase-high performance liquid chromatography on a semi-preparative C-18 Vydac column connected to a Waters (Millipore) HPLC system.

RESULTS and DISCUSSION

At the end of the AA-amyloid induction protocol significant quantities of amyloid were detected by Congo red (CR) staining (data not shown) and Th-T fluorescence (Fig.1). Mice like humans express two major isoforms of SAA (SAA1.1 and SAA2.1) during an acute-phase response but for mice only SAA1.1 can deposit as AA-amyloid (6). To determine if this was also the case in cell culture, J774 cells were incubated with AEF and either, purified SAA1.1 or SAA2.1 at 50 μg/ml (equivalent to their respective concentrations in plasma HDL-SAA), native HDL-SAA or HDL reconstituted with SAA1.1 or SAA2.1. Comparison of the purified SAAs showed that only SAA1.1 produced amyloid but in much reduced quantity (9% of HDL-SAA), while no amyloid was detected with SAA2.1. When purified SAA1.1 or SAA2.1 were first reassociated with HDL, the resulting amyloid-load with the reconstituted HDL-SAA1.1 was close to the amount observed for native HDL-SAA. Reconstituted HDL-SAA2.1 produced no amyloid. These data suggested that the pre-amyloidogenic conformation of SAA is critical for refolding and assembly into fibrils and a lipid microenvironment was likely necessary for this process.

Figure 1. Determination of SAA isoform preference and the effect of SAA delipidation on AA-amyloidogenesis. The conversion of delipidated SAA1.1, SAA2.1 and reconstituted HDL-SAA1.1 and HDL-SAA2.1 relative to native HDL-SAA (100%) was analyzed by Th-T fluorescence.

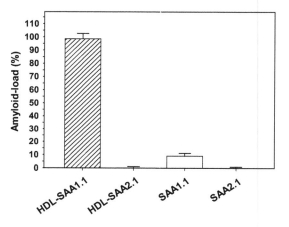

To examine whether HS is involved in the generation of the cell culture amyloid we tested the ability of LMW-hep, PVS and a synthetic peptide (27-mer) corresponding to the HS-binding site of SAA1.1 to inhibit amyloidogenesis (Fig. 2). Clinically relevant doses of LMW-hep administered to mice undergoing AA-amyloidosis has been reported to reduce tissue amyloid-loads (7) and we also found a similar effect in cell culture. PVS, a low molecular weight anionic polymer containing structural features of HS could achieve 50% and 100% inhibition at 0.5 μM and 9 μM, respectively. The anti-amyloid property of PVS has been demonstrated *in vivo* (8). We have previously identified a heparin/HS binding site on the C-terminal end of SAA1.1 (77-ADQEANRHGRSGKDPNYYRPPGLPAKY-103, 27-mer) (3) and have postulated that an interaction between this sequence and HS promotes SAA1.1 fibrillogenesis. To test this hypothesis, we attempted to block SAA1.1:HS interactions during amyloidogenesis by incubating the J774 cells with increasing concentrations of a synthetic 27-mer. Quantitation of the amyloid-loads at the end of the induction protocol revealed that the 27-mer was a profound inhibitor of amyloidogenesis, with an IC_{50} of 0.02 μM, which is 25-fold lower than for PVS. Also this effect was sequence specific. Scrambling the 27-mer's sequence (PLPAQGKPGPDHYARNDSYAKNRYERG), which destroys its HS binding activity (3) also caused a complete lost of inhibitory activity. Our data clearly demonstrates that SAA1.1:HS interactions are fundamental to AA-amyloidogenesis and suggests that peptides corresponding to other amyloid-precursor

HS binding sites may be a highly successful strategy in the prevention of these devastating amyloid-based diseases.

Figure 2. Natural and synthetic polymers inhibit amyloidogenesis in cell culture. Cells undergoing AA-amyloidogenesis were incubated with increasing concentrations of either low molecular weight heparin (LMW-hep, 3000 kD), polyvinylsulfonate (PVS), a synthetic peptide corresponding to the HS binding site of SAA1.1 (27-mer) or a 27-mer with its sequence randomized. At the end the induction protocol amyloid-loads were assayed by Th-T fluorescence.

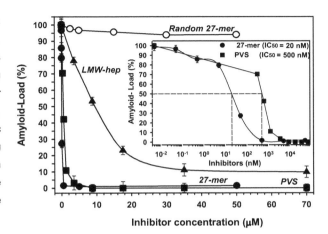

REFERENCES

1) Sipe J.D. (1994) Amyloidosis. *Crit. Rev. Clin. Lab Sci.* **31**, 325-354

2) Sipe J.D. and Cohen A.S. (2000) Review: History of the Amyloid Fibril. *J. Struct. Biol.* **130**, 88-98.

3) Ancsin J.B. and Kisilevsky R. (1999) *J. Biol. Chem.* **274**, 7172-7181.

4) Kluve-Beckerman B., Liepnieks J.J., Wang L. and Benson M.D. (1999) *Am. J. Pathol.* **155**, 123-133

5) LeVine H., III (1999) *Methods Enzymol.* **309**, 274-284

6) Meek R.L., Hoffman J.S. and Benditt E.P. (1986) *J. Exp. Med.* **163**, 499-510

7) Zhu H., Yu J. and Kindy M.S. (2001) *Mol. Med.***7**, 517-522

8) Kisilevsky R., Lemieux L.J., Fraser P.E., Kong X., Hultin P.G. and Szarek W.A. (1995) *Nat. Med.* **1**, 143-148

AN EM AUTORADIOGRAPHY AND IMMUNOFLUORESCENCE STUDY EXAMINING THE PATHWAY OF SERUM AMYLOID A THROUGH THE MACROPHAGE

S. Kinkley and R. Kisilevsky

Department of Pathology and Molecular Medicine, Queen's University, Kingston, Ontario Canada.

INTRODUCTION

Serum amyloid A (SAA) is an evolutionarily conserved acute phase protein whose plasma concentration increases 1000-fold during inflammation (1,2). The major biological function of AP-SAA still remains to be elucidated though there is growing evidence that SAA is involved in cholesterol transport (3). In this study we investigated the pathway that HDL/SAA followed through murine peritoneal macrophages to elucidate a putative role for SAA during inflammation. Using both immunofluorescence and EM autoradiography we showed that HLD/SAA was taken up rapidly by macrophages, proceeded through the cytoplasm to the peri-nuclear region, was then observed in the nucleus, later re-emerged to the cytoplasm and the plasma membrane where it was exocytosed into the extracellular environment. Through both of these techniques we were able to demonstrated that SAA has a finite association with macrophages that last approximately 1 hour, which is consistent with the half-life of SAA in the circulation. The observation that SAA is in the nucleus is to the best of our knowledge a novel finding and has implications that SAA may play a role in gene regulation.

MATERIALS AND METHODS

AP-SAA was collected and prepared as described in the literature (4). The HDL fraction of the plasma was then further purified by column chromatography using a Sephacryl S-300 column.

Peritoneal macrophages were obtained from inflamed and non-inflamed mice post CO_2 narcosis. Five ml of RPMI 1640 medium was then injected into the abdomen and massaged for approximately 1 min. The medium was retracted into the syringe, then passed over a Nitex 110 filter and centrifuged at 1,500 rpm for 5 minutes in a JA-17 rotor. The cells were then re-suspended and washed 3 more times in 1 ml of RPMI medium with 2% BSA and 5 mM HEPES (pH 7.2-7.4). A cell count was then determined using a hemocytometer.

Column purified AP-SAA was labelled by reductive methylation as described previously (5). The methylated SAA* reaction mixture was then dialysed against 3 changes of 6 M guanidine HCl at 4°C. An equivalent weight of HDL (1 mg) was added to the labelled SAA* and dialysed for 6 hours against 3, 1000 ml changes of TBS at 4°C. The dialysate was then passed over a pre-equilibrated (110 cm X 1 cm) Sephacryl S-300HR column in TBS at a rate of 10 ml/hour to separate the reconstituted HDL/SAA* from any un-associated SAA*. The radioactivity of the HDL/SAA* peak was then determined and the protein identity confirmed by PAGE.

The route of HDL/SAA taken in macrophages was determined by incubating the cells with HDL/SAA* for 0, 5, 10, 15, 30, 45 and 60 minutes. The cells in medium were then microcentrifuged at 8000 rpm, the medium aspirated and the cells fixed in 2% paraformaldehyde and 0.5% glutaraldehyde (pH 7.4) for 24 h before being embedded in Epon (Jembed 812 resin, Canemco Inc.). Confirmation of the route taken by the SAA was done by immunofluorescence. These experiments were repeated >15 times and were performed in culture as described above. Approximately 10,000 cells were plated on 5 mm coverslips (Fisher Scientific) and 15 ug of HDL/SAA, was added to each coverslip. The incubations were stopped by fixing the cells in 4% paraformaldehyde (pH 7.4) for 10 min. The cells were then pemeabilized in 10% Triton X 100 for 10 min. washed 3X in 1X PBS for 10 min. and blocked with 3% BSA at room temperature for 1 h. To detect the location of SAA within the macrophages 1 µl of affinity purified rabbit anti-mouse SAA primary antibody (1.7 mg/ml) was diluted 2000X in 3% BSA and added to the surface of each coverslip for 30 min. The cells were washed again 3X in 1X PBS for 10 min. A secondary antibody (goat anti-rabbit) conjugated with Alexa 488 (Molecular Probes) was diluted in a 1:750 ratio with 3% BSA and Phalloidon conjugated to TRITC (Sigma) was added to the mixture in a 1:400 dilution for 30 min. Following washes in PBS the excess wash was removed from the coverslips and 7 µl of Mowiol mounting solution (Calbiochem) supplemented with 0.6% Dabco reagent (Sigma) and 2 mg/ml of DAPI (Sigma) in a 1 to 50 dilution was added to a clean microscope slide for each coverslip. The cells and the location of SAA were then visualized at 40X and 63X objective on a deconvoluting epi-fluorescent light microscope (Leica Axiovert S-100) or at 100X on a Leica TCS SP2 multi-photon confocal microscope.

RESULTS

Both the EM autoradiography and Immunofluorescence experiments demonstrated similar and consistent results. These showed that SAA was taken up very rapidly by macrophages and that its association with these cells lasted approximately 60 min (Figure 1C). At 15 and 30 min SAA was observed either associated with the nuclear membrane or localized in the nucleus (Figure 1A and B).

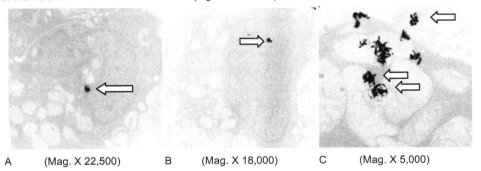

A   (Mag. X 22,500)   B   (Mag. X 18,000)   C   (Mag. X 5,000)

Figure 1- EM autoradiographs of cholesterol-loaded macrophages showing the location of SAA* at various time points after its uptake. A. Localization of SAA* to the nuclear membrane after 15 min. B. Localization of SAA* in the nucleus after 30 minutes. C. Localization of SAA* to the plasma membrane and in excretory vesicles after 60 minutes.

A statistical analysis of the grain-counts at the different time intervals indicated that there was a progressive significant shift in grains with time from the plasma membrane at t = 0 min, cell cytoplasm and perinuclear zone at t = 15 min, nucleus at t = 30 min, and excretory vacuoles and extracellular medium at t = 45-60 min.

To confirm the nuclear localization of SAA, we employed the immunofluorescence techniques described above. These cells were examined by confocal microscopy and a 3-D analysis was performed on these images using the Leica light confocal software. By looking at the position of the cross-hairs on the projected 2-D images the SAA was found to be localized within the nucleus rather than just associated with the nuclear membrane (color images not shown).

CONCLUSIONS

Through the use of both EM autoradiography and immunofluorescence techniques, we have been able to show the pathway of SAA through murine macrophages. SAA is taken up very rapidly by macrophages and it has a finite association with the macrophage lasting approximately 60 min before it is exocytosed back into the extracellular environment. We have also shown that between 15 and 30 minutes after its uptake SAA is localized to the nucleus, which is strongly suggestive that it may have a role in the regulation of certain macrophage genes during inflammation. This finding is to the best of our knowledge novel and it remains to be determined what genes these may be. Evidence for this possible role does exist in the literature. Elliot-Bryant et al., reported that the presence of SAA (50 or 100 μg/ml) resulted in increased proteoglycan synthesis and secretion by macrophages (6). She found that macrophages from azocasein treated mice responded to SAA with a 2-2.5 fold increase in cell associated dermatan sulfate and chondroitin sulfate and a disproportionate 6-fold increase in heparan sulfate compared to the normal macrophages (untreated group). More recently, Hatanaka et al., reported that SAA stimulated rapid expression and release of TNF-α and IL-8 mRNA in human neutrophils and monocytes (7). Each of these proteins is involved in inflammatory responses and is consistent with the idea that one of SAA's biological functions is related to the process gene regulation in inflammatory cells. These papers support the concept that SAA may be acting at the level of transcription.

REFERENCES

1.   Lowell, C. A., Stearman, R.S., and Morrow, J.F. Transcriptional regulations of serum amyloid A gene expression. *J. Biol. Chem.* 261,8453-61, 1986.

2.   Sipe, J.D., Rokita, H., and de Beer, F.C. Cytokine regulation of the mouse SAA gene family. 1993. Mackiewicz A, Kushner I, Baumann H eds. Acute Phase Proteins: Molecular Biology, Biochemistry, and Clinical Applications. 511-526 FL, CRC Press, Boca Raton.

3.   Tam, S.P., Flexman, A., Hulme, J. and Kisilevsky R. Promoting export of macrophage cholesterol: the physiological role of a major acute-phase protein, serum amyloid A 2.1. *J. Lip. Res.* 43, 410-20, 2002.

4.   Ancsin, J. B. and Kisilevsky, R. The heperan/heparan sulfate binding site on apo- serum amyloid A: implications for the therapeutic intervention of amyloidosis.. *J. Biol. Chem.* 274, 7172-81, 1999.

5.   Kisilevsky, R., Weiler, L., and Treloar, M. A. An analysis of alterations in ribosomal conformation using reductive methylation. *J.Biol.Chem.* 253, 7101-7108, 1978.

6.   Elliott-Bryant, R., Silbert, J.E., and Sugumaran G. Serum amyloid A, an acute-phase protein, modulates proteoglycan synthesis in cultured murine peritoneal macrophages. *Biochem. Biophys. Res. Comm.* 261, 298-301, 1999.

7.   Hatanaka, E., Furlaneto, C.J., Ribeiro, F. P., Souza, G.M., and Campa, A. Serum amyloid A-induced mRNA expression and release of tumor necrosis factor-alpha (TNF-α) in human neutrophils. *Immunol. Lett.* 91, 33-37, 2004.

CATHEPSIN B GENERATES NINE DIFFERENT AA AMYLOID PROTEINS BY ITS CARBOXYPEPTIDASE ACITVITY

C. Röcken[1], R. Menard[2], F. Bühling[1], J. Raynes[3], T. Kähne[1]

[1]Otto-von-Guericke-University, Magdeburg, Germany, [2]Biotechnology Research Institute, NRCC, Montreal, Canada, [3]Immunology Unit, Department of Infectious and Tropical Diseases, London School of Hygiene and Tropical Medicine, London, England

INTRODUCTION

AA amyloidosis develops in patients with chronic inflammatory diseases. The AA amyloid protein is derived from serum amyloid A (SAA), an acute phase protein synthesized and mainly secreted by the liver, and lacks one or two amino acids at the N-terminus, and between 15 and 83 amino acids at the C-terminus. Only a small proportion of the amyloid deposits enclose intact SAA. The majority of the amyloid proteins deposited in AA amyloid have undergone proteolysis. Patients with rheumatoid arthritis (RA) and AA amyloidosis show significantly increased amounts of proteolytic fragments of SAA in their serum compared to RA patients without amyloid [5]. Monocytes from healthy individuals completely degrade SAA, while monocytes from patients with amyloidosis produce a fragment similar to AA amyloid proteins. Similarly, activation of monocytes by interleukin-1β or interferon-γ hampers proteolysis *in vitro* and leads to partial degradation of SAA [6]. All these observations strongly suggest that AA amyloidosis is associated with or caused by defective SAA degradation in monocytic cells. As yet, little is known about the putative proteases and compartments, which may contribute to the formation of AA amyloid proteins. Cleavage of SAA to AA amyloid protein may occur intracellularly in macrophages [4], probably in an endosomal or lysosomal compartment. Macrophages are commonly found adjacent to amyloid deposits [9] and are able to bind and internalize SAA, forming amyloid deposits *in vitro*. They synthesize a broad range of proteases, which may process the precursor protein, and may also be involved in the degradation of amyloid deposits [3, 9]. The endosomal and lysosomal compartments are particularly rich in cysteine proteases, such as cathepsin (Cath) B. In the present study, we investigated the proteolytic action of recombinant CathB on SAA and AA amyloid proteins.

MATERIALS AND METHDOS

Recombinant serum amyloid A (rSAA) was obtained from Peprotech (London, Great Britain) and E64 from Bachem (Heidelberg, Germany). Recombinant CathB was generated as described elsewhere [1].

Unfixed splenic tissue containing amyloid was available from a patient with generalized AA amyloidosis. Using approximately 6 g of amyloidotic tissue, amyloid proteins were prepared by the water wash procedure [9]. A pellet was obtained with a whitish top layer, which contained amyloid proteins. This was stored at −20°C until further use. Samples from the pellet were de-lipidated and separated by gel filtration with sepharose CL-6B (Sigma-Aldrich, Deisenhofen, Germany). Pooled fractions of the elution peak were dialyzed against distilled water, concentrated and dissolved in guanidine HCl. The AA amyloid proteins were purified using high pressure liquid chromatography (HPLC) and a Jupiter 5u C5 300A column (250 x 4.60 mm; Phenomenex, Aschaffenburg, Germany). Peaks containing purified AA amyloid proteins were identified by SDS-PAGE, Western blotting, and mass spectrometry.

In vitro degradation experiments with recombinant human CathB were performed as follows: rSAA or AA amyloid proteins were dissolved in water to a concentration of 4 mg/ml and mixed in a ratio of 1:1 with reaction buffer (200 mM sodium acetate, 10 mM EDTA, and 5 mM DTT). Acetic acid was used to adjust to pH 5.2. Degradation was performed at 37°C and was started by the addition of CathB (0.1, 0.4 and 1.7 μM). The reaction was stopped by the addition of the cysteine protease inhibitor E64 (1.6 mM). Omission of cathepsin and incubation in the presence of 200 μM E64 served as a control.

Cleavage sites were determined using a Matrix-Assisted-Laser-Desorption/Ionisation-Time-Of-Flight-Mass-Spectrometer (MALDI-TOF-MS). Human SAA or AA amyloid proteins were incubated with either CathB as described above, and 0.5 μl aliquots were removed at different time points. The samples were subsequently co-crystallized with 0.5 μl sinapinic acid (20 mg/ml) in 70% acetonitrile on a SCOUT 384 -MALDI-Target. The mass spectrometry was performed on a MALDI-TOF-MS (Reflex III, Bruker Daltonics, Germany) in linear mode with internal calibration. The BioTools 2.0 software (Bruker Daltonics, Germany) was used for the annotation of the SAA fragments; the accepted mass tolerance was 100 ppm.

RESULTS

Degradation of rSAA by CathB

The first set of experiments aimed to investigate whether CathB is able to degrade SAA and whether the generated fragments resemble AA amyloid proteins found *in vivo*. The experimental conditions at which re-combinant human CathB cleaves SAA were optimised in preliminary experiments. Several protease concen-trations were chosen, at which rSAA (0.5 mg/ml) is cleaved stepwise after addition of active protease. Degra-dation was assessed by mass spectrometry.

CathB degrades rSAA completely at a concentration of 1.7 μM and less efficiently at concentrations below 0.4 μM. However, the cleavage products were identical irrespective of the concentration of CathB. Analysis of the degradation profile over time shows that CathB has mainly carboxypeptidase, some endoproteolytic and minor aminopeptidase activity. The cleavage sites between residues $Gly^{31}–Ser^{32}$, and $Val^{52}–Trp^{53}$ were found by mass spectrometry at low protease concentrations and no intermediates were detectable, suggesting endoproteolytic activity at these sites. Using different concentrations of CathB, we were able to identify further cleavage sites. Fragments ending at residues Gly^{99}, Arg^{87}, Gly^{86}, Trp^{85}, Asn^{83}, Ala^{82}, Ala^{78}, Ser^{76}, Glu^{74}, Phe^{69}, Arg^{67}, Gly^{50}, Lys^{46}, Tyr^{42}, Ala^{27}, Glu^{26}, Arg^{25}, Ser^{22}, and Ala^{20} are probably related to carboxypeptidase activity, while the fragment spanning residues Phe^3 to Arg^{47} may be generated from the Phe^3 to Val^{52} fragment by carboxypeptidase acitivity or may be an alternative endoproteolytic cleavage product. No degradation was observed in the absence of active protease or in the presence of E64.

Degradation of human AA amyloid proteins by CathB

We next investigated degradation of AA amyloid proteins obtained from a patient who had suffered from systemic AA amyloidosis. Following the extraction of AA amyloid proteins by the water wash procedure, gel filtration and purification by HPLC, we finally obtained a protein mixture of five AA amyloid proteins spanning residues Ser^2 to Asn^{64}, Ser^2 to Ile^{65}, Ser^2 to Gln^{66}, Ser^2 to Arg^{67}, and Phe^3 to Phe^{69}. CathB was able to degrade completely these AA amyloid proteins. Degradation by CathB showed primarily carboxypeptidase activity. No degradation was observed in the absence of active protease or in the presence of E64.

DISCUSSION

Despite the identification of numerous SAA-fragments created by a range of proteases, as yet no one type of protease has been identified, which has the capability to generate the diverse AA amyloid proteins found in human tissue. All these human AA amyloid proteins often differ from each other by only one or two amino acids. In further support of this notion, we have characterized AA amyloid proteins in two patients using mass spectrometry [8]. Previously, we found eight different AA amyloid proteins (Ser^2 to Tyr^{21}, Ser^2 to Asn^{64}, Ser^2 to Ile^{65}, Ser^2 to Gln^{66}, Ser^2 to Arg^{67}, Ser^2 to Ser^{76}, and Ser^2 to Gly^{86}) [9] and in this study five different AA amyloid proteins (Ser^2 to Asn^{64}, Ser^2 to Ile^{65}, Ser^2 to Gln^{66}, Ser^2 to Arg^{67}, and Phe^3 to Phe^{69}). In both cases, the AA amyloid proteins often differed only by a single amino acid. Following incubation of rSAA with CathB *in vitro*, we were able to create fragments of which 18 shared similarities with human AA amyloid proteins. In keeping with CathB's predominant carboxypeptidase activity, the cleavage products again differed only by one or two amino acids. As expected, CathB also showed weak endoproteolytic activity [2]. CathB could be a putative candidate responsible for the generation of most of the differently sized, C-terminally truncated AA amyloid proteins found in human AA amyloid deposits, as all these observations strongly suggest that AA amyloid proteins might be created by a carboxypeptidase with relatively low sequence specificity, such as CathB.

REFERENCES

1.   Bohne, S., et al. Cleavage of AL amyloid proteins by Cathepsin B, K and L. *J. Pathol.*, 203, 528-537, 2004.

2.   Chapman, H.A., Riese, R.J., and Shi, G.P. Emerging roles for cysteine proteases in human biology. *Annu. Rev. Physiol.*, 59, 63-88, 1997.

3.   Durie, B.G.M., et al. Amyloid production in human myeloma stem-cell culture, with morphologic evidence of amyloid secretion by associated macrophages. *New. Engl. J. Med.*, 307, 1689-1692, 1982.

4.   Kluve-Beckerman, B., Manaloor, J.J., and Liepnieks, J.J. A pulse-chase study tracking the conversion of macrophage-endocytosed serum amyloid A into extracellular amyloid. *Arthritis. Rheum.*, 46, 1905-1913, 2002.

5.   Migita, K., et al. Increased circulating serum amyloid A protein derivatives in rheumatoid arthritis patients with secondary amyloidosis. *Lab. Invest.*, 75, 371-375, 1996.

6.   Migita, K., et al. Impaired degradation of serum amyloid A (SAA) protein by cytokine-stimulated mono-cytes. *Clin. Exp. Immunol.*, 123, 408-411, 2001.

7.   Olsen, K.E., et al. What is the role of giant cells in AL-amyloidosis. *Amyloid*, 6, 89-97, 1999.

8.   Stix, B., et al. Proteolysis of AA amyloid fibril proteins by matrix metalloproteinases-1, -2, and -3. *Am. J. Pathol.*, 159, 561-570, 2001.

9.   Röcken, C., et al. A putative role for cathepsin K in degradation of AA and AL amyloidosis. *Am. J. Pathol.*, 158, 1029-1038, 2001.

A SERUM AMYLOID A PEPTIDE CONTAINS A LOW-pH HEPARIN BINDING SITE WHICH PROMOTES AA-AMYLOIDOGENESIS IN CELL CULTURE

Elena Elimova[1], Robert Kisilevsky[1,2] and John B. Ancsin[2]

Department of Biochemistry[1] and Pathology and Molecular Medicine[2], Queen's University, Kingston, Ontario Canada K7L 3N6

INTRODUCTION

We postulate that heparan sulfate (HS) promotes AA-amyloidogenesis by binding to serum amyloid A (SAA) and causing it to refold and assemble into AA-fibrils. A HS binding site is containined in residues 77-103 of SAA (1). A second HS binding site has been recently identified at residues 17-49, which differs from the first in that it is active only at acidic pH (data not shown). In a cell culture model of AA-amyloidogenesis (2) a peptide (33-mer) corresponding this later sequence could increase the amyloid-load by up to 180%. The pro-amyloidogenic activity of the 33-mer is sequence specific and dependent on His36. Furthermore, a 33-mer based on the SAA1.1 sequence promoted amyloidogenesis to a much higher level than the 33-mer for SAA2.1. The mechanism of action for the 33-mer is currently under investigation. The 33-mer blocks SAA binding to cells suggesting that this domain of SAA may be a binding site for an unidentified cell surface receptor. Our data are consistent with the post-nucleation phase of amyloidogenesis taking place extracellularly.

MATERIALS AND METHODS

The murine monocytic cell line J774A.1 (American Type Culture Collection, Manassas, VA) was cultured in RPMI (Sigma) media which contained 25 mM HEPES, 15% fetal bovine serum (FBS) and 50 µg/ml Penicillin-Streptomycin, at 37 C, 5% CO_2. The cell stocks were passaged every four days, and the media replaced every other day. Cells were seeded at a minimal density in 8-well chamber slides (Lab-Tek®, Nalge Nunc International, Naperville, IL) in 350 µl medium/well and allowed to reach about 80-90 % confluence (3 days), about 2.2×10^5 cells per well. To induce AA-amyloidogenesis, cells were treated for 24 h with 10 µg of amyloid enhancing factor (AEF) in the culture media, then the medium was removed and the cells rinsed with fresh media. To these cells 350 µl of medium was added containing 0.3 mg/ml high density lipoprotein associated SAA (HDL-SAA). In some experiments, SAA1.1 33-mer, SAA1.1 randomized 33-mer, SAA2.1 33-mer, or SAA1.1 33-mer with a H36 to A substitution were included at different concentrations thoughout the HDL-SAA treatments. At the end of the treatment period (7-days) the cells were either stained with Congo red to visualize the amyloid deposits, or the cells were dissolved in 1% NaOH and assayed for amyloid fibrils by thioflavin-T (Th-T) fluorescence (3). Fluorescence spectra of Th-T were acquired at 25°C with a Spectra Max Gemini 96-well plate reader. Cells were solubilized in 1% NaOH which was then neutralized (pH 7) and added to 6.25 µL of 2.5 mM ThT (Sigma) in PBS. Control spectra of Th-T and cell extract alone were determined. The emission spectra were collected by exciting samples at 440 nm (slit, 10 nm) and then monitoring emissions at 482 nm (slit, 10 nm). The J774 macrophages were incubated with HDL-SAA in the presence and absence of SAA1.1 33-mer and macrophage bound SAA was determined using Western blot analysis. Briefly, samples were boiled and resolved by SDS-PAGE using 12% gels. The gels were subsequently transferred to

Immobilon-P nylon membranes (Millipore) at 100V for 1 hour at 4°C in buffer containing 25 mM Tris, 192 mM glycine and 10% methanol, pH 8.3. The membranes were blocked in 6% skim milk powder and subsequently incubated with the SAA-specific antibody (developed in the lab) (1:1000 dilution of 1 mg/mL stock). The western blot was visualized using a HRP-conjugated goat-anti rabbit antibody (1:3000 dilution) and chemiluminescence (Amersham).

RESULTS AND DISCUSSION

Because PVS, LMW heparin, heparin and the 27-mer all inhibit amyloidogesis (2) it was expected that the 33-mer which contained a second heparin binding site would also act to prevent this process. Surprisingly, the 33-mer promoted amyloid deposition (Figure 1 and 2). By Congo red staining the co-incubation of 50 μM 33-mer with HDL-SAA, with or without AEF pre-treatment was observed to increase the amyloid-load in cell culture (data not shown). Using the Th-T fluorescence assay the 33-mer at 10 μM could increase the amyloid load by 180% in AEF pre-treated cultures (Figure 1). In the absence of pre-treatment AEF amyloid-load was increased by 60% (Figure 1).

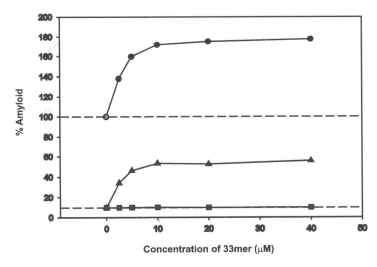

Figure 1: Th-T fluorescence assay of J774 macrophages incubated with SAA1.1 33-mer after AEF pretreatment (●), without AEF pre-treatment (▲) and randomized SAA1.1 33-mer without AEF pre-treatment (■).

Interestingly the substitution of H36 with A in the SAA1.1 33mer amyloid sequence significantly reduced its ability to stimulate amyloidogenesis, indicating that this histidine may be important in the promotion process (Figure 2). The enhancement curve of the 33-mer was shaped like a rectangular hyperbola indicating that this stimulation is saturable. It is likely that J774 macrophages can support only a given amount of amyloid and therefore the addition of high concentration of the 33-mer does not result in a further increase in amyloid deposition. The 33-mer may act to promote amyloidogenesis in manner similar to AEF. This hypothesis is supported by the observation that both AEF and the 33-mer have nearly identical amyloid promotion curves (unpublished data). Both of these curves are shaped like rectangular hyperbolas and demonstrate similar saturability.

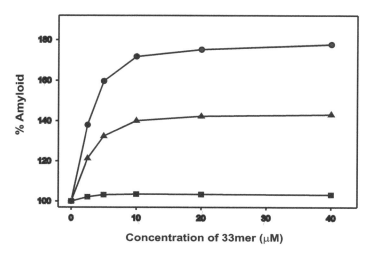

Figure 2: Th-T fluorescence assay of J774 macrophages incubated with SAA1.1 33-mer (●), SAA2.1 33-mer (■) and SAA 1.1 33-mer H36-A (▲) in the presence of AEF.

It is conceivable that the interaction with HS results in a conformational change in the 33-mer, perhaps to increase β-sheet conformation mimicking the conformational features of AEF. The 33-mer inhibits the binding of HDL-SAA to the J774 macrophages (data not shown).   These data also show that internalization of SAA may not be crucial to the process of amyloidogenesis since the 33-mer inhibits HDL-SAA internalization while promoting amyloidogenesis.   These conclusions are supported by the recent report from Barbara Kluve-Beckerman and co-workers in which they showed that fibroblast cells incubated with AEF could deposit AA-amyloid without internalization of SAA (4).

REFERENCES

1. Ancsin, J.B., and Kisilevsky, R. *J. Biol. Chem.* **274,**7172-7181, 1999.

2. Elimova E., Kisilevsky R., Szarek W.A. and Ancsin J.B. in Amyloid and Amyloidosis this volume

3. LeVine H., III *Methods Enzymol.* **309,**274-284, 1999.

4. Magy N., Liepnieks J.J., Benson M.D., and Kluve-Beckerman B. *Arthritis Rheum.* **48,**1430-1437, 2003.

MAPPING THE SURFACE RESIDUES ON MURINE HDL-ASSOCIATED SERUM AMYLOID A REVEALS SIGNIFICANT DIFFERENCES BETWEEN THE TWO MAJOR ISOFORMS

E. Elimova[1], R. Kisilevsky[1, 2] and J. B. Ancsin[2]

Departments of Biochemistry[1] and Pathology and Molecular Medicine[2], Queen's University, Kingston, Ontario Canada K7L 3N6

INTRODUCTION

During an acute-phase response mammalian species produce two very similar isoforms of serum amyloid A (SAA) associated with high density lipoprotein (HDL) (1). In mice only the 1.1 isoform is amyloidogenic, eventhough SAA1.1 and SAA2.1 have 91% sequence identity (2). We attempted to determine if differences in conformation existed between these two isoforms and whether this contributed to their difference in amyloidogenic potential. Applying conventional methods (crystallography, NMR) to examine SAA's 3-dimentional structure is complicated by its hydrophobic nature. To circumvent this problem we used three different chemical modifiers to explore the solvent exposed residues of native SAA associated with HDL. We investigated tryptophan (W), tyrosine (Y) and lysine (K) residues, which are found at identical positions in the two SAA isoforms. Preliminary analysis has revealed that significant differences in surface residue accessibility exist. One W, 2 Y and 4 K were modified for SAA1.1 compared with 0 W, 1 Y, and 3 K for SAA2.1. We were also able to identify the modified residues by mapping them to defined peptides generated by CNBr and proteolytic fragmentation. From these data we are able to conclude that there is a significant difference in conformation (or topology) between the two SAAs at residues 23-41.

MATERIALS AND METHODS

Plasma HDL-SAA concentrations were experimentally elevated in CD1 mice (Charles Rivers, Montreal, Quebec, Canada) by a subcutaneous injection of 0.5 ml of 2% (w/v) $AgNO_3$ which produces a sterile abscess. HDL-SAA was isolated by sequential density flotation as described previously (3). SAA surface W's were modified with hydroxy-nitrobenzyl bromide (Koshland's reagent) (4), Y's with tetra-nitro-methane (TNM) (5) and K's with trinitrobenzene sulfonate (TNBS) (6). Purified HDL-SAA was reacted with each of the modifying reagents individually then the unreacted reagent were removed by either gelfiltration on a Sepharose-G15 desalting column or dialysis. SAA1.1 and SAA2.1 were dissociated from HDL with 6 M guanidine-HCl and purified by reversed phase-high performance liquid chromatography (RP-HPLC) on a semi-preparative C-18 Vydac column connected to a Waters (Millipore) HPLC system (7). Purified SAAs were re-suspended in 0.1% TFA (Koshland's reagent and TNM modified SAA) or 20 mM Tris-HCl, 50 mM NaCl pH 7.5 (TNBS modified SAA) and the absorbance determined at the appropriate wavelength; Koshland's reagent (ε_{418nm}=18,000 $M^{-1}cm^{-1}$), TNM (ε_{360nm}=2,200 $M^{-1}cm^{-1}$) and TNBS (ε_{418nm}=13,100 $M^{-1}cm^{-1}$). Modified SAAs were fragmented by CNBr cleavage (7) and trypsin digestion, and the peptides produced were re-purified by RP-HPLC.

RESULTS AND DISCUSSION

Experimental data on the 3-dimentional structure of SAA has not been reported.   In this study chemical modification of specific residues using polar reagents has demonstrated that potentially important differences in the solvent exposed regions of SAA1.1 and SAA2.1 exist.   The three residues investigated in this study represent 15 out of 103 residues total (3 W, 6 Y and 6 K) and all are located at identical positions for the two isoforms. For SAA1.1, 9 of 15 residues were modified compared with only 5 of 15 residues for SAA2.1 (Fig. 1). These data clearly indicate that SAA1.1 and SAA2.1 have differences in 3-dimensional structure and/or topology on the HDL particle.

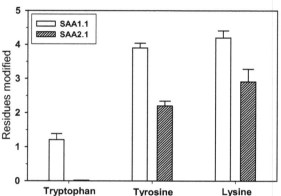

Figure 1. Determination of the number of W, Y and K residues modified by Koshland's reagent, TNM and TNBS, respectively. The number of residues modified was calculated based on the molar absorbpion co-efficient for each reagent per moles of SAA analyzed.

To identify which specific residues were modified, the SAAs were fragmented into defined peptides by a combination of chemical (CNBr) and enzymatic (trypsin) treatments.   Subsequent analysis of these peptides has identified 1 W and 3 Y, which are modified differently between SAA1.1 and 2.1 (Fig 2).   The K residues were not investigated.

```
SAA1.1   GFFSFIGEAFQGAGDMWRAYTDMKEAGWKDG DKYFHARGNYDAA RGPGG-50
SAA2.1   -----VH---------W--Y------NW-NS--Y-----Y---------


SAA1.1   VWAAEKISDARESFQEFFGRGHEDTMADQEANRHGRSGKDPNYYRPPGLPAKY-103
SAA2.1   -W--------G--A------------I---------------YY------D-Y
```

Figure 2. Protein sequence alignment for mouse SAA1.1 (gi:200911) and SAA2.1 (gi:200906).   Modified residues   are indicated (•) and a highly conserved sequence is shown in the box.

For SAA1.1 the three Ys (Y93, Y94 and Y103) at the C-terminus are solvent exposed compared with only two Ys (Y94 and Y103) for SAA2.1. Also W28, Y34 and Y41 are solvent exposed only for SAA1.1. Unexpectedly, the later two residues (Y34 and Y41) are located in a sequence (residues 32-44) that is invariant among SAAs and therefore would be expected to be identical in conformation for both SAAs. Furthermore, a synthetic peptide (33-mer) corresponding to residues 17-49 for SAA1.1 has been found to be a much strong promoter of

AA-amyloidogenesis in cell culture, than the same region of SAA2.1 (8). We propose that despite the sequence similarity between SAA1.1 and SAA2.1, their native conformations and/or topologies on HDL are significantly different and responsible, at least in part, for their divergent amyloidogenic potentials. In addition, we have identified two regions of SAA, which are exposed on the surface of HDL-SAA and potentially involved in binding surface receptors of macrophage with physiologic and amyloidogenic consequences.

REFERENCES

1) Hoffman J.S. and Benditt E.P. (1982) *J. Biol. Chem.* **257**, 10510-10517

2) Lowell C.A., Potter D.A., Stearman R.S. and Morrow J.F. (1986) *J. Biol. Chem.* **261**, 8442-8452

3) Havel R.J., Eder H.A. and Bragdon J.H. (1955) *J. Clin. Invest.* **34**, 1345-1353

4) Koshland D.E. Jr., Karkhanis Y.D. and Latham H.G. (1964) *J. Amer. Chem. Soc.* **86**, 1448-1450

5) Riodan J.F. and Vallee B.L. (1972) *Methods Enzymol.* **25**, 515-521

6) Drozdovskaya N.R., Kozlov I.A., Milgrom Y.M. and Tsybovski I.S. (1982) *FEBS Lett.* **150**, 385 -389

7) Ancsin J.B. and Kisilevsky R. (1999) *J. Biol. Chem.* **274**, 7172-7181

8) Elimova E., Kisilevsky R. and Ancsin J.B. *Amyloid and Amyloidosis* , this volume.

ANALYSES OF THE AMYLOID FIBRILS IN BOVINE AMYLOIDOSIS IN HOKKAIDO DISTRICT, JAPAN

M. Ueda[1], Y. Ando[2], K. Haraoka[3], T. Matsui[4], N. Takamune[6], S. Saito[7], M. Nakamura[2], H. Terazaki[3], T. Yamashita[1], S. Xuguo[2], J. Kim[2], T. Tateishi[2], S. Shoji[6], Y. Hoshii[5], T. Ishihara[5], M. Otagiri[7], M. Uchino[1]

[1]*Department of Neurology,* [2]*Department of Diagnostic Medicine, and* [3]*Department of Gastroenterology, Graduate School of Medical Sciences, Kumamoto University,* [4]*Obihiro University of Agriculture and Veterinary Medicine,* [5]*First Department of Pathology, Yamaguchi University School of Medicine,* [6]*Department of Biochemistry,* [7]*Biopharmaceutics Laboratory, Faculty of Pharmaceutical Sciences, Kumamoto University*

INTRODUCTION

It is well know that the bovine develops systemic amyloidosis and clinico-pathological findings on the bovine amyloidosis have been reported (1,4). Especially, the kidneys are the most commonly affected organ in the amyloidosis. In Hokkaido district, Japan, from 1992 to 2002, 619 of 394,166 aged bovines were diagnosed as amyloidosis by histopathological analyses. We analyzed the precursor protein of the amyloidosis in 5 frozen kidney samples of the bovines and clarified the characterization of bovine amyloidosis.

MATERIALS AND METHODS

We analyzed 5 frozen kidney samples of the bovine amyloidosis using immunohistochemical and biochemical methods.

Immunohistochemistry was performed on frozen sections of renal tissues using a rabbit anti-human AA serum given from First Department of Pathology, Yamaguchi University School of Medicine, and a sheep anti-human SAA antibody (CALBIOCHEM, Darmstadt, Germany) for detecting bovine serum amyloid A (SAA) / amyloid A (AA) protein.

Amyloid fibrils were extracted from 10g frozen kidney samples using a modified water extraction method (5). Thereafter, we performed SDS-PAGE, immunoblotting, and matrix-assisted laser desorption ionization/time-of-flight mass spectrometry (MALDI / TOF-MS) for detecting the amyloid precursor proteins.

RESULTS

Diffuse congophilic amorphous substances predominantly in the glomerulus of all the 5 bovine kidney samples. By immunohistochemistry, positive reaction for the rabbit anti-human AA serum and the sheep anti-human SAA antibody was observed in the lesions where Congo red staining was positive.

SDS-PAGE revealed that fibril proteins were visualized as two major bands of about 10 kDa and about 14 kDa in addition to a few minor bands. This result was in agreement with a former report (6, 8, 10)

Immunoblotting using the rabbit anti-human AA serum and the sheep anti-human SAA antibody revealed a single major band of 10 kDa and positive diffuse bands ranging from about 5 kDa to 10 kDa. However, the band of 14kDa was not detected by immunoblotting.

MALDI - TOF / MS for the extracted amyloid proteins disclosed different AA protein fragments. Ten different spectra were obtained, and three peaks with the molecular size of 10594.84, 9677.82 and 6438.30 Da were observed as major peaks. The peak of 14kDa was not also detected in MALDI-TOF / MS, in the same way as the result of immunoblotting.

DISCUSSION

We investigated 5 kidney samples of 619 amyloidosis bovine in Hokkaido district, Japan. All the samples showed amyloid deposits within the glomeruni and reacted anti-human SAA / AA antibody in the lesions where Congo red staining was positive. Previous reports showed incidence of amyloidosis ranging from 0.4 to 2.7% in slaughtered bovines (1,3). In Hokkaido district, a range of incidence was 0.15% to 2.8%, which was the same as the result of the previous reports.

In this study, our results suggest that amyloidosis occurring in aged bovines was secondary (AA) amyloidosis.

Previously, it was reported that SAA was partially degraded in C-terminal and the remaining N-terminal part formed amyloid fibril on the basis of amino acid sequence analysis (6,9). MALDI-TOF / MS of immunoprecipitated fibril proteins revealed that the molecular weight of intact SAA (14 kDa) was not detected, suggesting that amyloid fibrils mainly consisted of fragmented SAA. Three peaks with the molecular size of 10594.84, 9677.82 and 6438.30 Da were observed as major peaks, but cleavage sites predicted from these molecular weights did not completely correspond to those of previous report (6, 10). This suggests that some modifications for amino acids may occur in SAA or C-terminal sequences of SAA may not be very important for amyloid formation.

In summary, amyloidosis in aged bovines observed in Hokkaido district is secondary (AA) amyloidosis and different AA protein fragments are seen in the analysis. These results are considered to be important for elucidating the formation mechanism of the bovine amyloidosis.

REFERENCES

1. Johnson, K.H., Westermark, P., Sletten, K. and O'Brien, T.D., Amyloid proteins and amyloidosis in domestic animals, *Amyloid Int. J.Exp.Clin.Invest.,* 3, 270, 1996

2. Gruys, E., Amyloidosis in the bovine kidney, *Vet.Sci.Commun.* 1, 265, 1975

3. Monaghan, M., Renal amyloidosis in slaughter cattle in Ireland, *Irish. Vet. J.,* 36, 88, 1982

4. Johnson, R. and Jamison, K., Amyloidosis in six dairy cows, *J.Am.Vet.Med.Assoc.,* 185, 1538, 1984

5. Kaplan, B., Shtrasburg, S. and Pras, M., Micropurification techniques in the analysis of amyloid proteins, *J.Clin.Pathol.,* 56, 86, 2003

6. Rossevatn, K. et al., The complete amino acid sequence of bovine serum amyloid protein A (SAA) and of subspecies of the tissue-deposited amyloid fibril protein A, *Scand.J.Immunol.,* 35, 217, 1992

7. Husebekk, A. et al., Characterization of bovine amyloid proteins SAA and AA, *Scand.J.Immunol.,* 27, 739, 1988

8. Westermark, P. et al., Bovine amyloid protein AA: isolation and amino acid sequence analysis, *Comp.Biochem.Physiol.,* 85B, 609, 1986

9. Benson, M.D., DiBartola, S.P. and Dwulet, F.E., A unique insertion in the primary structure of bovine amyloid AA protein, *J.Lab.Clin.Med.,* 113, 67, 1989

10. Niewold, T.A., Tooten, P.C.J. and Gruys, E., Quantitation of fibrillar SAA in bovine AA-amyloid fibrils, in *Amyloid and Amyloidosis 1993*, Kisilevsky. R., Benson, M.D., Frangione, B., Gauldie, J., Muckle, T.J., and Young, I.D. Eds., Pathenon Publishing Group, New York,1993,140.

SERUM AMYLOID A (SAA) IN MAMMARY TISSUES WITH INFLAMMATORY PROCESSES AND IN MAMMARY CORPORA AMYLACEA

M.J.M. Toussaint[1], C. Hogarth[1,2], T.K.A. Nguyen[1,3], E. Loeb[1], J. Balciute[1,4], V. Vivanco[1,5], A.M. van Ederen[1], S Jacobsen[1,3,6], T.A. Niewold[3] and E. Gruys[1].

[1]Department of Pathobiology, Faculty of Veterinary Medicine, Utrecht University, Utrecht, The Netherlands, [2] Department of Veterinary Clinical Studies, University of Glasgow, Glasgow, Scotland [3]Animal Sciences Group, Lelystad, The Netherlands, [4] Department of Physiology and Pathology, Lithuanian Veterinary Academy, Kaunas, Lithuania, [5] Department of Animal Pathology, Veterinary Faculty, Zaragoza, Spain [6] Department of Clinical Studies, Royal Veterinary and Agricultural University, Copenhage, Denmark.

INTRODUCTION

During an acute phase reaction large quantities of different acute phase proteins can be found in blood plasma. One of these proteins is serum amyloid A (SAA). The liver is the predominant producer of SAA. Several isoforms of SAA are described for different species, some are found in a variety of extrahepatic tissues. In most mammals SAA4 appears to be the predominant isoform expressed extrahepatically (1). In the mammary gland a SAA3 has been described (2). It appears in milk during the dry period, has large quantities in colostrum and is elevated during mastitis (3).

General concern is on early diagnosis of mastitis. Acute phase proteins in milk such as SAA3, could be useful for this purpose (3). This study presents different approaches used to emphasis the value of SAA as early diagnostic of mastitis.

MATERIAL AND METHODS

Isoelectric focus (IEF) and Western blot (WB) techniques were performed routinely. IEF was used for a time curve of samples from cows intra-mammarily infected with *E. coli* 0:157 (30 cfu in one quarter). For detecting SAA isoforms in blood and milk on the WB after IEF, the biotinylated anti human antibody from a commercial kit (TP 802, Tridelta, Ireland) in 1:100 dilution was used, followed by Streptavidin-Horse Radish Peroxidase, 1:4000. (DAKO, Denmark), and visualization by incubating for 10 to 15 minutes with a TBS solution containing 0.5 mg/ml 3,3'-diaminobenzidin (Sigma-Aldrich, USA) and 0.02% of 30% hydrogen peroxide (Merck, Germany).

Amyloid fibrils were isolated from mammary corpora amylacea according to Pras *et al* 1968 (4). The corpora amylacea were purified after collection from frozen tissue samples corresponding to samples fixed for histology. This tissue was shown to be positive in Congo red stained sections. A part of the last pellet was kept in 6 M Guanidine/0.55 M TrisHCL pH 8.4 overnight and diluted 1:1 in sample buffer. This sample was put on a 15% SDS-urea gel. Thereafter the gel was blotted, using routine WB technique. The nitrocellulose membrane was stained with rabbit anti bovine AA (5) (1:1000) and goat anti rabbit alkaline phosphatase (DAKO, Denmark, 1:1000), followed by 4-nitro blue tetrazolium chloride and 5-bromo-4-chloro-3-indolylphosphate (NBT/BCIP, Roche, Germany).

Tissue sections from infected and control mammary glands were stained with hematoxylin & eosin and alkalic Congo red and with SAA monoclonal from Tridelta, using ABC method (Vector Laboratory, USA). Furthermore selections were stained with a DiG labeled probe for (milk) *SAA* (*in situ* hybridization using Roche Blocking reagent and NBT/BCIP as substrate for alkaline phosphatase). Alignment of the probe showed 100% homology with Bos taurus serum amyloid A protein mRNA (gi 23305877).

RESULTS

The milk SAA was shown to react in time more early after the experimental mammary *E. coli* infection than blood plasma acute phase SAA (Figure 1a,b) while it had an alkaline pl (>9) in contradiction to the plasma isotypes (pl around 6).

After isolation of the amyloid fibrils from corpora amylacea collected from the affected tissue from affected glands, and performing SDS electrophoresis followed by Western blot, one band of 14 kD reacting with the rabbit anti bovine antibody was found in (Figure 1c, lane 1). As can be seen in Figure 2A showing immunohistochemistry on tissue sections corpora located inside the acini (arrow) were positive for SAA.

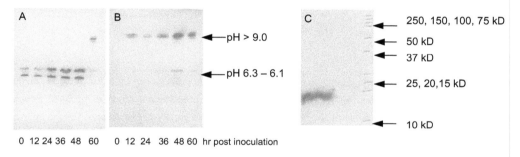

0 12 24 36 48 60   0 12 24   36 48 60 hr post inoculation

Figure 1. Western blot of IEF of time series of plasma (A) and milk (B) samples after inoculation with E.coli 0:157 (30 cfu in one quarter). The 60 hours sample from the milk was run on both gels to allow comparison between them. C) Western blot of isolated amyloid fibrils from corpora amylacea from affected udder (lane 1, M = molecular weight marker).

The monoclonal antibody from the commercial kit applied on bovine tissue sections (Figure 2B) gave comparable results as rabbit anti-bovine AA (not shown). It appeared to react with mastitis milk, mammary epithelium (Figure 2B, arrow) and corpora amylacea (Figure 2A, arrow) in tissue sections. The use of a probe against milk *SAA* resulted in local positive staining of the acinar epithelium (Figure 2C, arrow) indicating local production of SAA.

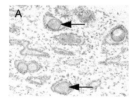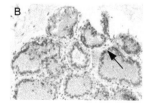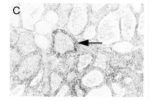

Figure 2. Corpora amylacea (A) inside mammary acini stained with anti SAA (immunohistochemistry using ABC). Mastitis udder (B) stained with anti SAA (immunohistochemistry using ABC). *In situ* hybridization of astitis udder (C) stained with milk *SAA* probe.

DISCUSSION

The findings presented here indicate milk SAA to be a marker for infectious mammary disorders. The milk isoform has a different pI, compared to the plasma isoform and it was measurable at an earlier time point compared to the blood plasma SAA. The milk isoform appeared to originate from mammary epithelium as shown by immunohistochemistry in combination with the *in situ* hybridization.

Moreover, during the study mammary corpora amylacea appeared to be positive for SAA. Isolation of amyloid fibrils from collected corpora revealed a SAA band on Western blot. Formerly has been shown that mammary corpora amylacea isolated from milk contain casein-related α-S2 protein (6). Further studies are required on these subjects.

REFERENCES

1 International Nomenclature Committee. Editorial Part 2 Revised nomenclature for serum amyloid A (SAA). Amyloid: Int. J. Exp. Clin. Invest. 1999, 6:67-70.

2 McDonald TL, Larson MA, Mack DR, Weber A. Elevated extrahepatic expression and secretion of mammary-associated serum amyloid A 3 (M-SAA2) into colostrum. Vet Immunol Immunopathol 2001, 83:203-211.

3 Eckersall PD, Young FJ, McComb C, Hogart CJ, Safi S, Weber A, McDonald T, Nolan AM and Fitzpatrick JL. Acute phase proteins in serum and milk from dairy cows with clinical mastitis. Vet. Rec. 2001, 148:35-41.

4 Pras M, Schubert M, Zucker-Franklin D, Rimon A, Franklin EC. The characterization of soluble amyloid prepared in water. Clin Invest. 1968, 47:924-33.

5 Boosman R, Niewold TA, Mutsaers CWAAM, Gruys E. Serum amyloid A concentrations in cows given endotoxin as an acute-phase stimulant. Am J Vet Res. 1989, 50:1690-4.

6 Niewold TA, Murphy CL, Hulskamp-Koch CA, Tooten PC, Gruys E. Casein related amyloid, characterization of a new and unique amyloid protein isolated from bovine corpora amylacea. Amyloid. 1999, 6:244-9.

ACKNOWLEDGEMENT

Part of this research was supported through a European Community Marie Curie Fellowship. The authors are solely responsible for the information published. It does not represent the opinion of the Community, and the Community is not responsible for any use that might be made of data appearing therein.

CHEMICAL TYPING OF PORCINE SYSTEMIC AMYLOID AS AA-AMYLOID

T.A. Niewold [1], C.L. Murphy [2], M.J.M. Toussaint[3], E. Gruys[3].

1. Animal Resources Development, Animal Sciences Group of Wageningen UR. P.O.Box 65, 8200 AB Lelystad, the Netherlands, E-mail: theo.niewold@wur.nl

2. University of Tennessee Graduate School of Medicine, Knoxville TN, USA

3. Department of Pathobiology, Faculty of Veterinary Medicine, Utrecht University, PO Box 80158, NL-3508 TD Utrecht, The Netherlands

ABSTRACT

Systemic amyloidosis is frequently reported in a wide variety of domestic animals, including avian, canine, feline, equine and ruminants. In most cases, the amyloid has been identified to be the AA-type. Porcine amyloidosis has been reported less frequently, in fact, descriptions are rare and refer to single cases (Jacob, 1971) . Furthermore, the amyloid has not been typed chemically thusfar. In the present report, we have extracted amyloid material from formalin fixed tissue sections. By subsequent amino acid sequencing, a N-terminal fragment was obtained which amino acid sequence showed a high degree of homology with known sequence of bovine and bovine AA-amyloid. This identifies porcine systemic amyloid as AA-amyloid. This is the first report on chemical identification of porcine amyloid. Furthermore, based on the partial sequence obtained, is was established that the amino acid sequence differs from the only porcine SAA sequences available in the public domain. The latter are derived from macrophages, and represent local or tissue-specific SAA (SAA3). This fact, and the similarity to bovine and ovine SAA1 suggests that pig AA-amyloid is derived from the systemic (hepatic) isoform of SAA. It is argued that this is a valuable addition to the sequences from various species already known, since pig SAA apparently has a low tendency to form amyloid. This will contribute to the elucidation of amyloidogenic factors by comparative sequence analysis.

CASE

A male pot bellied pig, 2.5 years was presented with a history of inappetite, kidney problems, emaciation, polyuria and polydypsia. The specific gravity of urine was 1.005, and proteinuria was found. Laparoscopic inspection of the peritoneum revealed various necrotic nodules (10 cm in diameter) of subperitoneal fat. In the mesenterium, smaller necrotic nodules occurred as well. Furthermore, amotility of the intestinal tract was observed. Kidney and adipose tissue needle biopsies were taken, and histologically, in the kidney severe interstitial fibrosis, tubular atrophy, a focally lymphoplasmacytic infiltrates and calcification were seen. In the adipose tissue, extensive necrosis and secondary inflammation and fibrosis were present.

The animal was diagnosed with chronic interstitial nephritis, and treated with Prepulsid and Prednison for a period of 9 months. The condition of the animal improved, but the during the last month the animal became lethargic and inappetent, and the animal died. At necropsy, the most significant findings were subacute pleuritis, pericarditis, a pale liver, kidney which appeared too small, pale and firm, with on cross-section having a thin cortex.

Histologically, Congo red staining demonstrated glomerular amyloid, and in the liver, extensive amyloid deposition in the space of Disse and focal calcification. Furthermore, interstitial amyloid was found in the pancreas. Adipose tissue revealed extensive necrosis and secondary chronic inflammation and fibrosis. The animal was diagnosed as suffering from the contracted form of renal amyloidosis.

Immunehistologically, a positive reaction was obtained using anti-AA antibodies directed against different species, the strongest with anti-canine AA (not shown).

SAMPLE PREPARATION

Amyloid material was extracted from formalin fixed tissue sections essentially according to Murphy et al., 2001. Briefly, four-micrometer-thick sections were cut from formalin-fixed paraffin-embedded kidney. Sections (24) were deparaffinized, and scraped from the slides into an Eppendorf tube containing 250 microliter of a 0.25 M tris(hydroxymethyl)aminomethane-hydrochloride, pH 8.0, 1.0 mmol/L disodium-EDTA solution, and 750 microliter of an 8 M guanidine hydrochloride solution. The tube was incubated at 37°C for 24-48 hours, and sonicated repeatedly until the solution was clear. The protein was reduced by addition of 10 microliter of 2-mercaptoethanol, and alkylated by addition of 30 microliter of iodoacetemide.

PROTEIN PURIFICATION

After centrifugation, the supernatant was subjected to purification by HPLC (Perkin-Elmer Model 200 injector, Danbury, Connecticut), using an Aquapore 300 C_8 (30 x 4 mm) column (Brownlee Columns, Perkin Elmer, Norwalk, CT). Proteins were separated by reversed phase liquid chromatography using an ABI model 140A solvent delivery system (Applied Biosystems), with linear gradient of 0.1% trifluoroacetic acid water-7% acetonitrile to 0.1% trifluoroacetic acid water-70% acetonitrile at a flow rate of 1 ml/min. Peaks were collected according to the absorbance at 220 nm using an ABI model 785A absorbance detector (Applied Biosystems). Fractions were concentrated in a Speed Vac sample concentrator, and on part of the sample, N-terminal sequencing was performed using an ABI model 494 Procise automated amino acid sequencer (Applied Biosystems).

Since the sequences obtained were exclusively of normal tissue constituents, the remainder of the material was subjected to N-terminal deblocking with *Pfu* pyroglutamate aminopeptidase (Takara Bio Inc. Otsu, Shiga, Japan). Peaks were resequenced after deblocking, and from one of the peaks a N-terminal sequence was obtained with homology to N-terminal sequences of bovine and ovine SAA (respectively SwissProt P35541, and P42819).

DISCUSSION

N-terminal amino acid sequencing of deblocked protein revealed a high degree of homology with known sequences of bovine and ovine AA-amyloid. The sequence differs from the only porcine SAA sequences available in the public domain (Neilan et al., 2003). The latter are derived from macrophages, and represent local or tissue-specific SAA (SAA3). This fact, and the similarity to bovine and ovine SAA1 suggests that pig AA-amyloid is derived from a systemic (hepatic) isoform of SAA. It is argued that this is a valuable addition to the sequences from various species already known, since pig SAA apparently has a low tendency to form amyloid. This will contribute to the identification of SAA intrinsic amyloidogenic factors.

LITERATURE

Jakob W. Spontaneous amyloidosis of mammals. Vet Path 8: 292-306, 1971

Murphy CL, Eulitz M, Hrncic R, Sletten K, Westermark P, Williams T, Macy SD, Wooliver C, Wall J, Weiss DT, Solomon A. Chemical typing of amyloid protein contained in formalin-fixed paraffin-embedded biopsy specimens.Am J Clin Pathol. 116:135-42, 2001.

Neilan JG, Kutish GF, Lu Z, Zsak A, Rock DL. Sequence analysis of African swine fever infected and non-infected porcine macrophage cDNA libraries. Genbank (2003) # GB47401

SERUM AMYLOID P COMPONENT IN MINK.*

*The work has been accepted for publication in Amyloid, the Journal of Protein Folding Disorders.

Lone A. Omtvedt[1,2], Tale N. Wien[1], Theresa Myran[1], Knut Sletten[2] and Gunnar Husby[1].

[1]Department of Rheumatology/ Institute of Immunology, Rikshospitalet, University of Oslo, Norway
[2]Department of Biochemistry/Biotechnology Centre of Oslo, University of Oslo, Norway

INTRODUCTION

In most species either C reactive protein (CRP) or serum amyloid P component (SAP) are acute phase reactants that increase up to a thousandfold following induction. They share similar subunit composition, calcium-dependent ligand binding capacity and extensive amino acid sequence homology (1). SAP and CRP have in general different calcium-dependent affinities for the ligands phosphorylethanolamine (PE) and phosphorylcholine (PC) (2). Therefore, affinity chromatography with either one of these ligands is often used to distinguish between CRP and SAP (3). The amyloid P component (AP), which is identical to circulating SAP is found in all forms of amyloid as extra-cellular material (4-5). Experimental AA amyloidosis in the mink is used as a model for the amyloid disease process, and it is thus important to characterise the different proteins involved in amyloid formation in this species. In the present work we have characterised SAP in mink.

MATERIAL AND METHODS

Animals; Blood was drawn from mink *(Mustela vison)* of both sexes from the research farm of the Norwegian School of Veterinary Science, Oslo, Norway. Acute phase serum from mink treated with bacterial lipopolysaccharide (LPS)(*Escherichia coli*, Bacto 3920-25, Difco Laboratories) was compared with serum from control mink (mink injected with 0.9% NaCl) (6).

Isolation and characterisation of serum pentraxins; Proteins were purified from mink serum by affinity chromatography, either using PE coupled ECH-sepharose 4B (Amersham Biosciences) or immobilised PC-gel (Pierce Biotechnology) (2). The purity of the sample was analysed by SDS-PAGE and N-terminal sequencing. Alkylated and non-alkylated SAP purified from control serum, was digested with trypsin, with staphylococcal V8 protease or with endoproteinase ASP-N (Roch) according to the manufacturer (7-8). Chemical fragmentation with CNBr was performed both on filter and in solution (9). Proteins and peptides were analysed by automatic Edman degradation and by mass spectrometry.

Database search; The elucidated sequence was used to search for homologous sequences using the search program FASTA in the protein database SWISSPROT (10). ClustalW was used to align the sequences (11).

RESULTS

SDS-PAGE of PE-purified mink protein showed a major protein band with an apparent molecular mass of 26 kDa, together with one minor band estimated to be 1/10 of the major band. SAP isolated from LPS-treated, untreated, male and female mink as well as the protein purified using a PC-column gave an identical two band pattern. The major SAP band did not stain with PAS/basic fuchsin, indicating there was little or no carbohydrate present. The molecular mass determined by mass spectrometry was 22,885kDa, consistent with the calculated mass of the amino acid sequence found.

Figure 1a.ESI-MS spectrum of SAP from mink, purified from serum.
Figure 1b.Tryptic peptides sequenced by ESI-MS

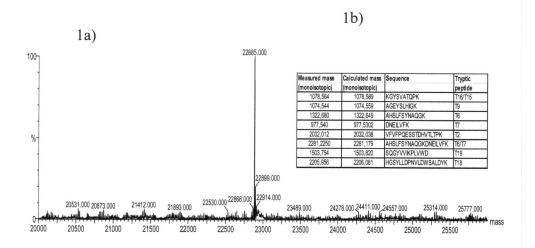

1b)

Measured mass (monoisotopic)	Calculated mass (monoisotopic)	Sequence	Tryptic peptide
1078,564	1078,589	KGYSVATQPK	T16/T15
1074,544	1074,559	AGEYSLHIGK	T9
1322,680	1322,649	AHSLFSYNAQGK	T6
977,540	977,5302	DNEILVFK	T7
2032,012	2032,038	VFVFPQESSTDHVTLTPK	T2
2281,2250	2281,179	AHSLFSYNAQGKDNEILVFK	T6/T7
1503,754	1503,820	SQGYVVIKPLVWD	T19
2205,956	2206,081	HGSYLLDPNVLDWSALDYK	T18

Fasta search in the protein database SWISSPROT showed that the elucidated sequence had the largest homology with the following sequences: The SAP sequences from human (73.0% homology), pig (73.5%), guinea pig (66.8%), rat (66.3%), Syrian hamster (63.7%), Armenian hamster (63.7%) and mouse (63.9%). Those sequences were aligned using the program ClustalW.

DISCUSSION

Of the two protein bands on the SDS-PAGE of mink, only the minor band was stained purple using the PAS staining method that specifically stains glycoproteins, and only the minor band changed its apparent molecular mass when digested with PNGaseF, indicating N-linked sugars bound to the protein. MS-spectrum of purified SAP shows that the molecular mass of the protein is 22,885 kDa, corresponding to the amino acid sequence found, giving room for no extra sugar attached to the molecule. The sum of all this indicates that there are two fraction of SAP in mink serum, where only the minor fraction is glycosylated. It is therefore the first time that a chiefly non-glycosylated SAP protein is found in a mammalian species. A non-glycosylated SAP does not seem to impede AA fibril formation, as mink readily form amyloid, both spontaneously and experimentally induced with LPS (6).

The mink protein purified with PC has an identical N-terminal sequence and SDS-PAGE profile as the PE purified protein. We must therefore conclude that the proteins purified with PC and PE are identical. Thus, mink SAP has CRP like properties, like FP, showing affinity for PC.

REFERENCES

1. Baltz, M. L., De Beer, F. C., Feinstein, A., and Pepys, M. B., Calcium-dependent aggregation of human serum amyloid P component, *Biochim Biophys Acta* 701 (2), 229-36, 1982.

2. Schwalbe, R. A., Dahlback, B., Coe, J. E., and Nelsestuen, G. L., Pentraxin family of proteins interact specifically with phosphorylcholine and/or phosphorylethanolamine, *Biochemistry* 31 (20), 4907-15, 1992.

3. Christner, R. B. and Mortensen, R. F., Binding of human serum amyloid P-component to phosphocholine, *Arch Biochem Biophys* 314 (2), 337-43, 1994.

4. Pepys, M. B. and Butler, P. J., Serum amyloid P component is the major calcium-dependent specific DNA binding protein of the serum, *Biochem Biophys Res Commun* 148 (1), 308-13, 1987.

5. Holck, M., Husby, G., Sletten, K., and Natvig, J. B., The amyloid P-component (protein AP): an integral part of the amyloid substance?, *Scand J Immunol* 10 (1), 55-60, 1979.

6. Wien, T. N., Omtvedt, L. A., Landsverk, T., and Husby, G., Characterization of proteoglycans and glycosaminoglycans in splenic AA amyloid induced in mink, *Scand J Immunol* 52 (6), 576-83, 2000.

7. Sletten, K., Natvig, J. B., Husby, G., and Juul, J., The complete amino acid sequence of a prototype immunoglobulin-lambda light-chain-type amyloid-fibril protein AR, *Biochem J* 195 (3), 561-72, 1981.

8. Austen, B. M. and Smith, E. L., Action of staphylococcal proteinase on peptides of varying chain length and composition, *Biochem Biophys Res Commun* 72 (2), 411-7, 1976.

9. Sletten, K. and Husby, G., The complete amino-acid sequence of non-immunoglobulin amyloid fibril protein AS in rheumatoid arthritis, *Eur J Biochem* 41 (1), 117-25, 1974.

10. Pearson, W. R. and Lipman, D. J., Improved tools for biological sequence comparison, *Proc Natl Acad Sci U S A* 85 (8), 2444-8, 1988.

11. Thompson, J. D., Higgins, D. G., and Gibson, T. J., CLUSTAL W: improving the sensitivity of progressive multiple sequence alignment through sequence weighting, position-specific gap penalties and weight matrix choice, *Nucleic Acids Res* 22 (22), 4673-80, 1994.

ACUTE-PHASE SERUM AMYLOID A (A-SAA) SYNTHESIZED BY HUMAN SYNOVIAL CELLS (SC) INDUCES SC MIGRATION

Y. Kumon, K. Arii[*], T. Suehiro[*], Y. Ikeda[*], H. Enzan[‡] and K. Hashimoto[*].

Department of Laboratory Medicine, [*]Department of Endocrinology, Metabolism and Nephrology, and [‡]Department of Pathology, Kochi Medical School, Kochi University, Kochi Japan

Human acute-phase serum amyloid A (A-SAA) has been considered to be mainly synthesized by liver, however, recent studies have shown that human A-SAA is synthesized extrahepatically and is present in many tissues including inflammatory synovium [1,2]. As the regulation of A-SAA synthesis by synovial cells (SC) and the physiological function(s) of A-SAA in the synovium remained unclear, the present study was undertaken to clarify 1) the control mechanism of A-SAA gene expression in SC, and 2) the A-SAA function, i.e. whether A-SAA affected SC migration or not.

MATERIALS AND METHODS

SC were obtained by enzymatic digestion of synovial membrane from rheumatoid arthritis patients at the time of total knee replacement [2]. The cultured SC were served for analysis of SAA1 (major A-SAA) mRNA expression using Northern blot analysis [2], for transcriptional regulation of SAA1 gene using luciferase reporter gene assay [3], and for SC migration assay [4]. The PCR product amplified from SAA1.1 cDNA was used as a probe for Northern blot analysis. The plasmid construct for luciferase assay contained 1465 bp (-1431/+34) of human SAA1.1 gene. SC migration assay was performed using a culture chamber of Chemotaxicell which has a polycarbonate membrane with pores of 8 microns in 24-well plates.

Immunohistochemistry on paraffin-embedded tissue sections of RA synovial membrane was performed using anti-human SAA1 monoclonal [5] and anti-human formyl peptide receptor like 1 (fPRL1) polyclonal [4]. A-SAA synthesis by SC was examined by immunoblot analysis of SC culture media [6].

RESULTS

With stimulation of SC by 10 ng/ml IL-1 plus 1 μM Dexamethasone (Dex), the increased levels of SAA1 mRNA expressed by SC were detectable after 8 hours, and continued to increase for up to 48 hours. Both of IL-1 and Dex were necessarily required to express optimal SAA1 mRNA by SC. SAA1 mRNA expression was upregulated synergistically by IL-1 and Dex in dose-dependent manner of IL-1 or Dex, each other. One μM corticosterone, 1 μM hydrocortisone (F) or 1 μM aldosterone also markedly upregulated SAA1 expression under the stimulation with 10 ng/ml IL-1. Ten ng/ml IL-6 or 10 ng/ml TNF-α alone did not detectably upregulate SAA1 mRNA expression by SC. These alterations of SAA1 mRNA expression by SC with cytokines and steroids were confirmed at the levels of transcription. The changes of SAA1 mRNA levels were identical to those of transcriptional activities. The secretion of A-SAA in SC culture media stimulated with 1 ng/ml IL-1 plus 1μM Dex was also confirmed.

To directly determine the influence of A-SAA on SC migration, we incubated SC with 50 μg/ml of apoSAAp, recombinant SAA1 and SAA2 hybrid protein, in the upper chamber and/or lower chambers (Figure 1). ApoSAAp induced SC migration and the major mode of SC migration by apoSAAp was chemotaxis, and chemokinesis was minor. To evaluate the effect of endogenous A-SAA on SC migration, we incubated SC with 10 ng/ml IL-1 plus 1μM Dex. The SC migration was upregulated by IL-1 plus Dex, and this migration of SC was inhibited by anti-SAA1 or anti-fPRL1.

The immunohistochemical examination of synovial membranes revealed that the surface SC were strongly, but the inner SC were weakly stained with anti-SAA1 and anti-fPRL1.

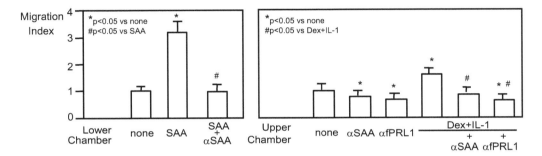

Figure 1. SC migration by A-SAA

Data are expressed as mean ± SD. Abbreviations are as follows; SAA, 50 μg/ml apoSAAp; Dex, 1 μM dexamethasone; IL-1, 10 ng/ml IL-1; α SAA, anti-A-SAA polyclonal ; αfPRL1, anti-formyl peptide receptor like 1 polyclonal.

DISCUSSION

We have shown that 1) SAA1 expression by SC was dependent on both IL-1 and glucocorticoids or mineralocorticoids; 2) A-SAA was synthesized by SC and was immunohistochemically identified in SC of inflamed synovium; and 3) A-SAA induced SC migration, which was mediated by A-SAA receptor, fPRL1.

It is worthy of note that the transcriptional regulation of human SAA1.1 gene in SC was almost similar to that in HepG2 cells, but is distinct from that in vascular smooth muscle cells [3]. This means that the efficient A-SAA synthesis by SC in human requires corticoid hormones like hydrocortisone, in addition to IL-1. IL-6 and TNF-α were found to be less effective in SAA1 expression by SC than IL-1. As one μM F used in this study is a physiological concentration, it would seem that SAA1 is usually synthesized and produced in inflamed joints. It was also interesting that A-SAA was immunohistochemically localized on the surface SC of synovial membrane. This might be the result of A-SAA production by the surface SC, because of dependence on highly concentrated IL-1 in the inflamed synovia.

The principle biological function of SAA has remained unknown. However, there are possible roles for A-SAA in the pathophysiology of joint inflammation; A-SAA induces migration of leukocytes [7], and upregulates the enzyme activities of secretory non-pancreatic phospholipase A2 [8] and matrix metalloproteinases [9], both of which are synthesized by SC and enhance joint destruction. We found herein that A-SAA was synthesized by SC and that A-SAA induced SC migration; the mode of SC migration was apparently by autocrine and/or paracrine mechanisms by means of fPRL1. Interestingly, anti-SAA1 and anti-fPRL1 downregulated SC migration in quiescent condition. Therefore, this phenomenon suggests that A-SAA is essential for SC migration. All of these findings suggest that A-SAA, present in inflamed joints, might contribute to the destruction of joints.

In conclusion, the locally synthesized A-SAA by SC induced SC migration via fPRL1, which was reportedly present on leukocytes and phagocytes [10]. However, the implication of fPRL1 in inflamed synovium remains unclear. Further studies how the locally synthesized A-SAA by SC affects joint destruction and how the regulation of fPRL1 on SC might therapeutically contributes to the control of joint inflammation will be needed.

REFERENCES

1. Urieli-Shoval, S. et al. Widespread expression of serum amyloid A in histologically normal human tissues: predominant localization to epithelium. *J. Histochem. Cytochem.,* **46**, 1377, 1998.

2. Kumon, Y. et al. Local expression of acute-phase serum amyloid A (A-SAA) mRNA in rheumatoid arthritis synovial tissue and cells. *J. Rheumatol.,* **26**, 758, 1999.

3. Kumon, Y. et al. Transcriptional regulation of serum amyloid A1 gene expression in human aortic smooth muscle cells involves CCAAT/enhancer binding proteins (C/EBP) and is distinct from HepG2 cells. *Scand. J. Immunol.,* **56**, 504, 2002.

4 Kumon Y. et al. Acute-phase serum amyloid A (A-SAA) is chemotactic for cultured human aortic smooth muscle cells. in *Amyloid and Amyloidosis 2001. The proceedings of the IXth International symposium on amyloidosis,* Bely, M. and Apathy, A. Eds., Budapest: PXP Elso Magyer Digit-lis Nyomda Rt., 2001, 96.

5. Kumon, Y. et al. A longitudinal analysis of alteration in lecithin-cholesterol acyltransferase and paraoxonase activities following laparoscopic cholecystectomy relative to other parameters of HDL function and the acute phase response. *Scand. J. Immunol.,* **48**, 419, 1998.

6. Kumon, Y. et al. Dexamethasone, but not interleukin-1 alone, upregulates acute-phase serum amyloid A (A-SAA) gene expression and production by cultured human aortic smooth muscle cells. *Scand. J. Immunol.,* **53**, 7, 2001.

7. Badolato, R. et al. Serum amyloid A is a chemoattractant: Induction of migration, adhesion, and tissue infiltration of monocytes and polymorphonuclear leukocytes. *J. Exp.Med.,* 180, 203, 1994.

8. Pruzanski, W. et al. Serum amyloid A protein enhances the activity of secretory non-pancreatic phospholipase A2. *Biochem. J.,* **309**, 461, 1995.

9. Mitchell, T.I. et al. The acute phase reactant serum amyloid A (SAA3) is a novel substrate for degradation by the metalloproteinases collagenase and stromelysin. *Biochim. Biophys. Acta.* **1156**, 245, 1993.

10. Le, Y., Oppenheim, J.J., and Wang, J.M. Pleiotropic roles of formyl peptide receptors. *Cytokine and Growth Factor Reviews,* **12**, 91, 2001.

SERUM AMYLOID A (SAA) INDUCES CYTOKINE PRODUCTION IN HUMAN MAST CELLS

K. Niemi[1], M.H. Baumann[1] and K.K. Eklund[2]

1) Protein Chemistry Unit, Institute of Biomedicine, University of Helsinki, Finland 2) Department of Medicine, Division of Rheumatology, Helsinki University Central Hospital, Finland

Secondary amyloidosis or AA amyloidosis is the most common form of amyloidosis worldwide. It occurs secondary to chronic inflammatory diseases such as rheumatoid arthritis, Crohns disease, Familial Mediterranean Fever or as a result of chronic infection (1). In AA amyloidosis the fibrillar part of amyloid deposits comprises of amyloid A (AA) which is derived from a circulating precursor protein, serum amyloid A (SAA) (2). SAA is an acute-phase protein and its level in the serum may be induced as much as 1000-fold during acute-phase response to inflammation, infection or trauma. SAA is mainly produced in the liver but recently the expression of SAA mRNA and SAA protein has also been demonstrated in a variety of other cell types (3), in the brain tissue of Alzheimer patients (4) and in inflamed rheumatoid arthritis synovium (5). SAA production is up-regulated by proinflammatory mediators, most importantly interleukins (IL)-1 and -6 and tumor necrosis factor (TNF)-α (6). The physiological function of SAA is not known but it has been shown to be capable of inducing the production of cytokines in THP-1 monocytes and lymphocytes (7,8).

In this study we have investigated the possible connection between inflammatory responding cells and the high concentration of circulating SAA in acute-phase inflammation. Here we show that SAA is cabable of inducing cytokine production in human mast cells and that this induction might be mediated through a G-protein coupled pathway.

MATERIALS AND METHODS

Human mast cells (HMC-1) were cultured in the presence of recombinant human apoSAA (rhSAA, 3-120 μg/ml). Cytokine concentrations in the cell culture supernatant after 24 hours were determined by ELISA.

RESULTS

As demonstrated in Fig 1 a significant and dose-dependent increase in cytokine production was observed in the presence of rhSAA. The highest rhSAA concentration used (120 μg/ml) increased the TNF-α production by 25-fold and IL-1β production by 38-fold. In the case of TNF-α this induction was comparable to that observed with the combination of PMA and Ca^{2+}-ionofore which is a very powerful stimulus of cytokine production in HMC-1 cells. Human albumin and heat-denatured rhSAA had no significant effect on cytokine production, suggesting that the effect of SAA is specific and that native conformation of SAA is required for the effect.

Interestingly, the incubation of rhSAA (60 μg/ml) together with PMA and Ca^{2+}-ionophore resulted in further stimulation of TNF-α production to more than 100-fold compared to the control, on average 2400 pg/ml/1×10^6 cells (data not shown). This suggests that the effect of SAA is mediated via a pathway other than calcium or

PKC-mediated signal transduction. Therefore we have also studied the effect of pertussis toxin (PTX), a known G-protein blocker, on SAA-induced cytokine production. According to our preliminary results, PTX seems to inhibit the cytokine production in mast cells.

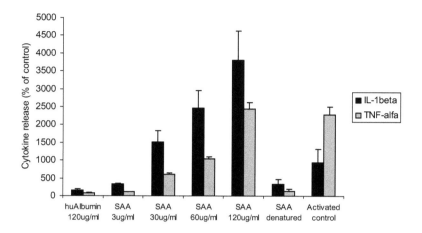

Figure 1. HMC-1 cell stimulation with rhSAA. HMC-1 cells were incubated in the presence of rhSAA for 24 hours at +37°C after which cytokine concentrations in the cell culture were determined with ELISA. Activation with the combination of PMA and Ca^{2+} ionophore was used as a positive control (activated control). Values of control (non-activated) cells are set to 100 %. Values are given as mean ± SEM.

DISCUSSION

Mast cells are powerful cells of the immune system which reside abundantly in connective tissue and in mucosal membranes. In addition to histamine, mast cells produce a wide variety of proinflammatory mediators such as various cytokines (9). Mast cells are also a rich source of tissue proteolytic activity (10). A pathogenic role for mast cells has been implicated in allergic diseases and recently also in inflammatory diseases, rheumatoid arthritis in particular (11). The distribution of mast cells in tissues (gut mucosa adjacent to blood vessels, skin and kidney) seems to be very similar to the tissue distribution of amyloid deposits in secondary amyloidosis. Increased density of mast cells has been found e.g. in kidney in renal amyloidosis (12).

SAA has been found to induce mast cell adhesion by a RGD-mediated manner (13) and to be chemotactic for mast cells by a PTX-sensitive signal transduction pathway (14). For the few cell types studied so far, the effect of SAA seems to be mediated by a G-protein linked receptor, FPRL1/LXA4R (formyl peptide receptor like 1 or lipoxin a_4 receptor). Up to now, the role of this receptor has been demonstrated in SAA-stimulated chemotaxis in phagocytes (15) as well as in SAA-induced IL-8 secretion from neutrophils (16). In addition, it has been verified that also the Alzheimer's amyloid β peptide (residues 1-42) and the prion peptide PrP106-126 utilize the same receptor (17).

The results of the present study show that SAA can induce significant production of cytokines in human mast cells, and that the native form of SAA is required for this induction. Also, according to our preliminary results, PTX inhibits the cytokine production. This is in a agreement with the hypothesis of a G-protein mediated signalling pathway of SAA. SAA-induced cytokine production in mast cells may have biological implications. Mast cell derived cytokines such as TNF-α can initiate local inflammatory reaction resulting in tissue and organ damage which in turn may also further increase the levels of SAA. Whether mast cell derived proteases are also involved in the degradation of SAA, in the process of amyloid formation and deposition in tissues is currently being studied by us.

REFERENCES

1. Rocken C, Shakespeare A. Pathology, diagnosis and pathogenesis of AA amyloidosis. *Virchows Arch* 2002 Feb;440(2):111-22.

2. Husebekk A, Skogen B, Husby G, Marhaug G. Transformation of amyloid precursor SAA to protein AA and incorporation in amyloid fibrils in vivo. *Scand J Immunol* 1985 Mar;21(3):283-7.

3. Urieli-Shoval S, Cohen P, Eisenberg S, Matzner Y. Widespread expression of serum amyloid A in histologically normal human tissues. Predominant localization to the epithelium. *J Histochem Cytochem* 1998 Dec;46(12):1377-84.

4. Liang JS, Sloane JA, Wells JM, Abraham CR, Fine RE, Sipe JD. Evidence for local production of acute phase response apolipoprotein serum amyloid A in Alzheimer's disease brain. *Neurosci Lett* 1997 Apr 4;225(2):73-6.

5. O'Hara R, Murphy EP, Whitehead AS, FitzGerald O, Bresnihan B. Acute-phase serum amyloid A production by rheumatoid arthritis synovial tissue. *Arthritis Res* 2000;2(2):142-4.

6. Jensen LE, Whitehead AS. Regulation of serum amyloid A protein expression during the acute-phase response. *Biochem J* 1998 Sep 15;334 (Pt 3)(Pt 3):489-503.

7. Patel H, Fellowes R, Coade S, Woo P. Human serum amyloid A has cytokine-like properties. *Scand J Immunol* 1998 Oct;48(4):410-8.

8. Furlaneto CJ, Campa A. A novel function of serum amyloid A: a potent stimulus for the release of tumor necrosis factor-alpha, interleukin-1beta, and interleukin-8 by human blood neutrophil. *Biochem Biophys Res Commun* 2000 Feb 16;268(2):405-8.

9. Gordon JR, Burd PR, Galli SJ. Mast cells as a source of multifunctional cytokines. *Immunol Today* 1990 Dec;11(12):458-64.

10. Miller HR, Pemberton AD. Tissue-specific expression of mast cell granule serine proteinases and their role in inflammation in the lung and gut. *Immunology* 2002 Apr;105(4):375-90.

11. Eklund KK, Joensuu H. Treatment of rheumatoid arthritis with imatinib mesylate: clinical improvement in three refractory cases. *Ann Med* 2003;35(5):362-7.

12. Toth T, Toth-Jakatics R, Jimi S, Takebayashi S. Increased density of interstitial mast cells in amyloid A renal amyloidosis. *Mod Pathol* 2000 Sep;13(9):1020-8.

13. Hershkoviz R, Preciado-Patt L, Lider O, Fridkin M, Dastych J, Metcalfe DD, Mekori YA. Extracellular matrix-anchored serum amyloid A preferentially induces mast cell adhesion. *Am J Physiol* 1997 Jul;273(1 Pt 1):C179-87.

14. Olsson N, Siegbahn A, Nilsson G. Serum Amyloid A induces chemotaxis of human mast cells by activating a pertussis toxin-sensitive signal transduction pathway. *Biochem Biophys Res Commun* 1999 Jan 8;254(1):143-6.

15. Su SB, Gong W, Gao JL, Shen W, Murphy PM, Oppenheim JJ, Wang JM. A seven-transmembrane, G protein-coupled receptor, FPRL1, mediates the chemotactic activity of serum amyloid A for human phagocytic cells. *J Exp Med* 1999 Jan 18;189(2):395-402.

16. He R, Sang H, Ye RD. Serum amyloid A induces IL-8 secretion through a G protein-coupled receptor, FPRL1/LXA4R. *Blood* 2003 Feb 15;101(4):1572-81.

17. Le Y, Oppenheim JJ, Wang JM. Pleiotropic roles of formyl peptide receptors. *Cytokine Growth Factor Rev* 2001 Mar;12(1):91-105.

THE OCCURRENCE OF PROTEIN-AA SPECIFIC ANTIBODIES IN EXPERIMENTALLY INDUCED MURINE REACTIVE AMYLOIDOSIS

Sofia Nyström and Gunilla T Westermark

Department of Cell Biology, Faculty of Health Sciences, Linköping University, Sweden

Reactive systemic amyloidosis is a complication to rheumatoid arthritis and other recurrent inflammatory disorders. Protein AA, an N-terminal cleavage product of the acute phase protein serum amyloid A (SAA), forms amyloid fibrils that are deposited in tissue as amyloid aggregates (1). A prolonged elevation of circulating SAA is a prerequisite for developing AA amyloidosis (2). There is a well established experimental model for AA amyloidosis in mice. Animals are repeatedly exposed to an inflammatory stimulus. Amyloid deposition will be quick and massive if an intravenous injection of protein, extracted from AA amyloid-laden mouse tissue, is given concomitantly with the first inflammatory stimulus (3). The protein extract is referred to as amyloid enhancing factor, AEF. Amyloid deposition is a dynamic process, and deposits have been shown to undergo resolution and relocation in experimental models and in humans as well, if the underlying disease is treated (4, 5). Not much is known about the mechanism(s) involved in amyloid resolution. There have been several studies showing resolution of amyloid plaques in Alzheimer's disease by triggering a humoral immune reaction against Aβ-protein (6). In an experimental model, resolution of light chain amyloidosis was accelerated, when animals were treated with amyloid-specific antibodies (7). There is evidence that auto-antibodies directed against protein AA occur in patients with rheumatoid arthritis (8). Since it is possible to promote amyloid regression by immunological modulation, we believe that the natural resolution of amyloid that occurs in reactive amyloidosis is mediated by an adaptive immunological mechanism.

This study confirms the presence of naturally occurring, native, AA-amyloid specific antibodies in murine experimental amyloidosis. We do also show that the native monoclonal AA-amyloid specific antibody B11, produced by conventional hybridoma-technique, when used for passive immunization, tends to decrease amyloid deposition in experimental reactive AA amyloidosis.

MATERIALS AND METHODS

Reactive amyloidosis was induced in a female out-bred NMRI mouse using silver nitrate and AEF. Silver nitrate injections were repeated once weekly, six times. After another five months, a silver nitrate booster was given and the animal was killed three days later. Spleen was collected. By conventional hybridoma technique, splenocytes were fused with non secreting sp2/0 myeloma cells. In the first step, ELISA technique was used for selection of clones producing AA-amyloid specific antibodies. ELISA 96 well microtiter plates were coated with protein AA ($10 \mu g$/ well) extracted from amyloid laden mouse liver. The next step of clone selection was immunohistochemistry on paraffin embedded tissue sections of mouse spleen, with deposits of AA-amyloid.

Hybridomas producing antibodies with AA-amyloid reactivity were sub cloned twice and the Ig-isotype was determined.

To investigate the effect on amyloid deposition of passive immunization with the native monoclonal AA-amyloid specific antibody B11, reactive amyloidosis was induced with silver nitrate and AEF in 20 female NMRI mice. Animals were divided into two groups. Prior to amyloid induction the experimental group (n=10) received 1 ml of B11 culture supernatant intra peritoneally (i.p.). Control group (n=10) received 1 ml culture supernatant i.p. of an irrelevant monoclonal antibody (anti human PC2) of the same isotype, IgM. Antibody culture supernatants were administrated three times. Silver-nitrate injections were repeated once after induction. On day 15, after amyloid induction, animals were killed. Spleens were collected and paraffin-embedded. Tissue sections, 7µm, were stained with Congo red and the amount of amyloid was estimated in polarized light.

RESULTS

The absorbance of B11 culture supernatant was 0.711 at 405 nm. Using immunohistochemistry, the B11 culture supernatant showed a sharp and specific immuno-reaction with murine AA amyloid in spleen, after pre-treatment of tissue sections with 20mM citrate buffer, pH6.0.

Two animals in the experimental group and two animals in the control group died during the experiment. To evaluate the effect on amyloid deposition of passive immunization with the native monoclonal AA-amyloid specific antibody B11, amyloid deposition in mouse spleen was graded I-V, table 1.

Table1. Grading of Congo positive amyloid deposits in mouse spleen

I	No amyloid
II	Very thin focal deposits at few follicles
III	Thin perifollicular amyloid at many follicles
IV	Moderate amyloid deposits around most follicles
V	Extensive amyloid deposits

In the experimental group no animal showed extensive amyloid deposits in spleen and no amyloid was found in three of eight animals. In the control group extensive amyloid deposits were found in four of eight animals and all animals showed some degree of amyloidosis.

DISCUSSION

Producing a hybridoma, B11, secreting AA amyloid specific antibodies, by using splenocytes from an animal suffering from reactive AA amyloidosis, proofs that there exists autoantibodies directed against AA amyloid or protein AA in mice. This study is still preliminary and absorption tests of antibody specificity are still lacking. But since immunoreactivity is restricted to amyloid deposits and the unspecific binding is low, we are convinced that the antibody is amyloid specific rather than SAA specific. The presence of amyloid specific autoantibodies may contribute to the natural resolution of amyloid deposits, occurring when inflammation ceases. It does also support the hypothesis that amyloid deposition may trigger an adaptive immune response.

We also present results indicating the possibility to decrease amyloid deposition by passive immunization with amyloid specific antibodies, in reactive amyloidosis in mice. It is a small study with a considerable falling off, 20%, in both the experimental and control group. The non planned deaths of animals took place late in the experiment and were probable due to repeated interventions. The decrease in amyloid deposition may be caused by a structural mechanism. Antibodies binding to the amyloid and interferes with the fibrillar structure, preventing growth of fibrils. Another mechanism of action may be that binding of antibody triggers the animal's immune response and macrophages are recruited to resolve the amyloid. However, it is not excluded that antibodies act by binding to circulating SAA or protein AA.

REFERENCES

1.   Sletten, K., Husby, G. (1974).The complete amino-acid sequence of non-immunoglobulin amyloid fibril protein AS in rheumatoid arthritis. Eur J Biochem., **41**, 117-25.

2.   Cunnane, G., Whitehead, A.S. (1999). Amyloid precursors and amyloidosis in rheumatoid arthritis. Baillieres Best Pract Res Clin Rheumatol., **13**, 615-28.

3.   Lundmark, K. Studies on Pathogenesis of Experimental AA Amyloidosis. (2001). Linköping University Medical Dissertation No 711.

4.   Waldenström, H. (1927). Uber das Entstehen und Verschwinden des Amyloids beim Menschen. Klin. Wschr., **47**, 2235-47.

5.   Wegelius, O. (1982). The resolution of amyloid substance. Acta Med Scand., **212**, 273-75.

6.   Monsonego, A, Weiner, H.L. (2003). Immunotherapeutic approaches to Alzheimer's disease. Science, **302**, 834-38.

7.   Hrncic, R., Wall, J., Wolfenbarge, D.A., Murphy, C.L., Schell, M., Weiss, D.T., Solomon, A. (2000). Antibody-mediated resolution of light chain-associated amyloid deposits. Am J Pathol., **157**, 1239-46.

8.   Maury, C.P., Teppo, A.M. (1988). Antibodies to amyloid A protein in rheumatic diseases. Rheumatol Int., **8**, 107-111.

9.   Kohler, G., Milstein, C. (1975). Continuous cultures of fused cells secreting antibody of predefined specificity. Nature, **256**, 495-97.

AGRIN IS A COMPONENT OF AA AMYLOID

T.N. Wien[1], R. Sørby[2], T. Landsverk[2], G. Husby.[1]

1. Department of Rheumatology / Institute of Immunology, Rikshospitalet, University of Oslo, Norway

2. Department of Basic Sciences and Aquatic Medicine, Norwegian School of Veterinary Science, Oslo, Norway

Agrin is a large heparan sulphate proteoglycan (HSPG) best known for its key role in acetylcholin receptor clustering in the neuromuscular synapse. The tissue distribution of agrin is widespread, and agrin has received increasing attention as new roles for this HSPG are disclosed.

HSPGs are found associated with all amyloids, and most studies have focused on perlecan. However, a glomerular basement membrane HSPG demonstrated in AA and AL glomerular amyloid ten years ago was later confirmed to be agrin (1). Recently, agrin accumulation was demonstrated in Aβ deposits (2) and *in vitro* agrin was shown to accelerate Aβ fibril formation as well as protect these fibrils from proteolysis (3).

We have previously demonstrated that splenic GAGs increase in experimental AA induced in the mink and that HS is the dominating splenic GAG in mink (4).

MATERIALS AND METHODS

Amyloid was induced in mink with lipopolysaccharide (LPS) (4). Amyloid in splenic sections was confirmed by Congo red birefringence. A sheep polyclonal antiserum against agrin from rat glomerular basement membrane (GR-14) (5) was kindly provided by Jacob van den Born (VU University Medical Center, The Netherlands). Reactivity to agrin in splenic cryosections was detected with the Vectastain ABC kit (Vector Laboratories) and visualized with diaminobenzidine (DAB).

RESULTS

Extensive splenic amyloid, characterized by positive Congo red staining with green birefringence in polarized light, was demonstrated in the LPS-treated animals. None of the controls showed any amyloid. The amyloid deposits were found in the germinal centres, the marginal zone enclosing the white pulp, the ellipsoidal walls and modest in the red pulp between the ellipsoids (Figure 1a). Immunohistochemical staining for agrin showed accumulation of this basement membrane HSPG in the splenic amyloid deposits congruent with the Congo red birefringence (Figure 1b). Furthermore, in all spleens strong agrin reactivity was demonstrated in nerve tissue and the inner subendothelial layers of arterioles, and weaker agrin reactivity was seen in connective tissue trabeculae.

Figure 1. Serial sections of amyloid mink spleen x 200.

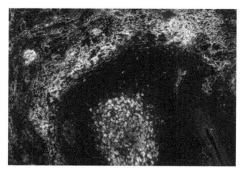
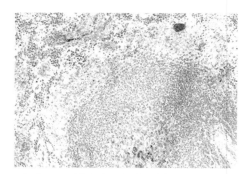

a. Congo red staining viewed under fluorescent light.   b. Immunohistochemical staining for agrin

CONCLUSION

We conclude that agrin is a component of AA amyloid also in this experimental mink model.

REFERENCES

1.   van den Born, J.et al., Distribution of GBM heparan sulfate proteoglycan core protein and side chains in human glomerular diseases, *Kidney Int.*, 43, 454, 1993.

2.   Verbeek, M.M.et al., Agrin is a major heparan sulfate proteoglycan accumulating in Alzheimer's disease brain, *Am. J. Pathol.*, 155, 2115, 1999.

3.   Cotman, S.L., Halfter, W., and Cole, G.J., Agrin binds to beta-amyloid (Abeta), accelerates abeta fibril formation, and is localized to Abeta deposits in Alzheimer's disease brain, *Mol. Cell Neurosci.*, 15, 183, 2000.

4.   Wien, T.N.et al., Characterization of proteoglycans and glycosaminoglycans in splenic AA amyloid induced in mink, *Scand. J. Immunol.*, 52, 576, 2000.

5.   Raats, C.J.et al., Differential expression of agrin in renal basement membranes as revealed by domain-specific antibodies, *J. Biol. Chem.*, 273, 17832, 10-7-1998.

AUTOFLUORESCENCE OF AA AMYLOID

T.N. Wien[1], R. Sørby[2], T. Landsverk[2], G. Husby.[1]

1. Department of Rheumatology / Institute of Immunology, Rikshospitalet, University of Oslo, Norway
2. Department of Basic Sciences and Aquatic Medicine, Norwegian School of Veterinary Science, Oslo, Norway

In our immunofluorescence investigations of basement membrane components in mink spleen, we encountered a problem of autofluorescence of AA deposits. The autofluorescence could thus not be clearly distinguished from the specific fluorescence obtained with the immunochemical reaction.

Many biological materials exhibit autofluorescence and this property is exploited in medicine and bioscience, e.g. in diagnostic endoscopies and in fluorescence activated cell sorting (FACS). Reports on autofluorescence of amyloid in the literature are limited and mainly describes Aβ (1).

We here report autofluorescence of AA amyloid with a pattern of excitation and emission that differs from what has been described for Aβ.

MATERIALS AND METHODS

Fluorescent properties were examined in splenic cryosections from 12 mink with lipopolysaccharide (LPS)-induced AA amyloidosis, 14 LPS-induced mink with no amyloid and 12 controls. We also examined cryosections from a murine spleen with AA amyloid as well as from four human amyloid cases of AA, AL and ATTR (lung, kidney, myocardium) all generously provided by Gunilla Westermark (Linköping University, Sweden) and paraffin sections of 2 cases of human AA amyloid (liver, lung, kidney) generously provided by Prof. Ole Petter Clausen (University of Oslo, Norway). Unfixed, acetone- or formol-calcium-fixed cryosections and deparaffinized sections were mounted in a non-fluorescent medium (polyvinyl alcohol, pH 8.2). The sections were examined in a Leica fluorescence microscope equipped with filter cubes with excitation range in the UV (A4), blue (L5) and green (N3 and TX2) wavelength areas (Table). Sections from the same organs were stained with Congo red and viewed under polarized and fluorescent light (2).

Table. Range of filter cubes used for fluorescence microscopy.

Leica filter cubes	Excitation range	Excitation filter	Dichromatic mirror	Suppression filter
A4	UV	360 ± 40	400	470 ± 40
L5	Blue	480 ± 40	505	527 ± 30
N3	Green	546 ± 12	565	600 ± 40
TX2	Green	560 ± 40	595	645 ± 75

www.leica-microsystems.com

RESULTS

Independent of fixation the amyloid deposits in 10 out of the 12 amyloid mink spleens exhibited extensive extracellular autofluorescence under green light excitation (TX2 and N3). The autofluorescence (Figure 1a) corresponded to the amyloid deposits of marginal zone, germinal centre, ellipsoids and red pulp, as verified by green birefringence under polarized light and Congo red fluorescence (Figure 1b). In spleens from control mink or LPS-injected mink that had not developed splenic amyloid, no extracellular autofluorescence could be demonstrated, only weak autofluorescence of a few scattered cells. Neither UV light excitation (A4), known to induce Aβ autofluorescence, nor blue light (L5) excitation induced autofluorescence in mink splenic AA amyloid. No autofluorescence was detected in the amyloid deposits of the murine or human cases examined.

Figure 1. Amyloid mink spleen viewed in green light excitation (TX2) x 200.

a. Autofluorescence of amyloid spleen, fixed in acetone, unstained.

b. Congo red fluorescence of amyloid spleen (same as in a), fixed in acetone and Congo red stained

CONCLUSION

We conclude that use of fluorophores and immunofluorescence calls for awareness of the autofluorescent properties of amyloid to avoid false positives. This property is apparently not common to all amyloids.

REFERENCES

1. Thal, D.R.et al., UV light-induced autofluorescence of full-length Abeta-protein deposits in the human brain, *Clin. Neuropathol.*, 21, 35, 2002.

2. Linke, R.P., Highly sensitive diagnosis of amyloid and various amyloid syndromes using Congo red fluorescence, *Virchows Arch*, 436, 439, 2000.

FIBRILLAR PROTEINS OCCURING IN THE NATURE CAN INDUCE AA AMYLOIDOSIS IN MICE

K. Lundmark[1], G.T. Westermark[2], A. Olsén[3], P. Westermark[4]

[1]Division of Pathology, Karolinska University Hospital - Huddinge, Sweden, [2]Division of Cell Biology, Linköping University, Sweden, [3]Division of Clinical Immunology, Göteborg University, Sweden, [4]Department of Genetics and Pathology, Uppsala University, Sweden

Secondary (AA) amyloidosis is a not uncommon complication of chronic inflammatory diseases. The insoluble protein fibrils with high content of β-pleated structure, consisting of amyloid protein A, are deposited in different organs. The precursor protein is serum amyloid A (SAA). Since there is not one specific amyloidogenic form of SAA identified in human, the reason why only a fraction of patients with longstanding inflammation and persistent high SAA plasma levels develops AA amyloidosis is not known.

AA amyloidosis can be induced in mice provoked by inflammatory challenge. The time for development of amyloid is dramatically shortened when the animals concomitantly receive the extract from a tissue with amyloid, which is named Amyloid Enhancing Factor (AEF). AEF is supposed to contain a seed for amyloid formation and nucleation is a widely accepted mechanism in amyloidogenesis. AA amyloid fibrils and synthetic amyloid-like fibrils were shown to serve as nucleus for fibril formation in experimental AA amyloidosis [1-3]. Similar results were obtained in studies with silk-derived fibrils [4].

Here we present nucleating features of two different naturally occurring amyloid-like protein fibrils in murine model of AA amyloidosis.

MATERIALS AND METHODS

We used two different fibrils with high content of β-pleated sheet: amyloid-like fibrils prepared from *Bombyx mori* silk [4], YMel *Escherichia coli* strain with amyloid-like fibers, curli, on the bacterial surface [5] and purified curli prepared as described [6]. All fibril preparations were evaluated with Congo red staining for green birefringence and electron microscopy for fibril formation.

We injected 5-10 weeks old FVB mice intravenously with 0.1 ml of sonicated fibril preparations together with inflammatory stimulation with a subcutaneous injection of 0.5 ml 1% silver nitrate. Control animals received 0.9% NaCl intravenously and 1% silver nitrate administrated exactly as the experimental groups. Subcutaneous injections of 0.1 ml 1% silver nitrate were repeated in all groups once weekly. The animals were sacrificed 16 days after first treatment. Congo red stained spleen smears and paraffin sections were evaluated in polarised light for typical green birefringence as described [3]. Probability of difference from control groups was determined using Fischer's exact test with INSTAT 2.01 software.

RESULTS

Silk and purified curli bound Congo red and showed green birefringence in polarised light, indicating amyloid-like organisation of the protein molecules. Electron microscopically both silk and curli preparations contained fine nonbranching fibrils (Fig. 1 a, b) similar to the native AA fibrils in the AEF preparation [1] (Fig. 1c).

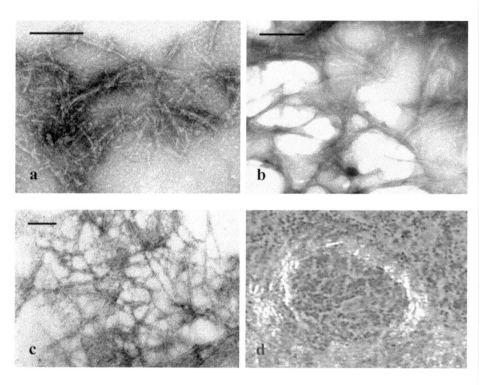

Fig.1. Electron micrographs of: **(a)** silk fibrils, **(b)** curli, **(c)** AA-fibrils extracted from murine amyloid. Fibril samples were negatively contrasted with uranyl acetate on Formvar coated copper grids. Bars: 200 nm. **(d)** Congo red stained amyloid deposit in the mouse treated with silk and 1% silvernitrate.

We found that all fibril preparations were sufficient to induce amyloidosis in experimental animals, and thus possessed amyloid enhancing activity (Table 1).

Table 1. Splenic amyloid deposits induced by treatment with different amyloid-like fibrils and 1%AgNO$_3$ compared to controls treated with 0.9% NaCl and 1%AgNO$_3$.

Treatment	Number of AgNO$_3$ injections	Duration of experiment	Incidence of mice with amyloid	Amyloid grade	P
Silk fibrils	3	16d	5/7	1+ to 3+	P=0.007
NaCl	3	16d	0/8	0	
YMel *E.coli*	3	16d	4/7	2+ t0 4+	P=0.03
NaCl	3	16d	0/8	0	
Purified curli	3	16d	3/8	1+ to 3+	P=0.2
NaCl	3	16d	0/7	0	

CONCLUSION

These results suggest that exposure (via ingestion or inhalation) to naturally occurring amyloid-like fibrillary proteins may introduce seeds that can start the nucleation in predisposed individuals, i.e. those with persistent high SAA production. This may offer an explanation why only a fraction of patients with longstanding inflammation develops AA amyloidosis. Hypothetically, environmental factors may be important risk factors not only for prion diseases, but also for AA amyloidosis and perhaps even other human amyloid diseases.

ACKNOWLEDGMENTS

This work was supported by Swedish Research Council (Project No. 5941), the Swedish Association against Rheumatism, the "Förenade Liv" Mutual Group Life Insurance Company, Stockholm, Sweden, Landstinget in Östergötland (Project No. 2003/050).

REFERENCES

1. Ganowiak,K., Hultman,P., Engström, U., Gustavsson, Å., and Westermark, P. (1994). Fibrils from synthetic amyloid-related peptides enhance development of experimental AA-amyloidosis in mice, *Biochem. Biophys. Res. Commun.* **199**, 306-12.

2. Johan,K., Westermark, G.T., Engström, U., Gustavsson, Å., Hultman, P., and Westermark P. (1998). Acceleration of amyloid protein A amyloidosis by amyloid-like synthetic fibrils. *Proc. Natl. Acad. Sci. USA,* **95**, 2558-63.

3. Lundmark, K, Westermark, G.T., Nyström, S., Murphy, C.L., Solomon, A., and Westermark, P. (2002) Transmissibility of systemic amyloidosis by a prion-like mechanism. *Proc. Natl. Acad. Sci. USA,* **10,** 6979-84.

4. Kisilevsky, R., Lemieux, L., Boudreau, L., Yang, D.S., and Fraser, P. (1999) New clothes for amyloid enhancing factor (AEF): silk as AEF. *Amyloid,* **6**, 98-106. 5. Olsén, A., Jonsson, A. and Normark, S. (1989) Fibronectin binding mediated by a novel class of surface organelles on Escherichia coli. *Nature,* **338**, 652-55.6. Collinson, S.K., Emödy, L., Muller, K.H., Trust, T.J., and Kay, W.W. (1991) Purification and characterization of thin, aggregative fimbriae from Salmonella enteritidis. *J. Bacteriol.,* **173**, 4773-81.

AMYLOID PRODUCTION BY HUMAN MONOCYTE CULTURES

B. Kluve-Beckerman, N. Magy, J. J. Liepnieks, and M. D. Benson

Indiana University School of Medicine and Veterans Affairs Medical Center, Indianapolis, IN, USA

INTRODUCTION

Amyloid formation by mouse peritoneal macrophages cultured in the presence of recombinant mouse SAA1.1 has been described previously (1). We now present human monocytes isolated from peripheral blood as an alternative cell culture model. In contrast to *in vitro* (test tube) studies, amyloid formation in macrophage and monocyte cultures occurs under physiologic conditions, *i.e.*, cells are maintained in neutral pH medium containing SAA at a concentration typical of that seen in acute phase serum (150 µg/ml). By using these model systems, events in AA amyloid pathogenesis, and importantly those initiating formation of an amyloid nidus, can be studied at the level of individual cells and before extracellular deposition and possible resorption have become confounding issues. Amyloid formation in macrophage and monocyte cultures is accelerated and augmented by, but not dependent on, the addition of amyloid-enhancing factor (AEF), and is accompanied by C-terminal cleavage of SAA. In agreement with previous findings from other laboratories, our data show nascent amyloid formation in lysosome-derived vesicles (2-5). Additionally, our observations suggest that release of these vesicles to the cell surface is an active metabolic process which does not directly result in cell death. The micrographs presented here depict monocytes engaged in various stages of amyloid formation.

MATERIALS AND METHODS

Human monocytes were isolated from peripheral blood by centrifugation on Histopaque and cultured on 8-well glass chamber slides or coverslips in RPMI medium containing 15% FCS. Amyloid-forming conditions were initiated on day 5 of culture. Mouse rSAA1.1, produced and purified as previously described (6), was added to the medium to establish an SAA concentration of 150 µg/ml. At this concentration, all of the SAA bound to HDL present in the FCS of medium. This was indicated by Sephacryl S-200 chromatography of SAA-containing medium and elution of SAA exclusively in fractions containing HDL. AEF, prepared as a glycerol extract of amyloidotic mouse spleen (7), was added to the medium of some cultures.

Monocytes incubated with Alexa-488-labeled SAA were observed via fluorescence scanning confocal microscopy (Perkin-Elmer UltraView) so that SAA binding, uptake and localization in and on living cells could be followed. To detect amyloid, cells were fixed and stained with Congo red and examined under polarized light to see birefringence. Other cultures were stained immunochemically to identify late endosomes and lysosomes (polyclonal antibody LAMP-1 from Santa Cruz), or perlecan, the core protein of heparan sulfate proteoglycan (monoclonal antibody from Zymed).

RESULTS AND DISCUSSION

Direct visualization of Alexa-488-SAA in living cultures shows SAA binds to and is internalized by monocytes. Less than half of the cells in culture take up SAA; the identity of this population is being investigated. Over time, SAA-containing vesicles having punctate or tubular structure appear to fuse forming large vesicular bodies (Figure 1).

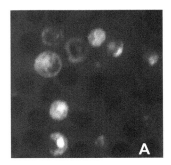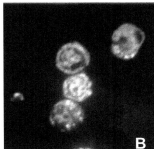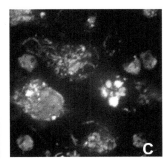

Figure 1. Confocal microscopy of monocytes showing binding and internalization of Alexa-488-SAA. Cells were cultured with Alexa-488-SAA for 4 hours and then with unlabeled SAA for 8 days to induce amyloid formation. Images are selected planar views through cells and were collected at the indicated time.

A.   30 minutes into pulse with Alexa-488-SAA.

B.   20 hours following pulse.

C.   8 days following pulse. Note big vesicular structures.

Amyloid, identified as Congo red-stained material, is first detected intracellularly in large vesicles. Over time, these vesicles are extruded from cells.   The nascent amyloid released in this process accumulates extracellularly around and on top of cells (Figure 2). Amyloid formation in and extrusion from monocytes does not directly result in cell death. Most likely, cells die when extracellular deposition becomes excessive.

Monocytes maintained under amyloid-forming conditions, but not control monocytes, demonstrate formation of large lysosomal (Lamp-1-staining) structures. If cells are not permeabilized prior to immunocytochemistry, only weak intracellular Lamp-1 staining is detected, whereas stronger extracellular staining associated with structures that look like disrupted vesicles is seen. Extracellular lysosomal staining of this nature is present only in cultures actively producing amyloid. We speculate that large lysosomal-derived vesicles transport nascent amyloid to the cell surface.

Monocyte cultures incubated with SAA without AEF for a period of two weeks exhibit little or no immunochemical staining for perlecan. Cultures treated in parallel, however, exhibit Congo red staining in intracellular vesicles and extracellularly, confirming that amyloid formation has occurred. This finding draws into question whether proteoglycans are required for initiation of amyloidogenesis, *i.e.*, for nidus formation. To further investigate this issue, experiments utilizing more sensitive immunofluorescent detection methods are currently being pursued. Monocytes incubated with either SAA and AEF or AEF alone, show strong perlecan staining, both peripherally at cell edges and in large intracellular vesicles. Extracellular amyloid deposits also stain strongly for perlecan. Experiments are in progress to determine the origin and significance of this perlecan. Since the AEF used in this study stains strongly for perlecan, it is reasonable to speculate that the perlecan detected in AEF-treated cultures was contributed by AEF, and this raises the possibility that proteoglycan is an active component of AEF. Another scenario warranting investigation is that robust amyloid formation in SAA/AEF-treated cultures triggers *de novo* perlecan synthesis by monocytes

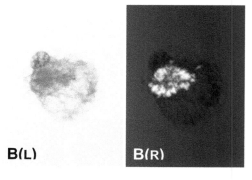

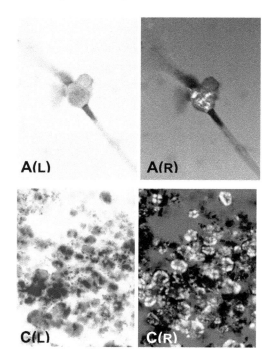

Figure 2. Amyloid in, on and around monocytes, as identified by Congo red staining (L) and birefringence (R).

A.   Nascent amyloid in vesicles (8 days with SAA).

B.   Nascent amyloid extruded from vesicle (8 days with SAA).

C.   Extensive amyloid deposition (14 days with SAA and AEF).

REFERENCES

1.   Kluve-Beckerman B, Liepnieks JJ, Wang L, Benson MD (1999). A cell culture system for the study of amyloid pathogenesis: Amyloid formation by murine peritoneal cells in the presence of recombinant serum amyloid A. *Am J Pathol* **155**, 123-133.

2.   Shirahama TS, Cohen AS (1975). Intralysosomal formation of amyloid fibrils. *Am J Pathol* **81**, 101-116.

3.   Takahashi M, Yokota T, Kawano H, Gondo T, Ishihara T, Uchino F (1989). Ultrastructural evidence for intracellular formation of amyloid fibrils in macrophages. *Virchows Archiv A* **415**, 411-419.

4.   Chronopoulos S, Laird DW, Ali-Khan Z (1994). Immunolocalization of serum amyloid A and AA amyloid in lysosomes in murine monocytoid cells: Confocal and immunogold electron microscopic studies. *J Pathol* **173**, 361-369.

5.   Miura K, Shirasawa H (1994). Amyloid A (AA) fibril formation in renal tubules occurs intracytoplasmically possibly at the site of membrane assembling structures. *Amyloid: Int J Exp Clin Invest* **1**, 107-113.

6.   Kluve-Beckerman B, Yamada T, Hardwick J, Liepnieks JJ, Benson MD (1997). Differential plasma clearance of murine acute phase serum amyloid A proteins SAA1 and SAA2. *Biochem J* **322**, 663-669.

7.   Axelrad MA, Kisilevsky R, Willmer J, Chen SJ, Skinner M (1982). Further characterization of amyloid-enhancing factor. *Lab Invest* **47**, 139-146.

EXPERIMENTAL AMYLOIDOSIS INDUCED BY FREUND'S ADJUVANT AND CASEIN IN TWO MOUSE STRAINS

R.Bradunaite, D.Povilenaite, L.Leonaviciene, D.Vaitkiene, A.Venalis

Institute of Experimental and Clinical Medicine of Vilnius University, Vilnius, Lithuania

INTRODUCTION

Amyloidosis is a disorder of protein folding in which normally soluble proteins are deposited extracellularly in the tissues as abnormal insoluble fibrils that accumulate and damage the structure and function of affected organs [1]. Approximately 45% of all generalized amyloidoses are secondary or AA amyloidosis [2,3]. AA amyloidosis can be induced in mice experimentally through injection of certain chemical or biological compounds, including casein, silver nitrate, Freund's adjuvant (FA), and lipopolysacharide [4,5]. These agents stimulate the production of proinflammatory cytokines, that mediate the inflammatory response by increasing the hepatic synthesis of serum amyloid A protein (SAA) [6]. This molecule serves as the precursor of the polypeptide that is deposited in the spleen, liver, and kidneys of affected animals as Congophilic, green birefringent fibrils, ie, AA amyloid.

Susceptibility to amyloid induction was compared in two strains of mice subjecting each group of BALB/c and C57BL/6 mice to multiple injections of casein (C) or complete FA.

MATERIALS AND METHODS

AA amyloidosis was induced in 67 BALB/c and C57BL/6 male mice in the models of sustained inflammation following C (5 times a week) or FA (once a week) subcutaneous (0.5 ml) injections for 3, 5 and 7 weeks. The animals were decapitated at 3 (1/3 of mice), 5 (the other 1/3) and 7 (the last 1/3) weeks post-injection (p.i.). The leukocyte count and the ESR were determined in the blood. The internal organs were examined macroscopically and weighted. The paraffin embedded sections from kidneys and spleen were stained with haematoxylin and eosin for identification of inflammatory changes. The amount of amyloid was assessed semi-quantitatively by polarizing microscopy after Congo red staining and was graded on a scale of 0 to 3 according to the density of amyloid masses seen under the microscope: 0 – no, 1+ – minimal, 2+ – moderate, and 3+ – heavy amyloid deposits.

The Student's t test was used to evaluate the statistical significance of the differences between the studied groups.

RESULTS

During the study, no mortality was observed in C-treated groups, whereas 13.3% of BALB/c and 6.67% of C57BL/6 mice perished during the first 3 weeks of experiment and one C57BL/6 mouse died two weeks later (5 weeks p.i.) in FA-treated groups. These animals were excluded from the final analysis.

Splenomegaly was revealed in the both mouse strains, but the relative weight of the spleen was 3 times higher in C-treated BALB/c than in C57BL/6 mice at 7 weeks p.i. (Table). ESR was the highest in C57BL/6 mice and the count of leukocytes in BALB/c mice that received C injections during 7 weeks.

BALB/c mice given FA had a progressive rise of amyloid deposits in the spleen, which reached the same level as in C-treated animals (Fig.). Traces of amyloid and minimal amyloid deposits were identified in 20% and 60% of FA-treated animals and in 33.3% and 66.7% of C given mice respectively. At 7 weeks p.i. C some more effectively than FA enhanced amyloid formation in the spleens of C57BL/6 mice: traces of amyloid were observed in 100% of animals given FA injections whereas 14.3% of C-treated mice had traces of amyloid, 57% - minimal, and 28.6% - moderate and heavy amyloid deposits. In the spleen of these mice the most average of amyloid deposits which significantly differed from the status at 3 weeks p.i. ($P < 0.01$) was observed (Fig.).

Table. The weight of organs and blood indices in BALB/c and C57BL/6 mice during the course of amyloidosis induced with Freund's adjuvant (FA) and casein (C)

Mouse strain	Subs-tan-ce	Duration of experi-ment	Relative weight of organs		Blood indices	
			Kidneys (g/kg^{-1})	Spleen (g/kg^{-1})	ESR (mm/h)	Leukocytes (10^9 L)
BALB/c	FA	3 weeks	1.37±0.08	1.65±0.14***	3.50±0.64$^+$	10.98±2.32
		5 weeks	1.84±0.09*	2.07±0.22***	2.00±0.71	8.17±0.47
		7 weeks	1.53±0.06	1.29±0.11***	***1.40±0.24	***7.29±0.53
	C	3 weeks	1.39±0.06	2.51±0.28***	4.17±0.87**	10.27±1.61
		5 weeks	1.48±0.06	3.17±0.26***	2.50±0.22^{++}	10.73±1.63
		7 weeks	1.58±0.06	5.17±0.11***	***3.33±0.21***	***21.98±2.43***
	Healthy mice		1.57±0.06	0.506±0.096	1.20±0.20	7.40±0.66
C57BL/6	FA	3 weeks	1.16±0.16	1.45±0.02***	5.75±1.37$^+$	7.60±1.04
		5 weeks	1.18±0.09	0.96±0.13**	$^{++}$1.75±0.25	***6.80±0.88
		7 weeks	1.15±0.07	0.86±0.10**	***1.50±0.29	$^+$5.63±0.72
	C	3 weeks	1.24±0.04	0.93±0.03***	6.33±0.40***	10.33±1.54
		5 weeks	1.19±0.05	1.92±0.20***	$^{++}$5.33±0.67***	***14.48±0.99***
		7 weeks	1.31±0.07	1.63±0.23***	***6.00±1.02^{++}	$^+$11.72±2.11*
	Healthy mice		1.38±0.12	0.376±0.02	1.33±0.33	6.40±0.15

Note: FA – Freund's adjuvant, C – casein, ESR – erythrocyte sedimentation rate. Symbols on the right - The differences are significant between normal mice and the test groups. Symbols on the left - The differences are significant between the FA- and C-treated groups on the same time of investigation (3. 5 or 7 weeks p.i.). * $P < 0.05$; ** $P < 0.01$; $^+P < 0.02$; $^{++} P < 0.002$; *** $P < 0.001$.

In the kidneys of C-treated BALB/c mice amyloid deposits were more prevalent than in FA-treated animals. Most of the amyloid deposited in C-challenged mice occured after 7 weeks p.i., where moderate amyloid deposits were revealed in 100% of mice. In C-treated animals the average amyloid deposition in the kidneys of BALB/c mice increased about twice in comparison with the status at 3 weeks p.i. ($P < 0.001$). Significant difference in amyloid deposition between FA- and C-treated groups ($P < 0.01$) was obtained (Fig.). The same have got in the kidneys of C-treated C57BL/6 mice (5 and 7 weeks p.i.). Middle amyloid deposits were found in 50% of animals treated with FA and in 71.4% - with C.

Amyloid was mainly confined to the marginal zone of the spleen and located in glomeruli, tubules and renal blood vessels walls of the kidneys.

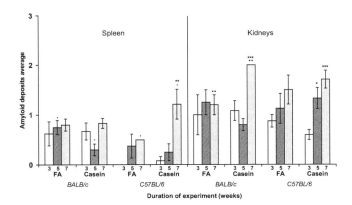

Figure. The average amyloid induction in the spleen and kidneys of BALB/c and C57BL/6 mice during the course of amyloidosis induced with Freund's adjuvant (FA) and casein (C).
The differences are significant between the average amyloid deposition in the groups after 3 weeks post-injection (p.i.) and the average induction of amyloid deposition after 5 or 7 weeks p.i.: * $P < 0.02$; *** $P < 0.001$. The diffirences are significant between FA- and C-treated group: • $P < 0.05$; •• $P < 0.01$
So, deposits of amyloid in the spleen and kidneys were found more frequently in those with C-treated than in those with FA-treated mice of C57BL/6 strain.

DISCUSSION

Our data indicate that both mouse strains are susceptible to amyloid induction. Amyloid deposition in BALB/c mice as well as in C57BL/6 mice was observed in kidneys and spleen, although C57BL/6 mice that received C developed amyloid deposits to a higher extent than did BALB/c and C57BL/6 mice, given FA. We think that FA-induced amyloidosis according to its course and pathological process, correspond to some degree the amyloidosis developing during RA. In the developed world chronic rheumatic and connective tissue diseases have been supposed to be the most frequent predisposing conditions to AA amyloidosis in man [7,8], with the predominant involvement of the spleen, liver, kidneys, and other parenchymal organs.

It is known that amyloidogenic stimulation in C-induced amyloidosis enhances the synthesis of proteoglycans, which are related to murine spleen reactive AA amyloid. This upregulation of proteoglycan expression precedes amyloid fibril formation [5]. The abundance of basement membrane glycosaminoglycan in the glomeruli is the main factor for renal AA deposition [9].

In conclusion our data indicate that different experimental conditions impose striking effects on the occurrence, frequency and distribution of amyloid deposits within the organs, and variation due to agent strain and mouse strain.

REFERENCES

1.   Pepys, M.B., *Amyloidosis*, Weatherall, D.J., Ledingham, J.G.G., and Warell, D.A., Eds., The Oxford Textbook of Medicine, 3rd ed., Oxford University Press, Oxford, UK, 1996, 2, 1512.

2.   Strege, R.J., Saeger, W., and Linke, R.P., Diagnosis and immunohistochemical classification of systemic amyloidoses. Report of 43 cases in an unselected autopsy series, *Virchows Arch.*, 433, 19, 1998.

3.   Hazenberg, B.P.S. and Van Rijswijk, M.H., Where has secondary amyloid gone?, *Ann. Rheum. Dis.*, 59, 577, 2000.

4.   Kisilevsky, R. and Young, I.D., *Pathogenesis of amyloidosis*. Bailliere's Clinical Rheumatology: Reactive Amyloidosis and the Acute-Phase Response, Husby G., ed., 8th London, Bailliere Tindall, 1994, 613.

5. Stenstad, T., Magnus, J.H., and Husby, G., Characterization of proteoglycans associated with mouse splenicc AA amyloidosis, *Biochem. J.*, 303, 663, 1994.

6. Sipe, J.D., Amyloidosis, *Crit. Rev. Clin. Lab. Sci.*, 31, 325, 1994.

7. Ishii, W. et al., Phenol Congo red staining enhances the diagnostic value of abdominal fat aspiration biopsy in reactive AA amyloidosis secondary to rheumatoid arthritis, *J. Intern. Med.*, 42, 400, 2003.

8. Wakhlu, A. et al., Prevalence of secondary amyloidosis in Asian North Indian patients with rheumatoid arthritis, *J. Rheumatol.*, 30, 948, 2003.

9. Mountz, J.D. and Hsu, H.C., Clinical features associated with correction of T-cell senescence: increased acute-phase response, amyloidosis and arthritis, *Dev. Comp. Immunol.*, 21, 509, 1997.

FIBRIN AND UBIQUITIN ENHANCE THE DEVELOPMENT OF CASEIN-INDUCED EXPERIMENTAL AA AMYLOIDOSIS IN C57BL/6 MICE

L.Leonaviciene, D.Povilenaite, R.Bradunaite, D.Vaitkiene, A.Venalis

Institute of Experimental and Clinical Medicine of Vilnius University, Lithuania

INTRODUCTION

Amyloidosis is a protein metabolism disorder where normally soluble proteins are transformed into insoluble fibrillar structures, which are deposited in the extracellular space of organs and tissues, thereby resulting in clinical disease [1]. The pathogenesis of secondary amyloidosis *in vivo* is not well understood. Much of our knowledge of the pathophysiology of this disease is derived from animal models [2] because reactive amyloidosis is the most common form in animals [3]. In the most prevalent experimental approach, amyloidosis is elicited after 2 to 3 weeks, during which susceptible mice are injected repeatedly, usually with azocasein or casein [2,4].

We have developed an experimental model of AA amyloidosis that involves C57BL/6 mice with casein (C)-induced amyloidosis and tried to enhance it by injections of dextran sulphate (DS), fibrin (F), and ubiquitin (Ub).

MATERIAL AND METHODS

48 male C57BL/6 mice were subjected to a conventional amyloid induction protocol receiving 0.5 ml subcutaneous injections of 12% vitamin free C for 5 days a week for period of 3 and 5 weeks. The animals were divided into 4 groups of 12 mice each. A control group (1st gr.) was injected the C only. In addition to C, the 2nd and 3rd groups received subcutaneous 1% DS and 5% F injections respectively once a week, and 4th group - a single intraperitoneal 1 µg/ml injection of Ub. The animals were decapitated at 3 (half of the mice) and 5 weeks (the other half) post-injection (p.i.). The leukocyte count and the ESR were determined in the blood. The internal organs were examined macroscopically and weighed. The indices obtained were compared with the indices for normal (healthy) animals.

The formalin-fixed spleen and kidney specimens were divided into two pieces and embedded in paraffin. The one set of slices was stained with haematoxylin-eosin to determine inflammation scores. Histological grading of amyloid in polarized light after Congo red staining of tissue slices was performed semiquantitatively using a scale of 0 to 3 according to the density of amyloid masses seen under the microscope: 0 – no, 1+ – minimal, 2+ – moderate, 3+ – heavy amyloid deposits.

The Student's t test was used to evaluate the statistical significance of the differences between the studied groups.

RESULTS

A post-mortem examination revealed splenomegaly ($P < 0.001$) in all the test groups in comparison with healthy animals (Table). The relative (3 weeks p.i.) weight of the spleen and kidneys in the 2nd and 4th groups respectively significantly differed from the control (C-treated) group. ESR in C+F- and C+Ub-treated groups markedly differed from the control and after 3 weeks stimulation was the highest in the C+F-treated group.

Table. Effect of dextran sulphate, fibrin, and ubiquitin on the weight of organs and blood indices in C57BL/6 mice with casein-induced experimental amyloidosis

Groups	Duration of experiment	Relative weight of organs		Blood indices	
		Kidneys (g/kg^{-1})	Spleen (g/kg^{-1})	ESR (mm/h)	Leukocytes (10^9 L)
1st	3 weeks	1.47±0.05	2.87±0.16***	4.00±0.45***	14.50±1.05***
(C)	5 weeks	1.23±0.05	2.42±0.14***	2.67±0.21**	15.83±2.35**
2nd	*3 weeks*	1.42±0.15	*2.32±0.15***	5.20±0.49***	17.60±2.31^{++}
(C+DS)	5 weeks	1.30±0.40	2.06±0.27***	5.60±1.43$^+$	19.40±3.29**
3rd	3 weeks	*1.39±0.05*	*2.71±0.20***	$^+$6.00±0.45***	18.47±1.52***
(C+F)	5 weeks	*1.16±0.05*	3.10±0.38***	$^+$3.83±0.31***	18.92±1.92***
4th	*3 weeks*	***1.12±0.06	2.27±0.23***	5.00±0.45***	19.30±3.59**
(C+Ub)	5 weeks	*1.12±0.06*	*2.27±0.21***	*3.67±0.33***	18.57±3.34**
Healthy mice		1.38±0.12	*0.376±0.02*	1.33±0.33	6.40±0.15

Note: C – casein, DS – dextran sulphate, F – fibrin, Ub – ubiquitin, ESR – erythrocyte sedimentation rate. Symbols on the right - The differences are significant between normal mice and the test groups. Symbols on the left - The differences are significant between the C-treated group and the other test groups after 3 and 5 weeks post-injection respectively: * P < 0.05; ** P < 0.01; $^+$ P < 0.02; $^{++}$ P < 0.002; *** P < 0.001.

The amount of amyloid deposited increased progressivelly with the duration of amyloidogenic stimulus. At 3 weeks p.i. traces and minimal deposits of amyloid in the spleen were observed in 33.3% of the mice treated with C+DS and C+Ub, and in 83.3% of the animals given C+F. 50% of the mice in both of the latter groups also had traces of amyloid deposits in the kidneys. The most extensive tissue amyloid deposits were seen in F+C-treated animals at 5 weeks p.i.: moderate (2+) (50%) and heavy (3+) deposits (50%) throughout the spleens, located perifolliculary in the marginal zone of the spleen follicles, were revealed in all the animals of this group.

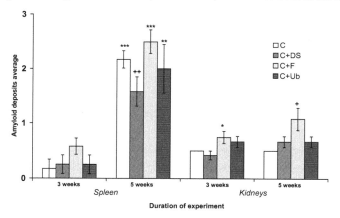

Figure. Effect of dextran sulphate, fibrin, and ubiquitin on average amyloid induction in the spleen and kidneys of C57BL/6 mice with casein-induced amyloidosis.
C – casein, DS – dextran sulphate, F – fibrin, Ub – ubiquitin. The differences are significant between the average induction of amyloid deposition in the groups after 3 weeks post-injection (p.i.) and the average induction of amyloid deposition after 5 weeks p.i.: ** P < 0.01;$^{++}$ P < 0.002; *** P < 0.001. The differences in the kidneys are significant between the C-treated group and C+F test group after 3 and 5 weeks p.i. respectively: * P < 0.05; $^+$ P < 0.02.

The incidence of AA was correlated with signs of inflammation at autopsy. 100% of the mice showed predominantly chronic renal lesions with glomerulonephritis and kidney tubular edema. Amyloid was deposited in the mouse kidneys predominantly in the arterial walls (66.7% of the animals), but also occured in the basement membrane and was located perifibrillary. Average amyloid deposition in the kidneys of C+F group was significantly higher than in the control C-treated group ($P < 0.02$) (Fig.). C+Ub resulted in similar incidence of amyloid accumulation in all the spleens, but in the kidneys 66.6% of mice had only traces of amyloid and 33.3% - minimal deposits. Glomerulonephritis, tubular edema, and tubular fibrinoid necrosis were found in 100%, 16,7%, and 50% animals respectively.

Summarizing the results obtained we can ascertain that the usage of F and Ub in addition to C to some degree enhances the formation of amyloid.

DISCUSSION

The formation of AA amylod fibril deposits is not well-understood but in the murine model of amyloidosis, deposits increase and are variable in respect to their extent and localization. The highest AA deposits are around the spleen lymphoid follicles [5], which our data also show. There is a greater increase in deposition when F and Ub are used together with casein. After 5 weeks, moderate and heavy amyloid deposits were found in these groups. Extensive AA amyloidosis was observed in 100% of the spleens whereas only

minimal AA deposition was found in the kidneys. It is known that amyloidogenic stimulation in casein-induced amyloidosis enhances the synthesis of proteoglycans, which are related to murine spleen reactive AA amyloid. This upregulation of proteoglycan expression precedes amyloid fibril formation [6]. The abundance of basement membrane glycosaminoglycan in the glomeruli is a main factor for renal AA deposition, however, this does not exclude the possibility that increased apoptosis in renal T cells plays a causative role for renal amyloidosis [7]. Since Ub demonstrates AEF activity *in vivo* and binds non-covalently to AA amyloid, some authors [8] suggest that Ub may indeed be 'fibril-AEF' and may play a crucial role in the pathogenesis of amyloidosis. Fibrin split products and fibrinopeptides also are the factors which enhance the permeability of the blood vessels and chemotaxis of the polymorphonuclear cells.

In conclusion, our studies demonstrate, that besides the casein which may be capable of forming amyloid fibrils, other inflammatory substances such as F and Ub can enhance its deposition in C57BL/6 mice. On the other hand, the higher pathology of the spleen and kidneys under treatment with F possibly reflects the formation of amyloid in these organs.

REFERENCES

1. Xing, Y. and Higuck, K., Amyloid fibril proteins, *Mech. Ageing Dev.*, 123, 1625, 2002.
2. Gruys, E. and Snel, F.W., Animal models for reactive amyloidosis, *Baillieres Clin. Rheumatol.*, 8, 599, 1994.
3. Berkow, R, *Amyloidosis*, In: Berkow, Fletcher (Eds), The Merck Manual of Diagnosis and Therapy, Merck Research Laboratories: Rahway, NJ 1992, 1052.
4. Gervais, F., Hebert, L., and Skamene, E., Amyloid-enhancing factor: production and response in amyloidosis-susceptible and -resistant mouse strains, *J. Leukoc. Biol.*, 43, 311, 1998.
5. Hutchinson, W.L. et al., AA amyloid deposition is complement independent, *IX International Symposium on Amyloidosis*, 2001, 39.
6. Stenstad T., Magnus J.H., and Husby G., Characterization of proteoglycans associated with mouse splenicc AA amyloidosis, *Biochem. J.*, 303, 663, 1994.
7. Alizadeh-Khiavi K. et al., Amyloid enhancing factor activity is associated with ubiquitin, *Virch. Arch.*, 420, 139, 1992.
8. Mountz, J.D. and Hsu, H.C., Clinical features associated with correction of T-cell senescence: increased acute-phase response, amyloidosis and arthritis, *Dev. Comp. Immunol.*, 21, 509, 1997.

LOWER PROPENSITY FOR AMYLOIDOGENESIS IN FEMALE MICE

S. Shtrasburg,[1], M. Pras[1], M. Dulitzky[2] , C. Pariente [2], R. Gal[3] , A. Mor[1] and A. Livneh[1]

Heller Institute of Medical Research[1] and Department of Obstetrics and Gynecology[2], Sheba Medical Center, Tel-Hashomer, Department of Pathology[3], Hasharon Hospital, Petach-Tikva, and Sackler Faculty of Medicine[1,2,3], Tel Aviv University, Israel.

INTRODUCTION

Susceptibility to amyloidosis is reduced in females as compared to males in several disease entities (1-3). Environmental or genetic factors may underlie this gender difference. The present study was undertaken to explore a possible lower propensity for amyloidogenesis in female mice and to elucidate hormonal associations of such a possible trend.

METHODS

Experimental design

Male and female Swiss mice, 8 to 18 weeks old, underwent induction of amyloidosis and the presence and amount of splenic amyloid was compared in the 2 sexes. The effect of sex hormones and adrenalin was determined by co-administration of hormones with amyloid induction and comparing the amount of splenic amyloid to control mice, receiving the solvents of the hormone alone.

Induction of amyloidosis

Amyloidosis was induced in the mice by two methods:

1. *Casein injection:* Daily subcutaneous (s.c.) injections of 0.5ml of 15% vitamin-free casein for 3 weeks (4).

2. *Enhanced induction by AEF:* Single i.v. injection of $1\mu g$ extremely active AEF (derived from acetone treated homo- genates of spleens of preamyloidotic mice), followed by 3, daily s.c. injections of 2% $AgNO_3$. Duration of the experiments varied between 2 to 6 days (5).

Quantification of amyloid deposition

Grading of amyloid in the spleens was performed using the crush and smear (C&S) technique, and a 5 grade scale (5).

Administration of hormones

Estradiol, water soluble, 0.5mg in 0.5 ml per day was injected i.p. to male mice. *Progesterone*, water soluble, 1mg in 0.5 ml per day was injected i.p. to male mice. *Testosterone*, in caster oil (C.O.), 2mg in 0.2 ml was injected i.m. daily, or on day 0 and day 2 as specified, to female mice. *Adrenalin*, 4 μg/0.2ml twice a day, was injected i.p. to female mice.

RESULTS

We found higher amounts of splenic amyloid in male mice (Table 1). Exposure of mice to female or male sex hormones failed to affect splenic amyloid deposition (Tables 2 and 3). In contrast, administration of adrenalin to female mice significantly attenuated amyloidogenesis (Table 4).-

Table 1: Amyloid deposition is less abundant in female mice.*

Gender	Positive mice of total	Mean amount of amyloid	P values
Male	35/35	3.93±0.33	
Female	34/35	3.27±0.86	0.000075

*Induction of amyloidosis was performed using the enhanced (AEF) protocol.
 Duration of the experiment – 6 days.

Table 2: Female sex hormones do not suppress amyloidogenesis in male mice.*

Type of experiment	Material injected	Mean amount of amyloid***
Study 1	Estradiol	3.54±0.44
Study 2	Progesterone	3.74±0.39
Study 3	Estradiol + Progesterone	3.80±0.44
Positive control	Cyclodextrin**	3.80±0.21

*Induction of amyloidosis was performed on 8 mice in each study or control
 group using the enhanced (AEF) protocol. Duration of the experiment – 6 days.
**Solvent of hormones
***No significant differences between study and control mice

Table 3: Exogenous testosterone does not promote amyloidogenesis in female mice.*

Type of experiment	Duration of experiment (days)	Material injected	Mean amount of amyloid***
Study 1	4	Testosterone	3.84 ±0.44
Control 1	4	Caster oil**	3.74 ±0.45
Study 2	3	Testosterone	3.62 ±0.37
Control 2	3	Caster oil	3.42 ±0.28
Study 3	2	Testosterone	0.45 ±0.55
Control 3	2	Caster oil	0.50 ±0.86

*Induction of amyloidosis was performed on ≥6 mice in each study or control
 group using the enhanced (AEF) protocol
**Solvent of testosterone
***No significant differences between study and control mice

Table 4: Adrenalin may reduce amyloidogenesis in female mice.*

Type of experiment	Duration of experiment (days)	Material injected (i.p.)	Amyloid grade (mean/group)	P values
Study 1 (Casein)	21	Adrenaline	3.00 ±1.60	
Control 1 (Casein)	21	Saline	3.96 ±1.18	0.047
Study 2 (AEF)	3	Adrenaline	1.80 ±1.01	
Control 2 (AEF)	3	Saline	2.33 ±0.98	NS

*Induction of amyloidosis was performed in ≥10 mice in each study or control group
 using casein injection

CONCLUSIONS

In the murine model, female amyloidogenesis is quantitatively less pronounced than that of males. Estrogen, progesterone (possible suppression in females) and testosterone (possible enhancement in males) contribute only negligibly to this finding. Increased production of other hormones, such as adrenalin, may cause gender differences by suppressing female amyloidogenesis.

These findings are consistent with human gender discrepancy in reactive amyloidosis, and may favor genetic over environmental factors in the protection of females from amyloidosis.

REFERENCES

1.   Gershoni-Baruch, R. et al. Male sex coupled with articular manifestations cause a 4-fold increase in susceptibility to amyloidosis in patients with familial Mediterranean fever homozygous for the M694V-MEFV mutation. *J. Rheumatol.,* 30, 308, 2003.

2.   Greenstein, A.J. et al. Amyloidosis and inflammatory bowel disease. A 50-year experience with 25 patients. *Medicine*, 71, 261, 1992.

3.   David, J., Vouyiouka, O., Ansell, B.M., et al. Amyloidosis in juvenile chronic arthritis: a morbidity and mortality study. *Clin. Exp. Rheumatol.*, 11, 85, 1993.

4.   Gruys, E. and Snel, F.W.J.J. Animal models for reactive amyloidosis. *Bailliere's Clin. Rheumatol.* 8, 599, 1994.

5.   Shtrasburg, S. et al. Extremely active murine amyloid enhancing factor. *Clin. Exp. Rheumatol.*, 14, 37, 1996.

IN VIVO INFLUENCE OF SERUM AMYLOID A 2.1 AND ITS ACTIVE DOMAINS ON MACROPHAGE CHOLESTEROL EXPORT

Shui-Pang Tam[1] and Robert Kisilevsky[1,2]

Departments of Biochemistry[1], and Pathology and Molecular Medicine[2], Queen's University, Kingston, Ontario Canada

INTRODUCTION

The accumulation of lipids, especially cholesterol, in several vascular cell types such as macrophages and smooth muscle cells, is a defining pathologic feature of atherosclerosis. This accumulation is likely due to an imbalance between the biosynthesis, uptake and esterification of cholesterol on the one hand, and, on the other hand, the de-esterification and export of unesterified macrophage cholesterol. Serum amyloid A (SAA) isoforms 1.1 and 2.1 are acute phase proteins produced by the liver in response to any acute tissue injury. Previously we have shown that SAA 2.1, but not 1.1, suppresses acyl-CoA:cholesterol acyl transferase (ACAT) and enhances cholesteryl ester hydrolase (CEH) activities in cholesterol-laden murine macrophages and in so doing shifts cholesterol into its transportable state increasing its rate of export (1). The present study further characterizes and defines the effect of SAA2.1 on cholesterol efflux from cholesterol-laden macrophages *in vivo*.

MATERIALS AND METHODS

To determine cholesterol export *in vivo* experiments were conducted as described and validated previously (1). Briefly, J774 macrophages were cholesterol-loaded with red blood cell (RBC) membranes and [³H]-cholesterol as described previously (1). Cells were washed four times with PBS/BSA and then detached from the culture dishes. Five millions cells in 200 μl DMEM were injected into mice through the tail vein. At various time points thereafter, approximately 25 μl of blood were collected from the tail vein of each animal into heparinized capillary tubes and then centrifuged for 5 min in an Adams Autocrit Centrifuge to separate red blood cells from plasma. Cholesterol efflux was determined by measuring the appearance of [³H]-cholesterol in plasma by scintillation counting. To study whether export of cholesterol from J774 cells to plasma is influenced by SAA2.1 or its domains, 24h after injection of cholesterol-loaded and labelled J774 macrophages into mice the same animals received, by iv injection, 200 μl of protein-free liposomes, or liposomes containing SAA1.1 or 2.1 or liposomes containing synthetic peptides corresponding to amino acid residues 1-20, 21-50, 51-80 and 74-103 of murine SAA2.1, respectively. The concentration of the injected SAA1.1 and 2.1 liposomes was 20 μM. Assuming a mouse total blood volume of 2 ml the final blood concentration was 2 μM. The liposomes containing the synthetic peptides were injected at a concentration of 5 μM resulting in a final blood concentration of 0.5 μM. At various time points following the second injection, approximately 25 μl of blood were collected from the tail vein of each animal and analyzed as described previously (1).

Peptides corresponding to amino acid residues 1-20, 21-50, 51-80 and 74-103 of the murine SAA2.1 protein sequence were synthesized by solid-phase peptide synthesis using a PE Applied Biosystems 433A peptide synthesizer. Liposomes were prepared by the cholate dialysis procedure of Jonas and co-workers (2).

RESULTS

We have previously demonstrated that the *in vivo* release of radiolabeled cholesterol from intravenously injected cholesterol-laden J774 macrophages is not a function of cell death nor of the destruction of these cells by an immune or ongoing inflammatory reaction but rather is dependent on a functional transporter (1).

To examine cholesterol export *in vivo* experiments were conducted as described previously (1). Five million [^{3}H]cholesterol-loaded J774 cells were injected intravenously into non-inflamed mice. After 24 h of equilibration, the animals were injected IV with medium alone (200 μl) or 200 μl medium containing protein-free liposomes or liposomes to a final concentration of 2 μM of SAA1.1 or 2.1. Cholesterol export was determined over a 96 h period by measuring the appearance of [^{3}H]cholesterol in plasma. Among these liposomes, only the liposomes containing SAA2.1 resulted in a 3-fold increase in cholesterol efflux to plasma of uninflamed animals (Figure 1).

The radioactivity released to plasma peaks 12 h after injection of liposomes and remains elevated for at least the next 60 h when compared to the other liposomes injected.

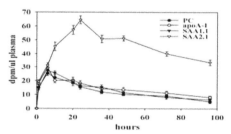

Fig.1. **Effects of pre-incubating cholesterol-laden J774 cells with protein-free liposomes or with liposomes containing either apoA-I, SAA1.1 or SAA2.1 on cholesterol efflux in uninflammed mice.**

To confirm that the active domains of SAA2.1 demonstrated in tissue culture are responsible for the increase in cholesterol efflux *in vivo*, liposomes to a final concentration of 0.5μM of synthetic peptides corresponding to amino acid residues 1-20, 21-50, 51-80 or 74-103 of murine SAA2.1 were each injected into uninflamed mice 24 h after the injection of five million [^{3}H]cholesterol-loaded J774 cells. As shown in Figure 2, only liposomes containing synthetic peptides corresponding to residues 1-20 and 74-103 of SAA2.1 caused significant increases (3 to 4-fold) of [^{3}H]cholesterol efflux to plasma. In each case the radioactivity peaks 16 to 20 h after injections of the liposomes and remains elevated for at least an additional 72 h (i.e. 96 h total).

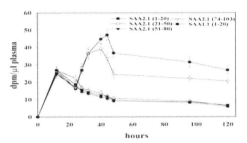

Fig. 2. Effects of liposomes containing various synthetic peptides of SAA2.1 on cholesterol efflux in vivo.

CONCLUSION

In our present work we have extended our previous *in vivo* studies by injecting radio-labeled cholesterol laden J774 cells into normal mice and allowing 24 h for these cells to establish themselves and for the equilibration of their cholesterol pools. Protein-free liposomes or those containing SAA1.1, 2.1 or short peptides spanning the entire length of isoform 2.1 were then injected IV. Radiolabeled cholesterol release was provoked only by liposomes containing SAA2.1 (see Figure 1) and SAA2.1 peptides 1-20 and 74-103 (see Figure 2). The cholesterol release was apparent within 6 h of the administration of isoform 2.1 or its relevant peptides, and peaked approximately16 h after the injection of these specific liposomes (i.e. approximately 40 h after the commencement of the experiment). These data are consistent with our previously published results (1), and the argument for the role of SAA2.1 in macrophage cholesterol mobilization *in vivo*

REFERENCES

1. Tam, S. P., Flexman, A., Hulme, J., and Kisilevsky, R. K. (2002) Promoting export of macrophage cholesterol: the physiological role of a major acute-phase protein, serum amyloid A 2.1. *J. Lipid Res.* **43:** 1410-1420.
2. Jonas, A., Kezdy, E., and. Wald, J. H. (1989). Defined apolipoprotein A-I conformations in recon-stituted high density lipoprotein discs. *J. Biol. Chem.* **264:** 4818-4825.

INFLUENCE OF MURINE SERUM AMYLOID A 2.1, AND ITS ACTIVE DOMAINS, ON HUMAN MACROPHAGE CHOLESTEROL EXPORT IN CELL CULTURE

Shui-Pang Tam[1] and Robert Kisilevsky[1,2]

Departments of Biochemistry[1], and Pathology and Molecular Medicine[2], Queen's University, Kingston, Ontario Canada

INTRODUCTION

Previously we showed that serum amyloid A (SAA) 2.1 inhibits acyl-CoA:cholesterol acyl transferase (ACAT) and stimulates cholesteryl ester hydrolase (CEH) activities in cholesterol-laden murine macrophages. This shifts cholesterol into its un-esterified form and in the presence of a cholesterol transporter and an extracellular acceptor there is a marked increase in its rate of efflux in culture and *in vivo*. We have also previously determined the domains of SAA2.1 that are responsible for its inhibitory effect on macrophage ACAT activity, its promotion of CEH activity and the enhancement of cholesterol efflux from cholesterol-laden macrophages.

The present study determined whether mouse SAA2.1 exerts these effects on a human derived monocytic cell line, THP-1. These monocytes were differentiated into macrophages by treatment with phorbol myristrate acetate (100nM). THP-1 macrophages, cholesterol loaded with red blood cell membrane fragments and then exposed to liposomes containing individual peptides that span the entire length of mouse SAA2.1.

MATERIALS AND METHODS

Human THP-1 macrophages (from ATCC) were cultured in 6-well tissue culture plates at one million cells per well and grown to 90% confluence in 2ml RPMI 640 medium supplemented with 10% fetal bovine serum. To load the cells with cholesterol, nearly confluent monolayers were washed 3 times with phosphate-buffered saline containing 2 mg/ml fatty acid free bovine serum albumin (PBS-BSA) and incubated for 5 h in RPMI supplemented with 5% lipoprotein-depleted serum (LPDS) (d > 1.25 g/ml) and 175 µg of RBC membrane cholesterol. For the purpose of pool equilibration of the added cholesterol, cell cultures were rinsed twice with PBS-BSA and incubated overnight in RPMI containing 5% LPDS. The relative activity of ACAT was determined in cholesterol-laden cells that had been cultured in medium containing no liposomes, protein-free liposomes or liposomes containing 0.5 µM synthetic peptides (final culture concentration) corresponding to amino acid residues 1-20, 21-50, 51-80 and 74-103 of murine SAA2.1, respectively. Following 3 h incubations with the above media, the cells were incubated for another 3 h period after addition of [^{14}C]-oleate (1,2). Cells were chilled on ice and washed twice with PBS-BSA and twice with PBS. After addition of [^{3}H]-cholesteryl oleate (6000 dpm/well) as an internal standard, the lipids were extracted from the cells and analysed by thin-layer chromatography as described previously (1,2). The radioactivity in appropriate spots was measured to determine the incorporation of radioactivity into cholesteryl esters as a measure of ACAT activity. To determine the rate of cholesteryl ester hydrolysis, 2 µg/ml of Sandoz 58-035 (an ACAT inhibitor) was added during liposomes incubation to prevent re-esterification of liberated [^{14}C]-oleate and free cholesterol. At various time points, cellular lipids were extracted and analyzed for cholesteryl ester radioactivity as described above. To examine cholesterol efflux in tissue culture, experiments were carried out as described previously (3). Peptides corresponding to amino acid residues 1-20, 21-50, 51-80 and 74-103 of the murine SAA2.1 protein sequence were synthesized by solid-phase peptide synthesis using a PE Applied Biosystems 433A peptide synthesizer. Liposomes were prepared by the cholate dialysis procedure of Jonas and co-workers (4).

RESULTS AND DISCUSSION

To map the domains in SAA2.1 that are responsible for modulating ACAT and CEH enzyme activities, we used 4 synthetic peptides corresponding to amino acid residues 1-20, 21-50, 51-80 and 74-103 of the murine SAA2.1 protein sequence and which span the entire sequence of this isoform. Residues 1-20 are closely related to residues 1-16 previously generated from the native protein by CNBr cleavage (5) and shown to profoundly inhibit macrophage ACAT activity in post-nuclear homogenates (6). Residues 21-50, 51-80 and 74-103 span an 80 residue peptide (residues 24-103) previously generated from the native protein by CNBr cleavage and which possessed the enhancing property for purified pancreatic neutral CEH (6). The incorporation of [^{14}C]oleate into cholesteryl ester was used as a measure of ACAT activity as described in "Materials and Methods". The relative ACAT activity was determined in cholesterol-laden cells that had been cultured in medium in the absence of liposomes or in the presence of protein-free liposomes or liposomes at 0.5 μM synthetic peptides corresponding to amino acid residues 1-20, 21-50, 51-80 and 74-103 of murine SAA2.1. Following 6 h incubations only the cells that had been exposed to liposomes containing synthetic peptides corresponding to amino acid residues 1-20 of SAA2.1 showed a 2.3-fold decrease in ACAT activity (Figure 1A), while other liposomes treatments had no significant effect on ACAT.

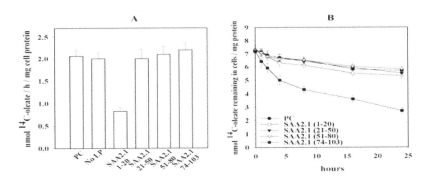

Fig. 1 Mapping the active domains in murine SAA 2.1 that are responsible for inhibiting ACAT and enhancing CEH activities

Experiments were also carried out to identify the domains in SAA2.1 that are responsible for enhancing CEH activity in J774 cells that had been pre-loaded and radio-labeled with cholesteryl esters (Figure 1B). This was performed using liposomes containing one of each of the four synthetic peptides of SAA2.1 noted above. These studies were done in the presence of Sandoz 58-035, an ACAT inhibitor, to prevent the re-esterification of liberated cholesterol and [^{14}C]oleate. Incubations proceeded for various times as indicated, following which the remaining quantities of [^{14}C]-labeled cholesteryl oleate in cells were measured to determine the rate of hydrolysis of cholesteryl ester. With re-esterification blocked, there were no significant differences in the rate of hydrolysis of [^{14}C]-labeled cholesteryl oleate in cells cultured in the presence of protein free liposomes or liposomes at 0.5 μM synthetic peptides corresponding to amino acid residues 1-20, 21-50 and 51-80 from the N-terminal of murine SAA2.1, respectively. However, an equivalent amount of liposomes containing the synthetic peptide corresponding to residues 74-103 of SAA2.1 caused a 3-fold increase in CEH activity.

To map the domains in SAA2.1 that are responsible for the increase in cholesterol export, THP-1 macrophages were loaded with cholesterol and labeled as described above. They were then incubated with, either HDL (10 µg/ml), or liposomes containing 0.5 µM synthetic peptides corresponding to amino acid residues 1-20, 21-50, 51-80 or 74-103 of murine SAA2.1 for 4 h. In some experiments, the above noted cells were exposed simultaneously to 0.5 µM (final culture concentration) synthetic peptides corresponding to amino acid residues 1-20 and 74-103 of murine SAA2.1 for 4 h. After the incubation period, the cells were washed three times with PBS/BSA to remove all the liposomes and the amount of [³H]-cholesterol radioactivity efflux to the medium containing 100 µg of HDL was measured at various time points as described above (3). Pre-incubation of cells with liposomes at 0.5 □M synthetic peptides corresponding to amino acid residues 21-50 or 51-80 of SAA2.1 did not cause any significant changes in the rate of [³H]cholesterol efflux into the medium containing the HDL. However, when cholesterol-laden THP-1 cells labelled with [³H]cholesterol were pre-incubated with liposomes at 0.5 µM synthetic peptides corresponding to amino acid residues 1-20 or 74-103 of SAA2.1 it was observed that 42.1 ± 3.9% and 35.6 ± 2.8% of total cellular [³H]-cholesterol were released into the medium when the cells were subsequently cultured in the presence of HDL. Under similar culturing condition, pre-incubation with the combination of these two synthetic peptides of SAA2.1 resulted in the export of 74.5 ± 3.6% of total cellular [³H]cholesterol to HDL (data not shown).

REFERENCES

1.   Oram, J. F., Mendez, A. J., Slotte, J. P. and Johnson, T. F. (1991) High density lipoprotein apolipoproteins mediate removal of sterol from intracellular pools but not from plasma membranes of cholesterol-loaded fibroblasts. *Arterioscler. Thromb.* **11:** 403-414.

2.   Mendez, A. J., Anantharamaiah, G. M., Segrest, J. P. and Oram, J. F. (1994). Synthetic amphipathic helical peptides that mimic apolipoprotein A-I in clearing cellular cholesterol. *J. Clin. Invest.* **94:** 1698-1705.

3.   Tam, S. P., Flexman, A. Hulme, J. and Kisilevsky, R. 2002. Promoting export of macrophage cholesterol: the physiological role of a major acute-phase protein, serum amyloid A 2.1. *J. Lipid Res.* **43:** 1410-1420.

4.   Jonas, A., Kezdy, E. and Wald, J. H. (1989) Defined apolipoprotein A-I conformations in reconstituted high density lipoprotein discs. *J. Biol. Chem.* **264:** 4818-4825.

5.   Ancsin, J. B. and Kisilevsky, R. (1999) The heparin/heparan sulfate-binding site on apo-serum amyloid A: implications for the therapeutic intervention of amyloidosis. *J. Biol. Chem.* **274:** 7172-7181.

6.   Ely, S., Bonatesta, R. Ancsin, J. B., Kindy, M. and Kisilevsky, R. (2001) The in-vitro influence of serum amyloid A isoforms on enzymes that regulate the balance between esterified and un-esterified cholesterol. *Amyloid J. Prot. Folding Disorders.* **8:** 169-181.

SYSTEMIC AA AMYLOIDOSIS IN THE COMMON MARMOSET

E. Ludlage,[1] C. L. Murphy,[2] S. M. Davern,[2] A. Solomon,[2] D.T. Weiss,[2] D. Glenn-Smith, S. Dworkin,[1] and K. G. Mansfield[1]

[1]*New England Primate Research Center, Harvard Medical School Southborough, MA, USA;* [2]*Human Immunology and Cancer Program, Department of Medicine, University of Tennessee Graduate School of Medicine, Knoxville, TN, USA; and* [3]*Department of Anthropology, University of California, Davis, CA, USA*

I. INTRODUCTION

AA amyloidosis develops in patients with chronic inflammatory or infectious diseases or other conditions associated with sustained elevation of SAA levels.[1] AA- related amyloid also has been found in other species,[2] including several types of primates.[3-5] Although the SAA levels in the non-human cases were not determined, it is important to note that post-mortem examination invariably revealed inflammation or infection involving at least one major site.

We now report the relatively high frequency of systemic amyloidosis among common marmosets (*Callithrix jacchus*) housed at the New England Primate Research Center. The fibrillar deposits had an anatomical distribution similar to that occurring in humans with this disorder and were identified immunohistochemically and by amino acid sequence analysis as AA. Interestingly, the mean SAA levels or extent of inflammatory-related pathology in the affected animals did not differ from marmosets without amyloid.

II. MATERIALS AND METHODS

A. ANIMALS

The 86 marmosets included in the study were >1-yr of age and had been euthanized during a 9-yr period between February 1991 and June 2000 due to cachexia.

B. IMMUNOHISTOCHEMISTRY

Tissue sections were immunostained using the ABC technique. The primary and secondary reagents included an anti-human SAA monoclonal antibody (mAb) and an affinity-purified horse anti-mouse IgG horseradish peroxidase conjugate, respectively.

C. SAA ASSAYS

Antemortem blood specimens were obtained by venapuncture and the separated serum frozen at -80°C. SAA concentrations were measured using a commercial ELISA kit.

D. AMYLOID EXTRACTION AND CHEMICAL CHARACTERIZATION

The amyloid was extracted from necropsy-derived marmoset liver by the method of Pras et al.[6] The resultant material was dissolved in 0.1% SDS buffer containing 0.1 M DTT and 8 M urea and electrophoresed on a 10% SDS/PAGE gel. For automated amino acid sequencing, the appropriate protein band was transferred onto a PVDF membrane and analyzed in a gas-phase sequencer.

III. RESULTS

A. INDEX CASE

Case #408-89 was a 13-yr-old female common marmoset with progressive weight loss and hepatomegaly; a liver biopsy revealed the presence of amyloid, as seen by light and polarizing microscopy. At post-mortem examination, similar deposits were found in the small intestine, kidneys, and adrenal glands. The fibrillar nature of this material was evidenced by electron microscopy and immunohistochemical studies of the liver showed that the amyloid was recognized by an anti-human SAA mAb.

To verify chemically that the hepatic amyloid was indeed AA in nature, fibrils were extracted from fresh-frozen tissue and subjected to SDS/PAGE where an ~8 kDa component was found. After transfer onto a PVDF membrane, the sequence of the first 37 residues of this protein was determined by Edman degradation and the remaining 36 by analyses of tryptic peptides. As shown in Figure 1, the protein was highly homologous in primary structure to that of human AA, differing only by 13 residues (not including the N-terminal R and S absent in the marmoset AA). Of note was the finding of 2 amino acids, Ala and Val, at position 52 of the marmoset AA that occurred at a ratio of ~10:1.

```
                        10         20         30         40         50
Marmoset AA     WFSFIGEAFG GARDMWRAYS DMREANYIGA DKYFHARGNY DAAQRGPGGA
   Human AA     RSF---L----D ---------- --------S ---------- ---K-----V

                        60         70
Marmoset AA     WAAKVISNAR EDIQQFLGHG AED
                 V
   Human AA     ---EA--D-- -N--R-F--G ---
```

Figure 1. Comparison of the N-terminal amino acid sequences of marmoset AA and human AA proteins. The residues are indicated by the one-letter code; (-); sequence identity.

B. RETROSPECTIVE ANALYSES

Review of tissue sections and clinical records of all 86 marmosets revealed that 15 (17%) had amyloid in one or more tissues. However, no significant differences in pathology or SAA levels were detected when comparing amyloid-bearing and non-amyloidotic animals.

IV. DISCUSSION

We have found that ~17% of common marmosets in our colony had amyloid deposits in various body tissues that, by immunohistochemical and amino acid sequence analyses, were identified unequivocally as AA in nature. The fibrils were comprised of the N-terminal 73 residues of the marmoset SAA protein and were homologous to human AA.

Although the development of systemic AA amyloidosis clinically or that induced experimentally typically is associated with chronic inflammation or infection and attendant, sustained elevation of the amyloidogenic SAA precursor molecule, these conditions did not differentiate affected from non-affected marmosets. Thus, we attribute the development of seemingly spontaneous AA amyloidosis in our animals to another factor (possibly genetic). Given the polymorphic nature of SAA in humans and other species,[1,7-11] and the fact that certain isotypes are preferentially deposited as amyloid,[12] it is possible that inheritable variations in the primary structure of this molecule among marmosets may be responsible for its amyloidogenicity, *e.g.*, those animals carrying a particular SAA allele that encodes the Val 62 variant would be prone to develop AA amyloidosis.

V. CONCLUSION

The discovery that common marmosets can spontaneously develop AA amyloidosis provides a unique opportunity to study in a primate model the pathogenesis of this disease and to develop therapies that can ameliorate or prevent this or other systemic amyloid-related disorders.

VI. ACKNOWLEDGEMENTS

This study was supported, in part, by USPHS Research Grant CA10056 from the National Cancer Institute and the Aslan Foundation. A.S. is an American Cancer Society Clinical Research Professor.

VII. REFERENCES

1. Benson, M., In: Scriver CR, Beaudet, A.L., Sly, W.S., Valle, D., eds. *The Metabolic and Molecular Bases of Inherited Disease,* 8th ed., McGraw Hill, New York, 2001, 5345.

2. Johnson, K., et al., Amyloid proteins and amyloidosis in domestic animals, *Amyloid: Int. J. Clin. Exp. Invest.,* 3, 270, 1996.

3.   Blanchard, J.L., Baskin, G.B., and Watson, E.A., Generalized amyloidosis in rhesus monkeys, *Vet. Pathol.,* 23, 425, 1986.

4.   Hubbard, G.B., et al., Spontaneous amyloidosis in twelve chimpanzees, Pan troglodytes, *J. Med. Primatol.,* 30, 260, 2001.

5. Slattum, M.M., et al., Amyloidosis in pigtailed macaques (Macaca nemestrina): pathologic aspects, *Lab. Anim. Sci.,* 39, 567, 1989.

6. Pras, M., et al., The characterization of soluble amyloid prepared in water, *J. Clin. Invest.,* 47, 924, 1968.

7. Boyce, J.T., et al., Familial renal amyloidosis in Abyssinian cats, *Vet. Pathol.,* 21, 33, 1984.

8. Husby, G., et al., Serum amyloid A (SAA): biochemistry, genetics and the pathogenesis of AA amyloidosis, *Amyoid: Int. J. Exp. Clin. Invest.,* 1, 119, 1994.

9. Westermark, G., et al., AA-amyloidosis. Tissue component-specific association of various protein AA subspecies and evidence of a fourth SAA gene product, *Am. J. Pathol.,* 137, 377 1990.

10. Booth, D.R. et al., SAA1 alleles as risk factors in reactive systemic AA amyloidosis, *Amyloid: Int. J. Clin. Exp. Invest.,* 5, 262, 1998.

11. Yamada, T., Okuda, Y., and Itoh, Y., The frequency of serum amyloid A2 alleles in the Japanese population, *Amyloid: Int . J. Clin. Exp. Invest.,* 5, 208, 1998.

12. *Benson, M., et al., Metabolism of amyloid proteins, Ciba Found Symp., 199, 104, 1996.*

THE INHIBITORY EFFECT OF MONENSIN ON MURINE AA AMYLOIDOSIS

T. Gondo[1], D. Cui[2], Y. Hoshii[2], H. Kawano[2], M. Takahashi[3], T. Iwata[3], T. Ishihara[2]

[1]*Department of Surgical Pathology, Yamaguchi University Hospital*
[2]*Department of Pathology, Yamaguchi University School of Medicine*
[3]*Division of Basic Laboratory Sciences, Faculty of Health Sciences, Yamaguchi University School of Medicine*

Monensin, Na^+ ionophore, blocks glycoprotein secretion (1, 2). In vitro, monensin blocks cytokine secretion of human monocytes. We examined the effect of monensin on experimental murine AA amyloidosis.

MATERIALS AND METHODS

Forty ICR female mice were used. All mice given amyloidogenic stimuli according to Ram's method were devided into 5 groups. Group A: Mice were once given 2.4mg monensin sodium salt *ip*. Group B: Mice were once given 0.48mg monensin sodium salt i*p*. Group D : Mice were daily given 0.2mg monensin sodium salt *po* for three weeks. Group C and group E were controls and mice were given saline. Experimental protocols are shown in Table 1 .

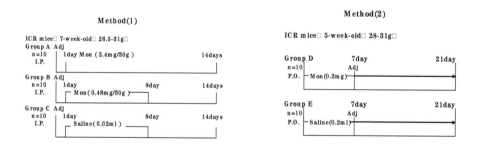

Table 1. Experimental protocols

For histological examination, resected spleens were fixed in 10% buffered formalin. Paraffin sections were stained with Congo red.

Morphometry: Amyloid deposits in sections from two or three slices of the spleen for each mouse were examined with confocal laser scanning microscopy (GB2000, OLYMPUS, Tokyo, Japan). Areas of amyloid deposits in each confocal images were determined with image analyzing soft Image-Pro (Figure 1). Average area of amyloid deposits in one slice of spleen were calculated.

RESULTS

Amyloid deposits in the spleen were observed 1/2 in group A, 7/8 in group B and 10/10 in group C. he average amylod area of one slice of the spleen(AAA) were $0.522 \times 10^6 \text{um}^2$ in group B and $1.33 \times 10^6 \text{um}^2$ in group C. In group B, amyloid deposits decreased as compared to group C (p=0.0041, Fig. 2). Amyloid deposits in the spleen were observed 10/10 in group D and 9/9 in group E. AAA were $0.729 \times 10^6 \text{um}^2$ in group D and $1.06 \times 10^6 \text{um}^2$ in group E. In group D, amyloid deposits decreased as compared to group E (p=0.044, Fig. 3).

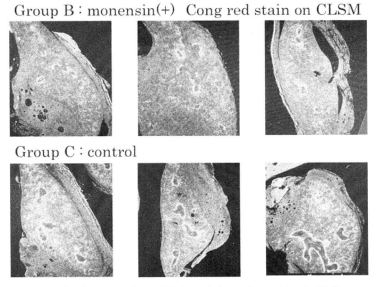

Figure 1. Confocal images of amyloid deposits in sections stained with Congo red

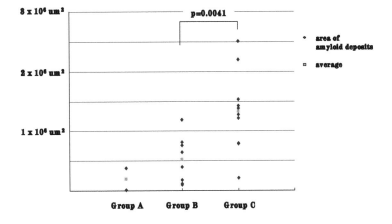

Figure 2. AAA in group A and B, C

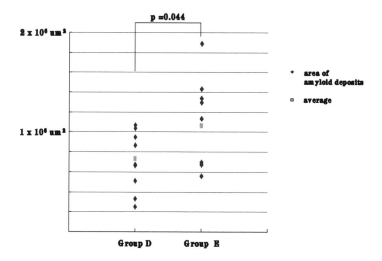

Figure 3. AAA in group D and E

DISCUSSION

The present data showed that monensin decreased amyloid deposits in murine AA amyloidosis. In preliminary experiment, monensin inhibited SAA secretion of murine hepatocyte in vivo as same as colchicine. It is assumed that monensin decreased murine amyloidosis by means of inhibiting SAA secretion of hepatocytes. The exact mechanism for inhibitory effect of monensin should be investigated.

REFERENCES

1. Callaway, T.R., et al. Secretion of a highly monensin-resistant *prevotella bryantii* subpopulation with altered outer membrane characteristics. *Appl. Environ. Microbiol.*, 65, 4753, 1999.
2. Nebbia, C., et al. Oxidative metabolism of monensin in rat liver microsomes and interactions with tiamulin and other chemotherapeutic agents: evidence for the involvement of cytochrome P-450 3A subfamily. D*rug Metab. Dispos.* 27, 1039, 1999.

AMYLOID-DETERMINING RESIDUES IN MOUSE SERUM AMYLOID A

B. Kluve-Beckerman, D. Marks, L. Wang, M. D. Benson, and J.J. Liepnieks

Indiana University School of Medicine and Veterans Affairs Medical Center, Indianapolis, IN, USA

INTRODUCTION

Mouse SAA1.1 (formerly SAA2) and SAA2.2 (formerly SAA CE/J) differ at only six of 103 positions (positions 6, 7, 11, 60, 63, and 101) (1, 2). SAA1.1 is highly amyloidogenic in mice and in macrophage and fibroblast cultures. In contrast, SAA2.2 does not form amyloid (3). Previous findings have indicated lack of SAA2.2 amyloidogenecity is dictated primarily by its structure (4). Three of the six differences between SAA2.2 and SAA1.1 are located in the N-terminal region. *In vitro* fibril formation has been achieved from peptides corresponding to the N-terminal portion of SAA1.1 (5), whereas recombinant truncated SAA lacking the first eleven residues fails to undergo *in vitro* fibril formation (6). Thus, this region is regarded as a determinant of fibrillogenesis. The aim of our study was to determine the amyloid-forming capability of various mutants of SAA1.1 and SAA2.2 in cell culture systems and thereby identify specific residues in SAA1.1 that correlate not only with fibrillogenesis but also with production of stable amyloid deposits.

MATERIALS AND METHODS

SAA1.1 and SAA2.2 cDNAs, cloned previously, were mutated via PCR-based *in vitro* mutagenesis methodology. Initially, residues in SAA2.2 were mutated to the corresponding residues in SAA1.1, and then SAA1.1 residues were changed to match those in SAA2.2. Mutated SAAs were produced in *E. coli* using the pET21a expression system (7). SAA proteins were purified from *E. coli* lysates via Sepharose CL-6B chromatography, chromatofocusing, and precipitation with ammonium sulfate (7). Mutations were confirmed by amino acid sequencing. Purified mutant SAAs were tested for amyloid-forming capacity in cell culture models (macrophage and fibroblast cultures) (8, 9). SAA1.1, given to identical cultures, served as a positive control.

RESULTS AND DISCUSSION

SAA2.2 proteins mutated to match SAA1.1 at one or more positions were tested for gain of amyloid-forming capacity. SAA2.2 mutant proteins identical to SAA1.1 at positions (6), (7), (6, 7), (6, 7, 11), (6, 7, 11, 60), (6, 7, 11, 63), or (101) failed to form amyloid. Note that two of the non-amyloidogeneic SAAs differ from highly amyloidogenic SAA1.1 at only two positions, either (63 and 101) or (60 and 101).

Residues in SAA1.1 were then changed to match those in SAA2.2; the mutant SAAs were added to cell cultures to determine if amyloid-forming capacity was maintained or lost. Mutating the residue at position 101 (Ala101 → Asp101) did not result in diminished amyloid formation. More surprisingly, amyloid formation was also demonstrated by SAA1.1 proteins mutated to SAA2.2 at position 6 (Ile6 → Val6) as well as position 101. However, when the residue at position 7 was also replaced (Gly7 → His7), the resultant SAA (identical to SAA2.2 at positions 6, 7, and 101) failed to form amyloid.

While the vast majority of SAA mutant proteins did not produce amyloid, many showed deposition in cell cultures (Figure 1).   These extracellular deposits, which did not stain with Congo red, were cleared from macrophage cultures over time, indicating susceptibility to proteolysis not demonstrated by SAA1.1-derived amyloid. Similar resorption of non-amyloid deposits was not observed in fibroblast cultures.

Table 1. Amyloid-formation in mouse macrophage cultures from mutated SAA proteins.

| SAA construct | \
Residue 6 | 7 | 11 | 60 | 63 | 101 | Amyloid |
|---|---|---|---|---|---|---|---|
| **SAA2.2** | V | H | L | G | A | D | − |
| mut 6 | I | H | L | G | A | D | − |
| mut 7 | V | G | L | G | A | D | − |
| mut 6,7 | I | G | L | G | A | D | − |
| mut 6,7,11 | I | G | Q | G | A | D | −* |
| mut 6,7,11,60 | I | G | Q | A | A | D | − |
| mut 6,7,11,63 | I | G | Q | G | S | D | − |
| mut 101 | V | H | L | G | A | A | − |
| **SAA1.1** | I | G | Q | A | S | A | + |
| mut 101 | I | G | Q | A | S | D | + |
| mut 6,101 | V | G | Q | A | S | D | + |
| mut 6,7,101 | V | H | Q | A | S | D | − |

* One preparation of this mutant formed tiny amounts of amyloid, but this result has not been reproduced.

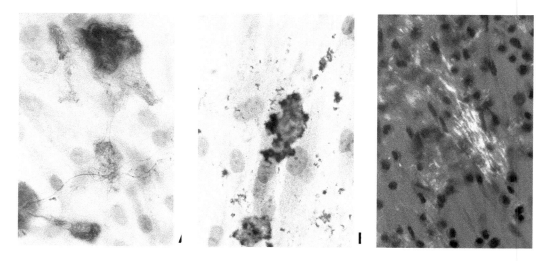

Figure 1. Deposition of SAA1.1 and SAA2.2 in cultures of human fibroblasts. Fibroblasts were cultured with SAA (250 µg/ml) for eight days and then stained immunochemically for SAA using monoclonal antibody F8J (home-made) or stained with Congo red.   Photographs were taken using a 60X oil objective.

A.   SAA1.1 deposits (dark areas), immunostained.

B. SAA2.2 deposits (dark areas), immunostained.

C. SAA1.1 deposits (light areas), stained with Congo red and viewed under polarized light.
    SAA2.2 deposits did not stain with Congo red.

CONCLUSIONS

Our results agree with those of other investigators who have shown the N-terminal region of SAA1.1 to be a determinant of amyloid fibrillogenesis. Our findings further suggest residues at other positions, namely Ala60 and/in Ser63, are critical to the generation of stable amyloid deposits.

REFERENCES

1. Yamamoto, K-I, Migita, S (1985). Complete primary structures of two major murine serum amyloid A proteins deduced from cDNA sequences. *Proc Natl Acad Sci USA* **82**, 2915-2919.

2. de Beer MC, de Beer FC, McCubbin WD, Kay CM, Kindy MS (1993). Structural prerequisites for serum amyloid A fibril formation. *J Biol Chem* **268**, 20606-20612.

3. Sipe JD, Carreras I, Gonnerman WA, Cathcart ES, de Beer MC, de Beer FC (1993). Characterization of the inbred CE/J mouse strain as amyloid resistant. *Am J Pathol* **143**, 1480-1485.

4. Wang L, Liepnieks JJ, Benson MD, Kluve-Beckerman B (2000). Expression of SAA and amyloidogenesis in congenic mice of CE/J and C57BL/6 strains. *Amyloid: Int J Exp Clin Invest* **7**, 26-31.

5. Westermark GT, Engstrom U, Westermark P (1992). The N-terminal segment of protein AA determines its fibrillogenic property. *Biochem Biophys Res Commun* **182,** 27-33.

6. Patel H, Bramall J, Waters H, de Beer M, Woo P (1996). Expression of recombinant human serum amyloid A in mammalian cells and demonstration of the region necessary for high-density lipoprotein binding and amyloid fibril formation by site-directed mutagenesis. *Biochem J* **318,** 1041-1049.

7. Kluve-Beckerman B, Yamada T, Hardwick J, Liepnieks JJ, Benson MD (1997). Differential plasma clearance of murine acute phase serum amyloid A proteins SAA1 and SAA2. *Biochem J* **322,** 663-669.

8. Kluve-Beckerman B, Liepnieks JJ, Wang L, Benson MD (1999). A cell culture system for the study of amyloid pathogenesis: Amyloid formation by murine peritoneal cells in the presence of recombinant serum amyloid A. *Am J Pathol* **155,**123-133.

9. Magy N, Liepnieks JJ, Benson MD, Kluve-Beckerman B (2003). Amyloid-enhancing factor mediates amyloid formation on fibroblasts via a nidus/template mechanism. *Arthritis Rheum* **48**, 1430-1437.

HYDROCORTISONE SUPPRESSES THE SECOND PHASE OF AMYLOIDOGENESIS IN A MOUSE MODEL

S. Shtrasburg[1] , M. Pras[1] , C. Pariente[2], R. Gal[3] and A. Livneh[1]

Heller Institute of Medical Research[1] and Department of Obstetrics and Gynecology[2], Sheba Medical Center, Tel Hashomer; Department of Pathology[3], Hasharon Hospital, Petah Tikva and Sackler Faculty of Medicine[1,3], Tel Aviv University, Israel.

INTRODUCTION

Steroid treatment and prevention were tried in human and animal AA amyloidosis, but with limited success and with conflicting results (1-4). The present study was undertaken to determine whether hydrocortisone (HC) inhibits amyloidogenesis in a mouse model.

METHODS

Experimental animals:

Male Swiss mice, 8 to 19 weeks old.

Induction of amyloidosis (5):

Amyloidosis was induced in mice using the enhanced method, in which one i.v. injection of $1\mu g$ of an extremely active amyloid enhancing factor (AEF), prepared by acetone extraction, according to the procedure of Shtrasburg et al, is followed by 3 once daily s.c. injections of 2% $AgNO_3$. Removal of mouse spleens for amyloid determination is usually carried out on day 6, three days after the last $AgNO_3$ injection. In the present study, in order to increase the sensitivity of the test, we used two shortened protocols, in which determination of splenic amyloid was carried out on days 3 and 4, instead of on day 6.

Administration of HC:

HC was administered intraperitoneally, in a daily mean amount of about 10mg/50g body weight (BW), in 0.5ml saline, either in a single dose or in divided doses (twice a day).

Control experiments

Control animals, of the same strain and age group, received the same amyloid induction protocol, but with 0.5ml saline alone, without HC.

Quantification of amyloid deposition:

Grading of amyloid in the spleens was performed using the Crush and Smear (C&S) technique, with a 5 grade scale (6).

RESULTS

HC was found to suppress the second phase of amyloidogenesis. Its best effect was observed in the experiments of the short duration (72 hrs), with up to 70% suppression (Table 1). Suppression was still observed in the 96 hrs experiments, but to a lesser extent (25%, Table 1). Divided HC doses retained their inhibitory effect at 72 hrs, but gave somewhat better protection at 96 hrs (34% versus 25%, Table 2).

Material injected	Daily Amount /mouse BW	Duration of experiment (hours)	Positive mice of total	Mean amount of amyloid	% Suppression	P values
HC	10mg /50g	72	5/6	0.80	68	P=0.0004
Saline	0.5ml	72	4/4	2.53		
HC	10mg /50g	96	7/7	2.60	34	P=0.002
Saline	0.5ml	96	8/8	3.96		

Table 1: Suppression of amyloidogenesis by a single daily dose of HC

Material injected	Daily Amount /mouse BW	Duration of experiment (hours)	Positive mice of total	Mean amount of amyloid	% Suppression	P values
HC	10mg /50g	72	7/12	0.83	69	P=0.0003
Saline	0.5ml	72	11/11	2.64		
HC	10mg /50g	96	16/17	2.56	25	P=0.004
Saline	0.5ml	96	17/17	3.41		

Table 2: Suppression of amyloidogenesis by divided doses of HC (twice a day)

CONCLUSIONS

HC was found to have a suppressive effect on the second stage of amyloidogenesis. It was effective even in a single daily administration. These findings support its use in clinical settings.

REFERENCES

1. Grayzel, H.G. et al. Amyloidosis – experimental studies. Part IX: The effect of corticotrophin (ACTH) and cortisone injections upon the production of amyloidosis in albino mice. *Exp. Med. Surg.*, 14, 332, 1956.
2. Cohen, A.S., Calkins, E., and Mullinax, P.F. Studies in experimental amyloidosis. III. The effect of cortisone administration on the incidence of casein-induced amyloidosis in the rabbit. *Arch. Int. Med.*, 110, 569, 1962.
3. Maxwell, M.H. et al. Corticosteroid therapy of amyloid nephritic syndrome. *Ann. Int. Med.*, 60, 539, 1964.
4. Fields, M., Laufer, A., and Polliack, A. Lysosomal enzyme studies in experimental amyloidosis of mice treated with cortisone. *Acta Path. Microbiol. Scand.*, 236, 45, 1973.
5. Shtrasburg, S. et al. Extremely active murine amyloid enhancing factor. *Clin. Exp. Rheumatol.*, 14, 37, 1996.
6. Shtrasburg, S., Gal, R., and Pras, M. Crush and smear technique for rapid detection and semiquantitation of amyloid deposition. *Biotech. Histochem.*, 66, 203, 1991.

TARGETING APOLIPOPROTEIN E FRAGMENTS IN AMYLOID DEPOSIT; A STUDY ON A MOUSE MODEL

Tomonosuke Someya, Toshiyuki Yamada, Mariko Kobayashi, Shinobu Fujita

Department of Pediatrics and Clinical Pathology, Juntendo University School of Medicine, and

Mitsubishi Kasei Institute of Life Sciences, Tokyo, Japan

BACKGROUND AND OBJECTIVES

Apolipoprotein E (apoE), as degraded fragments, is co-localized in amyloid deposits regardless of types. The in vivo targeting of the apoE, either diagnostic or therapeutic, has long been kept in mind. Recent establishment of human apoE knock-in mouse enabled experiments of utilization this approach. This study examined whether a monoclonal antibody specific to human apoE fragments is bound to and is accumulated in AA amyloid deposits in vivo in human apolipoprotein E knock-in mouse with AA-amyloidosis.

METHODS

A previously generated murine monoclonal antibody, YK-2, which recognizes degraded human apolipoprotein E in the amyloid deposits [1] and human apoE-knock-in mice [2] were utilized. The antibody was iodinated with I^{125} for the accumulation or clearance study and was biotinylated for the tissue study. AA amyloidosis was induced by a single intraperitoneal administration of amyloid enhancing and 3 times intraperitoneal injections of 0.5 ml Fruend complete adjuvant with a week interval. For the accumulation study, amyloidotic or control mice (n=5 each) were injected i.v. with the labeled antibodies and sacrificed at day 3 after injection. Then the radioactivity was counted in spleen, liver, kidney, and intestine. Blood samples were serially obtained for the clearance study. The presence of the biotinylated antibodies in the amyloid tissues were studied histochemically using peroxidase-conjugated avidin.

RESULTS

The antibodies accumulated in all the examined organs of amyloidotic mice with significant differences from non-amyloidotic mice at 72hrs after injection (Fig. 1). Plasma clearance of the antibody in amyloidotic mice was similar to that of non-amyloidtic mice. The histological examination revealed that the injected antibodies were localized in the amyloid deposits in the spleen of the amyloidotic mice.

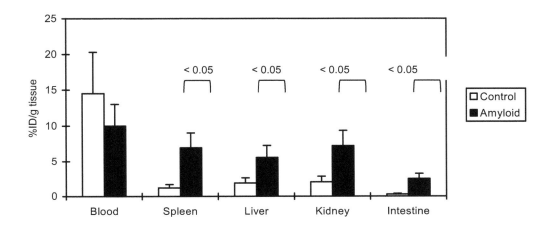

Figure 1. Accumulation of the injected antibody, YK-2 at day 3.

Data (accumulated activity) were shown as % of the injected dose per gram of organs.

CONCLUSIONS

These results suggest that apoE in the amyloid deposits was efficiently targeted by the present antibody. Further investigation utilizing this model is needed for the clinical application in the future.

REFERENCES

1) Yamada T, Kakihara T, Gejyo F, et al. A monoclonal antibody recognizing apolipoprotein E peptides in systemic amyloid deposits. Ann Clin Lab Sci 1994;24,243.

2) Hamanaka H, Fukui Y, Suzuki K, et al. Altered cholesterol metabolism in human apolipoprotein E4 knock-in mice. Hum Mol Genet 2000; 9, 353

SCINTIGRAPHY AT DIFFERENT TIME INTERVALS AFTER ADMINISTRATION OF ^{123}I LABELLED SERUM AMYLOID P COMPONENT (SAP) IN PATIENTS WITH AMYLOIDOSIS

B.P.C. Hazenberg[1], P.L. Jager[2], D.A. Piers[2], P.C. Limburg[1], M.N. de Hooge[2] and M.H. van Rijswijk[1]

Departments of Rheumatology (1) and Nuclear Medicine (2), University Hospital, Groningen, The Netherlands

INTRODUCTION

AA amyloidosis is sometimes seen in longstanding arthritis. Early detection is important for prognosis. Scintigraphy with ^{123}I labelled serum amyloid P component (SAP) may be helpful to detect amyloidosis in patients suspected to have amyloidosis (1).

OBJECTIVE

Visual assessment of SAP scans in patients with amyloidosis at different time intervals after administration of the tracer in order to obtain the optimum time interval for scanning.

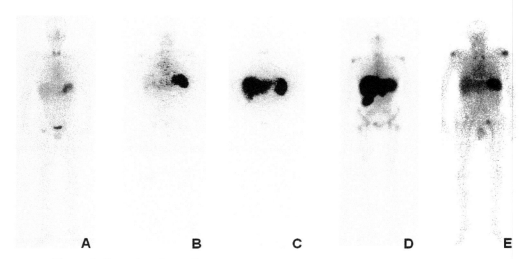

Figure 1. Examples of specific organ uptake on SAP scans (anterior view) after 24 hours. A Healthy control. B. Spleen. C. Liver and spleen. D. Liver and bone marrow. E. Liver and joints (shoulders).

PATIENTS AND METHODS

In this study 61 patients with systemic amyloidosis (27 AA type, 27 AL type, and 7 ATTR type), 6 patients with localised amyloidosis, and 8 controls were included. All patients and controls were treated in our university hospital.

A dose of 200 MBq ^{123}I labelled human SAP was injected intravenously in patients and controls as described (2). Scans were made 0.5, 4, 24, and 48 hours after administration of the tracer and scored visually by two blinded investigators. Scans of controls were used to obtain guidelines for maximum normal uptake. The optimum time interval for making a scan was studied.

RESULTS

Guidelines were formulated and used in a step-wise visual assessment of scans of patients with amyloidosis. In short: first anterior images assessing heart, liver, and lungs and then posterior images for spleen, bone marrow, kidneys, and adrenals. See figure 1 for some examples of organ uptake.

Table 1. Organ uptake of ^{123}I-SAP in AA and AL amyloidosis at different time intervals after injection.

Organ	AA (N=27)				AL (N=27)			
	0.5 h	4 h	24 h	48 h	0.5 h	4 h	24 h	48 h
Heart	0	0	0	0	0	0	0	0
Liver	2	2	2	2	13	13	14	14
Lung(s)	0	0	0	0	0	0	0	0
Spleen	22	24	24	23	17	20	22	19
Adrenal(s)	0	4	4	3	0	0	0	0
Bone marrow	0	0	0	0	5	4	6	2
Joints	0	2	3	1	1	2	4	3
Kidney(s)	6	9	14	11	3	5	6	3
Body total	22	**24**	**24**	23	20	22	**24**	22

Scans of 7 ATTR patients were most frequently positive (2 from 7) 24 hours after injection. Scans of 6 patients with localised amyloidosis were not informative. In 54 patients with AL and AA amyloidosis 42 scans (78%) were positive after 30 minutes, 46 scans (85%) after 4 hours, 48 scans (89%) after 24 hours, and 45 scans (83%) after 48 hours (see table 1). In only two patients (one AA and one AL type) the scan was negative after 24 hours whereas the scan was positive 4 hours after injection.

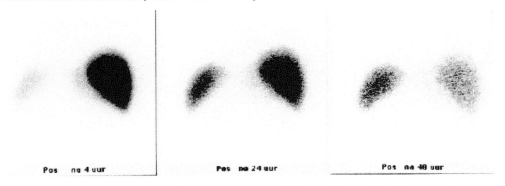

Pos na 4 uur   Pos na 24 uur   Pos na 48 uur

Figure 2. Differences in uptake of liver and spleen after 4, 24 and 48 hours. Posterior spot view.

In 7 AL patients a remarkable shift of uptake between liver and spleen was seen (figure 2): initially hepatic uptake was very intense, but after 4 to 24 hours splenic uptake increased gradually whereas hepatic uptake stabilised and seemed to decrease visually compared to splenic uptake. Hepatic uptake when present can be identified in almost all patients from 30 minutes after injection. On the contrary, kidney uptake can often not be identified until 24 hours after injection. The optimum moment of splenic uptake lies somewhere in between the uptake of liver and kidneys.

CONCLUSIONS

The optimum time interval for the scan was 24 hours after injection of the tracer. The scan was useful in identifying patients with AL and AA amyloidosis (in 89%). Intensity and time of uptake differed among organs. Liver uptake was seen first, followed by spleen, and later the kidneys.

REFERENCES

1.   Hawkins PN, Lavender JP, Pepys MB. Evaluation of systemic amyloidosis by scintigraphy with 123I-labeled serum amyloid P component. N Engl J Med 1990;323:508-13.

2.   Jager PL, Hazenberg BPC, Franssen EJF, Limburg PC, Van Rijswijk MH, Piers DA. Kinetic studies with iodine-123-labeled serum amyloid P component in patients with systemic AA and AL amyloidosis and assessment of clinical value. J Nucl Med 1998;39:699-706.

THE INCIDENCE, AND POLARIZATION OPTICAL CHARACTERISTICS OF PULMONARY ALVEOLAR CALCOSPHERITIES (CORPORA AMYLACEA) IN RHEUMATOID ARTHRITIS - A RETROSPECTIVE CLINICOPATHOLOGIC STUDY OF 210 AUTOPSY PATIENTS

Miklós Bély[1] and Ágnes Apáthy[2]

[1]*Department of Pathology, Policlinic of the Hospitaller Brothers of St.John of God and* [2]*Department of Rheumatology, National Institute of Rheumatology, Budapest, Hungary, H-1027 Budapest, Frankel L Street 17-19.*

INTRODUCTION

Alveolar microlithiasis is a rare pulmonary disease of unknown etiology. It is characterized histologically by concentrically laminated crystalloid calcified bodies (calcospherities, corpora amylacea) in the alveoli. Corpora amylacea (calcospherities) are eosinophilic, congophilic rounded, lamellated proteinaceous bodies, consisting mainly of calcium phosphate, and calcium carbonate. They are found in both normal and diseased lungs. Most of the patients are asymptomatic, but mild dyspnoe, restrictive lung disease, and cor pulmonale (with dramatic miliary roentgenographic changes) may also develop.

The **aim** of this study was to determine:

(1) the incidence of pulmonary microlithiasis in lungs of RA autopsy patients

(2) the size of corpora amylacea

(3) the histochemical and immunohistochemical characteristics of pulmonary corpora amylacea

(4) the stages of crystalloid formation

(5) the clinicopathological correlation of alveolar microlithiasis: the link with interstitial pneumonitis and/or interstitial fibrosis, the relation to congestive or restrictive cardiac insufficiency (based on the presence of heart failure cells in the alveoli).

PATIENTS AND METHODS

A randomized autopsy population of 210 in-patients (females 150, average age of 65.7 years; males 60, average age of 66.3 years at death) with *rheumatoid arthritis (RA)* was studied. Tissue samples of 338 lungs were available for histologic evaluation in 169 patients.

Pulmonary alveolar microlithiasis and interstitial pneumonitis, or fibrosis was histologically diagnosed post mortem. Histological evidence of chronic heart failure was determined by the presence of hemosiderin-laden alveolar macrophages. The correlation between pulmonary alveolar microlithiasis and interstitial pneumonitis, furthermore interstitial pneumonitis and/or interstitial fibrosis, or cardiac failure was determined by χ^2-test.

The tissue specimens were fixed in 8% formaldehyde solution and embedded in paraffin. Serial sections were cut and stained with HE or Congo red according to Romhányi, without alcoholic differentiation, and covered with gum arabic (1). The (AA) amyloid was determined and characterized histochemically by Congo red staining after pre-treatment with performate (85% HCOOH 8 ml + 30% H_2O_2 - 3 ml + 96% H_2SO_4 - 0,22 ml) according to

Romhányi (2) at 20°C for 1,3,5,10,15,20 or 25 sec, and with oxidation (5% $KMnO_4$ + 0,3% H_2SO_4, 1:1) induced degradation of amyloid deposits at 20°C for 30 sec, 1,2,3,4,5,6 or 10 min according to Wright et al. (3), and by streptavidin-biotin-complex/horseradish peroxidase immunohistochemical reactions (4,5). Autoperoxidase reaction was inhibited with 3% H_2O_2-methanol for 20 min at 20°C, and non-specific protein binding was inhibited by incubation in 5% human albumin solution for 20 min at 20°C. The slides were then incubated at 4°C for 12 hours with the primary antibody (anti-human amyloid A-component 1:100 [mono/DAKO MO759], anti-human amyloid P-component 1:200 [poly/DAKO], a-hu beta-2-Microglobulin 1:400 (po/DAKO A0072), followed by incubation with biotinylated second-antibody (Multilink; Biogenex) - 20 min at 20°C, and visualized by the streptavidin-biotin-complex/horseradish peroxidase - diaminobenzidine reaction 15 min at 20°C.

RESULTS

1. Alveolar microlithiasis was found in 16 (**4.73 %**) of 338 lung specimens from 169 RA patients.
2. The size of corpora amylacea ranged between 25 and 100μ.
3. The corpora amylacea were eosinophilic (Fig.1.a-b.), congophilic and birefringent (Fig.2.a-b.), showing different stages of crystalloid formation. They were staining positively with a-human beta-2-Microglobulin (Fig.3.a-b.). The beta-2-Microglobulin crystalloid corpora amylacea were resistant to KMnO4 oxidation for 30 sec-1 min, resistant/sensitive for 2 min (R<S), sensitive for 3 min; and resistant to performate pretreatment for 1-20 sec.
4. More or less pronounced interstitial pneumonitis was present in 124 (**36.7 %**), interstitial pneumonitis and/or interstitial fibrosis in 154 (**45.6 %**), and heart failure cells in 14 (**4.14 %**) of 338 lungs in 169 RA patients.
5. Interstitial pneumonitis (and/or fibrosis) was associated with alveolar microlithiasis in 16 of 124 (or 154) cases. Heart failure cells never accompanied to corpora amylacea. There was **no** significant correlation between pulmonary microlithiasis and interstitial pneumonitis (x^2=0.1830, p<0.94), or interstitial pneumonitis and/or fibrosis (x^2=-0.1727, p<0.5070), or congestive heart failure (x^2=0.0437, p<0.83). Analysis of the clinical data and post mortem findings **do not** suggest that pulmonary microlithiasis may have contributed to the development of interstitial pneumonitis, fibrosis, or heart failure.

DISCUSSION

Pulmonary alveolar microlithiasis is a progressive process, characterized by crystalloid formation of beta-2-Microglobulin. During this process the alveolar calcospherities progressively increase in number and size. The „maturation" of alveolar corpora amylacea is accompanied by deposition of minerals, namely calcium phosphates and carbonates.

The clinical significance of pulmonary alveolar microlithiasis is not known, but it does not seem to contribute to interstitial pneumonitis, fibrosis, or heart failure.

REFERENCES

1. Romhányi, G., Selective differentiation between amyloid and connective tissue structures based on the collagen specific topo-optical staining reaction with congo red. *Virchows Arch.*, 354, 209-222, 1971.
2. Romhányi, G., Selektive Darstellung sowie methodologische Möglichkeiten der Analyse ultrastruktureller Unterschiede von Amyloidablagerungen. Zbl. Allg. Pathol. Pathol. Anat., 123, 9-16, 1979.
3. Wright, J.R., Calkins, E., and Humphrey, R.L., Potassium permanganate reaction in amyloidosis. Laboratory Investigation, 36, 274-281, 1977.

4. Bély, M., and Apáthy, Á., Identification of Amyloid Deposits by Histochemical Methods of Romhányi: Applied Histochemistry. Systemic secondary (AA) amyloidosis in Rheumatoid Arthritis. Amyloid: J. Protein Folding Disord., 8, Suppl.2. (Guest Editor: M Bély) 177-182, 2001.

5. Bély, M., Differential diagnosis of amyloid and amyloidosis by the histochemical methods of Romhányi and Wright. Acta Histochemica, 105, 361-365, 2003.

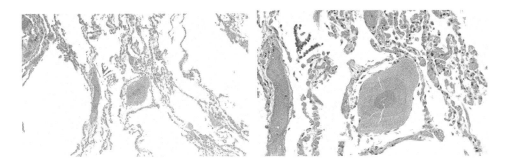

Fig.1.a-b. Lung. Corpus amylaceum, are rounded, eosinophilic, concentrically laminated crystalloid, calcified proteinaceous bodies, showing different stages of crystalloid formation. **(a)** HE, x125 **(b)** Same as Fig. 1a x200

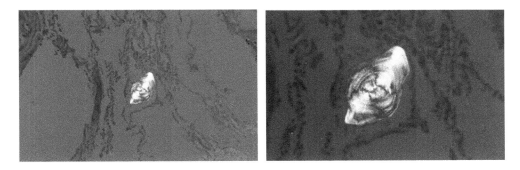

Fig.2.a-b. Lung. Corpus amylaceum are congophilic and concentrically lamellated bodies. **(a)** Congo red staining, without alcoholic differentiation, covered with gum arabic. Viewed under polarised light, x125 **(b)** Same as Fig. 2a x200

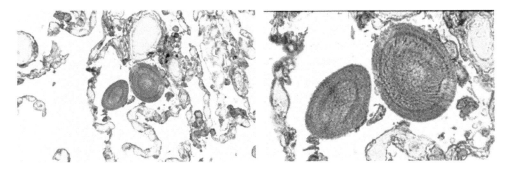

Fig.3.a-b. Lung. Corpora amylacea. The congophilic and birefringent proteinaceous calcospherities are positive for (a-hu) beta-2-Microglobulin. Negative for a-hu AA, or AL amyloid. **(a)** StrABC/HRP: anti-human (a-hu) beta-2-Microglobulin 1:400; [polyclonal antibody A0072; DAKO, Glostrup, Denmark], x125 **(b)** Same as Fig. 3a x200

PATIENT INFORMATION IN AA AMYLOIDOSIS – WHAT THE PATIENTS WANT

J.A. Hunter, E. Ellison, A. Francks, M. Gupta, R. Swann, N. Hanif

Rheumatology, Gartnavel General Hospital, Glasgow, G12 0YN, Scotland

In rheumatic diseases increasing levels of patient education are associated with improved skills and attitudes in addition to improved knowledge.[1] When patients know more about their condition, there is an improved sense of control with persisting reduction in depression and anxiety. Although patient education is probably best delivered by the multidisciplinary team, information leaflets alone can increase knowledge and improve compliance in several chronic conditions.[2]

AA amyloidosis is a rare but serious complication of chronic inflammatory disease. In the developed world rheumatoid arthritis (RA) is the commonest underlying condition. The consequences include nephrotic syndrome, renal impairment, resistant nausea and vomiting, and chronic diarrhoea and malabsorption. Many patients will have reached end stage renal failure five years after the diagnosis of amyloidosis.

Proteinuria is often intermittent initially, or attributed to other causes, such as drug therapy for inflammatory arthritis. Symptoms related to amyloidosis are often non-specific and the diagnosis, often delayed, depends on invasive biopsy procedures. The condition not readily compared to other common pathologies and is prognostically important to the patient. There are limited information resources for amyloid patients especially for those without internet access. We have therefore identified the perceived information needs of patients with AA amyloidosis, designed a patient information leaflet, and then assessed it by sending it to patients and their main carers.

Study 1. Perceived information needs - do they differ from other patients?

Methods: AA amyloid patients and unselected control attendees at the rheumatology clinic used the same questionnaire to assess previous information sources and the importance (5 point scale) of content and style of a leaflet for their condition. They were asked their views on information about: the definition of condition; its causes; epidemiology; its effects; investigations; available treatment; drugs, surgery and lifestyle changes that might help; support organisations; prognosis; legal, employment & financial issues; case histories as examples; pictures/diagrams; large print; colour; clarity of language; a glossary of medical terms; contact details; relevant internet sites; and support organisations.

Results: 12 of 15 AA amyloid patients replied (10 had RA). 1 carer also added his opinion. The 13 control subjects had RA(3), spondyloarthropathies (3), connective tissue diseases (4), and 1 each with gout, Henoch-Schoenlein purpura, and an undiagnosed polyarthropathy. The gender ratio differed between the amyloid and control groups (M:F 1:11 and 7:6 respectively), and the mean age in the amyloid group was higher (66 v 60years). 100% of amyloid patients gained information about their condition predominantly at the clinic compared with 62% of controls, who had gained knowledge through other health professionals (31%), leaflets (47%), the media (15%), support groups (15%), and through student contacts (15%) and the library.

Figure 1 shows the scores for the top 10 priority areas for a proposed leaflet for amyloid patients and controls. The controls usually attached greater importance to the priority areas.

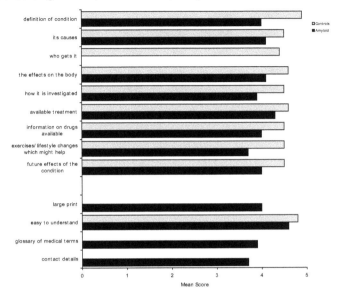

Figure 1. Histograms of mean scores for top 10 priorities for a leaflet (a mean score of 5 would indicate all respondents rated the item as "very important")

On the basis of the data a leaflet for AA amyloidosis was designed.

Study 2. Does the leaflet achieve standards expected by patients?

Methods: Leaflets and a new questionnaire were distributed to 16 AA amyloid patients and their carers, and 19 control patients and their carers (controls received the relevant Arthritis Research Campaign (ARC) leaflet. Subjects were asked to rate attributes of the leaflets out of 10.

Results: 12 of 16 amyloid patients and their carers responded, while 10 of 19 controls responded; they identified 7 responding carers. Figure 2 shows high scores for understandability and usefulness of different parts of the leaflet to amyloid patients and carers while figure 3 shows that patients and carers scored the general impression of the professionally printed ARC leaflets higher than the local AA amyloid leaflet.

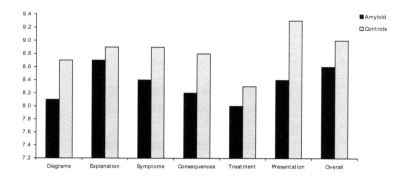

Figure 2. Histograms of mean scores (out of 10) for understandability and usefulness for amyloid patients and their main carers

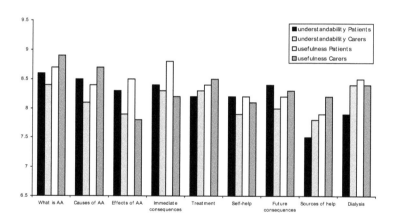

Figure 3. Histogram of combined patients' and carers' scores for general impressions of leaflets

Discussion: We have demonstrated that opinion can be gathered from patients about their needs for written information to allow the preparation of a leaflet for a rare condition. The leaflet then prepared met with little criticism and much welcome by patients and carers alike, and when assessed for usefulness, understandability and presentation scored with levels of acceptability close to those for leaflets produced by a national charity with an interest in patient education.

References

1.   Branch V.K. Lipsky K. Nieman T. Lipsky P.E. Positive impact of an intervention by arthritis patient educators on knowledge and satisfaction of patients in a rheumatology practice. Arthritis Care & Research, 12, 370, 1999

2.   Casey J. Johnson V. McLelland P.J. Impact of stepped verbal and written reinforcement of fluid balance advice within an outpatient haemodialysis unit : a pilot study. Human Nutrition & Dietetics,15, 43, 2002

ADVANCED GLYCATION END PRODUCTS AND ADVANCED OXIDATION PROTEIN PRODUCTION IN PATIENTS WITH AA AMYLOIDOSIS

R.Ryšavá, M.Kalousová *, T.Zima **, C.Dostál ***, M.Merta, V.Tesař

Department of Nephrology, *1ˢᵗ Institute of Medical Chemistry and Biochemistry, **Institute of Clinical Chemistry and ***Institute of Rheumatology, 1ˢᵗ Faculty of Medicine, Charles University, Prague, Czech Republic

The presence of chronic inflammation in the body results in the development of AA amyloidosis. Inflammation and mild or moderate forms of renal insufficiency are associated with increased oxidative and carbonyl stresses. Oxidative stress can be characterized as a dysbalance between free radicals and reactive oxygen on the one side and nitrogen species and antioxidants on the other side, in favor of free radicals [1]. Several experiments have shown increased oxidative stress in renal insufficiency (increased serum levels of lipoperoxidation products as well as increased levels of advanced oxidation protein products – AOPP) [2,3]. Patients (pts) with renal insufficiency have often decreased antioxidant defense (reduced enzyme activity and/or deficient cofactors). There is also a growing evidence for carbonyl stress in uremia and even in renal insufficiency, which is characterized as reactive carbonyl compounds overload (especially advanced glycation end products –AGEs).

We tested the hypothesis if the patients with AA amyloidosis and mild or moderate renal insufficiency do produce an increase of carbonyl/oxidative stress end products compared with pts suffering from rheumatoid arthritis (RA) without signs of AA amyloidosis and healthy controls (Co). We also investigated a new parameter of microinflammation - pregnancy associated plasma protein A (PAPP-A), which has a close relation to oxidative stress and is a new parameter of acute coronary syndrome.

PATIENTS AND METHODS

We investigated 17 pts with histologically proven AA amyloidosis (AA, mean age 58.8 ± 11.7 years; the underlying diseases were RA in 14 pts, eosinophilis fasciitis, Crohn´s disease and gout nephropathy in 1 pt resp; mean duration of underlying diseases was 15.6 ± 10.4 years) and 26 pts with RA without signs of AA amyloidosis (RA, mean age 62.7 ± 11.8 years; mean duration of RA was 13.36 ± 7.3 years). The stagging and activity score of RA were comparable in both groups. 20 healthy volunteers represented the control group (Co, mean age 54.9 ± 8.14 years).

We investigated AGEs, AOPP and PAPP-A as the parameters od carbonyl/oxidative stress, SAA (serum amyloid A), CRP and orosomucoid as the parameters of inflammation. TNF-α (tumor necrosis factor α), IL-18 (interleukin 18) and M-CSF (macrophage colony stimulating factor) were assessed as supplementary factors.

Determination of AGEs was based on spectrofluorimetric detection according to Henle and Munch in our modification. Determination of AOPP was based on spectrophotometric detection according to Witko-Sarsat in our modification. Other parameters were investigated by standard ELISA tests.

Statistics results are expressed as mean ± standard deviation; statistical significance was evaluated using ANOVA test and correlation using linear regression test.

RESULTS

Results are given in the table 1-3. We disclosed a significant positive correlation between AGEs and serum creatinine (r = 0.65) and PAPP-A (r = 0.59) resp, and a negative correlation between AGEs and GFR (r = -0.52). No correlation was found between AGEs and AOPP. AOPP did not correlate with any of investigated parameters. We disclosed a significant positive correlation between PAPP-A and the following parameters: serum creatinine (r = 0.59), SAA (r = 0.69), CRP (r = 0.49) and M-CSF (r = 0.76), and a negative correlation between PAPP-A and GFR (r = -0.94).

Table 1. Plasma concentrations of AOPP, AGEs and PAPP-A in the group of pts with AA, RA and healthy controls (Co)

	AA (n=17)	RA (n=26)	Co (n=20)	Statistical significance
AOPP (μmol/l)	145.4 ± 33.4	82.7 ± 37.3	81.8 ± 27.9	AA vs RA p<0.001 AA vs Co p<0.001
AGEs (AU)	5.8 ± 3.2 x10^5	3.7 ± 1.4 x10^5	2.7 ± 4.3 x10^5	AA vs RA p<0.001 AA vs Co p<0.001
PAPP- A (mU/l)	15.1 ±7.8	11.5 ± 3.5	11.2 ± 3.0	Ns

Table 2. Plasma concentrations of some (selected) cytokines and adhesion molecules in the group of pts with AA, RA and in healthy controls (Co)

	AA (n=17)	RA (n=26)	Co (n=20)	Statistical significance
TNF-α (pg/ml)	9.7± 5.6	21.1 ± 9.9	5.3 ± 0.6	RA vs AA p<0.001 RA vs Co p<0.001
M-CSF (pg/ml)	341.7 ± 242.6	680.9 ± 322.7	450.1 ± 71.8	RA vs AA p<0.001
IL-18 (pg/ml)	327.9 ± 206.9	345.6 ± 217.6	228.9 ± 146.2	Ns
ICAM-1 (ng/ml)	347.9 ± 108.3	420.4 ± 154.0	307.3 ± 56.4	RA vs Co p<0.05

Table 3. Plasma (and serum) concentrations of inflammatory and protein parameters, creatinine and GFR in the group of pts with AA, RA and healthy controls – (Co)

	AA (n=17)	RA (n=26)	Co (n=20)	Statistical Significance
SAA (μg/l)	80.7 ± 120.7	159.8 ±187.4	18.6 ± 32.2	RA vs Co p<0.05
Total protein (g/l)	63.0 ± 10.1	70.8 ± 4.5	72.8 ± 3.8	RA vs AA p<0.001 KO vs AA p<0.01
Albumin (g/l)	31.5 ± 9.0	39.8 ± 3.4	42.3 ± 1.9	RA vs AA p<0.001 KO vs AA p<0.001
CRP (mg/l)	17.0 ± 13.9	17.0 ± 19.5	2.7 ± 4.5	RA vs Co p<0.01 AA vs Co p<0.05
Orosomucoid (g/l)	1.5 ± 0.6	1,3 ± 0.5	0.76 ± 0,27	Ns
Glucose (mmol/l)	6.89 ± 2.69	4.71 ± 0.59	NA	RA vs AA p<0.001
Cholesterol (mmol/l)	5.99 ± 1.55	5.67 ± 0.85	NA	Ns
Creatinine (μmol/l)	218.5 ± 12.8	82.0 ± 15.0	87.4 ± 10.5	AA vs RA p<0.001 AA vs Co p<0.001
GFR (ml/s)	0.5 ± 0.3	1.2 ± 0.3	NA	RA vs AA p<0.001

Abbr:

NA   not assessed

GFR   glomerular filtration rate

DISCUSSION AND CONCLUSION

The finding of elevated plasma levels of AGEs and AOPP suggest a higher probability of tissue damage by carbonyl/oxidative stress in the group of pts with AA. The increase of plasma concentrations of AGEs is probably caused by decreased renal function in this group of pts as has been already described by other authors. Surprisingly no correlation was found between AOPP and other parameters; this fact likely mirrors a deeper disorder in the protein rather than in the glycid metabolism, both independent on the renal function. The plasma levels of PAPP-A - a new parameter of microinflammation, were increased in the AA group (however without statistical significance), this observation could be partly explained by the imparment of renal function in this group of pts (correlation of plasma levels of PAPP-A with serum creatinine and GFR resp). The correlation of plasma levels of PAPP-A with plasma levels of SAA and CRP resp could reflect the role of PAPP-A in the process of inflammation and could serve as a parameter of amyloidosis activity. The correlation of plasma levels of PAPP-A with M-CSF could represent the activation of macrophages and their role in the amyloid formation process.

Our (pilot) study, performed in order to provide a deeper insight into the pathogenesis of the process of amyloid formation, further strengthed our view of the importance of carbonyl/oxidative stress in the development of (AA) amyloidosis and also suggest that some new factors (as PAPP-A), described only recently, could play a role and/or serve a markers in the amyloidogenesis.

This study was supported by grant IGA MH CZ No NB/7035-3.

REFERENCES

1. Halliwell, B., Drug antioxidants effects: A basis for drug selection? *Drug*, 42, 569, 1991.
2. Miyata, T. et al., Alterations in nonenzymatic biochemistry in uremia: Origin and significance of "carbonyl stress" in long-term uremic complications. *Kidney Int*, 55, 389, 1999.
3. Witko Sarsat, V. et al., Advanced oxidation protein products as a novel marker of oxidative stress in uremia. *Kidney Int*, 49, 1304, 1996.

ETANERCEPT IN THE TREATMENT OF AA AMYLOIDOSIS IN JUVENILE RHEUMATOID ARTHRITIS-PRELIMINARY EXPERIENCE

H. Michels[1], R. Häfner[1], R.P. Linke[2]

[1] Rheumatic Childrens' Hospital, Garmisch-Partenkirchen, Germany
[2] Max-Planck-Institute of Biochemistry, Martinsried, Germany

INTRODUCTION

Amyloid A (AA) amyloidosis is a severe complication of chronic inflammatory disorders. It is characterized by extracellular deposition of amyloid in various tissues and organs, possibly leading to organ failure [4]. The major component of amyloid represent the amyloid fibrils which are derived from the circulating precursor serum amyloid A (SAA) in AA-amyloidosis. While chronic infections including tuberculosis were the most frequent underlying diseases in former times, these are today rheumatic diseases such as rheumatoid and juvenile rheumatoid arthritis (JRA), ankylosing spondylitis, Crohn's disease or familial Mediterranean fever. The treatment of AA-amyloidosis is directed against the inflammatory stimulus, here against the underlying disease JRA. In JRA, amyloidosis develops when methotrexate and other second line drugs fail. In this case, chlorambucil has often been used [1,5].

Etanercept has proven effective in JRA [2,3]. It is a fusion protein of two TNFα receptors connected by the Fc part of an IgG molecule. By blocking TNFα, and thus down-regulating the inflammatory pathways, Etanercept imitates the physiological process of a soluble TNFα receptor. Thus, it antagonizes the formation of SAA and may constitute a novel treatment of amyloid A amyloidosis [4]. Here, we have tried to answer two questions: (1) Can amyloidosis be halted? (2) Can organ functions be preserved or even restored? Here, we present our first experience with Etanercept.

PATIENTS & METHODS

Between 2001 and 2003, we have observed 11 patients with JRA and AA-amyloidosis, all of whom were now treated with Etanercept, 6 boys and 5 girls. Most of the patients had a systemic onset (Tab.1). All patients followed a rheumatic factor, negative polyarthritic course. One patient (JHU) died of septicemia during the observation period. The follow-up under Etanercept treatment was 1.9 ± 1.01 years.

In daily routines, CRP can be measured more easily compared with serum amyloid A (SAA). Since CRP parallels the SAA serum concentration, we used the serum CRP concentration to judge the amyloidogenic risk. A normal CRP concentration would mean amyloidogenesis is stopped. A high serum CRP concentration would demonstrate that the amyloidogenic process is up-regulated.

The hemoglobin concentration reflects a more complex situation. Low hemoglobin as a sign of "inflammatory anemia" can reflect a severe inflammation. On the other hand, a low hemoglobin can also reflect loss of blood, for instance as a result of the anti-rheumatic treatment with NSAIDs or as a consequence of intestinal amyloidosis. In case of amyloidosis, however, a normal hemoglobin concentration reflects a more favorable situation.

Table 1: Gender, onset type and course of JRA, age at onset of JRA and of amyloidosis, age at the beginning of the treatment with etanercept and the duration of the treatment with etanercept (Enbrel) are given. (Abbreviations: pauci, pauciarticular; RF, rheumatic factor; neg, negative; Poly, polyarthritis)

Name	Gender	JRA		Age (years) at				Duration Enbrel (Months)
	F – female M – male	Onset type	Course	Onset of JRA	Onset of Amyloidosis	Onset of Enbrel	Last Visit	
RBR	M	systemic	RFnegPoly	5.1	12.9	15	16	12
SLA	F	pauci	RFnegPoly	2.8	10.2	19.2	20.8	19
IJU	M	systemic	RFnegPoly	3.2	10.4	10.5	12.5	24
MHA	F	systemic	RFnegPoly	5.4	12.2	12.5	14.9	29
TEF	F	systemic	RFnegPoly	1.1	4.2	6.8	8.1	16
AOK	M	systemic	RFnegPoly	4.4	13.0	22.8	24.8	24
JHU	F	systemic	RFnegPoly	2.9	15.1	19.8	21.3 †	18
TPO	M	systemic	RFnegPoly	1.0	8.3	18.4	22.0	43
MGI	M	systemic	RFnegPoly	1.3	4.4	14.5	16.5	24
LTA	M	pauci	RFnegPoly	4.2	10.6	12.9	16.3	41
SDL	F	RFnegPoly	RFnegPoly	2.5	13.5	18.3	20.6	9

† deceased 3 January 2003 due to sepsis, multi-organ failure

The question as to whether or not organ functions could be preserved or restored, is being answered here only for the kidney function. As parameters, we have chosen the proteinuria during 24 hours and the serum creatinine concentration.

RESULTS

A normalization of CRP, as a sign for the arrest of amyloidogenesis, was observed in only two out of 11 patients (Tab.2). One of them had had a short duration of amyloidosis, while the amyloidosis had lasted for nine years in the other patient. In four patients the serum CRP concentration improved.

Table 2: The development of the parameters CRP, hemoglobin, creatinine, proteinuria and blood pressure before thetreatment with etanercept (Enbrel) and at the last visit under etanercept are described.

Name	CRP (mg/dl)		Hemoglobin (g/dl)		Creatinine (mg/dl)		Proteinuria (mg/24h)		Blood pressure (mmHg)	
	Before Enbrel	Last Visit	Before Enbrel	Last Visit	Before Enbrel	Last Visit	Before Enbrel	Last Visit	Before Enbrel	Last Visit
RBR	6.21	1.38	9.7	10.2	0.42	0.44	80	110	110/65	110/60
SLA	3.63	0.01	13.0	12.6	1.12	0.84	7400	2576	120/75	105/70
IJU	5.08	0.01	9.1	12.2	0.31	0.43	1019	160	100/60	110/60
MHA	8.29	3.11	5.5	7.2	0.81	0.85	120	123	115/70	110/70
TEF	7.11	11.64	11.4	12.3	0.45	0.42	594	221	105/65	95/70
AOK	4.68	0.80	7.8	9.0	0.87	2.29	2440	1547	120/80 [a]	120/70 [a]
JHU	2.68	4.04	8.6	9.0	2.01	1.99	5170	7236	115/70 [a]	130/85 [a]
TPO	2.89	0.92	10.7	13.7	0.60	0.77	364	1605	135/75 [a]	130/85 [a]
MGI	1.56	1.25	14.6	8.1	0.74	1.02	1186	1313	120/80 [a]	130/70 [a]
LTA	4.37	1.58	11.3	13.9	0.41	0.34	88	120	115/60 [a]	120/80
SDL	1.49	1.95	11.1	11.5	0.31	0.32	1568	1485	115/70	110/60

[a] antihypertensive treatment

Three patients showed an unchanged CRP, and the CRP even increased in two additional patients. Hemoglobin improved in six patients, remained unchanged in four and worsened in one patient. In the end, five patients out of the eleven demonstrated normal hemoglobin concentrations. Creatinine remained normal in five of the eleven patients. Two patients demonstrated an improvement. In the remaining four patients, the creatinine level remained unchanged in the pathological range or even increased. Proteinuria remained normal in three patients, improved in four, remained unchanged in two and worsened in two patients.

Since Etanercept showed no efficacy during nine months in the patient SDL, it was discontinued and replaced by Iinfliximab, with encouraging results: CRP normalized, and the proteinuria decreased markedly.

DISCUSSION

In JRA, amyloidosis develops when the treatment of the underlying disease (JRA) fails. Chlorambucil has been shown to improve the course of amyloidosis in JRA [1,5]. The toxicity of Chlorambucil, however, includes such severe side effects as leukemia and irreversible sterility in boys. Therefore, new forms of treatment with increased efficacy and reduced toxicity would be most desirable. Since TNFα plays an important role in the development of AA amyloidosis, it seemed reasonable to investigate the TNFα antagonist Etanercept as a new form of treatment. It turned out that about two-thirds of the patients took advantage of the therapy with Etanercept. In seven of the eleven patients, improvements could be achieved which were remarkable in part. However, one important goal, the normalization of CRP as a sign of an arrest of amyloidogenesis, was observed in only two patients. The proteinuria decreased in four of the eight patients with an elevated proteinuria prior to the Etanercept treatment. In cases with normal creatinine and no proteinuria at the beginning of the treatment with Etanercept, this condition could be maintained.

Taken together, our preliminary results demonstrate an improvement of several parameters in most of the patients. The course of amyloidosis, however, may be improved only in a smaller percentage. We have to remember that only about 30% of the patients with systemic JRA respond well to Etanercept. The earlier the treatment with Etanercept begins the more effective it may be. In cases where Etanercept does not work, it seems worthwhile to try another TNFα blocking agent, that is Iinfliximab. We need more patients and a long-term follow-up to attain more reliable conclusions.

REFERENCES

1.   David J. et al., Amyloidosis in juvenile chronic arthritis: A morbidity and mortality study. Clin Exp *Rheumatol* 11: 85, 1993.

2.   Lovell D.J. et al., Long-term efficacy and safety of etanercept in children with polyarticular-course juvenile rheumatoid arthritis: interim results from an ongoing multicenter, open-label, extended-treatment trial. Arthritis Rheum. 48, 218, 2003.

3.   Lovell D.J. et al., Etanercept in Children with Polyarticular Juvenile Rheumatoid Arthritis. *N.Engl.J.Med. 342,* 763, 2000.

4.   Merlini, G. and Bellotti, V, Mechanisms of disease: Molecular mechanisms of amyloidosis. *N.Engl.J.Med.* 349, 583, 2003.

5.   Schnitzer, T.J. and Ansell, B.M., Amyloidosis in juvenile chronic polyarthritis. *Arthritis Rheum.* 20 (suppl.),245, 1977.

EPIDEMIOLOGIC DESCRIPTION OF AMYLOIDOSIS DIAGNOSED IN UNIVERSITY HOSPITAL OF RENNES FROM 1995 TO 1999

C. Cazalets[1], B. Cador[1], N. Mauduit[2], O. Decaux[1], M.P. Ramée[3], P. Le Pogamp[4], M. Laurent[5], P. Jego[1], B. Grosbois[1]

1 : Service de Médecine Interne, Hôpital Sud, 16 boulevard de Bulgarie BP 90347 – 35203 RENNES Cedex 02 ; 2 Epidémiologie et hygiène Hospitalière 3 Service d'Anatomie et Cytologie Pathologique B , 4 Service de Néphrologie , 5 Département de Cardiologie, Hôpital Pontchaillou, rue Henri Le Guilloux, 35033 Rennes Cedex 9, France

1. INTRODUCTION

In France the descriptive epidemiology of amyloidosis is not well known. For this reason we decided to compile a register of amyloidosis in the University Hospital of Rennes (France).

2. MATERIAL AND METHODS

This retrospective study was performed on all the cases of amyloidosis diagnosed from 1st January 1995 to 31st December 1999. Diagnosis was assessed on positive red Congo staining by anatomo-pathology. Immunochemistry (anti SAA, anti-Kappa and Lambda light chain) was used to characterize the type of amyloidosis. Clinical data, staging and outcome of patients were analysed. Statistical analysis used SPSS 10.0, 1999 program.

3. RESULTS

3. 1. Incidence

Forty three amyloidoses were diagnosed during the five years of the study. The mean annual incidence was 8.6 cases per year. Five cases were diagnosed in 1995, 6 in 1996, 6 in 1997, 12 in 1998 and 14 in 1999.

3. 2. Demographic data

Twenty seven patients were female and 16 male (sex ratio 1.68). Mean age was 63.7 years ranging from 16 to 87 years.

3. 3. Sites of biopsy

In 42 patients (86 %) diagnosis was made on an affected organ: kidney (11), heart (7), gut (7), skin (5), gall bladder (4), synovium (3), urogenital tract (3), and trachea (2).

In seven patients (14%) diagnosis was made on an accessory site (minor salivary gland biopsy and/or bone marrow biopsy).

3. 4. Classification

The types of amyloidosis are summarized in table 1.

Among the 15 patients with an undetermined type of amyloidosis eleven patients were not studied as frozen samples were not obtained. In four patients immunochemistry tests were negative

3. 4. Extension

Thirty three patients (77 %) presented with a systemic amyloidosis

In ten patients (23%) amyloidosis was localized: gall bladder (3), tracheobronchial (2), skin (2), synovium (2) and eye lid (1).

3. 5. Survival

With a median follow-up of thirty six months median survival was 12 months (4 months in AA and 19 months in AL amyloidosis respectively). Twenty five patients have died. Causes of death are summarized in table 2.

4. DISCUSSION

Epidemiologic studies of amyloidosis are very rare. Kyle [1] has shown interesting results concerning the incidence of primary systemic amyloidosis in Olmsted County, Minnesota, 1950 through 1989. Estimated annual incidence was 8.9 new cases per million inhabitants corresponding to 2225 new cases each year in the United States. In our study we estimated that the annual incidence was 11.3 per million inhabitants for all types of amyloidosis and 5.2 per million inhabitants for AL amyloidosis. The incidence of transthyretin amyloidosis in previous studies was 10 to 20% of the incidence of AL amyloidosis. Regarding the reduction in chronic infectious diseases incidence of AA amyloidosis has dramatically decreased. As previously observed by Kyle [2] localized amyloidosis represented about 25 % of cases.

In our series we observed an increase in the annual incidence of amyloidosis. We are unable to affirm whether that increase is a coincidence or a reality. Diagnosis of amyloidosis was mainly performed on biopsy of affected organ. Bone marrow biopsy and minor salivary gland biopsy represented only 7 % of all biopsies and 14 % of positive biopsies. This fact emphasises the interest of biopsies of accessory sites as the sensitivity is very high : 90 % for bone marrow and abdominal fat aspiration (3), 52 to 72 % for abdominal fat aspiration (4), 86 % for minor salivary gland (5).

References

[1]   Kyle R, Linos A, Beard C, Linke R, Gertz M, O'Fallon W, et al. Incidence and natural history of primary systemic amyloidosis in Olmsted County, Minnesota, 1950 through 1989. Blood 79, 1817-22, 1992.

[2]   Kyle R, Greipp P. Amyloidosis (AL) Clinical and laboratory features in 229 cases. Mayo Clin Proc 58, 665-83, 1983.

[3]   Gertz M, Lacy M, Dispenzieri A. Amyloidosis: recognition, confirmation, prognosis, and therapy. Mayo Clin Proc 74, 490-4, 1999.

[4]   Duston M, Skinner M, Meenan R. Sensitivity, specificity, and predictive value of abdominal fat aspiration for the diagnosis of amyloidosis. Arthritis Rheum 32, 82-4, 1989.

[5]   Hachulla E, Janin A, Flipo R, Saile R, Facon T. Labial Salivary gland biopsy is a reliable test for the diagnosis of primary and secondary amyloidosis, A prospective clinical and immunohistologic study in 59 patients. Arthritis Rheum 36, 691-7, 1993.

Table 1. Classification of amyloidosis observed in 43 patients
20 AL Amyloidosis
· 15 Primary
· 5 Malignant plasmocytic dyscrasia
(4 Multiple Myeloma, 1 Waldnstroem's Macroglobulinemia)
7 AA Amyloidosis
4 Chronic Rheumatic Disease
(3 Rheumatoid Arthritis, 1 Psoriatic arthritis)
2 Chronic Infection
(1 tuberculosis, 1 chronic skin infection)
1 Covert inflammation
1 B2 microglobuline Amyloidosis
15 Undetermined amyloidosis

Table 2 Causes of death in 25 patients
16 Progression of Amyloidosis
· 10 Cardiac
· 5 Renal
· 1 Haemorrhage
2 Treatment toxicity
7 Non amyloidosis related

A HETEROZYGOTIC JAPANESE PATIENT WITH FAMILIAL MEDITERRANEAN FEVER (FMF) PYRIN M694I

Terazaki H[1], Okuda A, Katase K, Nakamura Mi[2], Ueda M, Kim JM, Haraoka K, Ando Y, and Sasaki Y[1].

[1]Department of Gastroenterology and Hepatology and [2]Department of Diagnostic Medicine, Graduate School of Medical Sciences, Kumamoto University, Kumamoto, Japan.

ABSTRACT

Although patients with familial Mediterranean fever (FMF) are often found in non-Ashkenazi Jewish, Armenian, Arab and Turkish people, only a few cases have been reported in Asian people.

We here report a 31-year-old man suffering from diarrhea and recurrent fever with pains in abdomen. For ten years, he had experienced repeated abdominal pain attacks once six months and underwent appendectomy at the age of 21-year-old. Laboratory examinations revealed that in the febril phase, leukocytosis and elevated CRP were recognized. Despite physical and laboratory examinations, infectious diseases, collagen disorders or malignancy was not listed. Although a colonoscopic examination revealed normal macroscopic findings, Congo red staining for mucosal biopsy specimen from the rectum showed amyloid deposits around vessels of propria mucosa and submucosa. Serum amyloid A levels were increased in the febril phase. DNA analyses by polymerase chain reaction (PCR) followed by cycle sequencing revealed a heterozygotic mutant (ATG to ATA) in codon 694 in exon 10 of FMF (pyrin) gene resulting in a substitution of isoleucine for methionine (M694I) in this patient. Based on these findings, he was diagnosed as FMF with amyloidosis. The symptoms have also been suppressed by treatment of oral colchicines. Although FMF had been thought to be a rare disease in Asian countries, such patients may be more often found if we suspect the disease in such cases.

INTRODUCTION

Familial Mediterranean fever (FMF) is an inherited inflammatory disease commonly found in Mediterranean populations such as Armenians, Arabs, Sephardic and Askenazi Jews, Turks and in other peoples of Mediterranean origin (1). In 1997, the gene causing FMF was cloned from the short arm of chromosome 16 which encodes pyrin. Patients with two mutations, either homozygotes or compound heterozygotes tended to have more severe conditions than heterozygotes (2). However, we here report that a Japanese patient with FMF with only a single pyrin mutation who developed amyloid depositions in the gastrointestinal organs.

A CASE REPORT

A 31-year-old man had been suffering from periodic fever and an abdominal pain since eight years ago. He was diagnosed as peritonitis and underwent the appendectomy. But he experienced abdominal pain once a year and admitted to several different hospitals. Every time, his laboratory examinations showed only leukocytosis and elevated CRP levels and he had been given antibiotics. Despite an uncertain diagnosis, his symptoms naturally disappeared within a week. On July 29, 2004, he was admitted to our hospital because of recurrent pain and continuous diarrhea. Chief complains: the patient complained of severe abdominal pain, vomiting, diarrhea.

Physical examinations:His temperature was 37.2°C. Pulse was 100/min, and blood pressure was 125/93 mmHg. The functions of the brain, neck, lungs, and heart were normal except for tachycardia. The abdomen was diffusely tender with increased bowel sounds. Rebound tenderness was not obtained in his abdomen. Laboratory examinations revealed that leukocytosis and elevated CRP were recognized. But infectious diseases, collagen disorders or malignancy were not listed.

Congo red staining for mucosal biopsy samples from the rectum revealed amyloid deposits around vessels of propria mucosa and submucosa (Fig. 1).

DNA analyses by polymerase chain reaction (PCR), followed by cycle sequencing revealed a heterozygotic mutant (ATG to ATA) in codon 694 in exon 10 of pyrin gene resulting in a substitution of isoleucine for methionine (M694I) in this patient. He was diagnosed as FMF with systemic amyloidosis. The symptoms have also suppressed by treatment of oral administration of colchicines.

DISCUSSION

We identified a Japanese FMF with amyloidosis. This mutation that caused amino acid replacement of methionine by isoleucine in position 694, seems to be identified mainly in Arabic people. In a study of 75 Arab patients with FMF, 41% were homozygous and compound heterozygous, and the main mutation M694I was detected in 63% of the alleles. Future, M694I was absent in Jews and Armenians and was detected in only 2% of the Turkish patients, who had many FMF patients. He was not shown the relation to these countries. And all Japanese FMF patients, whose mutation was a same position and a same amino acid replacement, were found in several area of Japan (3).   These areas were geographically distant, and a consanguineous relationship between these areas has not been identified. The presence of Japanese patients with the same mutant position in Pyrin gene suggested that there might be asymptomatic FMF gene carriers in Japan and some patients of FMF might develop AA amyloidosis as unknown origins.

CONCLUSIONS

1   We identified a heterozygotic Japanese FMF (Pyrin) M694I with amyloidosis.
2.   All Japanese FMF patients are all the same mutation Pyrin M694I.
3.   Not only homozygotes, but also heterozygous M694I gene carriers develop amyloidosis.

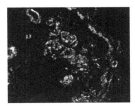

Figure 1. Congo red staining for mucosal biopsy samples from the rectum revealed amyloid deposits around vessels of propria mucosa and submucosa.

REFERENCE

1. Drenth JP, van der Meer JW. Hereditary periodic fever. N Engl J Med. 2001 Dec 13;345(24):1748-57.

2. Nir-Paz R, Ben-Chetrit E. Molecular diagnosis of familial Mediterranean fever. N Engl J Med. 2000 Jan 6;342(1):60.

3. Tomiyama N, Oshiro S, Higashiuesato Y, Yamazato M, Sakima A, Tana T, Tozawa M, Muratani H, Iseki K, Takishita S. End-stage renal disease associated with familial Mediterranean fever. Intern Med. 2002 Mar;41(3):221-4.

COMPARISON OF CLINICAL AND IMMUNOHISTOCHEMICAL FINDINGS IN PATIENTS WITH RHEUMATOID ARTHRITIS AND SECONDARY AMYLOIDOSIS

D. Povilenaite, V. Graziene, I. Butrimiene, A. Venalis

Vilnius University Institute of Experimental and Clinical Medicine; Rheumatology Center, Vilnius University, Lithuania

INTRODUCTION

About 3% to 20% of the patients with rheumatoid arthritis (RA) develop AA amyloidosis in the course of their disease depending upon the severity and duration of the arthritis in the population investigated [1]. Deposits of amyloid A protein (AA) are found in the walls of blood vessels and interstitial throughout the whole body. Amyloid is diagnosed by the typical green birefringence in polarized light of a tissue biopsy specimen stained with Congo red dye. Subsequent immunohistochemistry with anti-AA antibodies establishes the amyloid A nature of the amyloid involved [2,3]. Assessment of the extent of average amyloid deposition reveals from scarce to abundant extent in biopsies of tissues. Clinical amyloidosis is defined as the presence of symptoms or signs suggestive of visceral involvement by amyloid. Proteinuria by structural changes of renal basement membrane is often the first clinical presentation of generalised AA amyloidosis [4].

The aim of the study was to compare clinico-demographic data with localization sites and amount of AA amyloid deposits in biopsies of patients with RA.

MATERIALS AND METHODS

126 patients (111 women, 15 men) were included in the study (Table 1). AA amyloidosis was confirmed in biopsies of labial minor salivary gland (96 biopsies), gingiva (24), abdominal fat (3), kidney (1), and rectum (2) using Congo red staining according to Eastwood [5] with evaluation in polarized light and immunohistochemistry by using monoclonal Mouse Anti-Human Amyloid A clone mc1 (DAKO, Code M07501).

The amount of amyloid in biopsies was estimated semiquantitatively by grading of microscopically detected deposits: for scarce 1+ (15 patients), for moderate 2+ (55 patients), and for abundant 3+ (56 patients) groups.

Statistical analysis was performed using Student and χ^2 tests (P<0.05).

RESULTS AND CONCLUSIONS

No significant statistical differences for age of RA patients and RA disease duration (Table 1), anatomical and functional stage, RA seropositivity, ESR, α_2-globulin, γ-globulin, blood urea nitrogen, serum creatinine among three groups with different amount of amyloid in tissues were found on first detection of AA amyloid in biopsies (Table 2).

The proportions of patients with proteinuria (P=0.006) and microhaematuria (P=0.028) were statistically significantly higher in 2+ and 3+ patients groups as compared to 1+ group (Table 2).

AA amyloid deposits in basement membranes were found with same frequency in all three groups. AA amyloid deposits in the wall of blood vessels (P=0.037) and pericollagenic deposits (P<0.0001) were found significantly more frequently in the 2+ and 3+ grade groups, than in 1+ grade group (Figure 1).

Table 1. Characteristics of 126 rheumatoid arthritis (RA) patients with secondary amyloidosis in groups with different amount of amyloid in tissues

Amount of amyloid in tissues	Number of RA patients	Characteristic		
		Sex (Female/Male)	Age of RA patients, years, mean±SD (range)	RA disease duration, years, mean±SD (range)
scarce 1+	15	14/1	58.60±9.19 (39-71)	14.27±8.69 (2-30)
moderate 2+	55	46/9	57.42±11.84 (24-82)	13.05±8.12 (1.5-35)
abundant 3+	56	51/5	61.98±10.30 (39-81)	13.75±8.19 (2-30)

Table 2. Clinical characteristics and laboratory findings of rheumatoid arthritis patients with secondary amyloidosis in groups with different amount of amyloid in tissues

Characteristic	Amount of amyloid deposits in tissues			P value
	scarce 1+ (n=15)	moderate 2+ (n=55)	abundant 3+ (n=56)	
RF* positive, n (%)	10 (67)	33 (60)	32 (57)	0.7967
Anaemia, n (%)	7 (47)	35 (64)	41 (73)	0.1404
Proteinuria, n (%)	1 (7)	28 (51)	28 (50)	0.0060
Microhaematuria, n (%)	2 (13)	27 (49)	19 (34)	0.0283
ESR*, mm/h, mean±SD	37.57±21.49	49.12±21.50	52.22±19.72	0.0555
α_2-globulin, %, mean±SD	12.74±4.46	14.40±4.59	14.75±3.55	0.2526
γ-globulin, %, mean±SD	18.32±4.24	20.01±4.90	21.46±5.67	0.0886
Blood urea nitrogen, mmol/l, mean±SD	6.41±1.78	7.51±4.56	8.31±4.79	0.3130
Serum creatinine, μmol/l, mean±SD	87.86±17.38	105.38±61.02	120.38±69.84	0.2400

*RF=rheumatoid factor; ESR=erythrocyte sedimentation rate

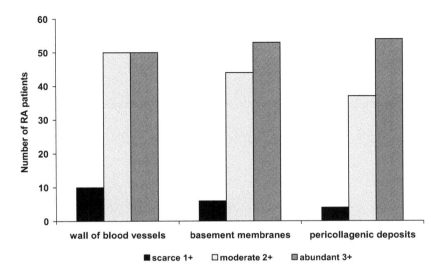

Figure 1. Localization sites of AA amyloid in tissues of rheumatoid arthritis (RA) patients in groups with different amount of amyloid in biopsies.

DISCUSSION

The median time from the onset of RA to the diagnosis of amyloidosis was 13-14 years in all patients groups. No clear correlation emerged between the amount of AA amyloid deposits in biopsies and the duration of RA up to diagnosis of amyloidosis. Only a half of the patients who have moderate and abundant deposits of amyloid in biopsies also have signs of clinical kidney dysfunction. Differences in the amount of AA amyloid in tissues might be explained by genetic factors associated with the synthesis of serum amyloid A protein (SAA).

REFERENCES

1. Hazenberg, B.P.C. and van Rijswijk, M.H., Where has secondary amyloid gone?, *Ann. Rheum. Dis.,* 59(8), 577, 2000.
2. Strege, R.J., Saeger, W., and Linke, R.P., Diagnosis and immunohistochemical classification of systemic amyloidoses, *Virchow Arch.,* 433, 19, 1998.
3. Röcken, C. and Shakespeare, A., Pathology, diagnosis and pathogenesis of AA amyloidosis, *Virchovs Arch.,* 440, 111, 2002.
4. Kuroda, T. et al., Comparison of gastroduodenal, renal and abdominal fat biopsies for diagnosing amyloidosis in rheumatoid arthritis, *Clin. Rheumatol.,* 21, 123, 2002.
5. Eastwood, H. and Cole, K.R., Staining of amyloid by buffered Congo red in 50% ethanol, *Stain Technol.,* 46, 208, 1971.

RENAL GLOMERULAR AL AND VASCULAR AA AMYLOIDOSIS IN A PATIENT WITH ANKYLOSING SPONDYLITIS

T. Pettersson, P. Anttila, T. Törnroth

Department of Medicine, Helsinki University Central Hospital, Helsinki, Finland

The case of a patient with a unique combination of renal vascular AA amyloidosis and predominantly glomerular AL amyloidosis is presented

CASE REPORT

History

A 59-year-old woman with a 41-year history of ankylosing spondylitis presented in 1998 with nephrotic-range proteinuria (3.6 g/24h). She had involvement of both spinal and peripheral joints, and had suffered from recurrent acute anterior uveitis. In spite of treatment with hydroxychloroquine, auranofin, sulphasalazine and intramuscular sodium aurothiomalate her inflammatory rheumatic disease had been continuously active and had led to spinal stiffness, cervical spine restriction and hip joint destruction. Bilateral hip arthroplasty had been performed. Serum creatinine concentration remained normal.

Kidney biopsy findings

Examination of a kidney biopsy specimen revealed small to medium-sized, mainly mesangial amyloid deposits in all 12 glomeruli. Massive transmural amyloid deposits were seen in most small blood vessels in the corticomedullary region. One larger artery showed patchy amyloid deposits in the muscular layer (media). Immunohistochemical characterization of the amyloid material using a panel of antisera revealed that the glomerular amyloid deposits stained with anti-lambda antiserum but not with anti-kappa or anti-AA antiserum, whereas most vascular amyloid deposits stained with anti-AA antiserum and others with anti-lambda antiserum but no deposits stained with both antisera (Figure 1).

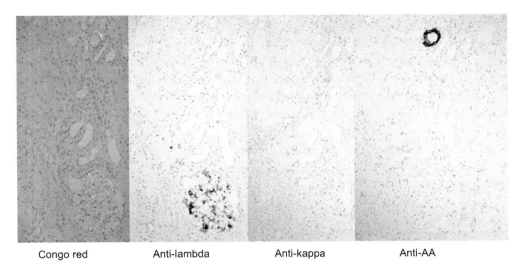

| Congo red | Anti-lambda | Anti-kappa | Anti-AA |

Figure 1. Glomerular AL-lambda and vascular AA-amyloidosis in a kidney biopsy specimen.

Protein studies

Immunofixation of serum showed a small (1.0-1.5 g/l) IgG-lambda type monoclonal immunoglobulin. No monoclonal protein was detected in the urine.

At detection of proteinuria and at follow-up, serum C-reactive protein concentration was within the normal range or slightly increased.

Other studies

Bone marrow biopsy showed a normal number of plasma cells. Practically all plasma cells stained with anti-lambda antiserum, whereas staining with anti-kappa antiserum was negative. No amyloid deposits were detected in the bone marrow specimen.

No osteolytic lesions were seen on skeletal radiographs. Biopsies of the gastric and duodenal mucosa showed amyloid deposits. A rectal biopsy was negative for amyloid. Further evaluation, including echocardiography, did not reveal any other clinically significant amyloid organ involvement.

Treatment

Over the following months there was a gradual increase in the concentration of the serum IgG-lambda paraprotein (maximal concentration 9.1 g/l). Five months after amyloidosis had been diagnosed, treatment with vincristine, doxorubicin and dexamethasone (VAD) was initiated. The patient received three courses of VAD followed by high dose melphalan with autologous stem cell reconstitution.

Follow-up

After chemotherapy, no monoclonal immunoglobulin has been detectable in her serum or urine. Her nephrotic syndrome has receded. At her latest visit to the hospital, four years after chemotherapy, the 24-hour excretion of protein was within normal limits. Her serum creatinine concentration has remained normal. Current treatment consists of low dose methotrexate and low dose prednisone.

CONCLUSIONS

There are a few previous reports of patients with more than one amyloid disease (1-4). This is, to the best of our knowledge, the first reported case where AL and AA amyloidosis have developed concurrently, not only in the same patient, but even in the same organ and the same compartment (artery) but in different locations. Our findings indicate that the two types of amyloid deposits are deposited independently.

Synthesis of acute phase reactants with development of AA amyloidosis, and clonal evolution of plasma cells with production of an amyloidogenic light chain leading to AL amyloidosis may both be linked to production of cytokines, particularly interleukin-6 in association with a chronic inflammatory disease.

REFERENCES

1. deSousa, M.M., Vital C., Ostler, D., Fernandes, R., PougetAbadie, J., Carles, D. and Saraiva, M.J. (2000). Apolipoprotein A1 and transthyretin as components of amyloid fibrils in a kindred with apoA1 Leu178His amyloidosis. *Am. J. Pathol.,* **156,** 1911-1917.
2. Fernandez-Alonso, J., Rios-Camacho, C., Valenzuela-Castano, A. and Hernanz-Mediano, W. (1994). Mixed systemic amyloidosis in a patient receiving long term haemodialysis. *J. Clin. Pathol.,* **47,** 560-561.
3. Isobe, T., Matsushita, T., Minakata, T., Takahashi, M., Hoshii, Y., Ishiwara, T. and Uchino, F. (1996). Coexistence of AL and Abeta2M amyloid in tissues of a patient with myeloma on hemodialysis. *Amyloid,* **3,** 41-43.
4. Röcken, C., Saeger, W. and Linke, R.P. (1993). Mehrere unterschiedliche Amyloidtypen bei einer 93-jährigen Patientin: Kasuistik eines Sektionsfalles. *Pathologe,* **14,** 42-46.

FAT TISSUE ANALYSIS BY CONGO RED METHOD AND BY AMYLOID A PROTEIN QUALIFICATION IN CLINICAL AA AMYLOIDOSIS

B.P.C. Hazenberg[1], J. Bijzet[1], P.C. Limburg[1], M.H. van Rijswijk[1], D. Garceau[2] and Fibrillex Amyloid Secondary Trial (FAST) Group

(1) Department of Rheumatology, University Hospital, Groningen, The Netherlands and (2) Neurochem Inc., 275 Armand-Frappier Boulevard, Laval, Quebec, Canada, H4S 2A

INTRODUCTION

Quantification of amyloid A protein can be used to detect AA amyloid in fat tissue (1). The method confirms the AA-type of amyloid and can be automated to become observer-independent.

OBJECTIVE

To study the sensitivity of amyloid A protein quantification and Congo red method in fat tissue of patients with clinical AA amyloidosis.

PATIENTS AND METHODS

Abdominal subcutaneous fat tissue of patients was analysed at the start of the phase II/III clinical trial with Fibrillex™, a GAG-mimetic drug. All patients had AA amyloidosis proven by a biopsy positively stained with Congo red as well as positive anti-AA immunohistology. All patients had renal symptoms, i.e. proteinuria of ≥1 g/day or diminished creatinine clearance between 20 and 60 ml/min.

Congo red stained slides (CR) were scored semi-quantitatively from 0 to 3+. Amyloid A protein in fat tissue was quantified by ELISA using murine monoclonal anti-human SAA antibodies Reu86.1 and Reu86.5 (Hycult Biotechnology, Uden, The Netherlands; Amyloid A protein reference range <12.6 ng/mg fat tissue) (1).

RESULTS

Fat tissue of 175 patients was available for analysis. In 156 patients ample fat tissue (at least about 60 mg) was available to be analysed with both methods. Congo red stained slides were positive in 144 (92%) and AA protein concentration was higher than 12.6 ng/mg fat tissue in 131 (84%) specimens. Both methods were concordant: median AA concentration rose from 1.0 ng/mg fat tissue for the negative Congo red specimens, to 64.5 ng/mg for Congo red 1+, to 345 ng/mg for 2+, and to 1146 ng/mg for 3+ positive specimens (figure 1).

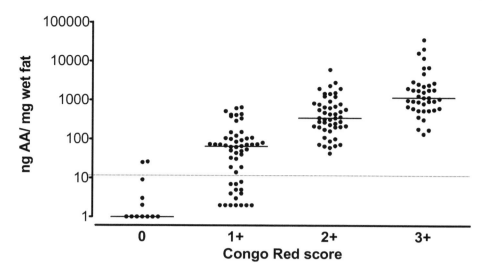

Figure 1. Detection of amyloid in fat tissue of 156 patients with clinical AA amyloidosis by Congo red stain (score 0-3+) and amyloid A protein quantification (ng/mg fat). The dotted line is the upper reference limit of controls without AA amyloidosis (12.6 ng/mg fat).

CONCLUSION

The Congo red method is more sensitive than quantification of amyloid A protein (92% vs. 84%) to detect amyloid in fat tissue of patients with clinical AA amyloidosis. However, good equipment and experience remain prerequisites for reliable results of the Congo red stain, especially when deposits are minute.

REFERENCES

1.   Hazenberg BPC, Limburg PC, Bijzet J, van Rijswijk MH. A quantitative method for detecting deposits of amyloid A protein in aspirated fat tissue of patients with arthritis. Ann Rheum Dis 1999; 58:96-102.

ACKNOWLEDGEMENTS

The authors like to thank the members of the **FAST Group** for their invaluable collaboration: H.B. Maiz (Tunisia), M.D. Benson (USA), I.Butrimiene (Lithuania), L.M. Dember (USA), H. Direskeneli (Turkey), A. Dispenzieri (USA), A. Filipowics-Sosnowska (Poland), P.D. Gorevic (USA), G. Grateau (France), A. Gul (Turkey), E. Hachulla (France), P.N. Hawkins (UK), B.P. Hazenberg (The Netherlands), J.A Hunter (UK), J.A. Jover (Spain), K. Kaarela (Finland), O. Lesnyak (Russia), A. Livneh (Israel), G. Merlini (Italy), J. Munoz Gomez (Spain), E.I. Nasonov (Russia), H. Ozdogan (Turkey), X. Puechal (France), I Rosner (Israel), X. Tena Marsa (Spain), J. Valverde Garcia (Spain) and P. Wiland (Poland).

ANEMIA OF CHRONIC DISEASE IN PATIENTS WITH AA-AMYLOIDOSIS: RELATIONSHIP TO POLYMORPHISMS OF THE INTERLEUKIN-1 GENE COMPLEX AND CIRCULATING CYTOKINE LEVELS

C.P.J.Maury[1],M.Liljeström[1],K.Laiho[2],S.Tiitinen[2], K.Kaarela[2] and M.Hurme[3]

[1]Department of Medicine, University of Helsinki and University Central Hospital,Helsinki,Finland
[2]Rheumatism Foundation Hospital, Heinola,Finland
[3]Department of Immunology,Tampere University and Tampere University Hospital, Tampere,Finland

Amyloidosis is a protein misfolding disorder leading to the accumulation in tissues of protein aggregates in a fibrillar form having a cross-β sheet conformation.Recent evidence shows that in contrast to previous views amyloid deposition is not a passive phenomenon, but may interfere with several cellular metabolic pathways.In both acquired and hereditary forms of amyloid diseases toxic protofibrillar intermediates have been identified and upregulation ofcellular oxidative mechanisms demonstrated.In the case of reactive amyloid A (AA) amyloidosis the receptor for advanced glycation end products (RAGE) is a target for the amyloidogenic form of the circulating precursor protein,serum amyloid A (SAA,ref.1).Since RAGE activation is associated with up-regulation of proinflammarory cytokines (2,3) a vicious cycle is generated with respect to AA amyloidogenesis.

Besides inducing   hepatic SAA production,proinflammatory cytokines are involved in the regulation of hematopoiesis (4-7).Here we show that anemic patients with AA amyloidosis have significantly elevated circulating levels of interleukin (IL)-1β and of IL-18 as compared with matched nonanemic amyloid patients, and that the occurrence of anemia in amyloidosis is associated with allele 2 (T) of the IL-1β-511 promoter gene.

SUBJECTS AND METHODS

We studied 54 adult patients (48 women ,6 men, median age 60 years, range 36-73 years) with seropositive, erosive RA fulfilling the 1987 revised criteria of the American College of Rheumatology.All patients had biopsy-proven amyloidosis;median duration of RA before   the diagnosis of amyloidosis was 20.5 years.Medication history included nonsteroidal anti-inflammatory drugs, glucocorticoids and conventional   disease modifying antirheumatic drugs, but not anti-tumor necrosis factor-α agents or IL-1 receptor antagonists.

Analysis of the polymorphisms of the IL-1 gene cluster was studied using polymerase-chain reaction-restriction fragment length assay as described (8).Blood samples from 400 adult blood donors obtained from the Finnish Red Cross Blood Transfusion Centre,Tampere, were used as a control panel in the IL-1 polymorhism studies. Plasma cytokine and acute phase protein levels were measured by immunoassays.Two-tailed Student's t-test for unpaired data, Mann-Whitney test,chi-square test, Correlation coefficient (r) and Spearman's rank correlation (r_s) were used in the statistical analyses as appropriate.Values are expressed as mean±SEM or as median ands ranges.

RESULTS

IL-1β and IL-18 levels were raised in the anemic patients with AA amyloidosis when compared with nonanemic patients matched with the anemic group with respect to duration of RA, sex, age and renal function, as well as with all patients with Hb≥110g/l (Table 1).

Table 1. Circulating IL-1β, IL-18 and CRP in patients with reactive AA-amyloidosis with or without anemia

Group	n	IL-1β pg/ml	IL-18 pg/ml	CRP μg/ml
Hb<110g/l	16	0.55± 0.17 ***	554± 83#	57 (22-110)#*
Hb>110g/l	16*	0.12± 0.02	359± 46	22 (<1-83)
Hb≥110g/l	38**	0.19 ±0.03	359± 25	28 (<1-83)

*Matched RA+A patients with respect to the anemic RA+A patients
**All RA+A patients with Hb ≥110g/l
*** P<0.05 v. matched patients and P=0.05 v. all anemic patients
# P=0.05 v. matched anemic patients and P<0.05 v.all anemic patients
#*P<0.001 v. matched patients and P<0.01 v. all anemic patients

In the anaemic amyloidosis patients circulating IL-1β correlated with IL-18 levels r=0.51, P<0.05) and circulating IL-18 with CRP levels (r_s=0.52, P<0.05).

The genotype 1-1 (C-C) of the IL-1β promoter gene at position 511 was less common among the anemic RA patients (19%) than among the nonanemic patients (70%; P<0.05). Conversely, the frequency of allele 2 (T) was increased (P<0.05) and that of allele 1(C) decreased (P<0.05) in anemic versus nonanemic amyloidosis patients .Compared with blood donors,nonanemic RA patients with amyloidosis had an increased allele 1 and a decreased allele 2 frequency (P<0.05).Circulating IL-1β levels, but not those of IL-18, tended to be higher among the allele 2 positive amyloid patients than among the allele 2 negative patients.

There were no significant differences in the genotype distribution or allele frequencies of the IL-1α-889 , IL-1β+3954 or IL-1Ra exon 2 polymorphic sites between anemic or nonanaemic amyloid patients. No significant differences were found in plasma ferritin, soluble transferrin receptor orerythropoietin levels between amyloidosis patients with or without anemia, though both ferritin and erythropoietin tended to be higher in the anemic patients.

DISCUSSION

The pathogenesis of anemia in reactive AA amyloidosis is complex. It may be related to such factors as the activity of the underlying inflammatory disease, therapy, amyloid deposition and amyloid protofibril-induced changes in metabolic pathways. The amyloid patients in this series had either normal or only mild renal dysfunction which argues against the possibility that renal anemia (associated with marked reduction in glomerular filtration rate) to any appreciable degree contributed to the anemia in these patients.Moreover, neither serum creatinine nor plasma erythropoietin significantly differed between the anemic and nonanemic patient groups. Seropositive, erosive RA was the underlying disease of all amyloid patients in this study, and medication included both nonsteroidal anti-inflammatory drugs, glucocorticoids and disease-modifying antirheumatic drugs, which all may have contributed to the development of anemia in the amyloid patients.However,the medication,including the use of cytostatics, was similar both in the anemic and nonanemic patient groups indicating that other factors were important in the pathogenesis of anemia in these patients.

We found elevated levels of the proinflammatory cytokines IL-1β and IL-18 in the circulation of anemic amyloid patients as compared with the nonanemic patients and demonstrate an inverse correlation between the cytokine and hemoglobin levels.These observations are in agreement with the current view on the key role of cytokines in the pathogenesis of ACD (4-7). It has been shown that IL-1β suppresses the colony formation of erythroid progenitors (4,5) and inhibits erythropoietin production (9). In addition, administration of IL-1 to mice induces anemia by suppressing CFU-E (10).IL-18, a member of the IL-1 family, has also been implicated in hematopoietic progenitor cell growth (11), but its role in erythropoiesis is unclear.The results of this study suggests that the effect may be in a direction similar to that of IL-1 and other proinflammatory cytokines.

We conclude that the occurrence of anemia in patients with reactive AA amyloidosisis associated with allele 2 (T) of the Il-1β -511 promoter gene and elevated levels of IL-1β and IL-18.The raised cytokine levels may generate a vicious cycle leading to accelerated amyloidogenesis, suppression of erythropoiesis and aggravation of the underlying inflammatory disorder.

REFERENCES

1. Du Yan S, Zhu H, Golabek A, Du H , Roher A , Yu J et al.Receptor-dependent cell stress and amyloid accumulation in systemic amyloidosis.*Nat Med* **6**:643,2000

2. Sousa MM, Du Yan S, Fernandes R, Guimaraes A, Stern D, Saraiva MJ.Familial amyloid neuropathy: receptor for advanced glycation end products-dependent triggering of neuronal inflammatory and apoptotic pathways.*J Neurosci* **21**:7576,2001

3   .Deane R, Du Yan S, Submamaryan RK, LaRue B, Jovanovic S, Hogg E et al.RAGE mediates amyloid-beta peptide transport across the blood-brain barrier and accumulation in brain.*Nat Med* **9**:907,2003

4. Maury CPJ , Andersson LC, Teppo AM, Partanen S , Juvonen E.Mechanism of anaemia in rheumatoid arthritis:demonstration of raised interleukin1β concentrations in anaemic patients and of interleukin 1 mediated suppression of normal erythropoiesis and proliferation of human erythroleukaemia (HEL) cells in vitro. *Ann Rheum* **47**:972,1988

5. Maury C P J , Juvonen E , Lähdevirta J ,Andersson L C .Mechanisms of anaemia and cachexia of chronic disease:evidence for a role of interleukin-1β and tumour necrosis factor-α. *Eur J Int Med* **2**:159,1991

6. Maury C P J, Lähdevirta J.Correlation of serum cytokine levels with haematological abnormalities in human immunodeficiency-virus infection.*J Intern Med* **227**:253,1990

7. Means RT Jr.Recent developments in anemia of chronic disease.*Curr Hematol Rev* **2**:116,2003

8. Hulkkonen J ,Vilpo J, Vilpo L,Koski T,Hurme M.Interleukin-1β, interleukin-1-receptor antagonist and interleukin-6 plasma levels and cytokine polymorphisms in chronic lymphocytic leukemia:correlation with prognostic parameters.*Haematologica* **85**:600,2001

9. Jelkmann W.Proinflammatory cytokines lowering erythropoietin production.*J Interferon Cytokine Res* **18**:555,1998

10. Furmanski P, Johnson CS.Macrophage control of normal and leukemic erythropoiesis.Identification of the macrophage derived erythroid suppressing activity as interleukin-1 and the mediator of its effect as tumor necrosis factor.*Blood* **75**:2328,1990

11. Ogura T,Ueda H ,Hosohara K, Tsuji R,Nagat Y,Kashiwamura S,Okamura H. Interleukin-18 stimulates hematopoietic cytokine a and growth factor formation and augments circulating granulocytes in mice.*Blood* **98**:2101,2001

INFLUENCE OF INFLAMMATION ON LIPID METABOLISM IN THE PATIENTS WITH RHEUMATOID ARTHRITIS

Y. Koseki, C. Terai, H. Kajiyama, M. Ito, H. Yamanaka, M. Hara, T. Tomatsu and N. Kamatani

Institute of Rheumatology, Tokyo Women's Medical University, Tokyo, Japan

Several lines of evidence show that cardiovascular disease (CVD) is a leading cause of death in patients with rheumatoid arthritis (RA). Standardized mortality ratios for patients with RA dying from CVD are reported to be 1.13 to 5.25 (1, 2). Mortality is not only linked to RA, but strongly linked to RA severity. Total cholesterol (TC) in active RA is reported to be reduced and TC in patients with active RA is reported to be lower than in those with inactive RA (3). Investigators consistently observe reduced high-density lipoprotein cholesterol (HDL-C) levels and frequently find reduced low-density lipoprotein cholesterol (LDL-C) levels in RA patients (3). Thus, the contribution of abnormal lipid metabolism to atherosclerotic change in RA is not clear.

Lipoprotein metabolism has been shown to be markedly altered during the inflammatory process in animal models (4) and in human inflammatory diseases (5). Serum amyloid A (SAA) is an apolipoprotein of HDL and a major acute phase protein that can increase to 1000 fold over normal level. During the acute phase response, SAA becomes the major apolipoprotein in HDL particles, and has been shown to change the metabolism and lipid composition of HDL-C (6). Thus SAA could be an important determinant of cholesterol metabolism in chronic inflammatory diseases like RA.

The purpose of this study is to reveal influence of inflammation on lipid metabolism in patients with RA.

PATIENTS AND METHODS

Objects of this study were 680 patients (118 males and 562 female) with RA who jointed in 3rd and 5th phase investigation of J-ARAMIS (Japanese large RA cohort study) investigation at October 2001 and October 2002. Among some 5000 patients who jointed J-ARAMIS investigation, consecutive RA patients who first visited our clinic during 1995-1997 were selected for this study and were followed prospectively. TC, HDL-C, LDL-C, apolipoprotein A-I (ApoA-I), and apolipoprotein A-II (ApoA-II) were determined as lipid profile tests, and erythrocyte sedimentation rate (ESR), CRP, and SAA were measured as markers of inflammation. Regular blood test, drugs used currently, body mass index (BMI), joint score, patient's visual analog scales (VAS) for pain, and patient's background were also evaluated. The relationship between lipid profiles and these factors were analyzed.

Mann-Whitney test, simple and multiple regression analysis were used for statistics.

RESULTS

Average age and duration of disease was 56.8 years and 9.4 years, respectively. Average number of swollen joints was 2.4. Average ESR, CRP, and SAA were 35.4 mm/hr, 1.28 mg/dl, and 72.3 µg/ml, respectively. The

RA patients were treated with disease modifying anti-rheumatic drugs (DMARDs) (63.4%), methotrexate (44.7%), and prednisolone (56.7%) with an average daily dose of 4.4 mg.

TC showed significant inverse correlation with each inflammatory markers; ESR (r=0.107, p=0.017), CRP (r=0.158, p=0.0004), and SAA (r=0.175, p<0.0001). The correlation coefficient is the highest between TC and SAA. Cholesterol levels were compared between the patients with CRP value in the highest and lowest quartile of CRP. HDL-C was significantly low in the highest CRP quartile (p=0.03), but LDL-C was not different between the highest and lowest CRP quartile (p=0.14).

Figure 1 shows the correlation analysis between TC and various variables. In simple regression analysis, TC associated with sex, age, number of swollen joints, patient's VAS, and all of 3 inflammatory markers, but not associated with BMI, hypertension, or any of the drugs. Multiple regression analysis showed any one of 3 inflammatory markers to be the second major determinant for TC next to age. Sex of the patients was the third determinant for TC (Figure 1).

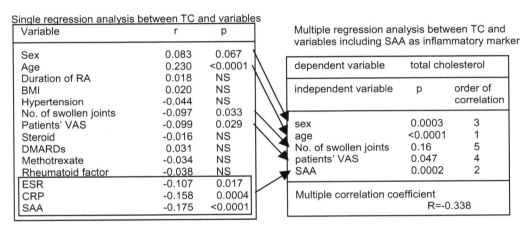

Single regression analysis between TC and variables

Variable	r	p
Sex	0.083	0.067
Age	0.230	<0.0001
Duration of RA	0.018	NS
BMI	0.020	NS
Hypertension	-0.044	NS
No. of swollen joints	-0.097	0.033
Patients' VAS	-0.099	0.029
Steroid	-0.016	NS
DMARDs	0.031	NS
Methotrexate	-0.034	NS
Rheumatoid factor	-0.038	NS
ESR	-0.107	0.017
CRP	-0.158	0.0004
SAA	-0.175	<0.0001

Multiple regression analysis between TC and variables including SAA as inflammatory marker

dependent variable	total cholesterol	
independent variable	p	order of correlation
sex	0.0003	3
age	<0.0001	1
No. of swollen joints	0.16	5
patients' VAS	0.047	4
SAA	0.0002	2

Multiple correlation coefficient
R=-0.338

Figure 1. Correlation analysis between total cholesterol and various factors in the patients with RA

To exclude the influence of age and sex on lipid metabolism, the correlation between the changes of lipid and inflammatory parameters between 2 different time point (October 2001 and October 2002) were analyzed. By using the differences of variables, the value of correlation coefficient increased up to over 0.4 (Table 1). This precise evaluation revealed that among 5 lipid profiles, ApoA-II correlated most significantly with all inflammatory markers. However, the correlation was the lowest in SAA compared to those in ESR and CRP.

Table 1. Correlation between the changes in lipid profiles and inflammatory markers of the patients with RA

	TC	HDL-C	LDL-C	ApoA-I	ApoA-II
ESR	r=-0.301	r=-0.265	r=-0.174	r=-0.395	r=-0.425
	p<0.0001	p<0.0001	p<0.0001	p<0.0001	p<0.0001
CRP	r=-0.321	r=-0.174	r=-0.175	r=-0.293	r=-0.396
	p<0.0001	p<0.0001	p<0.0001	p<0.0001	p<0.0001
SAA	r=-0.240	r=-0.116	r=-0.129	r=-0.226	r=-0.340
	p<0.0001	p=0.0058	p=0.0021	p<0.0001	p<0.0001

DISCUSSION

In this study, significant inverse correlation was observed between TC and inflammatory markers in RA patients. In fact, inflammation seemed to be one of the most important determinants of TC level in RA patients as shown in multiple regression analysis. Thus, the influence of inflammation on TC is stronger than that of sex difference. TC and HDL-C, not LDL-C were significantly low in the patients with high CRP, suggesting the decrease of TC was mainly caused by the depression of HDL-C. SAA synthesized in liver by stimulation of inflammatory cytokines (IL-1, IL-6, and TNFα) is secreted as apolipoprotein on HDL$_3$ particle. During acute phase response, excessive SAA displaces ApoA-I and ApoA-II from HDL. Since, ApoA-I is an important activating cofactor of lecithin-cholesterol acyltransferase, the relative lack of ApoA-I on HDL during inflammation disturb cholesterol esterification. This can account for the inverse correlation between SAA and HDL-C as well as TC.

While ApoA-I is contained in chylomicron, HDL$_2$ and HDL$_3$ particles, ApoA-II is present only on HDL$_3$ particles. This is the reason why among 5 lipid profiles, the changes in ApoA-II showed the strongest inverse correlation with the changes in each inflammatory marker. Among 3 inflammatory markers, only SAA could be directly involved in lipid metabolism, however, the correlation between lipid profiles and SAA were weaker than the correlation between lipid profiles and CRP or ESR (Table 1). This might be explained by the almost equal distribution of three major allotypes of *SAA1* gene (*SAA1.1, SAA1.5,* and *SAA1.3*) in Japanese population. SAA1 proteins produced by each *SAA1* allotypes have different amino acids in their 52nd and 57th positions of AA sequence. This region of SAA1 protein is assumed to be vital portion for the binding to cholesterol (7). Thus, different SAA1 protein might have different affinity to cholesterol. On the other side, no major polymorphism of CRP gene has been reported so far. This might explain that weaker correlation between the changes of SAA and lipid profiles that should be directly associated with SAA changes.

REFERENCES

1. Mutru, O. et al. Ten year mortality and causes of death in patients with rheumatoid arthritis, Brit. Med. J., 290,1797, 1985.
2. Wolfe, F. et al. The mortality of rheumatoid arthritis, Arthritis Rheum.,37, 481, 1994.
3. Lazarevic M.B. et al. Dyslipoproteinemia in the course of active rheumatoid arthritis, Semin. Arthritis Rheum., 22, 172, 1992.
4. Lindhorst, E. et al. Acute inflammation, acute phase serum amyloid A and cholesterol metabolism in the mouse. Biochim. Biophys. Acta, 1339,143, 1997.
5. Salazar, A. et al. Influence of serum amyloid A on the decrease of high density lipoprotein-cholesterol in active sarcoidosis, Atherosclerosis, 152, 497, 2000.
6. Pruzanski, W. et al. Comparative analysis of lipid composition of normal and acute-phase high density lipoproteins, J. Lipid. Res., 41,1035, 2000.
7. Liang, J.S. et al. Amino terminal region of acute phase, but not constitutive, serum amyloid A (apoSAA) specifically binds and transports cholesterol into aortic smooth muscle and HepG2 cells, J. Lipid. Res., 37, 2109, 1996.

SECTION 4

FAMILIAL AMYLOIDOSIS

MOLECULAR AND CELLULAR ASPECTS IN TRANSTHYRETIN AMYLOIDOSIS

M.J. Saraiva

Molecular Neurobiology, Institute for Cellular and Molecular Biology and ICBAS, University of Porto, Portugal

INTRODUCTION

Key questions in transthyretin (TTR) amyloidosis include pathways of fibril formation and consequences of TTR deposition. Molecular studies of intermediate species and detection of these intermediates "in vivo" are pivotal to answer these questions and for prospective therapeutical approaches. This review summarizes recent advances in our knowledge of intermediates of TTR fibrils and of tissue changes associated with TTR deposition.

1. INTERMEDIATES OF FIBRILLOGENESIS: "IN VITRO" CLUES

Examples of molecular studies of intermediates include the three-dimensional structure of aggressive mutants, such as Leu55Pro, which by x-ray diffraction revealed changes in the monomers compared to TTR WT by the disruption of beta-sheet strand D, which becomes a long loop connecting beta strands C and E. This variant packs into a tubular crystal formed by monomers with several channels running parallel to each other (1). That monomers are the building blocks of TTR fibrils was also evident in *in vitro* assembly analyses as observed by transmission electron microscopy (TEM), atomic force microscopy (AFM) and quantitative scanning transmission electron microscopy (STEM) for both TTR WT fibrils produced by acidification, and TTR Leu55Pro fibrils assembled at physiological pH. The morphological features and dimensions of TTR WT and TTR Leu55Pro fibrils were similar, with up to 300 nm long, 8 nm wide fibrils being the most prominent species in both cases. Other species were also evident; 4 – 5 nm wide fibrils, 9 - 10 nm wide fibrils and oligomers of various sizes. STEM mass-per-length (MPL) measurements revealed discrete fibril types with masses of 9.5 and 14.0 ± 1.4 kDa/nm for TTR WT fibrils and 13.7, 18.5 and 23.2 ± 1.5 kDa/nm for TTR Leu55Pro fibrils. These MPL values are consistent with a model in which fibrillar TTR structures are composed of 2, 3, 4 or 5 elementary protofilaments, with each protofilament being a vertical stack of structurally modified TTR monomers assembled with the 2.9 nm axial monomer-monomer spacing indicated by X-ray fibre diffraction data (2).

An amyloidogenic intermediate of TTR lacking the tetrameric fold has been generated by substituting/deleting mutants for the D-strand area (53-55) of TTR and 2 monoclonal antibodies (Mabs) raised against these structures had affinity to exposed separate cryptic epitopes, only expressed in TTR amyloid fibrils, but not in soluble TTR (non-mutated TTR or most common amyloidogenic mutants) (3). Besides cryptic epitopes present in TTR amyloid fibrils, these mabs also recognize soluble and insoluble TTR oligomers and a soluble tetrameric TTR mutant designed "in silico" based on the Leu55Pro structure to weaken dimer-dimer interactions. The structure adopted by one of these mutants, the Y78F TTR variant might resemble early intermediates consisting of modified tetramers that subsequently dissociate into monomers that polymerize into fibrils (4). Many other studies by biophysical methods such as NMR are trying to elucidate the structures of the monomeric intermediate species (5).

2. INTERMEDIATES OF FIBRILLOGENESIS: "IN VIVO" EVIDENCES

"in vivo" evidences for modified TTR intermediate species in plasma derive from studies using the above referred mabs against mutants for the D-stand area that expose cryptic epitopes. These amyloid-specific antibodies specifically recognize, in a direct enzyme-linked immunoassay (ELISA), plasma TTR from carriers of different mutations associated with FAP, both in asymptomatic individuals and patients; in contrast, they do not react with plasma TTR from healthy individuals, or carriers of non-pathogenic mutations. Thus, this antibody recognizes TTR conformations in plasma that express cryptic epitopes shared by amyloidogenic TTR variants associated with FAP, not present among non-pathogenic TTR molecules (6).

"in vivo" evidence for the presence of TTR intermediates in tissues, in the form of non-fibrillar aggregates was recently demonstrated in nerves from asymptomatic carriers of TTR V30M, in individuals without amyloid deposition and without reduction in number of fibers when compared to normal individuals in the stage prior to loss of unmyelinated and myelinated fibers, and major nerve fiber degeneration (FAP 0); despite the absence of Congo red birefringence (the hallmark of amyloid) TTR was present in all 12 cases investigated as revealed by immunohistochemistry and by immunocitochemistry with an anti-TTR antibody (7). It was hypothesized that TTR might deposit in a non-fibrillar, or pre-fibrillar form in early stages of FAP, before assembling into mature amyloid fibrils, that give the characteristic green-birefringence by Congo red staining, as observed in subsequent stages of FAP.

3. TISSUE CHANGES ASSOCIATED WITH TTR DEPOSITION

Tissue changes associated with TTR deposition have been investigated at the RNA and protein levels in tissues from asymptomatic carriers of TTRV30M (FAP 0) and from FAP patients. Microarray analyses revealed changes in a number of different genes; among them, genes related to matrix remodelling as described in this Symposium (Sousa et al.). Previous work on signalling cascades "in vivo" showed up-regulation of interleukins, inducible nitric oxide synthase and activation of caspase 3 in nerves from FAP 0 individuals and subsequent FAP stages (8); in a cell culture system using a Schwanoma cell line incubated with TTR non-fibrillar aggregates, caspase-3 activation was inhibited by antibodies to the soluble part of the receptor for advanced glycation end products (RAGE); thus, at the membrane level, at least 1 receptor is involved in signalling cascades triggered by the presence of TTR deposits. At the nuclear level, at least the NF-kB transcription factor is involved; it has been clearly observed to be overexpressed and translocated into the nucleus in FAP tissues (9). In summary, inflammation, oxidative stress and matrix remodelling occur in FAP tissues; the signaling cascades behind these events need clarification.

4. CLOSING REMARKS

Trends in TTR amyloidosis at the molecular and cellular levels encompass the interplay between inflammation, oxidative stress, matrix remodelling which can be either a consequence of toxicity from intermediate species/amyloid, or be players in TTR deposition (see figure 1). Efforts should be taken towards elucidating these aspects to advance towards therapeutical approaches.

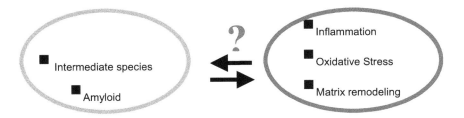

Figure 1 – Trends in TTR amyloidosis at the molecular and cellular levels

REFERENCES

1. Sebastião, M.P., Saraiva, M.J., Damas, A.M. (1998) The crystal structure of amyloidogenic Leu55 Pro transthyretin variant reveals a possible pathway for transthyretin polymerization into amyloid fibrils. *J. Biol. Chem.* **273**, 24715-24722.

2. Cardoso, I., Goldsbury, C., Muller, S.A., Olivieri, V., Wirtz, S., Damas, A.M., Aebi, U., Saraiva, M.J. (2002) Transthyretin fibrillogenesis entails the assembly of monomers: A molecular model for *in vitro* assembled transthyretin amyloid-like fibrils. *J. Mol. Biol.* **317**, 687-699.

3. Goldsteins, G., Persson, H., Andersson, K., Olofsson, A., Dacklin, I., Edvinsson, A., Saraiva, M.J., Lundgren, E . (1999) Exposure of cryptic epitopes on transthyretin only in amyloid and in amyloidogenic mutants. *Proc. Natl. Acad. Sci. U. S. A.* **96**, 3108-3113.

4. Redondo, C., Damas, A.M., Olofsson, A., Lundgren, E., Saraiva, M.J. (2000) Search for intermediate structures in transthyretin fibrillogenesis: soluble tetrameric Tyr78Phe TTR expresses a specific epitope present only in amyloid fibrils. *J. Mol. Biol.* **304**, 461-470.

5. Liu, K., Kelly, J.W., Wemmer, D.E. (2002) Native state hydrogen exchange study of suppressor and pathogenic variants of transthyretin. *J. Mol. Biol.* **320**, 821-32.

6. Palha, J.A., Moreira, P., Olofsson, A., Lundgren, E., Saraiva, M.J. (2001) Antibody recognition of amyloidogenic transthyretin variants in serum of patients with familial amyloidotic polyneuropathy. *J. Mol. Med.* **78**, 703-707.

7. Sousa, M.M., Cardoso, I., Fernandes, R., Guimarães, A., Saraiva, M.J. (2000) Deposition of transthyretin in early stages of familial amyloidotic polyneuropathy: evidence for toxicity of non-fibrillar aggregates. *Am. J. Pathol.* **159**,1993-2000.

8. Sousa, M.M., Yan, S.D., Fernandes, R., Guimarães, A., Stern, D., Saraiva, M.J.M. (2001) Familial amyloid polyneuropathy: RAGE-dependent triggering of neuronal inflammatory and apoptotic pathways. *J. Neurosc.* **21**, 7576-7586.

9. Sousa, M.M., Yan, S.D., Stern, D., Saraiva, M.J. (2000) Interaction of the receptor for advanced glycation end products (RAGE) with transthyretin triggers nuclear transcription factor kB (NF-kB) activation. *Lab. Investigation* **80**, 1101-1110.

GENOTYPE-PHENOTYPE CORRELATIONS AND MANAGEMENT IN TRANSTHYRETIN (TTR) RELATED AMYLOIDOSIS

Y. Ando

Department of Diagnostic Medicine, Graduate School of Medical Sciences, Kumamoto University, Kumamoto, Japan.

INTRODUCTION

Familial amyloidotic polyneuropathy (FAP) is a fatal hereditary amyloidosis and it has been identified amyloidogenic mutated transthyretin (ATTR), apolipoprotein A-I (AApoA-I) and gelsolin (Agel) as FAP amyloidogenic proteins. Among qualified 100 point of mutation or deletion in TTR gene, 13 mutations are nonpathologic forms. Other abnormal TTRs induce FAP which can be classified into several phenotypes, such as neuropathic, oculoleptomeningeal, and cardiac types. Several types of ATTR mutants do not show neuropathy although they are classified into FAP.

LEPTOMENINGEAL TYPE

Attention has recently been focused on (oculo)leptomeningeal form of FAP induced by several types of TTR mutations, such as FAP ATTR Y114C. Cerebral amyloid angiopathy (CAA) and ocular amyloidosis are often characteristic clinical features in those types of FAP. CAA is characterized by the amyloid deposition in the media and adventima of medium and small arteries, arterioles, and occasionally veins of the cortex and leptomeninges. Typical clinical CNS manifestations include cerebral infarction and hemorrhage, hydrocephalus, ataxia, spastic paralysis, convulsion and dementia. Although amyloid deposits in the meningo-cerebrovascular system were thought to be the cause of those clinical CNS symptoms, precise mechanism remains to be elucidated.

AUTONOMIC DYSFUNCTION AND TREATMENT

Of various clinical symptoms, those in the autonomic dysfunction group have the greatest effect on patients' lives, because they restrict the daily life of FAP patients. To evaluate various types of autonomic dysfunction, we employed several methods, some of which we originally developed. We previously reported the relationship between cerebral blood flow and orthostatic hypotension as determined by echo-ultrasonography. Reverse flow in the common carotid artery and the vertebral artery was always noted when patients stood and experienced clinical symptoms such as syncope or a feeling of faintness. We evaluated morphological changes induced by autonomic dysfunction in the venules of the finger tip by near-infrared spectrophotoscopy in FAP patients. Macroscopic abnormalities of the venules, such as tortuosity, irregular venous caliber and microaneurysm-like change were observed in FAP patients with autonomic dysfunction by the examination. The method demonstrated in this study may serve as a useful tool for screening the autonomic dysfunction and amyloid deposits around the vessels. Table 1 shows the summary of autonomic function tests and treatments in FAP patients. Several treatments were effective to improve quality of life of FAP patients.

THERAPEUTIC STRATEGIES FOR FAP

Since liver transplantation have several problems and can not prevent ocular and leptomeningeal amyloid deposition, we must develop a new essential therapy for ocluoleptomeningeal type of FAP and FAP ATTR V30M. Here, I describe new therapeutic approaches for FAP that our group has recently investigated.

1. Cr^{3+}

Stabilizing TTR in tetrameric form is one of the most important therapeutic strategies for preventing amyloid deposition in tissues of FAP patients. NSAID derivatives are one of the candidate agents that may be used to stabilize TTR in tetrameric form. It is well known that several metals play an important role in amyloid formation in various forms of amyloidosis. We have tested the effect of different metals on amyloid formation of TTR in vitro. It is interesting that of the various metals, Cr^{3+} showed the strongest stabilizing effect on the tetrameric form of both the wild-type and the variant form of TTR. Moreover, Cr^{3+} suppressed the amyloid formation in vitro. Because Cr^{3+} is now widely used as a health food, administering it to FAP patients presents no problem.

2. (*trans, trans*)-1-bromo-2,5-bis-(3-hydroxycarbonyl-4-hydroxy)styrylbenzene (BSB)

BSB, a novel tool for detecting amyloid deposits in vitro and in vivo, was examined the possibility for preventing amyloid fibril formation of FAP. A quartz crystal microbalance examination revealed that BSB had significant affinity for the amyloid fibrils purified from vitreous FAP samples. An electron microscopic analysis of a TTR solution incubated at pH 4 for 5 days revealed that amyloid fibrils of TTR in the absence of BSB were straight and smoothly configured. In comparison, TTR amyloid fibrils in the presence of BSB were immature, had irregular bead-like shapes, and had much larger diameters, suggesting that BSB had an inhibitory effect on amyloid formation.

3. Gene therapy

Antisense or ribozyme therapy would work at the mRNA level and may be a nonsense therapy as an essential FAP therapy because TTR is a rapid turnover protein and cannot easily be regulated by mRNA levels. We chose targeted TTR gene repair, and constructed three different types of SSOs. Because HepG2 cells have the wild-type TTR gene and secrete TTR, we tried to convert the wild-type TTR gene to the variant TTR gene. Ten% of the wild-type gene was converted to the variant gene with 74 mer SSOs embedded in 0.5% atelocollagen. Next we injected SSOs intrahepatically into transgenic mice that possessed the mice TTR Met30 gene to repair the TTR gene. By this treatment, 8.7 % of the TTR gene was converted into the wild-type TTR gene.

4. Antibody therapy

Recently domino liver transplantation using FAP patient's liver has been frequently performed. For the second recipient, a variant TTR is an alien protein, so in several patients' serum, autoantibody for a variant TTR can be detected. In such cases, the variant TTR levels become extremely low levels than FAP patients. As Alzheimer disease, antibody therapy using a variant TTR injection may induce autoantibody to decrease serum variant TTR levels and amount of amyloid deposition in the tissues.

REFERENCES

1. Ando,Y., New therapeutic approaches for familial amyloidotic polyneuropathy (FAP), *Amyloid* 10, S55, 2003

2. Ando, Y. et al., A novel tool for detecting amyloid deposits in systemic amyloidosis in vitro and in vivo, *Lab.Invest.*, 83, 1751, 2003

3. Nakamura, M. et al., Targeted conversion of the transthyretin gene in vitro and in vivo, *Gene.Ther.*, 2004 in press.

THE BINDING OF DIETHYLSTILBESTROL TO TRANSTHYRETIN – A CRYSTALLOGRAPHIC MODEL

E. Morais-de-Sá [1], P.J.B. Pereira [1], M.J.Saraiva [2,3] and A.M.Damas [1,3]

[1]*Unidade de Estrutura Molecular* & [2]*Unidade de Amiloide, Instituto de Biologia Molecular e Celular, Universidade do Porto, Rua do Campo Alegre, nº823, 4150 Porto, Portugal*
[3]ICBAS- Instituto de Ciências Biomédicas de Abel Salazar, Universidade do Porto, *Largo Prof. Abel Salazar, nº2, 4099-003 Porto, Portugal*

INTRODUCTION

Transthyretin (TTR) is a plasma protein which is known to form fibrillar extracellular deposits in patients suffering from three main types of pathologies, namely Senile Systemic Amyloidosis (SSA), Familial Amyloidotic Cardiomyopathy (FAC) and Familial Amyloidotic Polyneuropathy (FAP).

Dissociation of the TTR tetramer is probably essential in the mechanism of amyloid formation and since the binding of small molecules to the TTR central channel stabilizes the tetramer, several compounds are being studied as a prospect for a non invasive therapeutic approach (1,2). Diethylstilbestrol (DES) is a synthetic estrogen that was shown to be a competitive inhibitor for thyroid hormone binding to TTR and therefore it may have a stabilizing effect over the protein quaternary structure (3). The crystallographic structure of TTR, first determined by C.Blake and collaborators, revealed a tetramer formed by four identical subunits, each of them consisting of two four stranded β-sheets (4).

In the present study, the three-dimensional structure of the TTR-DES complex was determined by X-ray crystallography with the aim of identifying the key features of DES that are responsible for its ability to bind TTR.

MATERIALS AND METHODS

TTR was incubated overnight with a ten fold molar excess of DES, and crystals were grown by the hanging drop vapour diffusion method, at 14 ºC, using as precipitant agent a solution that consisted of 2.4 M ammonium sulfate, 7% glycerol and 0.2 M sodium acetate, pH 5.4.

Diffraction data were collected using synchrotron radiation at the ESRF-Grenoble. The crystal belongs to space group $P2_12_12$ with unit cell dimensions a=42.3 Å, b=85.4Å, c=63.4Å and diffracted to 1.8 Å resolution. The structure was solved by the molecular replacement method and refined to a final R_{factor} of 18.5% and R_{free} of 21.5%.

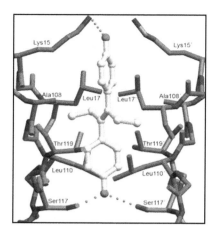

Figure 1. Interactions of DES with TTR

RESULTS AND DISCUSSION

As expected from the reported data about DES ability to compete for thyroid hormone binding to TTR, the structural data shows that the molecule binds within the thyroxine binding site. However, it binds deeper towards the center of the tetramer in comparison with thyroxine, inducing a conformational change in Ser117. In fact, one of DES hydroxyl groups is hydrogen bonded to two Ser117 residues, linking the two dimers and probably contributing for the stabilization of the tetrameric structure.

The interactions established between DES and TTR in the thyroxine binding site are presented in Fig 1 for one of the binding sites. Apart from the interaction at the center of the tetramer with Ser117, DES establishes hydrogen bonds at the channel entry with Lys15 and several dimer-dimer contacts appear as a result of DES binding in the interface between the two dimers. One of the interesting features of DES are its ethyl groups that fit securely in the middle halogen binding pocket and thereby participate in several hydrophobic interactions. This pocket accepts the halogen atoms from the natural ligand thyroxine when it binds TTR (5).

The structural information concerning the TTR-DES complex will be used to assist in the design of DES derivatives with increased affinity and selectivity for TTR.

ACKNOWLEDGMENTS

This work has been supported by grants POCTI/35735 and POCTI/44821/2002 from FCT, Fundação para a Ciência e Tecnologia, Portugal. The authors would like to thank Paul Moreira for his excellent technical assistance in the preparation of recombinant transthyretin and acknowledge the European Synchrotron Radiation Facility for the use of beam line ID14-EH3 and for the technical assistance given by the ESRF staff.

REFERENCES

1. Hammarstrom, P., Wiseman, L., Powers, E.T. and Kelly, J.W., Prevention of transthyretin amyloid disease by changing protein misfolding energetics. *Science,* 299, 713, 2003.
2. Klabunde, T,Petrassi, H.M.,Oza, V.B., Raman, P., Kelly, J.W. and Sacchetti, J.C., Rational design of potent human transthyretin amyloid disease inhibitors. *Nature Struct. Biol.* 7, 312, 2000.

3. Ishihara, A., Sawatsubashi, S., Yamauchi, K. Endocrine disrupting chemicals: interference of thyroid hormone binding to transthyretins and to thyroid hormone receptors. Mol Cell Endocrinol. 199, 105, 2003.

4. Blake, C. C., Geisow, M. J., Oatley, S. J., Rerat, B., and Rerat, C. Structure of prealbumin: secondary, tertiary and quaternary interactions determined by Fourier refinement at 1.8 Å. *J Mol Biol* 121, 339, 1978.

5. Paz, P. d. I., Burridge, J. M., Oatley, S. J., and Blake, C. C. F. *The design of drugs to Macromolecular targets* (Beddel, C. R., Ed.), John Wiley, New York, 1992.

LIVER TRANSPLANTATION IN ATTR Met30 AND OTHER VARIANTS: TIME TO SHIFT THE PARADIGM?

A.J. Stangou[1,3], N.D. Heaton[1], M Rela[1], J. O'Grady[1], C.J. Mathias[2], H.J.B. Goodman[3], H.J. Lachmann[3], A. Bybee[3], D. Rowzcenio[3], J. Joshi[3], R.S. Williams[1], M.B. Pepys[3], P.N. Hawkins[3]

[1]Institute of Liver Studies, King's College Hospital, London, United Kingdom
[2]The National Hospital for Neurology and Neurosurgery, London, United Kingdom
[3]Centre for Amyloidosis and Acute Phase Proteins, Royal Free and University College Medical School, London, United Kingdom

The outcome and survival following liver transplantation for FAP vary substantially between different mutations, with best results reported for FAP TTR Met30. Indications for OLT for FAP comprise genotypic confirmation of TTR variant, relevant symptoms, objective evidence of developing FAP disease. We describe the diversity in clinical features, disease course and outcomes amongst TTR variants in the U.K series, and assess whether current standard criteria for patient selection for OLT are applicable uniformly across all TTR mutations.

Between 1993 and 2003, 54 FAP patients underwent OLT (40 with the TTR Met30 variant, 6 Ala60, 3 Pro52, 1 Thr84, 1 Ala71), and 3 with the TTR Tyr77, Glu84 and Leu33 variants had combined heart/liver transplant. Initial presentation in all TTR Met30 cases was with sensorimotor neuropathy, and at later stages varying degrees of autonomic involvement. SAP scintigraphy identified visceral amyloid deposits in each case. Subclinical cardiac autonomic dysfunction was common finding, but echocardiography was normal in 36 and demonstrated significant cardiac amyloid in only 4 cases. Mean IVS thickness was 11.2mm before, and 10.8mm on follow-up post OLT. Of the 4 cases with abnormal echo pre-OLT, 3 improved and one remained unchanged. One other patient developed cardiac arrhythmia at 10 years post OLT requiring pacing, and a 65 yo patient had a sudden cardiac death with no history of heart failure. Objective and subjective recovery of autonomic features with improvement in bowel, bladder and cardiovascular function was documented in 28/40 patients within 6mo-2yrs post OLT, and stabilisation in the remaining, whereas peripheral nerve function was much slower and was only observed in patients with early disease at OLT. Overall survival for TTR Met30 was 80%.

By contrast, in non-Met30 variants, presentation was mainly with dysautonomia, and peripheral neuropathy was almost invariably a late manifestation. With the exception of cases with known family history, however, a definitive diagnosis of variant FAP had been commonly obtained by the time peripheral neuropathy was evident. All such cases had significant cardiac amyloid disease on echocardiography. SAP scintigraphy identified visceral deposits in most, but was negative in all TTR Pro52 cases. Following OLT, all non-Met30 cases with severe cardiac amyloidosis preoperatively, developed rapid increase in mean IVS thickness from 14mm to 20mm within 3-6 months post OLT (p<0.05). The full echo findings corroborated by histology in 4 patients were suggestive of progressive amyloid deposition.

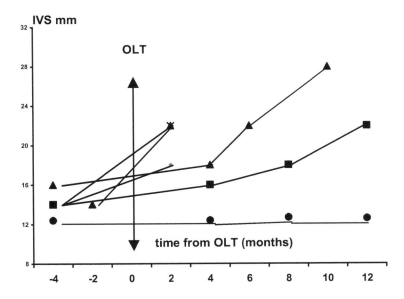

Graph1. Rapidly progressive amyloid cardiomyopathy post OLT in 5/6 TTR Ala60 cases

The TTR Ala60 and Pro52 patients underwent OLT, with IVS and LVPW thicknesses at the time of OLT of 13mm and 12mm respectively (NR 9-11 +/_ 1 mm). Both patients, having previously maintained stable parameters, have recently developed progressive amyloid cardiomyopathy, albeit in much slower rate than previous cases, with increase in IVS and LVPW thickness to 15 and 17 mms, at 2 and 7 years respectively. Echocardiographic appearances are completely consistent with severe cardiac amyloid.

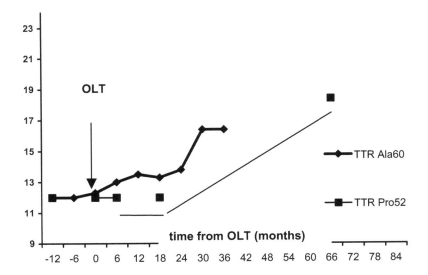

Graph 2: Late progression of cardiac amyloidosis in 2 non-Met 30 cases, at 2 and 6 years post OLT.

A further 3 patients who had been identified as apparently asymptomatic gene carriers at family screening (1 TTR Ala60, 1 Glu84 with 1 TTR Pro52), had early amyloid cardiomyopathy in the absence of objective evidence of neuropathy at diagnosis. During monitoring of those cases to determine timing for OLT when relative symptoms or evidence of disease progression were evident, cardiac amyloid progression supervened in all by the time subclinical evidence of neuropathy was obtained in 2/3 cases, therefore rendering them too advanced for isolated OLT.

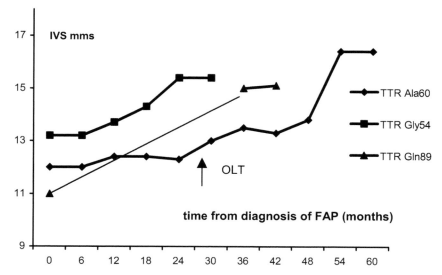

Graph 3: Echo evidence of progression of cardiac amyloid with normal neurophysiology in 3 asymptomatic TTR non-Met30 cases diagnosed at family screening

No evidence of recurrent amyloidosis in the cardiac graft was seen in any of the recipients of combined transplants. Neuropathy features and general wellbeing have improved in 2 cases, deteriorated in one, or remained unchanged in the rest. Overall survival in OLT for TTR non-Met30 was only 30%.

Results of OLT for FAP in TTR Met30 patients who present with neuropathic symptoms are excellent with best outcome in those with early disease. The poor results of FAP for non-Met30 were compounded by certain adverse effects such as progressive cardiomyopathy post OLT which may be mutation-dependent. We have previously suggested that continuing deposition of wild-type TTR on a pre-existing template of variant TTR in the heart may underlie this phenomenon. The severity of cardiac amyloid load at the time of OLT, may have a role in determining the rate progressive cardiomyopathy occurs post OLT, and our experience suggests that this may happen even if only microscopic deposits are present in the myocardium at OLT, albeit in a much slower rate. Patients with non-Met30 who present with advanced cardiac amyloid are not suitable candidates for OLT, but may be considered for combined cardiac/liver grafts.

Although awaiting onset of symptoms is a valid selection criterion in ATTR Met30, our findings make a case for considering pre-emptive OLT before even the earliest cardiac amyloid deposition has occurred in patients with high-penetrance non-Met30 amyloidogenic TTR variants. Coordination and collaboration between centres and the FAPWTR is essential in identifying all such variants and record outcomes of transplantation for systematic evaluation of results. Family screen of these cases and close follow-up with low threshold for frequent histological sampling, neurophysiology studies, and SAP scintigraphy is vital for selection of early cases likely to benefit from OLT.

NON-TRANSTHYRETIN HEREDITARY AMYLOIDOSES

Merrill D. Benson, M.D.

Department of Pathology and Laboratory Medicine, Indiana University School of Medicine and Richard L. Roudebush Veterans Affairs Medical Center, Indianapolis, Indiana USA

In general I think we should discourage the use of "non-" in our attempts to classify diseases, drugs, therapeutics, etc., including the fashionable terms "non-steroidal" anti-inflammatory drugs, "non-surgical" treatments, "non-invasive" procedures. It is not just that "non" carries an aura of negativity, but it suggests that we, except for a specific subject, combine everything that is not that specific subject into one group, classification, pot, and/or pigeon-hole. This is true for the non-transthyretin hereditary amyloidoses. Each and every one of these diseases is a fascinating subject which can well stand on its own without admitting that it is not related to transthyretin. Transthyretin may have been the first type of hereditary amyloidosis to be characterized and, indeed, certainly has the greatest number of disease associated mutations. However, several other types of hereditary amyloidosis have graced the history of our subject, and, with advancement in our knowledge of pathogenesis of β-fibril formation and thoughts on potential therapeutics, each has added to the world of amyloidosis.

Table 1

Mutant proteins other than transthyretin associated with autosomal dominant systemic amyloidosis.

Protein	Mutation	Clinical Features	Geographic Kindreds
Apolipoprotein AI	Gly26Arg	PN[a], Nephropathy	United States
	Leu60Arg	Nephropathy	United Kingdom
	Trp50Arg	Nephropathy	United Kingdom
	del60-71 insVal/Thr	Hepatic	Spain
	del70-72	Nephropathy	South Africa
	Leu75Pro	Hepatic	Italy, United States
	Leu90Pro	Cardiomyopathy, Cutaneous	France
	Arg173Pro	Cardiomyopathy, Cutaneous, Laryngeal	United States
	Leu174Ser	Cardiomyopathy	Italy
	Leu178His	Cardiomyopathy, Laryngeal	France
Gelsolin	Asp187Asn	PN[a], Lattice corneal distrophy	Finland, United States, Japan
	Asp187Tyr	PN[a]	Denmark, Czech
Cystatin C	Leu68Gln	Cerebral hemorrhage	Iceland
Fibrinogen Aα	Arg554Leu	Nephropathy	Mexico
	Glu526Val	Nephropathy	United States
	4904delG	Nephropathy	United States
	4897delT	Nephropathy	France
Lysozyme	Ile56Thr	Nephropathy, Petechiae	United Kingdom
	Asp67His	Nephropathy	United Kingdom
	Trp64Arg	Nephropathy	France
	Phe57Ile	Nephropathy	Canada
Apolipoprotein AII	Stop78Gly	Nephropathy	United States
	Stop78Ser	Nephropathy	United States
	Stop78Arg	Nephropathy	United States, Russia

[a]PN = Peripheral Neuropathy

Table 1 lists the mutant proteins other than transthyretin that we now recognize to cause systemic amyloidosis. The various mutations associated with disease that were known prior to the present symposium are listed along with some of the clinical features which set these diseases apart. When looked at in the aggregate these six proteins which are associated with systemic "non" transthyretin amyloidosis represent approximately 30 different mutations which makes our attempts at specific diagnosis, at the very least, interesting.

Table 2.

Historical Perspective and Present Day Classification

History	Name	Protein Characterization	Present
Meretoja 1969	Finnish — Lattice Corneal Dystrophy/FAP IV	Gelsolin 1990	Gelsolin
Van Allen 1969	Iowa/FAP III	Apo AI G26R 1989	Apo AI
Gudmundsson 1972	HCHWA	Cystatin C 1986	Cystatin C
Alexander 1975 Mornaghi 1982	Familial Renal Amyloidosis	Fibrinogen Aα chain 1993	Fib. Aα
Lanham 1982 Zalin 1991	"Ostertag" Nephropathic	Lysozyme 1993	Lysozyme
Weiss and Page 1972	"Ostertag"	Apo AII 2001	Apo AII

For the student of history, Table 2 lists the present day protein classification of these diseases and, where possible, the verified incident clinical description of the disease. These diseases are fascinating. Some, such as fibrinogen Aα-chain, cystatin C, and gelsolin, involve primarily one organ system. Others, such as apolipoprotein AI and lysozyme often give varied pictures of organ dysfunction and tease us with the idea that it might be possible to link a specific amino acid mutation's position in the peptide chain to the clinical manifestation of the disease. Certainly there is much yet to be learned about how protein structure and function may be related to specific tissue factors that dictate when, where, and how rapidly β-pleated sheet fibrils find a permanent resting place and invariably impact on organ function and life itself.

Finally, a few words on the diagnosis of hereditary amyloidosis: Many diseases can be diagnosed by specific laboratory tests whether by tissue biopsy, histochemistry, serum protein analysis, or DNA analysis. In the case of the hereditary amyloidoses we are often limited to the histological demonstration of amyloid and have few other laboratory procedures to more definitively diagnose the problem and, therefore, suggest therapeutic intervention. DNA testing is only commercially available for transthyretin. Mutations in apolipoprotein AI, apolipoprotein AII, lysozyme, fibrinogen Aα-chain, cystatin C, and gelsolin can be detected by specific tests, but only in laboratories and facilities dedicated to the study of amyloidosis. This leaves the diagnostician with only the knowledge of the various diseases to focus on one specific diagnosis which may then be verified by specific testing. In Table 3 I have tried to list some of the factors that we can use to settle on a specific diagnosis in the realm of hereditary amyloidosis, or at least point the way to a more definitive laboratory test.

Table 3.

Aids For Diagnosis

Organ Involvement	Think
1. Vitreous Opacities	TTR
2. Neuropathy	TTR, ApoAI (G26R)
3. Restrictive Cardiomyopathy	TTR, ApoAI (C-terminal)
4. Kidney Failure	ApoAI, FibAα, ApoAII, Lys
Slow	ApoAII, ApoAI, Lys
Rapid	FibAα
5. Hepatic Amyloid	ApoAI, Lys
6. Dermal	ApoAI (C-terminal)
7. Laryngeal	ApoAI (C-terminal)
8. Lattice Corneal Dystrophy	Gelsolin

FAMILY HISTORY HELPS BUT NOT CONCLUSIVE.

In summary it is important to be as knowledgeable as we can about the various forms of hereditary amyloidosis, including not only the transthyretin amyloidoses, but also the "non" transthyretin amyloidoses. We should realize that study of each disease entity has much to offer, not only to the care of patients afflicted with these diseases and their families, but also what knowledge of each type of disease can add to our understanding of the pathogenesis of all the protein deposition diseases.

FIBRINOGEN A α-CHAIN AMYLOIDOSIS: CLINICAL FEATURES AND OUTCOME AFTER HEPATORENAL OR SOLITARY KINDEY TRANSPLANTATION

A.J. Stangou[1], H.J. Lachmann[2], H.J.B Goodman[2], A. Bybee[2], D Rowzcenio[2], G Tennent[2], S.O. Brennan[3], J.G O'Grady[1], N.D. Heaton[1], M. Rela[1], M.B. Pepys[2], P.N. Hawkins[2]

[1]Institute of Liver Studies, King's College Hospital, London, United Kingdom
[2]Centre for Amyloidosis and Acute Phase Proteins, Royal Free and University College Medical School, London, United Kingdom
[3]Molecular Pathology Laboratory, Canterbury Health Laboratories, Christchurch, New Zealand

Fibrinogen amyloidosis due to mutations in the fibrinogen A α-chain gene (Afib) is the most frequently diagnosed form of hereditary amyloid disease in the U.K. We review its clinical features, course of the disease and outcome after either solitary kidney transplant or hepatorenal transplantation.

We have identified the fibrinogen A α-chain Glu526Val variant in 38 patients who had presented with renal disease in the 3rd-8th decade (median age 57 years). The disease has variable penetrance, remarkable tropism for the kidney, and rapid progression from initial presentation with features of proteinuria, hypertension and mild renal impairment, to end stage renal failure and dialysis dependance within just 1-5 years. Serum amyloid P component (SAP) scintigraphy showed renal amyloid in all, and splenic deposits in most cases. Renal histology comprising near replacement of the glomeruli by amyloid without any interstitial or vascular involvement is virtually pathognomic. *(Picture 1)*

 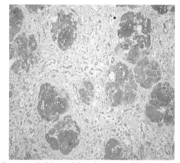

Picture 1: Enlarged glomeruli, no interstitial or vascular amyloid on histology and immunohistochemistry

Fibrinogen is exclusively synthesised by the liver but hepatic involvement is extremely rare and liver amyloid deposits were present only in 2 cases. No patient has had clinical peripheral neuropathy or cardiac amyloid, but many patients had coronary atherosclerosis, even before progression to ESRF. Nine patients have received isolated renal transplantation as a form of renal replacement therapy without simultaneous transplantation of the liver, and five had combined liver and kidney transplants (LKT).

Eleven kidney grafts have overall been utilised for solitary renal transplantation in 9 patients; of those, 5 grafts in 4 patients were lost due to renal amyloid recurrence within 3-7 years from transplantation, (picture 2), a further 2 grafts have severe amyloid recurrence and compromised function, and 3 grafts were lost to primary non-function/ Chronic allograft nephropathy. Only one out of the eleven renal grafts continues to have normal function without scintigraphic evidence of recurrent amyloid so far, at 11 months post renal transplant.

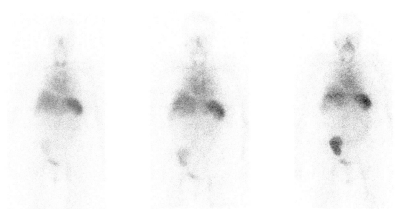

Picture 2: Evolution of amyloid in transplanted kidney over 5 years

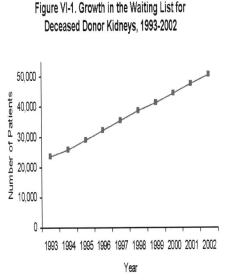

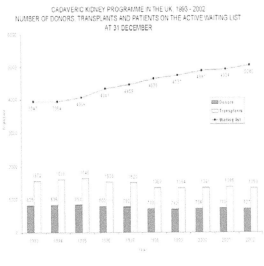

Source: *2003 OPTN/SRTR Annual Report,* Table 5.1.

Graph: UNOS and U.K Transplant Registry data
Indicating extreme organ shortage, and long waiting times for renal transplantation

Five patients have undergone combined liver and kidney transplantation as a potentially curative treatment. One patient had LKT 7.5 years ago for the indication of liver failure in association with hepatic amyloidosis (one of the 2/43 cases of Afib with liver involvement) and 2 previous failed renal transplants. Following review of the treatment options for AFib in the face of the extremely disappointing results of isolated renal transplantation, a further four patients had combined liver kidney transplant in the past 4 weeks-8 months, as the most rational approach with potentially best long term outcome prospects. Two patients received LKT pre-emptively, before commencement of renal replacement therapy, and two had been on haemodialysis and on the renal transplant list for 3-4 years. Transplantation procedures and post-operative course were uneventful in the 2 cases who received pre-emptive LKT, ITU stay was short and uncomplicated and overall hospital stay less than 3 weeks. In the 2 patients who had been on long term dialysis prior to transplantation, surgery and post transplant course was very challenging. Both patients had evidence of significant multivascular atheromatous disease, poor quality tissue and impaired healing with wound dehiscence in one case, multiple septic episodes and in addition features of dysautonomia. The overall outcome of hepatorenal transplantation has been excellent and all 5 patients are alive with functioning grafts. The longest term patient remains extremely well with normal kidney and liver function and no scintigraphic evidence of amyloid at 7.5 years *(picture 3)*. In the most recently transplanted 4 patients, amyloid scintigraphy at 6 months showed no evidence of amyloid deposition. In two of the hepatorenal transplant cases so far, at 6 years and 8 months post- LKT respectively, we have confirmed by mass spectrometry that the amyloidogenic variant fibrinogen has been completely eliminated from the plasma. Two of the explanted Afib livers were successfully used in domino procedures for transplantation in 2 patients with HCV cirrhosis and hepatocellular cancer.

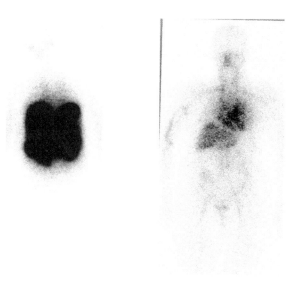

Picture 3: Baseline SAP scan showing extensive liver amyloid (left), and normal (right) scan at 5 years, following combined liver and kidney transplant

The clinical and biochemical outcomes after hepatorenal transplantation for fibrinogen amyloidosis define the liver as the source of production of the circulating amyloidogenic protein, and liver transplantation as a potentially curative treatment option. The very encouraging results of hepatorenal transplantation, in the context of great shortage of kidney grafts, with 5023 CRF patients in the U.K currently on the waiting list for kidney transplant, strongly support the combined liver/kidney transplant approach for all eligible younger patients. Issues for consideration include selection of patients for LKT, the role of pre-emptive transplantation and the inclusion of Afib amyloidosis and domino Afib in the Familial Amyloid Polyneuropathy World Transplant Registry (FAPWTR).

A NOVEL COMPLEX INDEL MUTATION IN THE FIBRINOGEN Aα CHAIN GENE IN AN ASIAN CHILD WITH SYSTEMIC AMYLOIDOSIS

A.Bybee[1], H.G.Kang[2], I.S.Ha[2], M.S.Park[3], H.I.Cheong[2], Y.Choi[2], J.A.Gilbertson[1], M.B.Pepys[1] and P.N.Hawkins[1]

[1]*Centre for Amyloidosis and Acute Phase Proteins, Royal Free and University College Medical School, London, United Kingdom*
[2]*Department of Pediatrics, Seoul National University Medical School, Clinical Research Institute, Seoul National University Hospital, Seoul, Korea;* [3]*Department of Pediatrics, Sungkyunkwan University School of Medicine, Seoul, Korea*

Systemic amyloidosis in early childhood is extremely rare, and is usually of AA type complicating JIA and other chronic inflammatory diseases. We report the molecular basis of amyloidosis that presented with asymptomatic proteinuria in an otherwise healthy Korean girl aged 7 years, who progressed to end-stage renal failure associated with amyloid hepatomegaly within 2 years. A renal biopsy showed enlarged glomeruli virtually replaced by amyloid, but no interstitial or vascular involvement, an appearance identical to that seen in more than 40 Caucasian patients we have evaluated with hereditary fibrinogen A α chain Glu526Val amyloidosis. The amyloid deposits in fixed sections stained specifically with antibodies to fibrinogen, and we identified a frameshift deletion-insertion (indel) mutation in one allele of her fibrinogen A (chain gene (FGA NM_000508.2, c.1636_1650del, 1649_1650insCA). This encodes a partly novel peptide and a premature stop signal (517_522delinsQSfsX548), similar to the two previously reported amyloidogenic point deletions at positions 522 and 524. The mutation was absent in samples verified to be from her parents, indicating that it had occurred de novo. No mutations were identified in the other genes known to be associated with hereditary renal amyloid. This is the first case of AFib characterized in an Asian individual, and the distinctive renal histology offered a strong clue to the diagnosis. The disease is potentially curable by combined hepatorenal transplantation.

PHENOTYPE, GENOTYPE AND OUTCOME IN HEREDITARY ApoAI AMYLOIDOSIS

[1]PN Hawkins, [1]A Bybee, [1]HJB Goodman, [1]HJ Lachmann, [1]D Rowczenio, [1]JA Gilbertson, [2]J O'Grady, [2]N Heaton, [2]A Stangou, [1]MB Pepys.

[1]Centre for Amyloidosis and Acute Phase Proteins, Royal Free and University College Medical School, London, United Kingdom
[2]Institute of Liver Studies, King's College Hospital, London, United Kingdom

We report the phenotype, genotype and clinical course of 17 patients with hereditary apolipoprotein AI amyloidosis evaluated in our centre during the past 5 years. The apoAI variants were Gly26Arg in 4 cases, Trp50Arg in 2 cases, Leu60Arg in 5 cases, Arg173Pro in 3 cases, and Del70-72 in one case. Two new variants Leu64Pro and Ala175Pro were identified. There was substantial heterogeneity in the age of onset and breadth and combination of clinical manifestations among patients with the same and different variants, but notable observations included: a variable degree of neuropathy in 3 out of 4 patients with Gly26Arg but not in those with any other variants; cardiac involvement in some patients with Leu60Arg and Arg173Pro; early onset of bruising and hoarseness but late onset of renal and cardiac amyloid in patients with Arg173Pro; compelling evidence for cerebral amyloid angiopathy in a patient with Leu60Arg; and ocular involvement with Leu60Arg and Del70-72. The newly identified variant Leu64Pro presented in an Italian man with renal failure in middle-age associated with major but silent deposits in the liver and spleen, and Ala175Pro presented in a young British man with laryngeal and possibly testicular involvement, but no visceral or cardiac deposits. Four patients had renal transplants 1, 6, 14 and 23 years ago, two patients had combined kidney and heart transplants 3 and 10 yrs ago, and two had liver transplants 2 and 6 years ago, and all continue to have excellent graft function. Despite recurrence of amyloid histologically in many of the grafts, the clinical outcome of solid organ transplantation has been remarkably good in this series.

PIGMENTED CILIARY EPITHELIUM CELLS SYNTHESIZED TRANSTHYRETIN IN THE RABBIT EYE

T. Kawaji,[1] Y. Ando,[2] M. Nakamura,[2] K. Yamamoto,[2] E. Ando,[1] M. Fukushima,[1] A. Hirata,[1] and H. Tanihara,[1]

[1]Department of Ophthalmology and Visual Science, and the [2]Department of Diagnostic Medicine, Graduate School of Medical Sciences, Kumamoto University, 1-1-1 Honjo, Kumamoto 860-8556, Japan

INTRODUCTION

Transthyretin (TTR)-related familial amyloidotic polyneuropathy (FAP) is an autosomal dominant inherited disorder characterized by systemic accumulation of polymerized mutated TTR in the peripheral nerves and other organs. In addition to systemic clinical manifestations, various ocular symptoms are commonly recognized in TTR-related FAP. Among these, vitreous opacity and glaucoma are the most serious and result in severe visual disturbances (1,2) . The main source of plasma TTR has been documented to be the liver, but the retinal pigment epithelium (RPE), the choroid plexus of the brain, and the visceral yolk sac endoderm are also known to synthesize and secrete TTR (3).

Most FAP patients with secondary glaucoma have vitreous opacities, dandruff-like substances on the lens surface or pupillary margin, and pigment deposition in the chamber angle (2). This evidence led to the common belief that glaucoma associated with FAP is caused by two mechanisms: First, deposition of perivascular amyloid in conjunctival and episcleral tissue may contribute to the elevated intraocular pressure (4,5). Second, ocular amyloid deposition originating from the vessels and the RPE may result in obstruction of the aqueous outflow route (5). However, we doubted this second hypothesis because we had several FAP patients who had glaucoma with severe amyloid deposition on the pupil and pupil fringe with little or no vitreous opacity (2). These data suggested that TTR synthesis sites other than the RPE, sites close to the chamber angle, may exist in the ocular tissues.

In the present study, we clarified sites of TTR synthesis in ocular tissues in addition to the RPE.

MATERIALS AND METHODS

Animals

Japanese adult albino male rabbits were used in this study.

Tissue Preparation

The eyes were cut circumferentially at the cornea to remove the cornea and lens to make posterior cups, and samples were embedded in paraffin. These sections were used for in situ hybridization and reverse transcription-polymerase chain reaction (RT-PCR) analysis. Forty eyes were cut circumferentially at the sclera (1 mm posterior from the limbus) to make anterior cups without RPE and posterior cups. These fresh tissues were used for real-time quantitative RT-PCR analysis.

In Situ Hybridization

In situ hybridization was performed with a tyramide signal amplification system with fluorescein-labeled probes according to the manufacturer's instructions with a slight modification.

Isolation of RNA from Targeted Tissue Areas and Qualitative RT-PCR

To obtain pure CPE without RPE, we used the Pinpoint Slide RNA Isolation System II and isolated RNA from specific paraffin-embedded tissue areas on slides.

RT-PCR was performed with SuperScript One-Step RT-PCR with Platinum Taq according to the manufacturer's protocols.

Real-time Quantitative RT-PCR

Total RNA was prepared from fresh tissue samples by using AquaPure RNA Isolation Kit. Real-time one-step PCR for TTR mRNA in rabbits was performed by using a LightCycler thermal cycling system.

RESULTS

In Situ Hybridization

In situ hybridization assays of the rabbit eye sections with the antisense probe revealed the presence of hybridization signals for TTR mRNA not only within the RPE cells but also the CPE cells (Fig. 1A, C). The signals were distributed uniformly and abundantly within the cytoplasm of all RPE and CPE cells. Control sections hybridized with the sense probe showed no hybridization signal (Fig. 1B, D).

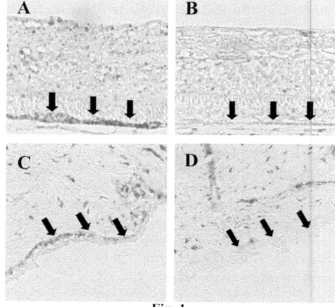

Fig. 1

Isolation of RNA from Targeted Tissue Areas and Qualitative RT-PCR

We isolated RNA from specific paraffin-embedded tissue areas on slides. RT-PCR revealed expression of TTR mRNA in CPE cells. The expression level in the RPE cells was higher than that in the CPE cells.

Real-time Quantitative RT-PCR

Real-time quantitative RT-PCR analysis with the LightCycler revealed that the TTR mRNA level in CPE cells was approximately one-third of that in the RPE cells.

DISCUSSION

Our in situ hybridization analysis revealed TTR expression not only in RPE cells, as previously reported, but also in CPE cells, as we expected. Because the average density of TTR expression signals in CPE cells was weak, we confirmed that this signal was positive by using RT-PCR. Moreover, we isolated RNA from targeted tissue areas in which CPE cells were included and RPE cells were excluded by using the Pinpoint Slide RNA Isolation System II. Our RT-PCR analysis convincingly confirmed the presence of TTR expression in CPE cells.

Quantitative analysis revealed significant production of TTR in CPE cells, with real-time quantitative RT-PCR demonstrating that the level of TTR mRNA expression in CPE cells was about one-third of that in RPE cells. These results suggest that TTR in aqueous humor and vitreous humor is predominantly derived from retinal vessels and RPE, but that a significant proportion of the TTR is produced by CPE cells, especially the TTR in the aqueous humor. TTR synthesized by CPE cells may cause amyloid deposition at the pupillary margin and angle chamber, which would result in glaucoma. Although we had several FAP patients who had glaucoma with amyloid deposition on the pupil and pupil fringe but with little or no vitreous opacity, we believe that CPE cells, but not RPE cells, may predominantly contribute to amyloid formation.

REFERENCES

1. Kimura, A. et al., Secondary glaucoma in patients with familial amyloidotic polyneuropathy, *Arch.Ophthalmo.*, 121, 351, 2003

2. Futa, R. et al., Familial amyloidotic polyneuropathy: ocular manifestations with clinicopathological observation, *Jpn.J.Ophthalmol.*, 28, 289, 1984

3. Cavallaro, T. et al., The retinal pigment epithelium is the unique site of transthyretin synthesis in the rat eye, *Invest.Ophthalmol.Vis.Sci.*, 3, 497, 1990

4. Nelson, G.A., Edward, D.P., Wilensky, J.T., Ocular amyloidosis and secondary glaucoma, *Ophthalmology*. 106, 1363, 1999

5. Silva-Araujo, A.C., Tavares, M.A., Cotta, J.S., Castro-Correia, J.F., Aqueous outflow system in familial amyloidotic polyneuropathy, Portuguese type, *Graefes.Arch.Clin.Exp.Ophthalmo.*, 231, 131, 1993

UPREGULATION OF THE EXTRACELLULAR MATRIX REMODELING GENES, BIGLYCAN, NEUTROPHIL GELATINASE-ASSOCIATED LIPOCALIN AND MATRIX METALLOPROTEINASE-9 IN FAMILIAL AMYLOID POLYNEUROPATHY

[1] *M. M. Sousa,* [2] *J.B. Amaral,* [2] *A. Guimarães and* [1,3] *M. J. Saraiva*

[1] *Molecular Neurobiology, Institute for Cellular and Molecular Biology;* [2] *Hospital Geral de Santo António;* [3] *ICBAS, University of Porto, Portugal.*

INTRODUCTION

In the peripheral nervous system (PNS) of familial amyloid polyneuropathy (FAP) patients, transthyretin (TTR) aggregates and fibrils occur extracellularly; in latter stages of disease progression, nerves present complete axonal loss and endoneurial contents are replaced by amyloid, collagen, Schwann cells and fibroblasts. Changes in proteoglycan type and distribution could account for the derangement of collagen and the altered physical properties of tissues with TTR deposition; however, information is lacking concerning the molecular events that underlie tissue remodelling in FAP. The identification of genes differentially expressed during the course of FAP is needed. However, the amount of mRNA that is possible to obtain from nerve biopsies is limiting to perform gene expression studies. Labial salivary gland (SG) biopsies have been reported as a useful tool for the diagnosis of amyloid neuropathy (1); this tissue presents TTR deposition that correlates with deposition in the PNS and represents a minimally invasive method (1). In the present study we investigated differentially expressed genes related to ECM remodelling in SG biopsies aiming at further elucidating the molecular effects of TTR deposition in FAP.

METHODS

Subjects: General characterization of SG biopsies from FAP patients consisted on TTR immunohistochemistry and Congo red staining. Sural nerve biopsies from FAP patients, asymptomatic carriers and controls were available at HGSA, Porto (2).

Immunohistochemistry: Primary antibodies were: anti-TTR Mab39-44 (3), anti- MMP-9 (Chemicon, 1:1000), anti-NGAL (Dr. Borregaard, University of Copenhagen, 1:1000) and anti-biglycan (Dr Roughley, Mc Gill University, Montreal, 1:500).

Microarray analysis: SG from 4 FAP patients and 4 control individuals were hybridised to the U95A oligonucleotide array according to Affymetrix's recommendations; a gene was considered to have modified expression if it averaged $\geq \pm 2.5$ fold change.

RT-PCR: Total RNA was isolated from additional SG biopsies from 3 FAP patients and 3 control individuals, and subjected to RT-PCR with the Superscript II kit (Invitrogen).

Digestion of TTR aggregates and fibrils by MMP-9: TTR aggregates and fibrils were obtained following Cardoso *et al* (4). Active MMP-9 (Calbiochem) was added overnight at 37°C to (i) soluble TTR and (ii) day 0

TTR aggregates and daily (i) during 6 days to day 0 TTR aggregates and (ii) during 9 days to 6 days TTR fibrils. In parallel experiments, serum amyloid P component (SAP, Calbiochem) was incubated with day 0 TTR aggregates for 2 hours at 37 °C after which MMP-9 was added as described above.

RESULTS

Differential gene expression in salivary gland biopsies with TTR deposition: Two genes differentially expressed coding extracellular matrix (ECM)-related proteins were identified, namely biglycan and neutrophil gelatinase-associated lipocalin (NGAL), both upregulated in FAP (2.7±0.9 and 6.7±3.4 times, respectively).

Upregulation of the ECM-related proteins, biglycan, NGAL and MMP-9 in tissues with TTR deposition: Microarray results were confirmed by RT-PCR (data not shown). At the protein level, IHC analysis revealed that SG biopsies from FAP patients (n=5) had increased biglycan and NGAL immunostaining when compared to normal individuals (n=6) (Figure 1). NGAL, was originally purified from neutrophils but is expressed in a variety of tissues (5). Since NGAL exists as a complex with matrix metalloproteinase-9 (MMP-9) modulating its activity by protecting it from degradation (6), we carried out IHC for MMP-9. Upregulation of NGAL in FAP tissues (n=5) was accompanied with an increased immunostaining for MMP-9 (Figure 1).

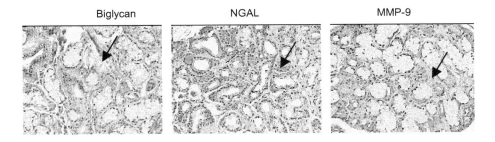

| Biglycan | NGAL | MMP-9 |

Figure 1. IHC for biglycan (left), NGAL (center) and MMP-9 (right) of SG from FAP patients.

To establish if the events observed in SG are relevant in the PNS, we performed IHC analysis for biglycan; NGAL and MMP-9 in nerve biopsies of different stages of FAP progression. In a stage prior to amyloid deposition, biglycan staining was already markedly increased. In the case of NGAL and MMP-9, FAP nerves presented an increased staining only after presence of amyloid deposits (not shown).

It has been previously reported that MMP-9 is able to degrade amyloid β (7). We tested if MMP-9 might also be able to degrade TTR: *in vitro*, MMP-9 degraded both TTR aggregates and fibrils (data not shown); however, in the presence of SAP, the ability of MMP-9 to degrade TTR fibrils was clearly diminished.

DISCUSSION

The present study represents the first assessment of molecular events related to ECM remodeling upon TTR deposition in FAP. SG biopsies are a sensitive means to investigate molecular events related to TTR deposition during the pathogenesis of FAP; the genes found differentially expressed in FAP SG biopsies were also altered in the nerve. Biglycan has been reported in other amyloid deposits such as AA amyloidosis and light-chain deposition (8, 9). The observation that biglycan is also associated with TTR fibrils emphasizes the importance of this proteoglycan in amyloid disorders. Similarly to biglycan, NGAL and MMP-9, were also increased in FAP relatively to control samples. However, this was only evident in stages where fibrilar TTR was already present. We show that MMP-9 is able to degrade both TTR aggregates and fibrils obtained *in vitro*. However, in FAP 0 nerves, MMP-9 levels were maintained unaltered relatively to controls, suggesting that *in vivo* TTR aggregates are not removed possibly due to the presence of other tissue factors. In later stages of the disease, increased levels of this enzyme should lead to fibril degradation. Our data suggests however that as *in vivo* TTR fibrils are bound to other components such as SAP, proteolysis by MMP-9 is prevented. Upregulation and/or activation of metalloproteinases in sites of early aggregate deposition may serve as a future therapeutic target. Our observations raise the possibility that abnormalities in basement membrane may be an early event, and play an integral part in the FAP amyloidogenesis process.

ACKNOWLEDGEMENTS

We thank Rui Fernandes and Rosana Moreira (IBMC, Porto) for tissue processing. This work was supported by grants from POCTI program and Gulbenkian Foundation, Portugal.

REFERENCES

1. Lechapt-Zalcman, E., Authier, F. J., Creange, A., Voisin, M. C., Gherardi, R. K. (1999) Labial salivary gland biopsy for diagnosis of amyloid polyneuropathy. *Muscle Nerve* **22**, 105-107.
2. Sousa, M. M., Du Yan, S., Fernandes, R., Guimarães, A., Stern, D. and Saraiva, M.J. (2001). J. Neurosci. , **21**, 7576-7586.
3. Goldsteins, G., Persson, H., andersson, K., Olofsson, A., Dacklin, I., Edvinsson, A., et al. (1999). Proc. Nat.l Acad. Sci. U.S.A., **96**, 3108-3113.
4. Cardoso, I., Goldsbury, C. S., Muller, S. A., Olivieri, V., Wirtz, S., Damas, A. M., Aebi, U., and Saraiva, M. J. (2002). Transthyretin fibrillogenesis entails the assembly of monomers: a molecular model for in vitro assembled transthyretin amyloid-like fibrils. *J. Mol. Biol.* **317**, 683-695.
5. Friedl, A., Stoesz, S. P., Buckley, P., and Gould, M. N. (1999). Neutrophil gelatinase-associated lipocalin in normal and neoplastic human tissues. Cell type-specific pattern of expression. *Histochem. J.* **31**, 433-441.
6. Yan, L., Borregaard, N., Kjeldsen, L., and Moses, M. A. (2001). The high molecular weight urinary matrix metalloproteinase (MMP) activity is a complex of gelatinase B/MMP-9 and neutrophil gelatinase-associated lipocalin (NGAL). Modulation of MMP-9 activity by NGAL. *J. Biol. Chem.* **276**, 37258-37265.
7. Backstrom, J. R., Lim, G. P., Cullen, M. J., and Tokes, Z. A. (1996) Matrix metalloproteinase-9 (MMP-9) is synthesized in neurons of the human hippocampus and is capable of degrading the amyloid-beta peptide (1-40). *J. Neurosci.* **16**, 7910-7919.
8. Moss, J., Shore, I., and Woodrow, D. (1998) An ultrastructural study of the colocalization of biglycan and decorin with AA amyloid fibrils in human renal glomeruli. *Amyloid* **5**, 43-48.
9. Stokes, M. B., Holler, S., Cui, Y., Hudkins, K. L., Eitner, F., Fogo, A., and Alpers, C. E. (2000) Expression of decorin, biglycan, and collagen type I in human renal fibrosing disease. *Kidney Int.* **57**, 487-498.

HOW IMPORTANT IS THE ROLE OF COMPACT DENATURED STATES ON AMYLOID FORMATION BY TRANSTHYRETIN?

J. R. Rodrigues and R. M. M. Brito*

Departamento de Química, Faculdade Ciências e Tecnologia, Universidade de Coimbra, 3004-535 Coimbra, Portugal; and Centro de Neurociências de Coimbra, Universidade de Coimbra, 3004-517 Coimbra, Portugal. *Corresponding author: brito@ci.uc.pt

Amyloid fibril formation and deposition have been associated with a series of diseases, including Alzheimer´s, Spongiform encephalopathies, and several systemic amyloidosis. In most of these amyloid diseases, it has been shown that the normal precursor protein, due to proteolysis, mutation or molecular environment stress, undergoes misfolding, leading to molecular species with a high tendency for ordered aggregation into amyloid. However, the structural nature of these amyloidogenic intermediates is the subject of debate.

Transthyretin (TTR) is a homotetrameric plasma protein with a high percentage of beta-sheet. TTR has been implicated in diseases such as Familial Amyloid Polyneuropathy (FAP) and Senil Systemic Amyloidosis (SSA). The current view on the mechanism of amyloid formation by TTR implies tetramer dissociation and monomer partial unfolding (for a recent review see (1)). However, very little is known about the structure of the amyloidogenic intermediates or the extent of monomer unfolding required for amyloid formation. Limited proteolysis in the fibrillar state and monoclonal antibody binding to highly amyloidogenic TTR mutants suggested loss of structure in beta-strands C and D and also in loop DE (2).

Here, based on molecular dynamics (MD) unfolding simulations of TTR, we propose that compact denatured states may play a central role as amyloidogenic species. From our simulations it is not clear how partially or locally unfolded species could provide the framework for protofibril formation. Alternatively, our simulations seem to indicate that in order to unfold beta-strands C and D and the loop DE, and expose beta-strands A and B for subunit interaction and aggregation, a global unfolding event is required. Thus, amyloid formation by TTR does not seem to be mediated by a local structural fluctuation or a local partial unfolding event, but by a global denaturing process of the subunit beta-sandwich. If in fact, compact denatured states play a central role on amyloid formation, the good agreement between amyloidogenic potential and protein conformational stability, observed for several proteins, could be more easily explained.

MOLECULAR DYNAMICS UNFOLDING SIMULATIONS OF TTR

The structural nature of the molecular species responsible for amyloid formation in TTR is not known. Several authors have suggested partial unfolding of the monomeric species as a key event in TTR amyloidogenensis (1, 3, 4, 5). However, the extent of this unfolding event is not known. In order to explore the unfolding pathways of monomeric TTR, we performed multiple thermal-induced unfolding simulations of WT- and L55P-TTR. Figure 1 shows several structures along four different molecular dynamics (MD) simulations, at 500 K, of one WT-TTR subunit. The structures at the left (Fig. 1) represent the X-ray crystal structure of the WT-TTR subunit (pdb entry 1tta) with its well characterized beta-sandwich topology formed by beta-sheet CBEF on one side and beta-sheet DAGH on the other side of the sandwich (6).

A general qualitative analysis of Figure 1 clearly shows different protein unfolding pathways, when comparing the four MD simulations. This is perfectly consistent with the current view on protein folding which implies that in an ensemble of protein molecules not all the individual members follow the same folding or unfolding pathway. In all the simulations the protein looses a significant amount of secondary structure, but maintains a general globular collapsed topology. In some instances the structure expands but this is due to conformational fluctuations of the highly mobile N- and C-terminus.

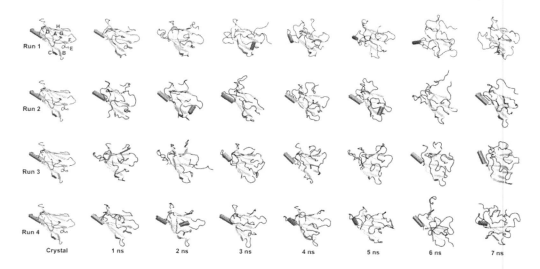

Figure 1. Secondary structure ribbon representations of the WT-TTR subunit along four different molecular dynamics (MD) unfolding simulations. Beta-strands, alpha-helices and turns and coil are represented by arrows, cylinders and tubes, respectively. The simulated time along the MD trajectories is indicated at the bottom of the figure. The simulations were performed with the progam NAMD (7), at 500 K, using periodic boundary conditions and an integration time step of 2 fs. Bond lengths between hydrogen and heavy atoms were constrained with the SHAKE algorithm. Long range electrostatic interactions were computed using the Particle Mesh Ewald method. Initial coordinates for the protein were taken from pdb entry 1tta (6). The only difference in the setup of each one of the four runs resides in the assignment of initial atomic velocities.

A detailed analysis, not possible here, of the trajectories in Figure 1 reveals several interesting aspects. The native alpha-helix is lost early in some of the simulations, but refolds and unfolds several times. Additionally, non-native alpha-helices transiently form in the regions of residues 7, 62 and 100. More importantly, concerning the persistence of the beta-sandwich topology, we can conclude that beta-strands E, B and A are the most stable, followed by beta-strands F, G and C, and beta-strands H and D are the least stable. This indicates that a core formed by beta-strands E, B and A is maintained until very late in most of the unfolding simulations. In our simulations, protein unfolding is accompanied by a significant loss in the number of native contacts, by a significant increase in the solvent accessible surface area, but by a moderate increase in the radius of gyration of the protein. It is however very interesting to note that total unfolding of beta-strands D and C and loop DE, and their displacement from the core of the beta-sandwich is not observed, even late in the simulations. Thus, in order to build a monomer topology compatible with the experimental data obtained for the fibrillar state, it seems that a global denaturing event is required because partial unfolding and displacement of beta-strands D and C is not a commonly observed event in the unfolding simulations.

CONCLUSIONS

Although not much is known about the identity of the amyloidogenic intermediates in most of the amyloid diseases, many molecular researchers in the field, including us, have used the expression "partially unfolded species" to describe the amyloidogenic intermediates. In the present report, we put forward new ideas, based on the results of MD unfolding simulations, concerning the structural nature and role of the partially unfolded species in amyloid formation, in general, and in TTR amyloidogenesis, in particular.

The MD unfolding simulations reported here clearly show that to unfold and displace beta-strands D and C and the loop DE from the core of the TTR beta-sandwich, as required by the subunit topology observed in the fibrillar state (8), a global denaturing event is required. Displacement of strands D and C from an intact beta-sandwich formed by strands BEF and AGH was not observed in any of the unfolding simulations performed with WT- or L55P-TTR. Thus, amyloid formation by TTR could be mediated by global unfolding of TTR subunits with low conformational stability, as we have previously suggested (5). This mechanism would not require the presence of a partially unfolded intermediate in the unfolding pathway of TTR, but the amyloidogenic molecular species would be found among the ensemble of monomeric compact denatured states of TTR. This interpretation allows immediate clarification of the correlation observed between the conformational stability of the TTR subunits and the amyloidogenic potential for several TTR variants (5).

ACKNOWLEDGMENTS

This work was supported in part by grant POCTI/BIO/35685/2000, FCT, Portugal. We acknowledge the computer resources provided by the Centre of Computational Physics, Universidade de Coimbra, Portugal.

REFERENCES

1. Brito, R.M.M., Damas, A.M, Saraiva, M.J. (2003) *Curr. Med. Chem.- Immun., Endoc. & Metab. Agents*, **3**, 349-360.
2. Goldsteins, G., Persson, H., Andersson, K., Olofsson, A., Dacklin, I., Edvinsson, A., Saraiva, M.J., Lundgren, E. (1999) *Proc. Natl. Acad. Sci. USA* **96**, 3108-3113
3. Lai, Z., Colon, W., Kelly, J.W. (1996) *Biochemistry* **35**, 6470-6482.
4. Quintas, A., Saraiva, M.J., Brito, R.M.M. (1999) *J. Biol. Chem.* **274**, 32943-32949.
5. Quintas, A., Vaz, D.C., Cardoso, I., Saraiva, M.J., Brito, R.M.M. (2001) *J.Biol.Chem.* **276**, 27207-27213.
6. Hamilton, J.A., Steinrauf, L.K., Braden, B.C., Liepnieks, J., Benson, M. D., Holmgren, G., Sandgren, O., Steen, L. (1993) *J. Biol. Chem.* **268**, 2416-2424.
7. Kalé, L., Skeel, R., Bhandarkar, M., Brunner, R., Gursoy, A., Krawetz, N., Phillips, J., Shinozaki, A., Varadarajan, K., Schulten, K. (1999) *J. Comp. Physics* **151**, 283-312.
8. Serag, A.A., Altenbach, C., Gingery, M., Hubbell, W.L., Yeates, T.O. (2002) *Nat.Struct.Biol.* 9, 734-739.

THE X-RAY CRYSTALLOGRAPHIC STRUCTURE OF THE AMYLOIDOGENIC VARIANT TTR TYR78PHE

R. Neto-da-Silva[1], P. J. B. Pereira[1], M. J. M. Saraiva[2,3] and A. M. Damas[1,3]

[1]*Molecular Structure Unit and* [2] *Molecular Neurobiology Group, Institute for Molecular and Cell Biology, Universidade do Porto, Rua do Campo Alegre, nº 823, 4150 Porto, Portugal.*
[3]*ICBAS - Instituto de Ciências Biomédicas Abel Salazar, Universidade do Porto, Largo Prof. Abel Salazar, nº2 4099-003 Porto, Portugal.*

INTRODUCTION

The systemic deposition of transthyretin (TTR) under the form of insoluble fibrils is the hallmark of TTR amyloid related diseases. Senile Systemic Amyloidosis (SSA) is characterized by the presence of protein deposits containing the wild-type (WT) protein, whereas a point mutation in the amino acid sequence is the biochemical marker for Familial Amyloidotic Polyneuropathy (FAP) (1).

TTR is composed of four identical subunits, which assemble forming a well-defined channel where the thyroxine binding sites are located. Each subunit comprises 8 β-strands (A-H) that assemble non-covalently in a β-sandwich, with the strands CBEF facing the strands DAGH. The single α-helical element encompassed in each monomer is located between strands E and F. The dimer is composed of one pair of eight-stranded β-sandwiches, which are held together through interactions between the strands HH' and FF'. The tetramer is formed when two dimmers associate through contacts between the AB loop of one dimer and the H strand of the other dimer (2). Several mutations have been identified in TTR, most of which lead to protein aggregation in the form of amyloid deposits. The Tyr to Phe exchange in the position 78 of the amino acid sequence generates a highly amyloidogenic TTR variant (2). This protein harbors the interesting particularity of reacting in its soluble tetrameric form with a monoclonal antibody raised against structures presenting the amyloidogenic characteristics. Hence, the Tyr78Phe variant may resemble an early intermediate in the fibrillogenesis cascade (3). Individuals carrying this mutation suffer from peripheral neuropathy, cardiomyopathy, carpal tunnel syndrome, and skin amyloidosis (4) (5).

In order to broaden our knowledge about the structural features responsible for TTR misfolding and aggregation, we have crystallized and solved the three dimensional structure of the TTR Tyr78Phe variant.

MATERIALS AND METHODS

Crystals of the recombinant TTR Tyr78Phe were grown by vapour diffusion methods. The protein concentration in the drop was 10mg/ml, and the reservoir solution was composed of 2.0 M ammonium sulfate, 7% glycerol and 0.2 M sodium acetate, pH 4.8. X-ray diffraction data were collected at the European Synchrotron Radiation Facility (ESRF) to a maximum resolution of 1.55 Å. Crystals belong to space group P2₁2₁2, with unit-cell parameters a = 42.72 Å, b = 86.08 Å, c = 63.80 Å, α = β = γ = 90°. A 98% complete data set was obtained with an overall R_{merge} of 7.2%. The structure was solved using the initial phases from the Thr119Met variant (PDB code 1F86), and subsequently refined to a final R_{factor} of 19.8% (R_{free} 23.9%) (6).

RESULTS AND DISCUSSION

As it has been reported for most TTR variants, the final refined model obtained for TTR Tyr78Phe displays a high overall homology to the structure of the WT-protein deposited in the PDB. Residue 78 is located in the single α-helix present in each monomeric subunit of the TTR homotetramer. In the native protein the hydroxyl group of Tyr78 establishes important hydrogen bonds with main-chain atoms of Pro24 and Ile26 and with the side-chain of Asp18 (Fig.1A). These are residues encompassed in the TTR AB loop, which is thereby intimately associated to the core of the molecule.

The Tyr to Phe exchange in the variant protein leads to the abrogation of these interactions. A new water molecule is observed near the AB loop, favouring a close-to-native hydrogen bond network within this region. Subtle structural differences in relation to the WT-protein are also observed in the α-helix, the site where the mutation occurs. The novel contacts established by some residues belonging to the α-helix and its vicinity, namely Lys80, Leu82, Gly83, Ile84 and Ser85, may account for these differences (Fig.1B).

Despite of the new observed interactions, the mutation impairs important contacts between the α-helix and the AB loop. As a consequence, they render looser structures, with a weaker association between the loop and the body of the molecule. We hypothesize that this may underlie the highly amyloidogenic behaviour observed for this protein.

(A)   (B)

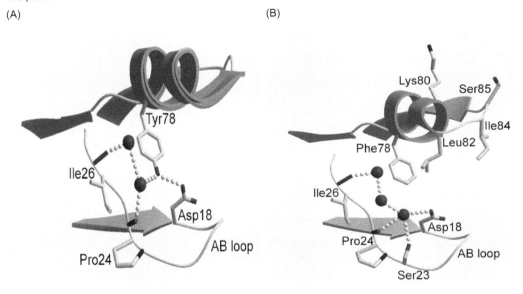

Figure 1. The interactions observed in the WT- and Y78F-TTR in the region of the mutation. (A) In WT-TTR the hydroxyl group of Tyr78 establishes key contacts with residues comprised in the AB loop, providing a relevant association between two regions located far apart in the protein structure. (B)In the Tyr78Phe variant the contacts between the AB loop and the protein core are lost. An additional water molecule ensures near-native hydrogen bonds within the AB loop region, preventing its complete displacement. The residues shown in the α-helix and its vicinity are involved in contacts not present in WT-TTR.

ACKNOWLEDGMENTS

This work has been supported by grants POCTI/35735 and POCTI/44821/2002 from FCT, Fundação para a Ciência e Tecnologia, Portugal. The authors would like to thank Paul Moreira for his excellent technical assistance in the preparation of recombinant transthyretin and acknowledge the European Synchrotron Radiation Facility for the use of beamline ID14-EH3 and for the technical assistance given by the ESRF staff.

REFERENCES

1. Saraiva, M.J.M. Molecular genetics of Familial Amyloidotic Polyneurpathy, *J. Periph. Nerv. Sys.,* 1, 179, 1996.
2. Blake, C. C., Geisow, M. J., Oatley, S. J., Rerat, B., and Rerat, C. Structure of prealbumin: secondary, tertiary and quaternary interactions determined by Fourier refinement at 1.8 Å. *J Mol Biol* 121, 339, 1978.
3. Redondo, C. et al., Search for intermediate structures in transthyretin fibrillogenesis: soluble tetrameric Tyr78Phe TTR expresses a specific epitope present only in amyloid fibrils, *J. Mol. Biol.,* 304, 461, 2000.
4. Magy, N. et al., A transthyretin mutation (Tyr78Phe) associated with peripheral neuropathy, carpal tunnel syndrome and skin amyloidosis, *Amyloid,* 10, 29, 2003.
5. Anesi, E. et al., Therapeutic advances demand accurate typing of amyloid deposits, *Am J Med.,* 111, 243, 2001.
6. Sebastião, M.P. et al., Transthyretin stability as a key factor in amyloidogenesis: X-ray analysis at atomic resolution, *J. Mol. Biol.,* 306, 733, 2001.

AN INTERACTIVE STRATEGY FOR THE ACCURATE CHARACTERIZATION OF TRANSTHYRETIN VARIANTS USING COMBINED PROTEOMIC AND GENOMIC TECHNOLOGIES

S.R. Zeldenrust,[1] H.R. Bergen, III,[2] A.I. Nepomuceno,[2,3] M.L. Butz,[4] D.S. Snow,[2] P.J. Dyck,[5] P.J.B. Dyck,[5] C.J. Klein,[5] J.F. O'Brien,[6] S.N. Thibodeau[4] and D.C. Muddiman[2,7]

[1]Department of Hematology,[2]W.M. Keck FT-ICR Mass Spectrometry Laboratory, Mayo Proteomics Research Center, [3]Department of Chemistry, Virginia Commonwealth University, Richmond, Virginia 23284 [4]Department of Molecular Genetics, [5]Perhipheral Neuropathy Research Center, Department of Neurology [6]Department of Laboratory Medicine and Pathology,[7]Department of Biochemistry and Molecular Biology, Mayo Clinic College of Medicine, Rochester, Minnesota 55905

BACKGROUND

Transthyretin associated hereditary amyloidosis (ATTR) is an inherited disease in which mutations in the primary structure of transthyretin (TTR, prealbumin) lead to the extracellular polymerization of insoluble protein fibrils resulting in organ failure and ultimately death. We have developed an integrated approach to characterize the TTR gene and protein in which the initial assay is the analysis of intact plasma TTR by ESI mass spectrometry. Positives are reflexed for DNA sequence analysis while real-time PCR confirms polymorphisms. A schematic of this protocol is shown in Figure 1.

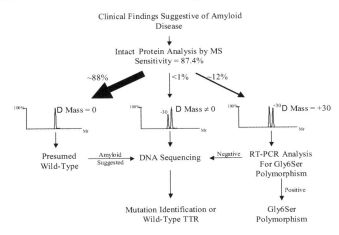

Figure 1. Flow diagram for the clinical analysis of TTR variants. The majority of samples will show no mass difference and can be presumed to be wild type (left branch). If a clinician still suspects amyloid, the archived buffy coat can be analyzed for the presence of mutations by sequencing the TTR exons. The presence of a variant (middle branch) will invoke DNA sequencing of the four TTR exons to determine the exact mutation. The presence of a +30 variant with a high ratio relative to the wild type is probably the Gly6Ser polymorphism and will be analyzed via real-time PCR (right branch). A single peak at +30 would be indicative of homozygous Gly6Ser. The percentages of each branch are based on the mutation incidence in the general population and do not take into account any referral bias.

EXPERIMENTAL

A. Protein Analysis

Samples from patients diagnosed with amyloid, polyneuropathy and control patients were analyzed for the presence of a variant in TTR. TTR protein was purified with an immunoaffinity resin prepared with 20 mm POROS AL resin. Plasma (20 mL) from archived plasma samples was analyzed. The purified TTR was reduced with 10 ml of 100 mM tris(2-carboxyethyl) phosphine (TCEP) at 55°C for 15 min and analyzed via mass spectrometry. A Sciex 150 single quadrupole MS was utilized for the analysis. The appearance of two peaks (or a single peak shifted in mass indicative of a homozygous variant) including the wild-type mass of 13,761 Da was indicative of the presence of a variant and was reflexed for DNA sequence analysis.

B. Genomic Analysis

DNA sequencing was performed after amplification of the four TTR exons with 250 ng of genomic DNA with 1 U AmpliTaq GOLD polymerase (Applied Biosystems). Forward and reverse PCR products for each exon were utilized in subsequent sequencing reactions. BigDye[â] Terminator v1.1 Cycle Sequencing kit (Applied Biosystems) was utilized for sequencing at one-half the recommended volume. An ABI 3100 capillary sequencer with uncoated 36 cm capillaries was utilized to read the TTR DNA sequence. ABI Prism 3100 Data Collection software (v 1.1) was used to analyze the data.

RESULTS

A.   Reduction of TTR with TCEP

Reduction of the purified TTR greatly reduces the complexity of the spectra as shown in Figure 2.

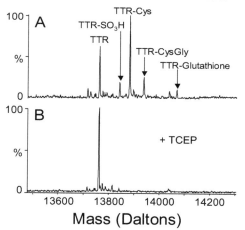

B. Mass spectral analysis of TTR samples

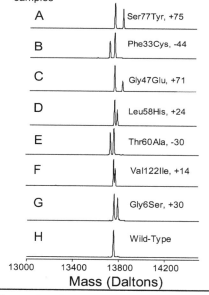

Figure 2. Mass spectra of immunoaffinity purified TTR before (A) and after (B) reduction with tris(2-carboxyethyl)phosphine. The multiple adducts are removed and the entire signal collapses into the free TTR increasing the signal-to-noise ratio.

Figure 3. Transformed spectra of TCEP reduced TTR. The measured mass difference and the actual variant are shown on the right.

C. Identification of a novel TTR variant

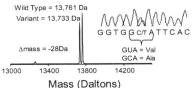

Wild Type = 13,761 Da
Variant = 13,733 Da

GGTGG C/T A TTCAC

GUA = Val
GCA = Ala

Δmass = -28Da

13000   13400   13800   14200

Mass (Daltons)

Figure 4. The transformed spectra of immunoaffinity purified and reduced TTR of a novel variant (Val94Ala). The inset shows reverse DNA sequence showing the substitution of a T for a C in the DNA sequence of codon 94. The expected mass shift for this variant is −28 Da and is identical to the measured mass shift.

D. Precise mass measurement of TTR variants with FT-ICR MS.

Patient TTR Variant Analysis using LC-dual-ESI-FT-ICR-MS

Sample	Protein Variant	DNA Variant	Mass Shift (Da)	Monoisotopic Neutral Mass Observed	Monoisotopic Neutral Mass Theoretical	MMA$_{int}$ (ppm)	Error (Da)
A	G6S	GGT - AGT	(+30)	13782.914	13782.899	1.1	0.02
B	T60A	ACT - GCT	(-30)	13722.867	13722.878	-0.8	-0.01
C	G6S	GGT - AGT	(+30)	13782.927	13782.899	2.0	0.03
D	E54G	GAG - GGG	(-72)	13680.846	13680.867	-1.5	-0.02
E	S77Y	TCT - TAT	(+76)	13828.914	13828.920	-0.4	-0.01
F	G47E	GGG - GAG	(+72)	13824.889	13824.909	-1.4	-0.02
G	S77Y	TCT - TAT	(+76)	13828.914	13828.920	-0.4	-0.01

E. FT-ICR MS analysis of a cis double TTR variant

Figure 5. The theoretical isotopic distribution of the [M+11H+]11+ charge state for a wild-type individual (Top Right) and a patient thought to be wild-type obtained on a 7 Tesla LC-dual-ESI FTICR. The monoisotopic mass for the theoretical and experimental +11 charge state is indicated. The mass measurement accuracy was -72 ppm. This difference in mass was sufficient to determine this individual was not wild-type. DNA sequencing results confirmed that this patient contained two variants on the same allele (G6S and V30A). A small family pedigree of the double cis-variant family is indicated.

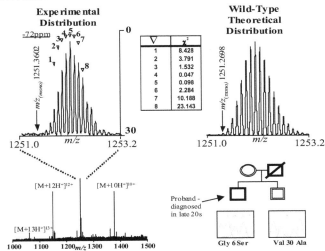

Experimental Distribution

Wild-Type Theoretical Distribution

∇	2
1	8.428
2	3.791
3	1.532
4	0.047
5	0.098
6	2.284
7	10.188
8	23.143

Proband - diagnosed in late 20s

Gly 6 Ser   Val 30 Ala

DISCUSSION

All mass spectral determinations were corroborated by DNA sequence analysis. Additionally, all Gly6Ser polymorphisms were correctly called based on the +30 mass shift and an equal relative abundance of the +30 polymorphism relative to wild-type TTR.

CONCLUSION

We describe a two-tiered approach to TTR mutation analysis. The first stage involves mass spectral analysis. Positives are subsequently reflexed for DNA sequencing or, in the case of the Gly6Ser polymorphisms, real-time PCR analysis by the LightCycler. This reflexive method is efficient and cost effective since it eliminates the necessity of sequencing most samples and allows us to screen for the familial forms of amyloidosis in a broader patient population in a timely fashion. Using our overall approach, all variants were correctly identified. FT-ICR and DNA sequencing combined was able to identify a cis-double mutation with a +30-28=+2 Da shift. Furthermore, a novel variant, Val94Ala was also identified.

CODEPOSITION OF TRANSTHYRETIN AND IMMUNOGLOBULIN LAMBDA LIGHT CHAIN IN SENILE CARDIAC (ATTR) AMYLOIDOSIS

J.J. Liepnieks(1), M.D. Benson(1,2)

(1) Department of Pathology and Laboratory Medicine, Indiana University School of Medicine (2) Richard L. Roudebush Veterans Affairs Medical Center, Indianapolis, Indiana USA

INTRODUCTION

Transthyretin (TTR) is the major component of amyloid fibrils in familial TTR amyloidosis and senile cardiac amyloidosis. In familial TTR amyloidosis, the fibrils contain a mutant TTR (1). The disease is usually characterized by peripheral and autonomic polyneuropathy, but the heart and other organs may be affected. In senile cardiac amyloidosis, the fibrils contain normal TTR (2). Deposits are present predominantly in the heart and blood vessel walls, with approximately 25% of the population over 80 years of age affected. We report a case of senile cardiac amyloidosis where an immunoglobulin lambda light chain was codeposited with normal TTR in the amyloid.

PATIENT

A 77 year old White male presented with restrictive cardiomyopathy. An endomyocardial biopsy stained with Congo red and exhibited green birefringence under polarized light indicating amyloid. Immunoelectrophoretic analysis of serum and urine showed no monoclonal immunoglobulin component. Immunohistochemical analysis of the biopsy showed the deposits stained positive with TTR. Genetic analysis of the patient's DNA showed no mutation in his TTR gene. A diagnosis of senile cardiac amyloidosis was made. The patient died five years later, and cardiac tissue was obtained for biochemical analysis.

METHODS

Cardiac tissue was homogenized in 0.1M sodium citrate, 0.15M sodium chloride and centrifuged at 12,000 RPM (3). The pellet was repeatedly homogenized and centrifuged before exhaustive dialysis against water and lyophilization. The isolated fibrils stained with Congo red and exhibited green birefringence under polarized light. Fibrils were solubilized in 6M guanidine hydrochloride, 0.5M Tris pH8.2 containing 10mg dithiothreitol/ml and 1mg EDTA/ml, alkylated with iodoacetic acid, and centrifuged. The supernatant was chromatographed on a Sepharose CL6B column equilibrated and eluted with 4M guanidine hydrochloride, 50mM Tris pH8.2. Pooled fractions were exhaustively dialyzed against water and lyophilized. The Sepharose CL6B pool was digested with trypsin and fractionated by reverse phase HPLC on a Synchropak RP8 column eluted with an acetonitrile gradient. HPLC peaks were subjected to Edman degradation analysis.

Residues	Sequence
#67-89+?	1 5 10 15 20 S-G-N-T(A)S(L)T-I-S-G-L-Q-A-E-D-E-A-D-Y-Y-C(S)...
#104-110	1 5 V-T-V-L-G-Q-P-K
#111-129	1 5 10 15 A-A-P-T-V-T-L-F-P-P-S-S-E-E-L-Q(A)(N)(K)
#130-147+?	1 5 10 15 A-T-L-V-C-L-I-S-D-F-Y-P-G-A-V-T-V-A...

Figure 1. Lambda light chain peptides present in trypsin digest of Sepharose CL6B pool from guanidine hydrochloride solubilized cardiac amyloid fibrils. Parentheses denote residues not completely verified.

RESULTS

Analysis of the Sepharose CL6B pool demonstrated that the major component in the cardiac fibrils was TTR. Tryptic peptides from residues 10-126 of TTR were identified all containing normal TTR sequence. In addition, peptides from an immunoglobulin lambda light chain were present (Figure 1). Two peptides starting with residues 111 and 130 of the constant region, and two peptides starting with residues 67 and 104 of the variable region were found. The molar amount of the light chain peptides was less than 10% than that of the TTR peptides.

DISCUSSION

Analysis of the cardiac amyloid showed that while the major protein component in the fibrils was normal TTR, a monoclonal immunoglobulin lambda light chain was also codeposited as a minor constituent in this case of senile cardiac amyloid. The variable region sequence of the light chain is most homologous with the lambda II light chain subgroup. These findings raise the possibility that small amounts of light chain fibrils could act as a nidus or AEF and promote fibril formation from normal TTR in some cases of senile cardiac amyloidosis. The known increasing prevalence of monoclonal immunoglobulin proteins with age might be a factor in the high prevalence of senile cardiac amyloidosis seen in aged individuals.

REFERENCES

1. Benson MD, Amyloidosis. In: Scriver CR, Beaudet AL, Sly WS, Valle D, editors. *The Metabolic and Molecular Bases of Inherited Disease, 8th Ed.* New York: MCGraw Hill; 2001, p. 5345-5378.
2. Westermark P, Sletten K, Johansson B and Cornwell III GG (1990). Fibril in senile systemic amyloidosis is derived from normal transthyretin. *Proc Natl Acad Sci USA* **87**: 2843-2845.
3. Nichols WC, Liepnieks JJ, Snyder EL and Benson MD (1991). Senile cardiac amyloidosis associated with homozygosity for a transthyretin variant (Ile-122). *J Lab Clin Med* **117**: 175-180.

MACROSCOPIC ASSESSMENT OF SKIN VENULES IN PATIENTS WITH FAMILIAL AMYLOIDOTIC POLYNEUROPATHY (FAP) USING NEAR-INFRARED SPECTROPHOTOSCOPY: AS A MARKER OF AUTONOMIC DYSFUNCTION

K. Obayashi[1], Y. Ando[2], T. Miida[1], M. Nakamura[2], T. Yamashita[3], M. Ueda[3], K. Haraoka[4], Terazaki, H.[4], and M. Okada[1]

[1]Division of Clinical Preventive Medicine, Department of Community Preventive Medicine, Niigata University Graduate School of Medical and Dental Sciences, 1-757 Asahimachi, Niigata 951-8510, Japan
[2]Department of Diagnostic Medicine, [3]Department of Neurology, and [4]Department of Gastroenterology, Graduate School of Medical Sciences, Kumamoto University 1-1-1 Honjo, Kumamoto 860-0811, Japan

INTRODUCTION

Patients with FAP show various clinical symptoms reflecting peripheral circulatory disturbances. Orthostatic hypotension is one of the most common and serious complications in these patients (1). The integral role of the arterioles in the regulation of blood pressure to orthostatic stress is well established (2). Although change in the function of venules may also be found, it has not been well examined in those patients.

Recently, attention has been focused on near-infrared spectrophotoscopy as a non-invasive tool to assess the venules (3). In this study, we evaluated morphological changes in the skin venules of the fingers in patients with FAP using near-infrared spectrophotoscopy, and discussed the utility of this method.

METHODS

Ten FAP patients (6 women and 4 men; aged 33-54 years) and 10 healthy volunteers (4 women and 6 men; aged 29-42 years) were examined. The patients took no drugs at least for 24 hours before the examination. None of the healthy volunteers had vascular diseases or autonomic dysfunction.

The shapes, diameters, and venous oxygenation indices (VOIs) of those vessels before and after deep inspiration were evaluated using a near-infrared spectrophotoscopy system (ASTRIM, Sysmex Co., Japan). Subjects put the second finger of the left hand on Astrim. Near-infrare rays uttered by light source irradiated their fingers, and the pictures of skin venules in the finger were recorded by charge-coupled device camera.

Blood pressure was measured with the patients and healthy volunteers in both a supine and a 70 degrees upright position by using a tilting table, and the degree of orthostatic hypotension was defined by levels of the reduced systolic blood pressure.

All the data are expressed as mean ± SD. The significance of the data was determined by the unpaired Student's t-test. P values less than 0.05 were considered significant. The correlation between the degree of orthostatic hypotension and vasoconstriction of the venules was investigated by the Pearson's parametric coefficient.

RESULTS

Six of 10 FAP patients showed morphological abnormalities of the venules, such as tortuosity, irregular venous caliber, and microaneurysm-like change (Fig.1). The baseline diameter in the venules and the percent increase in VOI after deep inspiration were not significantly different between healthy volunteers and FAP. However, percent vasoconstriction of the venules after deep inspiration was significantly lower in the FAP patients than that in the healthy volunteers ($p<0.05$). A linear negative correlation was observed between the degree of blood pressure change and the degree of vasoconstriction of the venules in the FAP patients ($p<0.05$, $r=-0.68$).

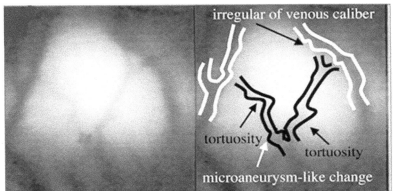

Fig. 1 Morphological abnormalities of the skin venules in a 33-year-old male FAP patient.

DISCUSSION

We have demonstrated morbid changes on the venules in the finger macroscopically and non-invasively in the FAP patients. Interestingly, these changes seemed to be very similar to those found in the eye conjunctiva of the patients (4).

Three major factors, amyloid deposition around the vessels, the dysfunction of smooth muscles, and dynamic changes in blood velocity accompanied by orthostatic hypotension may contribute to the phenomenon of the venules. However, the amount of perivascular amyloid deposition in the venules is usually smaller than that of the arterioles in FAP. Moreover, these impairments are shown not only in FAP patients but also in MSA patients despite there being no amyloid deposition. Therefore, the effect of amyloid deposition on the venules' abnormality may be small. Concerning the possibility of smooth muscle dysfunction, the effect may also be very small because FAP patients do not have any abnormalities of the smooth muscles in the venules. In addition, similar impairments could not be seen in progressive muscular dystrophy patients who showed dysfunction of smooth muscles by the same examination. Thus, it is reasonable to consider that the major factor in these impairments is dynamic changes in blood velocity accompanied by orthostatic hypotension. We could not directly evaluate the effect of changes in venus pressure during the provocative stimuli on the vasoconstriction of venules. However, it is well documented that orthostatic hypotension induces dynamic changes in blood velocity and repeated such condition for a long time may lead to the morphological changes on the venules in the finger of patients with autonomic dysfunction (3). In addition, degree of vasoconstriction of venules was correlated with the degree of blood pressure change, suggesting that autonomic dysfunction should strongly influence on the vasoconstriction of venules. It has been known that a significant correlation between retinal vessel and systemic autonomic nerve function was observed (5, 6). However, it has not been well examined in the correlation between the peripheral venules in the extremities and systemic autonomic nerve function. This report is the first to document that a significant correlation between the peripheral venules in the extremities and systemic autonomic nerve function is observed, and evaluation of the venules demonstrated in this study may be useful for screening the peripheral autonomic dysfunction.

REFERENCES

1. Ando, Y., Suhr, O., Autonomic dysfunction in familial amyloidotic polyneuropathy, *Amyloid* 5, 288, 1998.

2. Kontos, H.A., Richardson, D.W., Norvell, J.E., Mechanisms of circulatory dysfunction in orthostatic hypotension, *Trans.Am.Clin.Climatol.Ass.,* 87, 26, 1976.

3. Hall, J.W., Pollard, A., Near-infrared spectrophotometry: a new dimension in clinical chemistry, *Clin.Chem.,* 38, 1623, 1992.

4. Ando, E. et al., Ocular microangiopathy in familial amyloidotic polyneuropathy, type I, *Graefes.Arch.Clin.Exp.Ophthalmol.,* 230, 1, 1992.

5. Lanigan, L.P., Clark, C.V., Hill, D.W., Retinal circulation responses to systemic autonomic nerve stimulation, *Eye,* 2, 412, 1988.

6. Steinle, J.J., Pierce, J.D., Clancy, R.L., G. Smith, P., Increased ocular blood vessel numbers and sizes following chronic sympathectomy in rat, *Exp.Eye Res.,* 74, 761, 2002.

INCIDENCE OF TRANSTHYRETIN VAL122ILE AMYLOID MUTATION IN AFRICAN-AMERICANS BORN IN INDIANAPOLIS, INDIANA, USA

M.D. Benson(1,2), M. Yazaki(1), T. Yamashita(1), K. Hamidi Asl(1)

(1) Department of Pathology and Laboratory Medicine, Indiana University School of Medicine (2) Richard L. Roudebush Veterans Affairs Medical Center, Indianapolis, Indiana USA

INTRODUCTION

Hereditary transthyretin amyloidosis is associated with greater than 80 different mutations in transthyretin. While the most common type of transthyretin amyloidosis in the world (Val30Met) is found in large pedigrees in Northern Portugal, Northern Sweden, Japan, and smaller families in United States, Europe, and Australia, one particular TTR mutation (Val122Ile) is found predominantly in Americans of African descent (1-3). This mutation causes late-onset restrictive cardiomyopathy (after age 60). In previous reports allele frequencies of 0.014 (148 subjects) and 0.020 (1,688 subjects) were found (4,5). Those studies, however, used DNA samples obtained for genetic analysis of other diseases in various geographic areas. The present study determines the gene frequency in newborns in a mid-American city (Indianapolis, Indiana).

METHODS

Cord blood samples were collected at the time of birth from 1,973 subjects at the county/city hospital. Mothers were categorized only as to ethnicity. DNA was isolated by standard techniques. The Val122Ile mutation was detected by SSCP and confirmed by PCR-RFLP.

RESULTS

Race	No. Tested	No. Positive	% Positive	Allele Frequency
African-American	1,000	30[*]	3%	.0155
Caucasian	453	2	0.4%	.0022
Hispanic	490	0		
Other (Asian, etc.)	30	0		

[*]One homozygous for Val122Ile

CONCLUSION

In a population selected only by ethnicity and birth in an urban community the finding of 3% incidence of the TTR Val122Ile mutation associated with late-onset cardiac amyloidosis is similar to previous reports of studies on preselected DNA samples, but lower than the 3.9% found in one study (5). The amyloidosis caused by this mutation is often clinically mistaken for other types of cardiac disease in the African-American population. Demonstration of the high incidence of this mutation in African-Americans is important for the education of individuals giving care to this population.

REFERENCES

1. Gorevic PD, Prelli FC, Wright J, Pras M and Frangione B (1989). Systemic senile cardiac amyloidosis-identification of a new prealbumin (transthyretin) variant in cardiac tissue: Immunologic and biochemical similarity to one form of familial amyloidotic polyneuropathy. *J Clin Invest* **83**: 836-843.

2. Nichols WC, Liepnieks JJ, Snyder EL and Benson MD (1991). Senile cardiac amyloidosis associated with homozygosity for a transthyretin variant (Ile122). *J Lab Med* **117**: 175-180.

3. Saraiva MJM, Sherman W, Marboe C, Figueeira A, Costa P, De Freitas AF and Gawinowicz MA (1990). Cardiac amyloidosis: report of a patient heterozygous for the transthyretin – isoleucine 122 variant. *Scan J Immunol* **32**: 341-346.

4. Jacobson DR, Reveille JD and Buxbaum JN (1991). Frequency and genetic background of the position 122 (Val Ile) variant transthyretin gene in the Black population. *Am J Hum Genet* **49**: 192-198.

5. Jacobson DR, Pastore R, Pool S, Malendowicz S, Kane I, Shivji A, Embury SH, Ballas SK and Buxbaum JN (1996). Revised transthyretin Ile-122 allele frequency in African-Americans. *Hum Genet* **98**: 236-238.

AN HISTORICAL CASE OF PRIMARY AMYLOIDOSIS WAS RE-DIAGNOSED AS ATTR (V30M)

R.P. Linke [1], W. Schlote [2], L. Gerhard [3], P. Winter [4], K. Altland [4]

[1] *Max-Planck-Institut für Biochemie, Martinsried, Germany*

[2] *Neuropathologie der Universität, Frankfurt a.M., Germany*

[3] *Institut für Neuropathologie, Essen, Germany*

[4] *Institut für Humangenetik, Giessen, Germany*

[1] *Correspondence: Am Klopferspitz 18a, D-82152, Martinsried, Germany;*

E-mail: linke@biochem.mpg.de;

ABSTRACT

An unusual systemic amyloidosis with visceral and cerebral amyloid deposits was reported by W. Krücke. However, the amyloid protein of this patient has never been classified according to the chemical nosology. Subsequent citations referred to this disease as "primary amyloidosis", indicating a monoclonal gammopathy as the underlying disease. Here, this disorder was re-diagnosed using immunohistochemistry and gene analysis. The results show the identification of the hereditary amyloidosis of transthyretin-origin (ATTR) due to the M30V mutant.

INTRODUCTION

Before the advent of the new nosology of the amyloid diseases, which is based on chemistry (1), unusual amyloid disorders had been published. One of such cases was reported by Krücke in 1950 (2). From the time of Krücke until the present, this unusual disease was considered as a special form of "primary amyloidosis", a term which has been later applied and restricted to diseases of immunoglobulin origin without overt monoclonal gammopathy. Therefore, this case was preliminarily catalogued without the identification of the protein of origin. Fortunately, the retrieval of a series of Krücke's original paraffin blocks enabled us to reassess this disease using modern technology. Our re-investigation resulted in a clear and precise diagnosis, which is discussed here.

CASE HISTORY

The female patient (GNE) died at the age of 38 years in 1948 after two years of progressive weakness, weight loss, cardiomyopathy, vomiting and obstipation, amenorrhoea, and hair loss. The neurologic examination showed signs of sensomotoric polyneuropathy of both feet. Her gait became insecure and she suffered from a retention of urine. Five days before her death she became somnolent and lost her consciousness (1).

MATERIAL AND METHODS

Paraffin blocks of several organs including brain (see below) were available from patient GNE for this re-evaluation. Paraffin-embedded tissue sections (4 μm thick) were prepared and staining was performed with Congo red according to Puchtler et al. (3). The stained sections were evaluated for the presence of amyloid in bright light, in polarized light (3) and in fluorescent light (4). The protein of origin of this amyloid was identified by applying a set of antibodies that are able to distinguish 12 different amyloids in tissue sections, through the application of the peroxidase-anti-peroxidase technique of Sternberger as described (4-8). Each antibody was controlled by a tissue section containing the respective known amyloid class. Gene analysis of the transthyretin (TTR) gene was performed on formalin-fixed and in paraffin-embedded tissue. DNA was extracted from a formalin-fixed and in paraffin-embedded liver, the TTR-exons were amplified and sequenced. In addition, restriction length polymorphism with Nsi I was performed on exon 2.

RESULTS

Diagnosis of amyloid: Amyloid was present in all major organs such as the liver, spleen, kidney, adrenal glands, skin, intestine, neural tissue, muscles, but also in the plexus chorioidicus, the leptomeningeal vessels of the brain and the spinal cord. In particular amyloid was present in the vessels of the sub-arachnoidal space, in the pellucid septum and in Ammon's horn.

Immunohistochemistry: Amyloid of the leptomeningeal vessels of the brain and the spinal cord, amyloid of the liver and spleen all showed a very strong reaction with the antiserum directed against the amyloid of transthyretin origin. No other reaction was significant with one exception, the antibody directed against the λ-light chain derived amyloid (ALλ) reacted as well. In contrast to the uniform staining with anti-ATTR, the light chain antibody stained only partly, while most amyloid remained unstained by this antibody indicating a reaction against a non-amyloidotic constituent.

Gene analysis: Gene analysis of all three exons of the TTR gene showed a single point mutation only in exon 2 on one strand, in the codon 30 (GTG/ATG) leading to the amino acid exchange of Val 30 to Met. No other mutation was noticed. Restriction length polymorphisms corroborated this finding.

DISCUSSION AND CONCLUSIONS

From the data presented, the disease of patient GNE with systemic amyloidosis was established as belonging to the hereditary group of amyloidoses of TTR-origin (ATTR), although no family history was reported. This mutation (V30M) represents the most common form of hereditary amyloidosis. There are more than 80 different amyloid-promoting mutations known (9).

This diagnosis is based on the identification of the amyloid protein of TTR origin and on the nucleotide sequence. Of particular interest are the cerebral symptoms that were interpreted by her physicians as either lues cerebri, brain tumor or multiple sclerosis, since the diagnosis of amyloidosis was only made in tabula (at autopsy).

Whether the cerebral symptoms are due to the amyloid deposits is not clear. In favor of this interpretation are the findings of nodular amyloid deposits in the wall of small cerebral blood vessels which could have caused the small vicinal brain infarctions and those of the spinal cord. There were also no histological signs of lues cerebri nor those of multiple sclerosis. Leptomeningeal amyloid in systemic amyloidosis of transthyretin origin has been published by several groups.

ACKNOWLEDGEMENTS

Supported by the Deutsche Forschungsgemeinschaft, Bonn-Bad Godesberg (grant Li 247/ 12-3) and the Max-Planck-Institut für Biochemie, Martinsried, Germany (Prof. Dr. R. Huber). For technical assistance we thank J. Lindermeyer and M. Bandmann.

REFERENCES

1. Glenner, G.G.:Amyloid and amyloidosis. The β-fibrilloses; *New Engl. J. Med.* 302, 1283 and 1333, 1980.

2. Krücke, W. : Das Zentralnervensystem bei generalisierter Paramyloidose; *Arch. Psychiatr. Neurolog;* 185, 165, 1950.

3. Puchtler, H., Sweat, F., and Levine, M.: On the binding of Congo red by amyloid; *J. Histochem. Cytochem.* 10, 355, 1962.

4. Linke, R., P.: Highly sensitive diagnosis of amyloid and various amyloid syndromes using congo red fluorescence, *Virchows Arch. Path. Anat.* 436, 439, 2000

6. Linke, R.P.: Monoclonal antibodies against amyloid fibril protein AA. Production, specificity and use of immunohistochemical localization and classification of AA-type amyloidosis. *J. Histochem. Cytochem.* 32, 322, 1984.

7. Linke, R.P., Gärtner, V., and Michels, H.: High sensitivity diagnosis of AA amyloidosis using congo red and immunohistochemistry detects missed amyloid deposits; *J. Histochem. Cytochem.* 43, 863, 1995.

8 Schröder, R., Nennesmo, I., and Linke, R.P.: Amyloid in a multiple sclerosis lesion is clearly of Aλ-type; *Acta Neuropath,* 100, 709, (2000).

9. Connors, L.H., et al.; *Amyloid: J. Protein Folding Disord.* 10, 160, 2003.

ECHOCARDIOGRAPHIC ABNORMALITIES IN TRANSTHYRETIN (TTR) VI22I ALLELE CARRIERS

Tagoe CE, Schwartzbard A, Shah A, Buxbaum JN, Salman L, Feliciano E, Labat M, Mehra D, Malendowicz S, Hanna G, Pierre S, Stedman MR, Roswell R, Fiore LD, Brophy MT, Gagnon D, Jacobson DR

New York Harbor VA Healthcare system and New York University Schoolf of Medicine, New York, NY, USA, Scripps Research Institute, La Jolla, CA, and VA Boston Healthcare System and Boston University School of Medicine, Boston, MA, USA

The transthyretin (TTR) V122I allele is associated with cardiac amyloidosis. Nearly all affected patients have had Black African ancestry. The TTR V122I allele is carried by 3.9% of African-Americans, with equal prevalence throughout the U.S.A. (1). Most clinical information about allele carriers comes from individual case reports, and little is known about carriers' disease penetrance or early disease manifestations. It is also known that cardiac amyloidosis is more prevalent in African-Americans than in Caucasian Americans (fig. 1) (2). In a previous study, we have shown that the excess prevalence of cardiac amyloidosis in Africans is cardiac ATTR, largely V122I ATTR (3). While these studies have established the association of the TTR V122 gene with clinical disease, the allele's penetrance in causing clinical disease, and the early findings in allele carriers on noninvasive testing, are unknown.

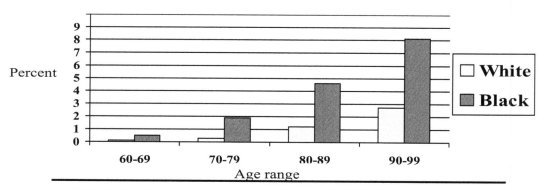

Figure 1. Age-dependent prevalence of cardiac amyloidosis in a large autopsy study.

Mortality from cardiovascular disease in African-Americans is 50% greater than in White Americans. Congestive heart failure (CHF) is also more common, occurs at a younger age, and is associated with higher mortality in Afircan-Americans. CHF in African-Americans also has less association with coronary artery disease, and more with hypertension and with apparently "idiopathic" causes. Furthermore, death rates for ischemic heart disease are greater in Arican-Americans than in Caucasian-Americans. Various possible explanations for these observations have been studied, including racial differences in vascular autonomic reactivity, D-dimer levels, endothelium-mediated vasodilation, platelet Ca++ stores and turnover, insulin-induced venodilation, endothelin levels, and ACE polymorphisms. We postulate that an undiagnosed, treatment-resistant cardiomyopathy common in African-Americans could contribute to this population's excess cardiovascular disease as compared to other groups, and that TTR V122I amyloidosis could be that undiagnosed cardiomyopathy. The objective of this study is therefore to investigate the early echocardiographic findings in TTR V122I allele carriers without a known personal or family history of cardiac amyloidosis.

METHODS

To study TTR V122I allele carriers' early non- invasive findings, we performed a matched cohort study nested in a cohort of 354 African-American males > age 60, without a personal or family history of amyloidosis, referred for echocardiography for any reason. DNA was isolated from buccal swabs on each subject. Multiplex allele-specific PCR was performed to determine allele status. The bottom band of 238 bp in lanes 5-8 indicates the presence of the variant in these samples; the top band at 291 bp is a positive control for the success of the PCR reaciton (figure 2). Positives were confirmed by another PCR reaction and restriction analysis (figure 3). Each TTR V122I allele carrier was matched for age, hypertension, diabetes, and coronary artery disease with two controls negative for the V122I allele. Echocardiograms on 12 TTR VI22I heterozygotes (mean age 74 years) and 18 controls were read by two cardiologists blinded to allele status. Generalized linear models were used to analyze the data with generalized estimating equations to adjust for correlated data. Data are presented on 12 subjects and 18 controls for whom the echocardiographic analysis is complete (table 1).

RESULTS

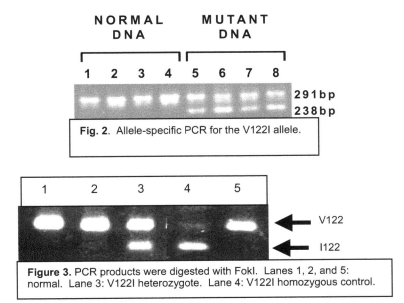

NORMAL DNA   **MUTANT DNA**

| 1 | 2 | 3 | 4 | 5 | 6 | 7 | 8 |

291bp
238bp

Fig. 2. Allele-specific PCR for the V122I allele.

| 1 | 2 | 3 | 4 | 5 |

← V122

← I122

Figure 3. PCR products were digested with FokI. Lanes 1, 2, and 5: normal. Lane 3: V122I heterozygote. Lane 4: V122I homozygous control.

354 subjects were enrolled (age range 60-98, mean age 73). 18/354 subjects were heterozygous for TTR V122I (5.1%, which is equal to the population frequency of 3.9%). No V122I homozygotes were found. The mean age of the TTR V122I heterozygotes was 72, identical to the mean age of the entire group. The mean age of the 12 patients with complete echocardiography data to date is 74.

Table 1. Echocardiography data

Variable	Subjects	Controls	p value
Intraventricular septum thickness	1.39 mm	1.15 mm	.03
Posterior wall thickness	1.33 mm	1.10 mm	.01
Deceleration time	139 msec	194 msec	.003
Left atrial size	4.7 cm	4.1 cm	.048

Our data indicate that among African-American males over age 60 referred for echocardiography for any reason, carriers of TTR VI22I have several specific echocardiographic abnormalities that are known to be associated with diastolic dysfunction. The abnormalities seen in the V122I carriers are similar to the abnormalities previously described in patients with cardiac amyloidosis of various types. We conclude that TTR V122I allele is a common cause of heart disease in African-Americans, i.e., is highly penetrant in causing cardiac dysfunction, and also that TTR V122I appears likely to contribute to the excess cardiac disease seen in African-Americans.

REFERENCES

1. Jacobson DR, Pastore R, Pool S, Malendowicz S, Kane I, Shivji A, Embury SH, Ballas SK, Buxbaum JN. Revised transthyretin Ile 122 allele frequency in African-Americans. Hum Genet. 1996 Aug;98(2):236-8.

2. Buck FS, Koss MN, Sherrod AE, Wu A, Takahashi M. Ethnic distribution of amyloidosis: an autopsy study, Mod Pathol. 1989 Jul;2(4):372-7.

3. Jacobson DR, Pastore RD, Yaghoubian R, Kane I, Gallo G, Buck FS, Buxbaum JN. Variant-sequence transthyretin (isoleucine 122) in late-onset cardiac amyloidosis in black Americans. N Engl J Med. 1997 Feb 13;336(7):466-73.

FAMILIAL AMYLOIDOTIC POLYNEUROPATHY WITH SEVERE RENAL INVOLVEMENT IN ASSOCIATION WITH TRANSTHYRETIN Gly47Glu IN DUTCH, BRITISH AND AMERICAN FAMILIES

E.B. Haagsma [1], P.N. Hawkins [2], M.D. Benson [3], H.J. Lachmann [2], A. Bybee [2] and B.P.C. Hazenberg [4]

Department of Gastroenterology and Hepatology (1), and Rheumatology (4), University Hospital, Groningen, The Netherlands; National Amyloidosis Centre (2), Royal Free Hospital, Royal Free and University College Medical School, London, UK; and Department of Pathology and Laboratory Medicine (3), Indiana University School of Medicine, Indianapolis, USA

INTRODUCTION

Familial amyloidotic polyneuropathy (FAP) is an autosomal dominant disorder associated with more than 80 different transthyretin (TTR) mutations. The clinical features of FAP are broad and variable, but knowledge of the pattern and natural history of disease associated with particular mutations nevertheless offers the best guidance for management of individual patients, including the role and timing of treatment by orthotopic liver transplantation.

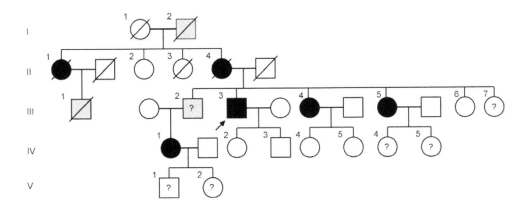

Figure 1. Pedigree of the Dutch family. Darkened square or circle means affected individual. A grey shaded square or circle means probable carrier of the mutation. The arrow denotes the propositus Case A III-3. The question mark denotes family members not tested for the mutation.

OBJECTIVE

FAP in association with TTR Gly47Glu has been described previously in an Italian kindred (1). We report here its phenotype in seven further patients from Dutch, British, and American (Finnish) families.

PATIENTS AND METHODS

DNA analysis showed the presence of the TTR Gly47Glu mutation in all seven affected individuals (five Dutch, one British and one American). Clinical evaluation of the patients was focused on organ function of heart, kidneys, gastrointestinal tract, eyes, and nerves. SAP scintigraphy was performed in five patients.

RESULTS

The pedigree tree with the five patients of the Dutch kinship (figure 1) showed the characteristic autosomal dominant type of inheritance.

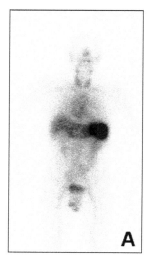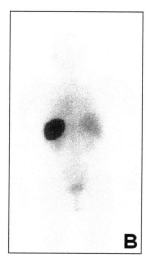

Figure 2. SAP scintigraph (24 hours after administration of SAP) of the propositus Case A III-3. Anterior view (A) and posterior view (B) with intensive uptake of the spleen (++) and slightly increased uptake of the kidneys (+). Beside the increased uptake of the spleen and kidneys the expected normal picture of blood pool activity in heart and major blood vessels is visible as well as some radioactive degradation products in stomach and urinary bladder.

Four women and three men were studied with a median age at first symptoms of 49 years (range 33-69 years). First symptom was nausea and weight loss (2 patients), CTS (2 patients), erectile impotence, oedema, and bladder dysfunction (one patient each). Autonomic neuropathy was seen in six patients (three severe), motor neuropathy in three and some signs of sensory neuropathy in all seven. Left ventricular wall thickness was increased in all seven (ranging from 12 to 19.5 mm). None of the patients showed proteinuria, but creatinine clearance was decreased in more than half of the cases (median value 54 ml/min, range 10-95 ml/min). Weight loss was seen in five patients (two severe) and disturbed bowel motility in four patients (the two patients with severe weight loss had the most disturbed bowel motility. No vitreous opacities were seen.

Only four patients were deemed to be sufficiently fit to undergo orthotopic liver transplantation, and clinical deterioration was generally rapid. SAP scintigraphy showed no specific uptake in one patient, only splenic uptake in three patients and uptake in spleen and kidneys in one patient (figure 2).

CONCLUSIONS

Characteristic clinical features of FAP associated with TTR Gly47Glu included amyloid cardiomyopathy and autonomic failure but, unusually, severe renal failure was present in more than half of the cases. Only four patients were deemed to be sufficiently fit to undergo orthotopic liver transplantation, and clinical deterioration was generally rapid. These observations support early intervention with orthotopic liver transplantation in patients with FAP associated with TTR Gly47Glu.

REFERENCES

1.   Pelo E, Da Prato L, Ciaccheri M, Castelli G, Gori F, Pizzi A, Torricelli F and Marconi. Familial amyloid polyneuropathy with genetic anticipation associated to a gly47glu transthyretin variant in an Italian kindred. Amyloid: J. Protein Folding Disord 2002; 9:35-41.

CLINICAL DIAGNOSIS OF FAMILIAL TRANSTHYRETIN AMYLOIDOSIS BY AUTOMATED DATA ACQUISITION AND PROCESSING WITH QUADRUPOLE ORTHOGONAL TIME-OF-FLIGHT MASS SPECTROMETRY

M.E. McComb,1,2 A. Lim,1,2,5 R. Théberge,1,2,5 T. Prokaeva,4,5 L.H. Connors,2,5 M. Skinner4,5 and C.E. Costello1,2,3,5

1Mass Spectrometry Resource; Departments of 2Biochemistry, 3Physiology and Biophysics, and 4Medicine; and 5Amyloid Treatment and Research Program, Boston University School of Medicine, Boston, MA 02118 USA

INTRODUCTION

Transthyretin (TTR) is a 13.7 KDa transport protein found predominantly in plasma where it binds and transports retinol-binding protein and thyroxine. Amino acid substitution in TTR is hypothesized to destabilize TTR and cause it to form an intermediate that self associates into amyloid fibrils. Familial transthyretin amyloidosis (ATTR), is associated with the deposition of the TTR variants as amyloid fibrils in various tissues and organs. A large number of TTR variants have been identified, the majority being amyloidogenic, using traditional gene based sequencing methods and newer mass spectrometry (MS) methods. Since the only effective treatment of ATTR is liver transplantation, the correct clinical diagnosis is critical. Our laboratory, in collaboration with the Amyloid Treatment and Research Program, has developed an extensive sample purification/analysis protocol for the MS characterization of TTR; this is shown schematically in Figure 1.A. We have thus characterized a number of TTR variants of clinical significance using MS for ATTR identification and peptide mapping and tandem mass spectrometry (MS/MS) for protein sequencing of ATTR variants and identification of post-translational modifications. Here we explore the application of automating the MS characterization of ATTR using on-line information dependent acquisition (IDA) capillary LC MS/MS for initial sequence analysis followed with automated data-base searching.

EXPERIMENTAL

TTR was immunoprecipitated from the serum of patients and purified by centrifugation and reversed phase HPLC. Proteolytic digestions were performed. MALDI-TOFMS was performed using a Finnigan MAT Vision 2000 mass spectrometer. Nanospray MS and on-line capillary LC-MS with information dependent acquisition (IDA) MS/MS were performed on an Applied Biosystems QSTAR QoTOF mass spectrometer. coupled with an LC Packings capillary LC system. Separation was on 300 µm capillary columns packed in-house with Michrom Magic C18 phase using standard water:ACN gradients. MS and MS/MS were acquired in the positive polarity mode with resolution > 1:9,000 (fwhm) and better than 25 ppm mass accuracy (external calibration). Information dependent acquisition (IDA) was used to obtain MS/MS spectra of peptides as they eluted from the LC system. MS peaks which exceeded a threshold of 10 counts per second were subjected to MS/MS using preset collision energies proportional to the *m/z* value of the precursor (ca. 18-60 V, lab frame). Data from IDA-LC-MS/MS experiments were analyzed using the Mascot (www.matrixscience.com) search engine and the PRO-ID (Applied Biosystems) search engine with custom user programmed databases .

RESULTS and DISCUSSION

Mass spectrometry (MS) is becoming an important tool in the field of clinical research including the application for the diagnosis and characterization of ATTR. MS analysis of intact TTR provides an indication on the possibility of a variant by yielding the mass difference between the wild type and variant TTR. An example is shown here in Figure 1.B. for the ATTR A97S variant indicated with a mass difference of +16 Da from the intact mass of the TTR protein. Intact mass analysis, alone, may be misleading. TTR is known to contain post-translational modifications on Cys10 (*e.g.*. sulfonation and cysteinylation, as indicated in Figure 1.B). In addition, other PTMs may occur, *e.g.,* oxidation, which may account for intact protein mass shifts. Additionally, this mass difference does not define the location of the variant.

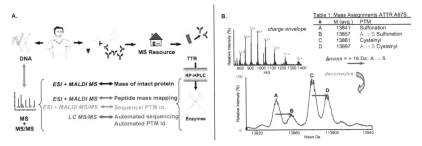

Figure 1. A. Sample preparation scheme for the isolation, purification and characterization of TTR from human serum. MS is performed on intact TTR and proteolytic digests of TTR using ESI MS, MALDI MS and LC MS/MS. MS results are compared with DNA analysis results for confirmation. **B.** ESI MS analysis of intact TTR, showing the charge envelope and deconvolved spectrum indicating Cys10 PTMs and a potential variant A97S. Average masses for peak assignments are shown in Table 1 (inset).

 Peptide mass mapping provides for a more conclusive identification and location of a potential variant. Figure 2.A. shows mapping MALDI TOF MS and LC MS results of the A97S variant indicating 100% sequence coverage and an accurate monoisotopic mass shift of 15.99 Da localized on the T10 peptide. This approach is adequate for confirmatory analysis of MS results with those obtained with DNA sequencing methods. For a true clinical diagnosis, though, it is necessary to sequence the individual peptides that exhibit mass shifts relative to peptide digests of unmodified wt protein in order to locate and conclusively identify the variant. The same holds true for additional forms of TTR which arise due to post-translational modification of the expressed protein. MS/MS via nanospray or MALDI are suitable for this purpose given some prior information about a particular sample and some idea as to where a variant may occur (*e.g.*. peptide mass mapping and gene results). However, being operator dependent, manual collection and interpretation of data is time consuming and requires expertise.

 Proteomics developed MS/MS methods such as IDA-LC-MS/MS offers advantages to the clinical applications of mass spectrometry by automating the data acquisition phase. Data are shown here for the A97S ATTR protein characterized in this manner using IDA LC-MS/MS. Figure 2.B. shows the principle of the experiment and the resulting TIC, SIC and MS/MS data from the normal and variant peptides. Each T10 peptide yielded complete sequence information allowing the unequivocal assignment of the A97S variant.

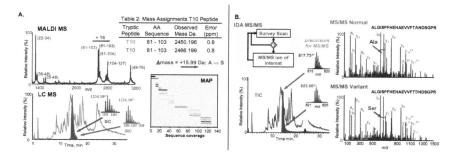

Figure 2. A. MS peptide mass mapping of a tryptic digest of TTR. Mass differences observed in MALDI MS and LC MS are indicative of the variant A97S located on T10 (seq. 81-103). Accurate masses obtained from the deconvolved spectra (spectra and SIC inset) are shown in Table 2 (inset). **B.** LC IDA MS/MS obtained on a tryptic digest of TTR. MS/MS was obtained automatically on the precursor peaks for the [M+3H]3+ charge state of each T10 peptide. Complete sequence coverage on the normal T10 and variant (A97S) T10 peptides allowed conclusive variant identification.

As an LC MS/MS experiment may yield >300 MS/MS spectra, some of which may require complex interpretation, a proteomics approach was employed to simplify and facilitate the analysis. LC MS/MS data were subjected to database searches using Mascot and PRO-ID search engines. Conclusive variant identification was automatically obtained with statistically significant confidence scores in all cases where a quality data set was submitted. However, erroneous results could be obtained. These included false positive or negative results and were observed if the database interrogation was not performed with due care.

CONCLUSIONS

Mass spectrometry affords the direct characterization of a protein for the study of a protein-based disease including variant analysis and PTM characterization. While intact mass analysis and peptide mass mapping may indicate the possibility of a variant, only MS/MS provides for conclusive unambiguous determination. This process is greatly facilitated with automated LC IDA MS/MS and automated analysis.

ACKNOWLEDGMENTS

This work was supported by NIH grants P41-RR10888 (CEC) and S10-RR15942 (CEC), the Gerry Foundation (MS), and the Young Family Amyloid Research Fund (MS). We thank Applied Biosystems for their support with the PRO-ID software.

REFERENCES

1.   Connors, L.H., Lim, A., Prokaeva, T., Roskens, V.A. and Costello, C.E., Tabulation of Human Transthyretin (TTR) Variants, 2003, *Amyloid: J. Protein Folding Disorders*, 10, 160, 2003.
2.   Lim, A., Prokaeva, T., McComb, M.E., O'Connor, P.B., Théberge, R., Connors, L.H., Skinner, M., Costello, C.E., Characterization of Transthyretin Variants in Familial Transthyretin Amyloidosis by Mass Spectrometric Peptide Mapping and DNA Sequence Analysis, *Anal. Chem.*, 74, 741, 2002.

LIVER TRANSPLANTATION FOR FAMILIAL AMYLOIDOTIC POLYNEUROPATHY DOES NOT PREVENT DISEASE PROGRESSION IN A MAJORITY OF PATIENTS

S. Zeldenrust.[1], R. Perri[2], D. Brandhagen[2], C. Rosen[3], J. Poterucha[2], M. Gertz[1], R. Wiesner[2] and G. Gores[2]

Department of Medicine, Divisions of Hematology[1] and Gastroenterology[2] and Transplant Surgery[3], Mayo Clinic, Rochester, MN, 55905

ABSTRACT

<u>Aim</u>: To determine the outcome in patients transplanted for FAP.

<u>Methods</u>: We reviewed the medical records of all patients undergoing OLT for FAP at our institution between 11/96 and 12/02. Information collected included: sex, mutation, age, clinical manifestations, and patient and allograft survival.

<u>Results</u>: We evaluated 12 males with a mean age of 58 years (range 47-68). Six received liver, and six patients liver and heart transplants. TTR mutations included MET-30 (n=4), ALA-60 (n=4), TYR-77 (N=2), GLY-42 (n=1) and LYS-89 (n=1). Mean time from symptoms to diagnosis was 3.4 years (0-10). Patients were listed for OLT an average of 10.8 months (0-60) after diagnosis. Mean time from listing to OLT was 8.2 months (1-18). Five patients died after OLT, during a mean follow-up of 3.5 years (1-6). Survival was 100% (1 year) and 92% (3 year). Symptoms worsened after OLT in seven, and improved or stabilized in five. Following OLT, neuropathy improved in four patients, worsened in seven and was unchanged in one. Prior to OLT, all twelve patients had an increased IVST on echocardiogram. Post-OLT cardiac symptoms improved in six (five had cardiac transplantation), worsened in three, and were unchanged in three (one had cardiac transplantation). Post-OLT echocardiograms were available in 4 of 6 patients who had not received a cardiac transplant. Echocardiograms were performed an average of 29.3 months (12-36) post-transplant and IVST improved in one, worsened in two, and remained the same in one.

<u>Conclusions</u>: The benefits of OLT in FAP, particularly in non-MET-30 patients, remain indeterminate.

INTRODUCTION

Potentially curative liver transplantation for transthyretin-associated amyloidosis (ATTR) was first reported in 1990 in Sweden. Since that time, over 600 patients have received orthotopic liver transplantation (OLT) for ATTR. Early reports indicated the procedure was well-tolerated and resulted in significant clinical improvement and amyloid regression. However, subsequent reports have shown that not all patients improve after OLT, with some patients experiencing progressive neuropathy and cardiac deposits. Thus the benefit of OLT in ATTR remains unclear, especially in patients with mutations other than V30M. We examined the outcomes in patients undergoing OLT for ATTR at our institution in order to provide more information on the benefit of OLT for this disease.

AIMS

This study was performed to determine the outcome, including the status of clinical manifestations, in patients undergoing OLT for ATTR. In addition, predictors of outcome in patients undergoing OLT for ATTR were determined.

METHODS

A retrospective chart review of all patients undergoing OLT for ATTR at Mayo Clinic Rochester from 11/96-12/02 was perfromed. Symptoms attributable to ATTR were assessed at the initial evaluation, time of OLT, and last follow-up. Information collected included: symptom duration, age at diagnosis and OLT, sex, TTR mutation, pre and post-OLT clinical manifestations, modified body mass index [mBMI = (weight/[height^2]) x serum albumin (g/L)], and patient and allograft survival. Neurological involvement was assessed by patient symptoms of autonomic and peripheral neuropathy. Cardiac involvement was assessed by serial echocardiograms. Intraventricular septal thickness (IVST) was measured on all patients undergoing echocardiogram.

PATIENTS

12 male patients underwent orthotopic liver transplantation for ATTR. The mean age was 57 years (44-67). 6 received liver transplant alone, while 5 underwent combined liver/heart and 1 liver/heart/kidney transplant. The initial clinical manifestations were neurological (n=7), cardiac (n=4) and gastrointestinal (n=1). The mean time from symptoms to diagnosis was 3.4 years (range 0-10), with a mean time of diagnosis to OLT of 18.8 months. See Table 1 for individual patient characteristics.

Table 1. Patent Characteristics

Pt.	Mutation	Multiple Tx	Disease Duration	Follow-up	Neuro Status	Cardiac Status
1	Met-30		2 years	66 months	Improved*	Improved*
2	Met-30		4 years	40 months	Progressed*	
3	Tyr-77	Kidney/Heart	5 years	35 months	Stable*	
4	Ala-60	Heart	6 years	40 months	Progressed	
5	Ala-60		3 years	33 months	Progressed	Progressed*
6	Ala-60		2 years	12 months	Progressed	Progressed*
7	Lys-89	Heart	2 years	49 months	Progressed*	
8	Met-30		5 years	48 months	Progressed*	Progressed*
9	Tyr-77	Heart	1 year	4 months	Improved	
10	Gly-42	Heart	0 years	75 months	Improved	
11	Met-30		10 years	24 months	Progressed	
12	Ala-60	Heart	1 year	73 months	Improved	

Shaded rows indicate deceased patients. * indicates subjective data (echo, EMG) available.

RESULTS - CARDIAC

The IVST increased in all patients pre-OLT. All heart transplant recipients had non-Met30 mutations. Of the 4 post-OLT echocardiograms obtained in non-heart transplant recipients Mean time post-OLT 32 months (14-41); 1 was improved, 2 showed worsening and 1 was unchanged

RESULTS - NEUROLOGICAL

7 patients had progression of their neurological symptoms, while the remaining 5 patients had improved or stable symptoms. Of the 7 with progressive disease, subjective evidence of improvement (EMG) was available for 3 patients; the remainder were classified according to subjective findings. EMG data was available for 2 of the 5 patients with stable or improved symptoms, with the others classified by subjective findings. Of the 4 patients with the Met30 mutation, 3 had progressive neurological symptoms following OLT, while 1 had improved symptoms, with EMG data verifying improved nerve conduction.

PREDICTORS OF OUTCOME

Variable	Improved or Stabilized (n=5)	Worsened (n=7)	p
Disease duration (months)*	49.2 +/- 48.1	67.6 +/- 30.0	0.485
mBMI at time of transplant*	933 +/- 112	941 +/- 152	0.918
Age (years)*	51.8 +/- 6.7	60.3 +/- 5.8	0.050
Met 30 mutation	1/4	3/4	n/a
non-Met 30	4/8	4/8	n/a

* denotes mean ± standard deviation.

PATIENT AND ALLOGRAFT SURVIVAL

The mean follow-up was 3.8 years. Patient survival at 1 year was 100% and 92% at 3 years. There were five deaths. One patient received a second OLT for hepatic artery thrombosis.

Predictors of Survival

Variable	Survivors	Non-survivors	p
Age (years)*	54.0 +/- 6.6	60.6 +/- 7.0	0.147
Disease duration (months)*	48.3 +/- 39.5	76 +/- 33	0.209
mBMI*	920 +/- 114	962 +/- 163	0.656
Met 30	2/4	2/4	n/a
non-Met 30	5/8	3/8	n/a

* denotes mean ± standard deviation.

SUMMARY

Symptoms progressed despite OLT in a majority of our patients with FAP. The majority of our patients undergoing OLT for FAP had non-Met30 mutations. All cardiac transplant recipients had a non-Met30 mutation. The mBMI, disease duration, and Met30 mutation were not predictive of outcome or survival after OLT. Further studies are needed to better define the benefit of OLT for FAP, particularly for non-Met30 patients

CONTINUED DEPOSITION OF WILD-TYPE TRANSTHYRETIN IN CARDIAC AMYLOID AFTER LIVER TRANSPLANTATION

J.J. Liepnieks(1), M. Yazaki(1), M.D. Benson(1,2)

(1) Department of Pathology and Laboratory Medicine, Indiana University School of Medicine (2) Richard L. Roudebush Veterans Affairs Medical Center, Indianapolis, Indiana USA

INTRODUCTION

Transthyretin (TTR) amyloidosis is the most prevalent type of hereditary systemic amyloidosis. The majority of the over 80 mutations in the TTR gene result in amyloid deposition in peripheral nerve that leads to progressive polyneuropathy. Amyloid cardiomyopathy, however, is the most common cause of death. Since plasma TTR is produced mainly in the liver, liver transplantation has been employed as a promising therapy for patients with hereditary TTR amyloidosis to stop synthesis of plasma variant TTR. Several studies of patients after liver transplantation have shown no further progression in neurological symptoms including autonomic dysfunction in many patients (1-4). A few clinical studies, however, have reported progression of cardiac amyloidosis and vitreous amyloidosis still occurs after liver transplantation (5-8). In order to assess the possible contribution of wild-type TTR to continuing amyloid deposition after liver transplant, we investigated the relative amounts of wild-type and variant TTR in the cardiac amyloid of patients who underwent liver transplant compared to that of patients without liver transplant.

METHODS

Amyloid fibrils were isolated from cardiac tissue by repeated homogenizations in sodium citrate/sodium chloride, sodium chloride, and water (9). The water washes and final pellet were exhaustively dialyzed against water and lyophilized. Fibrils were solubilized in 6M guanidine hydrochloride, 0.5M Tris-HCl pH8.2 containing 10mg dithiothreitol/ml and 1mg EDTA/ml, alkylated with iodoacetic acid, and centrifuged. The supernatant was chromatographed on a Sepharose CL6B column equilibrated and eluted with 4M guanidine hydrochloride, 50mM Tris-HCl pH8.2. Pooled fractions were exhaustively dialyzed against water and lyophilized. Sepharose CL6B pools were digested with trypsin and fractionated by reverse phase HPLC on a Synchropak RP8 column eluted with an acetonitrile gradient. Peaks were analyzed by Edman degradation.

RESULTS

The relative amounts of variant and wild-type TTR in cardiac amyloid were determined from the recoveries of tryptic peptides containing the variant or wild-type TTR residue. Eight hearts from patients without a liver transplant were analyzed (Table1). Seven from Val30Met, Thr60Ala, Ile84Ser, and ΔVal122 patients

Table 1. TTR Composition of Cardiac Amyloid in Non-Liver Transplant Patients.

Patient	TTR Mutation	% Variant TTR	% Normal TTR
1	Val30Met	56%	44%
2	Val30Met	60%	40%
3	Thr60Ala	60%	40%
4	Ile84Ser	60%	40%
5	Ile84Ser	63%	37%
6	Ile84Ser	65%	35%
7	ΔVal122	70%	30%
8	Val122Ile	50%	50%

Table 2. TTR Composition of Cardiac Amyloid in Liver Transplant Patients.

Patient	TTR Mutation	Time Post-Transplant	% Variant TTR	% Normal TTR
9	Ala25Ser	1 Month	60%	40%
10	Val30Met	2 ½ Years	35%	65%
11	Thr60Ala	5 ½ Years	20%	80%

contained more variant (56-70%) than wild-type (30-44%) TTR in the amyloid. A Val122Ile heart contained approximately equal amounts of variant and wild-type TTR. Three hearts from patients who underwent liver transplant were analyzed (Table 2). The relative amount of wild-type TTR in cardiac amyloid increased with time post-transplant from 40% after one month to 65% after 2 ½ years to 80% after 5 ½ years.

DISCUSSION

Analysis of the cardiac amyloid in paients without liver transplant indicates that the deposition of variant TTR predominates over wild-type TTR in hereditary TTR amyloidosis except for the Val122Ile patient where approximately equal amounts of variant and wild-type TTR were present. Analysis of the cardiac amyloid in patients with liver transplant indicates that the relative amount of wild-type TTR increases with time after transplant. These results suggest that plasma wild-type TTR continues to deposit in cardiac amyloid after synthesis of plasma variant TTR is halted by liver transplant.

REFERENCES

1. Holmgren G, Ericzon B-G, Groth C-G, Steen L, Suhr O, Andersen O, Wallin BG, Seymour A, Richardson S, Hawkins PN and Pepys MB (1993). Clinical improvement and amyloid regression after liver transplantation in hereditary transthyretin amylodosis. *Lancet* **341**: 1113-1116.
2. Suhr OB, Holmgren G, Steen L, Wikstrom L, Norden G, Friman S, Duraj FF, Groth C-G and Ericzon B-G (1995). Liver transplantation in familial amyloidotic polyneuropathy. *Transplantation* **60**: 933-938.
3. Ikeda S-I, Takei Y-I, Yanagisawa N, Matsunami H, Hashikura Y, Ikegami T and Kawasaki S (1997). Peripheral nerves regenerated in familial amyloid polyneuropathy after liver transplantation. *Ann Intern Med* **127**: 618-620.

4. Bittencourt PL, Couto CA, Farias AQ, Manchiori P, Massarollo PCB and Mies S (2002). Results of liver transplantation for familial amyloid polyneuropathy type I in Brazil. *Liver Transpl* **8**: 34-39.

5. Ando Y, Ando E, Tanaka Y, Yamashita T, Tashima K, Suga M, Uchino M, Negi A and Ando M (1996). *De novo* amyloid synthesis in ocular tissue in familial amyloidotic polyneuropathy after liver transplantation. *Transplantation* **62**: 1037-1038.

6. Dubrey SW, Davidoff R, Skinner M, Bergethon P, Lewis D and Falk RH (1997). Progression of ventricular wall thickening after liver transplantation for familial amyloidosis. *Transplantation* **64**: 74-80.

7. Stangou AJ, Hawkins PN, Heaton ND, Rela M, Monaghan, M, Nihoyannopoulos P, O'Grady J, Pepys MB and Williams R (1998). Progressive cardiac amyloidosis following liver transplantation for familial amyloid polyneuropathy. *Transplantation* **66**: 229-233.

8. Munar-Qués M, Salvá-Ladaria L, Mulet-Perera P, Solé M, López-Andreu FR and Saraiva MJM (2000). Vitreous amyloidosis after liver transplantation in patients with familial amyloid polyneuropathy: Ocular synthesis of mutant transthyretin. *Amyloid Int J Exp Clin Invest* **7**: 266-269.

9. Nichols WC, Liepnieks JJ, Snyder EL and Benson MD (1991). Senile cardiac amyloidosis associated with homozygosity for a transthyretin variant (Ile-122). *J Lab Clin Med* **117**: 175-180.

DEPOSITION AND PASSAGE OF TRANSTHYRETIN THROUGH THE BLOOD-NERVE BARRIER IN RECIPIENTS OF FAMILIAL AMYLOID POLYNEUROPATHY LIVERS

M. M. Sousa [1], J. Ferrão [2], R. Fernandes [1], A. Guimarães [3], J. B. Geraldes [2], R. Perdigoto [2], L. Tomé [2], O. Mota [2], L. Negrão [4], A. L. Furtado [2] and M. J. Saraiva [1, 5].

[1] *Molecular Neurobiology, Institute for Cellular and Molecular Biology, Porto;* [2] *Transplantation Department, University Hospitals of Coimbra* [3] *Neuropathology, Hospital Geral de Santo António, Porto;* [4] *Neurology Department, University Hospitals of Coimbra and* [5]*ICBAS, University of Porto, Portugal.*

INTRODUCTION

Given the fact that approximately 90% of plasma transthyretin (TTR) is produced by the liver, liver transplantation (LT) has been used to abolish the production of most of the mutant TTR in familial amyloid polyneuropathy (FAP). As the shortage of donor livers has been the rate-limiting factor in the expansion of LT, domino liver transplantation (DLT), has been performed; in this procedure FAP patients receive and sequentially donate grafts to other recipients. However DLT involves specific ethical problems namely, the graft of a liver producing a pathogenic protein.

Cytotoxicity of early non-fibrillar TTR aggregates has been previously found in asymptomatic FAP patients (FAP 0) (1, 2); despite the absence of amyloid fibrils, FAP 0 nerves show signs of neuronal stress presenting increased expression of pro-inflammatory cytokines, oxidative stress-related molecules (inducible nitric oxyde synthethase and 3-nitrotyrosine- 3-NT) and apoptosis (1, 2).

The purpose of this work was to study the presence of *de novo* amyloid deposition and toxicity in the skin and nerve of recipients of FAP livers up to 7 years after DLT.

MATERIALS AND METHODS

Participants: Livers from 15 FAP patients were transplanted to 15 recipients with chronic liver disease or with liver tumors. DLT recipients had no clinical history or signs of FAP. 4 DLT recipients (TD) performed a skin biopsy before the transplant; skin biopsies were collected either up to 2 years after DLT (n=6), between 2 and 5 years (n=6) or 6 years after DLT (n=3). Nerve biopsies were performed in recipients that were subjected to DLT for more than 5 years. Control skin biopsies were performed in FAP patients (n=5) and non-FAP patients (n=4) submitted to conventional LT 3 years before.

Immunohistochemistry (IHC): Primary antibodies used were: mouse monoclonal anti-TTR Mab39-44 (3) and anti-3 nitrotyrosine (3-NT, Chemicon, 1:1000).

Electron microscopy (EM) and immunogold TTR labeling: Rabbit anti-human TTR (DAKO, 1:100) was added for 1 h at room temperature; control grids were incubated with preabsorbed anti-human TTR.

Electromyography (EMG): Motor responses were recorded with surface electrodes (belly-tendon montage); the radial (wrist-digit I) and median (wrist-digit III) nerves sensory responses were recorded antidromically through near-nerve needle electrodes after superficial stimulation.

RESULTS

De novo amyloidosis in recipients of domino liver transplantation (DLT): We analysed skin biopsies from recipients of FAP livers 1 to 7 years after DLT. Up to 2 years after DLT, none of the skins from recipients presented TTR deposition; 3 to 5 years after DLT, deposition was variable and did not correlate with time of transplant; 6 to 7 years after DLT amyloid deposition was evidenced in all the cases (Table 1).

years after DLT	TTR/CR
1 (n=3)	- -
2 (n=3)	- -
3 (n=1)	+ +
4 (n=2)	+ -
5 (n=1)	- -
5 (n=1)	+ -
5 (n=1)	+ +
6 (n=2)	+ +
7 (n=1)	+ +

Table 1. Deposition of TTR and amyloid (CR) in skin biopsies performed in recipients of FAP livers 1 to 7 years after DLT

The amount of deposited TTR was generally scarce both in non-fibrillar and fibrillar forms. One of the recipients where biopsy was performed 6 years after DLT, presented TTR deposition comparable to the characteristic TTR deposition in the skin of FAP patients. Presence of TTR amyloid fibrils in the skin of DLT recipients, was further confirmed by TTR immunoEM (Figure 1)

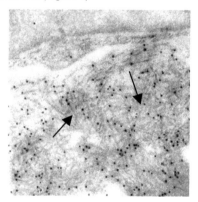

Figure 1 . Anti-TTR immunoEM of a skin biopsy performed 3 years after DLT.
Toxicity of *de novo* TTR deposition: 3-NT IHC in skin biopsies from DLT recipients with fibrillar (DLT +/+) and non-fibrillar TTR deposition (DLT -/+), was generally not increased relatively to normal skin biopsies and to non-FAP individuals submitted to LT (data not shown). Therefore, the amount of TTR deposition in the majority of the DLT recipients does not seem sufficient to promote increase in oxidative stress in target organs.

Nerve pathology in recipients of DLT: Sural nerve biopsies from 6 recipients of FAP livers that underwent DLT in the preceding 5 to 7 years, were performed. Five of the biopsies had no TTR deposits (Table 2); one of the biopsies, performed 6 years after DLT, disclosed a small deposit of TTR amyloid fibrils in the epinerve (Table 2). This TTR positive nerve did not reveal any 3-NT immunostaining (data not shown). To further evaluate the occurrence of polyneuropathy in DLT recipients, EMG was performed 5-7 years after transplant. The only DLT recipient that showed TTR deposition in the nerve, did not present any signs of polyneuropathy (Table 2). Two of the DLT recipients showed signs of polyneuropathy that developed post-transplantation (Table 2). However, the morphometric changes were not related to FAP but were most probably the consequence of other pathologies causing sensory axonopathy, such as diabetes and alcoholism.

Table 2. Deposition of TTR and morphometric study of nerve biopsies from recipients of FAP livers and EMG

Yrs post DLT	TTR/CR		MF/mm²	UF/mm²	EMG prior DLT	EMG at time of biopsy
5	-	-	5368	39738	No PNP	Sensory axonal NP
5	-	-	7331	83136	NA	NA
5	-	-	3571	88330	NA	No signs of PNP
7	-	-	3525	48340	NA	No signs of PNP
6	-	-	4691	52951	No PNP	Sensorymotor changes
6	+	+	4976	44503	No PNP	No signs of PNP

evaluation five to seven years after DLT. MF- myelinated fiber; UF- unmyelinated fiber; PNP- polyneuropathy; NP- neuropathy; NA- non-available.

DISCUSSION

De novo amyloidosis occurs in the skin and nerve of recipients of FAP livers after DLT. Sites related to TTR deposition did not display signs of oxidative stress such as increase in 3-NT epitopes probably due to the reduced amount of deposited TTR. The fibrillogenesis pathway might be accelerated in adults either due to the enhancement of mechanisms involved in amyloid formation or to the downregulation of inhibitory pathways. In individuals where the liver is the sole source of mutated TTR, the protein is able to deposit in the connective tissue of the epinerve and skin, suggesting that the plasma pool of the protein is sufficient for PNS involvement in FAP. DLT using FAP livers should continue to be considered an experimental procedure under careful surveillance and recipients of FAP livers should be followed up using an extensive neurological and pathological examination.

REFERENCES

1. Sousa, M. M., Cardoso, I., Fernandes, R., Guimaraes, A. and Saraiva, M. J. (2001) Deposition of transthyretin in early stages of familial amyloidotic polyneuropathy: evidence for toxicity of nonfibrillar aggregates .Am. J. Pathol., **159**, 1993-2000.

2. Sousa, M. M., Du Yan, S., Fernandes, R., Guimarães, A., Stern, D. and Saraiva, M.J. (2001) Familial amyloid polyneuropathy: receptor for advanced glycation end products-dependent triggering of neuronal inflammatory and apoptotic pathways. J. Neurosci. , **21**, 7576-7586.

3. Goldsteins, G., Persson, H., Andersson, K., Olofsson, A., Dacklin, I., Edvinsson, A., Saraiva, M. J., and Lundgren, E. (1999) Exposure of cryptic epitopes on transthyretin only in amyloid and in amyloidogenic mutants. Proc. Nat.l Acad. Sci. U.S.A., **96**, 3108-3113.

ACKNOWLEDGEMENTS

This work was supported by Portuguese grants from POCTI Program (FCT), Gulbenkian Foundation and SAUDE XXI Program (Health Ministry).

FAMILIAL AMYLOIDOSIS: RECENT NOVEL AND RARE MUTATIONS IN A CLINICAL POPULATION

L.H. Connors[1,2], T. Prokaeva[2], H. Akar[2], M. Metayer[2], P. Smith[2], A. Lim[2,3], R. Théberge[2,3], C.E. Costello[1,2,3], M. Skinner[2,4]

[1]Department of Biochemistry, [2]Amyloid Treatment and Research Program, [3]Mass Spectrometry Resource, and [4]Department of Medicine at Boston University School of Medicine, Boston, MA, USA

The hereditary forms of systemic amyloidosis (AF) represent a diverse group of autosomal dominant diseases that occur much less frequently than immunoglobulin light chain amyloidosis (AL).[1,2] Establishing a diagnosis of AF can be challenging when the phenotype is similar to that seen in some patients with AL and an accurate family history of disease is absent. This report details several new and unusual mutations associated with AF identified at our center since the IX International Symposium on Amyloidosis.

MATERIALS AND METHODS

Biopsy proof of amyloid disease was demonstrated with Congo red in all cases described. AF testing was pursued when a diagnosis of AL was inconclusive and included serum transthyretin (TTR) variant screening by isoelectric focusing (IEF), genetic mutation analyses (direct DNA sequencing, RFLP) of TTR (exons 2-4), apolipoprotein AI (exons 3,4), apolipoprotein AII (exons 3,4), fibrinogen Aα (exon 5), lysozyme (exons 2, 4) and gelsolin (exon 4), and mass spectrometric (MS) characterization (ESI, MALDI) of variant TTR proteins. [3]

RESULTS AND CONCLUSION

Total evaluations, new patients, and new cases of AF seen at the Boston University Amyloid Clinic from January 2000 to April 2004 are shown in Table 1. Genetic abnormalities associated with AF were identified in 64/659 (9.7%) of new patients. The specific types of AF in this group are profiled in Table 2. AF associated with TTR (ATTR) was diagnosed in 54/64 (84%) patients and the majority of this group (44/54) had one of the more frequently occurring pathologic TTR mutations including TTR-Ala60 (9), Met30 (8), His58 (6) and Ile122 (6). In the group of 10 patients with non-TTR type disease, AF was found in association with ApoAI (4), fibrinogen Aα (3), lysozyme (2) or ApoAII (1).

Table 1. BU Amyloid Clinic Population

	2000	2001	2002	2003	2004	TOTAL
Total Evaluations	389	415	435	408	124	1771
New Patients	161	159	174	120	45	659
New AF Cases	17	12	22	10	3	64

Table 2. AF Population (N=64) Profile

AF Type	N
TTR	54
Apolipoprotein AI (ApoAI)	4
Apolipoprotein AII (ApoAII)	1
Fibrinogen Aα	3
Lysozyme	2

Clinical and demographic information on 20 patients with rare and novel (bolded italics) forms of AF are presented in Table 3. The abbreviations for phenotype listings are as follows: H = heart; GI = gastrointestinal; N = neuropathy; E = eye; and K = kidney. Eight rare forms of TTR (Arg10, Asn18, Glu18, Cys33, Leu41, Pro49, Leu64, Thr81) and 1 novel mutation (Ala94) were identified in 10 patients. Cardiomyopathy was the main clinical feature in all but the TTR-Leu41 case. Non-TTR mutations were found in 10/64 (16%) of the patients with AF and included ApoAl-Arg26, ApoAl-Pro75, fibrinogen Aα-Val526, lysozyme-Arg64, and a new ApoAll-Leu78 mutation identified in a woman with renal involvement.

Table 3. Rare and Novel AF variants

AF Type	Variant	Sex	Age	Phenotype
TTR	Arg10	M	75	H
	Asn18	F, M	57, 43	H
	Glu18	M	47	H, GI, N
	Cys33	F	51	H, E, N, K
	Leu41	F	44	E
	Pro49	M	66	H, N
	Leu64	M	62	GI, N, H
	Thr81	M	68	H
TTR	*Ala94*	*M*	*50*	*H, N, K*
ApoAl	Arg26	M, F, F	42, 56, 40	K, GI, (N)
	Pro75	M	71	K, GI
Fibrinogen Aα	Val526	F, F, M	55, 59, 61	K
Lysozyme	Arg64	F, F	52, 46	GI, (K)
ApoAll	*Leu78*	*F*	*45*	*K*

Two novel variants/mutations (TTR-Ala94 and ApoAll-Leu78) were demonstrated using a combination of genetic and biochemical techniques (Figures 1 and 2). IEF screening of serum from a 50 year old Turkish man with cardiomyopathy, neuropathy and renal involvement demonstrated wild type and variant forms of TTR (Figure 1a). Direct DNA sequencing of TTR exons 2-4 revealed heterozygosity at the second base position in codon 94 (exon 4) and corresponded to TTR-Val94Ala (Figure 1b). This TTR mutation was confirmed by RFLP using the restriction enzyme Bsm I (Figure 1c). Post-digestion analysis showed the occurrence of 2 fragments (201bp and 72bp) as well as uncleaved product (271bp). No reaction occurred with DNA containing only the wild type sequence. Mass spectrometric examination of TTR immunoprecipitated from serum demonstrated near equal amounts of the wild type protein and a variant form (Figure 1d). The variant TTR had a mass shift of -28 compared to normal protein in both the unmodified and S-sulfonated structures. This data was consistent with a valine to alanine residue substitution in the variant protein. In the second case, initial IEF screening of serum from a 45 year old woman of Portuguese and Phillipino descent with renal disease was negative for a TTR variant (Figure 2a). Genetic mutation analysis of exon 4 of the ApoAll gene showed heterozygosity at the second base position in codon 78 (Figure 2b). The coincident occurrence of guanine and thymine at this location corresponded to the wild type protein with a stop codon and a variant ApoAll containing a carboxy-terminus extension of 21 amino acid residues. This mutation was confirmed using the restriction enzyme Dra III which digests wild type DNA, but not the mutated sequence (Figure 2c). DNA containing a different mutation in codon 78 was also run as a positive control.

In conclusion, patients with AF comprise ~10% of our total amyloid clinic population. The majority of patients with AF had the TTR-associated type of familial amyloidosis (84%). Cardiac dysfunction was often featured in the TTR-associated forms of AF, while renal involvement was common in the non-TTR types. Two new mutations associated with AF, TTR-Ala94 and ApoAll-Leu78, are highlighted.

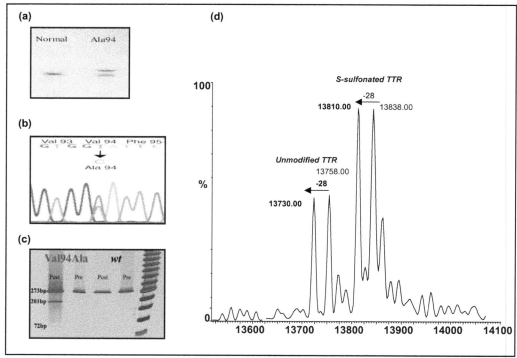

Figure 1. TTR-Ala94 analyses by a) IEF, b) direct DNA sequencing, c) RFLP and d) ESI MS.

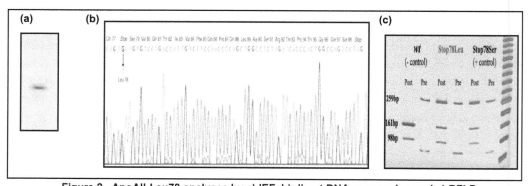

Figure 2. ApoAII-Leu78 analyses by a) IEF, b) direct DNA sequencing and c) RFLP.

This research was supported by the Young Family Amyloid Research Fund and the Gerry Foundation.

REFERENCES

1.   Falk, R.H. and Skinner, M., The systemic amyloidoses: an overview, *Advances in Int. Med.*, **45**, 107-137, 2000.

2.   Benson, M.D., Amyloidosis, in *The Metabolic and Molecular Bases of Inherited Disease*, Scriver, C.R., Beaudet, A.K., Sly, W.S. and Valle, D., Eds., McGraw-Hill, New York, 2001, pp. 5345-5378.

3.   Théberge, R., Connors, L.H., Skinner, M., Skare, J. and Costello, C.E., Characterization of transthyretin mutants from serum using immunoprecipitation, HPLC/electrospray ionization and matrix-assisted laser desorption/ionization mass spectrometry, *Anal. Chem.*, **71**, 452-9, 1999.

RENAL APOLIPOPROTEIN A-I AMYLOIDOSIS (AAPO-A-I) ASSOCIATED WITH A NOVEL MUTANT LEU64PRO

C. L. Murphy,[1] S. Wang,[1] K. Weaver,[1] M.A. Gertz,[2] D.T. Weiss,[1] and A. Solomon[1]

[1]*Human Immunology and Cancer Program, Department of Medicine, University of Tennessee Graduate School of Medicine, Knoxville, TN, USA and* [2]*Mayo Clinic, Rochester, MN, USA*

I. INTRODUCTION

The apolipoprotein (apo) A-I-associated amyloidoses are an autosomal dominant familial disorder characterized by the pathologic deposition of N-terminal fragments of mutated apo A-I molecules as fibrils within vital organs throughout the body.[1] In this hereditary disorder, 11 different sequence alterations have been detected in the amyloidogenic precursor protein that resulted from point mutations, deletions, and a deletion/insertion in the apo A-I gene.[2] We now report that the fibrils extracted from the kidney of a patient with renal amyloidosis contained yet another apo A-I mutation, the substitution of proline for leucine at position 64 (Leu64Pro). Further, we demonstrated that this alteration resulted from a one-base transition in codon 64 of the subject's apo A-I gene.

II. CASE REPORT

The patient is a 58 year-old male of Italian ethnic background with marked proteinuria and edema. A renal biopsy revealed extensive amyloid deposition within the kidney.

III. MATERIALS AND METHODS

A. DNA ANALYSES

Genomic DNA was prepared from peripheral blood leukocytes. Amplification of the first 214 bases of the 700bp comprising exon 4 of the apo A-I gene was achieved using primers and reaction conditions specified by Hamidi Asl et al.[3]

B. AMYLOID EXTRACTION AND CHEMICAL CHARACTERIZATION

Amyloid fibrils were extracted from 4 μm-thick formalin-fixed, paraffin-embedded kidney sections and the protein was purified by reverse-phase HPLC and digested with trypsin under conditions described by Murphy et al.[4] For tandem mass spectrometric (MS/MS) identification of peptides, samples were separated by reverse phase HPLC and the effluent directed into an ion-trap mass spectrometer. Isolation and analysis of plasma apo A-I was performed as previously described.[2]

C. IMMUNOHISTOCHEMISTRY

Tissue was immunostained using the ABC technique. The primary reagent was a murine anti-human apo A-I mAb and the secondary, a biotinylated goat anti-mouse antibody.

IV. RESULTS

A. IDENTIFICATION OF THE APO A-I MUTATED GENE

Sequencing of the first 215 bp of apo A-I exon 4 revealed a thymine (T) to cytosine (C) transition in the second base of codon 64; this T→C alteration would result in the substitution of Pro for Leu at position 64 in the product encoded by the mutant gene, *i.e.*, CTC (Leu) vs. CCC (Pro). The patient was heterozygous for this mutation, as evidenced in RFLP analysis by the presence of both variant and wild-type populations.

B. CHARACTERIZATION OF RENAL DEPOSITS

Passage of the amyloid extract through the reverse phase HPLC column yielded 4 peaks. Each was digested with trypsin and the resultant HPLC-isolated peptides were analyzed in the ion-trap mass spectrometer. Here it was determined that the proteins in one of the isolates encompassed positions 11-23, 28-40, 41-59, 46-59, 62-77, and 89-96 of the apo A-I molecule (Figure 1). Genetic studies had predicted that the Leu→Pro substitution occurred at position 64; this finding was confirmed in MS/MS analyses where it was found that the 62-77 tryptic peptide had the Pro residue, as evidenced from an anomalous M_r of 1916.50 vs. the expected M_r of 1932.80 for the wild-type 62-77 molecule (that has a Leu residue at position 64). Substantiation of the apo A-I nature of the renal amyloid deposits also was obtained immunohistochemically. Interestingly, chemical analyses of the plasma apo A-I revealed that only the wild-type Leu was present at position 64 (the mutant Pro residue was not detected).

```
                  10          20          30          40          50
Apo A-I   DEPPQSPWDR VKDLATVYVD VLKDSGRDYV SQFEGSALGK QLNLKLLDNW
Amyloid   [        ]---------- -[       ]--- ---------- ----------
                                    ▼
                  60          70          80          90         100
Apo A-I   DSVTSTFSKL REQLGPVTQE FWDNLEKETE GLRQEMSKDL EEVKAKVQPY
Amyloid   --------- --P------ ------         --- ------
```

Figure 1. Comparison of the amino acid sequence of the N-terminal portion of apo A-I with that of the apo A-I-related 96-residue amyloid component deposited in the patient's kidney. Note the Leu64Pro alteration in the amyloid at position 64. (---), sequence identity; [ ], peptides not identified.

V. DISCUSSION

Our studies using mass spectrometric technology have shown that the pathologic fibrillar deposits found in the kidney of a patient with renal amyloidosis were composed of an ~96 residue N-terminal fragment of apo A-I. Analyses of genomic DNA revealed a single base transition at nucleotide 1331 (T→C) in one Apo A-I allele that resulted in the replacement of leucine by proline at position 64 (Leu64Pro). The Leu64Pro substitution was confirmed through MS/MS analyses of a tryptic peptide encompassing residues located at positions 62-77 of the apo A-I molecule.

The structural basis of apo A-I amyloidogenicity and factors responsible for selective tissue deposition (kidney, liver, etc.) are presently unknown. Particular alterations in primary structure, *i.e.*, point mutations resulting in an increased positive charge, neutral substitutions, deletions and/or insertions, may render the molecule unstable and thus prone to form fibrils. Given the α-helical nature of apo A-I,[5] it is noteworthy that, among the now known

12 amyloidogenic variants, 7 have involved substitutions or deletions of the hydrophobic amino acid leucine and, in 4 cases (including the present), the introductions of the α-helix disruptive proline residue.

VI. CONCLUSION

Amyloid that is deposited mainly in the kidney can be a dominant feature in certain of the familial apo A-I-related amyloidoses.[1] This pathologic manifestation now has been associated with at least 5 mutations involving the amyloidogenic precursor protein and can occur in individuals of diverse ethnic backgrounds. Although the diagnosis of this type of amyloid may be made immunohistochemically using an anti-apo A-I antibody, in order to establish unequivocally the nature and, in most cases, the presence of the mutation in the amyloidogenic precursor protein, chemical analysis of fibrils extracted from the deposits is required. Further, the demonstration of an alteration in the apo A-I gene, through analysis of the genomic DNA or its products, serves to confirm the inheritable nature of this disorder.

VII. ACKNOWLEDGEMENTS

This study was supported, in part, by USPHS Research Grant CA10056 from the National Cancer Institute and the Aslan Foundation. A.S. is an American Cancer Society Clinical Research Professor.

VIII. REFERENCES

1. Benson, M.D., The metabolic and molecular bases of inherited disease. *Amyloidosis*, 8th ed., McGraw Hill, New York, 2001, 5345.

2. Coriu, D., et al., Hepatic amyloidosis resulting from deposition of the apolipoprotein A-I variant Leu75Pro, *Amyloid: Int. J. Clin. Exp. Invest.,* 10, 215, 2003.

3. Hamidi Asl, K., et al., A novel apolipoprotein A-1 variant, Arg173Pro, associated with cardiac and cutaneous amyloidosis, *Biochem. Biophys. Res. Commun.*, 257, 584, 1999.

4. Murphy, C.L., et al., Chemical typing of amyloid protein contained in formalin-fixed paraffin-embedded biopsy specimens, *Am. J. Clin. Pathol.*, 116, 135, 2001.

5. Borhani, D.W., et al., Crystal structure of truncated human apolipoprotein A-I suggests a lipid-bound conformation, *Proc. Natl. Acad. Sci. USA,* 94, 12291, 1997.

HEREDITARY AMYLOIDOSIS IN A SPANISH FAMILY ASSOCIATED WITH A NOVEL NON-STOP MUTATION IN THE GENE FOR APOLIPOPROTEIN AII

D. Rowczenio, J.A. Gilbertson, A. Bybee, D. Hernández[1] and Hawkins PN

Centre for Amyloidosis and Acute Phase Proteins, Royal Free and University College Medical School, London, United Kingdom
Haematology Department, La Paz Hospital, Madrid, Spain[1].

We report here a Spanish family with dominantly inherited amyloidosis presenting in middle age with slowly progressive renal involvement. Renal biopsy showed substantial glomerular, interstitial and vascular amyloid deposits, which did not stain with a panel of antibodies to SAA, kappa and lambda light chains, fibrinogen A ᾱchain, lysozyme or apoAI. No mutations were identified in the genes for these latter three proteins, but all affected members of the kindred had a TGA>AGA substitution in the gene for apolipoprotein AII (ApoA2; NM_001643.1 c.359T>A) encoding a novel non-stop change (p.X101Argext21) at the position previously annotated as codon 78. Substitution of this stop codon by an arginine residue is predicted to result in extension of translation to the next in-frame stop codon 60 nucleotides downstream. The novel residue is therefore followed by the same 20-residue *C*-terminal peptide extension that has previously been reported in 3 other families with hereditary apoAII amyloidosis, although these have been due to different nucleotide substitutions (p.X101Argext21(TGC>CGA), p.X101Serext21 and p.X101Glyext21). The frequency and penetrance of amyloidogenic apoAII mutations in the wider population is unknown, and sequencing of the apoAII gene should therefore be considered in patients with slowly progressive renal amyloidosis in whom definitive evidence of AL or other types cannot be obtained.

HEREDITARY RENAL AMYLOIDOSIS IN A GERMAN FAMILY ASSOCIATED WITH FIBRINOGEN A α CHAIN Glu540Val

A. Bybee[1], M. Hollenbeck[2], E.R. Debusmann[2], D. Gopau[1], J.A. Gilbertson[1], H.J. Lachmann[1], M.B. Pepys[1] and P.N. Hawkins[1]

[1]*Centre for Amyloidosis and Acute Phase Proteins, Royal Free and University College Medical School, London, United Kingdom*
[2]*Nephrologische Klinik, Knappschaftskrankenhaus, Osterfelder Str. 155a, D-46242 Bottrop, Germany*

A German woman presented with proteinuria and haematuria aged 49, and was found to have amyloidosis on renal biopsy. Her father died due to renal failure aged 47, having developed symptoms 2 years previously. Her biopsy showed extensive amyloid localized solely to the glomeruli, a characteristic distribution that suggested fibrinogen A (chain amyloidosis. DNA analysis confirmed that the patient was heterozygous for a novel missense mutation in the fibrinogen A (chain gene (FGA NM_000508.2; c.1706A>T) encoding the variant Glu540Val, and the renal amyloid deposits stained specifically with antibodies to fibrinogen. Her sister also had the variant and was found to have asymptomatic proteinuria and similar histological findings on renal biopsy. The mutation was present in two of the sisters' three children, none of whom have any evidence of renal dysfunction. None of the subjects with fibrinogen A (chain Glu540Val had evidence of neuropathy, cardiac amyloidosis or any sign of extra-renal amyloid involvement other than in the spleen, which was imaged on SAP scintigraphy in both the proband and her sister. This newly discovered mutation is in the same region of exon 5 that contains the five other known amyloidogenic fibrinogen A (chain mutations, and the clinical and histological picture in this family was indistinguishable from that associated with the prevalent fibrinogen A (chain variant, Glu526Val.

SYSTEMIC ALys AMYLOIDOSIS PRESENTING WITH GASTROINTESTINAL BLEEDING

C. Röcken[1], K. Becker[2], B. Stix[1], T. Rath[5], T. Kähne[3], M. Fändrich[4], A. Roessner[1], F.W. Albert[5]

Institutes of Pathology of the [1]Otto-von-Guericke-University, Magdeburg, Germany, and [2]Westpfalz-Klinikum GmbH, Kaiserslautern, Germany; [3]Institute of Experimental Internal Medicine, Otto-von-Guericke-University, Magdeburg, Germany; [4]Institute of Molecular Biotechnology, Jena, Germany; [5]Department of Internal Medicine III, Westpfalz-Klinikum GmbH, Kaiserslautern, Germany.

ABSTRACT

We report on a patient who, at 36-years of age, was found to suffer from systemic ALys amyloidosis primarily affecting the gastrointestinal tract and kidney. Biopsy specimens from the gastrointestinal tract and kidney were used for histological, immunohistochemical and biochemical studies. Genomic DNA was extracted from peripheral blood leukocytes and analyzed by polymerase chain reaction and sequencing. Extended amyloid deposits were found in the upper gastrointestinal tract and entire colon, leading to erosions and gastrointestinal bleeding. Amyloid was found in the kidney resulting in mild proteinuria. Immunohistochemical and biochemical studies identified lysozyme as the amyloid protein. DNA analyses of the lysozyme gene revealed a T to C transversion at the first position of codon 112, indicating a replacement of Trp by Arg at residue 112. This mutation was not found in the lysozyme genes of the patient's mother and daughter. To the best of our knowledge, this is the seventh report and the first German patient with the exceedingly rare ALys amyloidosis. ALys amyloidosis initially and commonly presents with gastrointestinal and renal symptoms.

CASE REPORT

A 36-year-old Caucasian man was admitted to hospital for a kidney biopsy. Six months prior to the current admission, the patient was admitted to another hospital for recurrent abdominal pain. Gastroduodenal endoscopy and colonoscopy showed gastric bleeding. Amyloid was found in biopsies taken from the upper gastrointestinal tract and colon (see below). In his past medical history, the patient suffered from knee pain for 20 years and was followed-up by an orthopedist. The patients mother was 61-years old and had no history of gastrointestinal or renal disease. His father had died at age 46 years supposedly from tuberculosis. The patient had a stepsister and a 7-week-old daughter.   Physical examination on admission showed no clinical signs of heart or kidney failure, and neurological deficits were not found. The patient was normotensive (120/80 mmHg), in sinus rhythm. ECG was normal. A mild normocytic anaemia (RBC 3.6 x 1012/l, Hb 10.8 g/dl) was found and levels of urea (15 mg/dl) and creatinine (1.3 mg/dl) were in the upper range of normal. The serum creatinine increased during hospital stay up to 1.7 mg/dl. Electrolytes and serum electrophoresis were normal. Protein-clearance was 190 mg/24 hours, in keeping with mild glomerular proteinuria, and creatinin clearance was impaired (46 ml/min). X-ray of the chest, lumbar spine and abdomen, ultrasound of the heart, and lung function tests were normal. An enlarged spleen (16.6 cm x 8.4 cm) was found in the ultrasound of the abdomen and confirmed by CT scan. Oesophago-gastro-duodenoscopy and colonoscopy showed small red spots in the rectal mucosa. A kidney biopsy was taken.

RESULTS

Biopsies taken from the duodenum and colon showed extended amyloid deposits in the lamina propria as shown by green birefringence after Congo red staining [6]. Atrophy of the glands was noted in the duodenal biopsies. Extravasates of erythrocytes were commonly found in the biopsies and was probably related to an increased tissue fragility. CD68$^+$ macrophages were found adjacent to or within the amlyoid deposits. Glomerular and vascular amyloid deposits were also found in a kidney biopsy consistent with a systemic ALys amyloidosis. The amyloid deposits were homogeneously immunoreactive with an antibody directed against lysozyme (polyclonal; DAKO; Denmark). Anti-fibrinogen and anti-transthyretin showed some immunostaining of the interstitum lining amyloid, whilst the center of the deposits was immunonegative. No immunostaining was observed with antibodies directed against AA amyloid, gelsolin, β2-microglobulin, λ- and κ- light chain (all DAKO).

Peripheral blood leucocytes were obtained from the patient, the patient's mother and daughter. Following extraction of genomic DNA, exons 1 to 4 of the lysozyme gene were analyzed by PCR and sequenced using the primers listed in Table 1. A T to A transversion at the first position of codon 112 was found in the patient, indicating a replacement of Trp by Arg at residue 112. This mutation was not found in the lysozyme genes of the patient's mother and daughter.

Table 1: The list of primers used for the amplification of exons 1 - 4 of the lysozyme gene.

Exon	forward (5`→ 3`)	reverse (5`→ 3`)
1	CAG TCA ACA TGAAGG CTC TCA T	GTT CCA TAC GTA GCT AAT TCT CT
2	AGT CAC TTA GTG TTG CTG TTT	ACC AGA TTG GTC AAA TAT TAG
3	ATT CCT TAC CAC CTG TCT TTC A	TGC TCT CAT TGT ATA TAT CAA CAG
4	TCG GCA TTC TAT GCT CTA CTGA	TGA CGG ACA TGT CTG TTT TGA C

DISCUSSION

To the best of our knowledge, we describe here the first German patient with systemic amyloidosis related to the deposition of lysozyme (ALys amyloidosis). Lysozyme is a ubiquitous bacteriolytic enzyme present in external secretions, polymorphs and macrophages, which in its native, "wild-type" form does not form amyloid, unless a gene mutation leads to the exchange of an amino acid in the mature protein. Until now, four different point mutations of the lysozyme gene have been linked to ALys amyloidosis, of which all were associated with the exchange of a single amino acid residue, i.e. Ile56Thr, Phe57Ile, Trp64Arg, and Asp 67His (Table 2)[3-5, 7, 8]. In searching for the underlying germline mutation we sequenced exons 1 through 4 of the lysozyme gene and found that our patient is heterozygous for a point mutation in codon 112 of exon 4 of the lysozyme gene denoting a replacement of Trp by Arg at residue 112. This mutation was not found in the lysozyme genes of the patient's mother and daughter. The mutation, in codon 112, has not been previously described and may therefore cause amyloidosis in our patient. Mutations of the lysozyme gene leading to an exchange of an amino acid can destabilize the tertiary structure of the protein, promoting fibrillogenesis [2].

The presence of CD68-immunoreactive macrophages within or adjacent to the amyloid deposits of our patient is an intriguing observation, as macrophages synthesize and secrete lysozyme, as well as many other proteases. However, whether the presence of macrophages is the cause or effect of amyloid deposition and whether they contribute to proteolysis of the amyloid protein(s) requires further investigations.

Table 2. Hereditary ALys amyloidoses (modified according to [1])

Mutation	Clinical features	Geographic Kindreds
Ile56Thr	Nephropathy, dermal petechiae, gastrointestinal bleeding	England
Phe57Ile	Nephropathy	Italian-Canadian
Trp64Arg	Nephropathy, ocular and oral sicca syndrome, Gastrointestinal bleeding	France
Trp64Arg	Gastrointestinal bleeding	Italy
Asp67His	Nephropathy, hepatic and gastrointestinal bleeding	England
Trp112Arg	Gastrointestinal bleeding, Nephropathy	Germany (present case)

REFERENCES

1. Benson, M.D. et al. A new human hereditary amyloidosis: The result of a stop-codon mutation in the apolipoprotein AII gene. *Genomics, 72*, 272-277, 2001.

2. Booth, D.R. et al. Instability, unfolding and aggregation of human lysozyme variants underlying amyloid fibrillogenesis. *Nature*, 385, 787-793, 1997.

3. Gillmore, J.D. et al. Hereditary renal amyloidosis associated with variant lysozyme in a large English family. *Nephrol Dial.Transplant*, 14 , 2639-2644, 1999.

4. Harrison, R.F. et al. 'Fragile' liver and massive hepatic haemorrhage due to hereditary amyloidosis. *Gut*, 38, 151-152, 1996.

5. Pepys, M.B. et al. Human lysozyme gene mutations cause hereditary systemic amyloidosis. *Nature*, 362, 553-557, 1993.

6. Puchtler, H., Sweat, F., and Levine, M. On the binding of Congo red by amyloid. *J.Histochem.Cytochem.*, 10, 355-364, 1962.

7. Valleix S. et al. Hereditary renal amyloidosis caused by a new variant lysozyme W64R in a French family. *Kidney Int.*, 61, 907-912, 2002.

8. Yazaki M., Farrell S.A., and Benson M.D. A novel lysozyme mutation Phe57Ile associated with hereditary renal amyloidosis. *Kidney Int.*, 63, 1652-165, 2003.

SECTION 5

CENTRAL NERVOUS SYSTEM AMYLOIDOSIS

PRESENILIN-INDEPENDENT AND JLK-SENSITIVE γ-SECRETASE PATHWAY: A HOPE FOR ALZHEIMER'S DIESASE THERAPY?

F. Checler

Institut de Pharmacologie Moléculaire et Cellulaire, UMR6097 du CNRS, 660 Route des Lucioles, Sophia-Antipolis, 06560 Valbonne, France

BACKGROUND

Alzheimer's disease (AD) is characterized by the abnormal cortical deposition of a set of peptides referred to as amyloid β peptides (Aβ). These peptides derive from the proteolytic processing of a large transmembrane precursor called βAPP by two enzymes, namely β- and γ-secretases (Fig.1) that release the N- and C-terminus of Aβ, respectively (for review see **1**). Although the exact etiology of AD remains questionable, several lines of biochemical, anatomical and genetic evidence suggest that the Aβ peptides at least contribute to the pathogenesis of AD. Along with this hypothesis, it can be postulated that blockade of secretases should hopefully lead to the reduction of Aβ production and therefore, slow down or eventually prevent AD neurodegenerative process. In this context, a series of works aimed at identifying the nature of secretases have led to the unequivocal identification of the β-secretase as BACE (β-site APP Cleaving Enzyme). This new aspartyl protease appears mainly responsible for cerebral β-secretase activity as its genetic depletion abolishes the production of Aβ in mice models of AD (for review see **2**). The nature of putative γ-secretase candidates is more discussed but it appears clearly that at least part of the γ-secretase-mediated Aβ production involves a multi-proteic complex in which the presenilins (PS) play a significant role (for review see **3**). PS1 and PS2 are two related proteins which, when mutated give rise to aggressive forms of AD and it has been demonstrated that it is associated to an exacerbated production of Aβ42, the more pathogenic species of the amyloid β peptides.   On the other hand, the knock out of PS1 and PS2 abolishes the production of Aβ in cells overexpressing the βAPP precursor (**4**). These data have led to the design of inhibitors of γ-secretase that interact physically with PS and prevent Aβ production. Altogether, these results have led to the proposal that presenilins would correspond to a new type of aspartyl protease bearing the γ-secretase activity (**5**).

IS TARGETING γ-SECRETASE HOPELESS?

While BACE depletion appears mostly innocuous for the mice, the knockout of PS is lethal in embryo suggesting that either Aβ exerts a vital embryonic function in mammals or, alternatively, that PS-dependent γ-secretase targets other proteins involved in important cell functions. Although the exact function of Aβ is still a matter of discussion, it has been clearly shown, by both genetic and pharmacological PS inactivation, that PS-dependent γ-secretase targets other substrates and that this leads to drastic side-effects, not only at the embryonic stage, but also at the adulthood. Among others, Notch is one of the proteins that undergo intramembranous PS-dependent γ-secretase-like cleavage. This cleavage releases an intracellular fragment called NICD (<u>N</u>otch <u>I</u>ntra<u>C</u>ellular <u>D</u>omain, (**6**) that acts as a transcription factor (Fig.1). Inhibition of NICD production triggers deleterious effects at both embryonic stage and at the adulthood. Therefore, the blockade of γ-secretase cleavage is likely to be associated to physiopathological effects apparently extremely difficult to overcome, particularly if PSs correspond to the only γ-secretase activity.

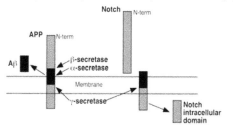

Figure.1: βAPP and Notch undergo similar γ-secretase-like cleavages

EVIDENCE OF A PRESENILIN-INDEPENDENT γ-SECRETASE PATHWAY

Recent data indicate the existence of a clear PS-independent Aβ production. Thus, we have demonstrated that PS-deficient fibroblasts produce endogenous Aβ40 and Aβ42 (Fig.2). These two products can be detected both intracellularly and in secretion medium (**7**). It should be emphasized here that previous works concerning the inhibition of Aβ production in PS-deficient cells were carried out with cells over-expressing mutated βAPP while our work concerns endogenous production of Aβ. In order to avoid any possible contamination by Aβ present in cell serum/medium or in our precipitating antisera, we measured Aβ from cells cultured with synthetic serum devoid of any animal protein (Prolifix S6) and immunoprecipitation of Aβ was carried out with the IgG-purified fraction of antisera. This procedure fully confirmed the generation of Aβ by cells devoid of presenilins (Fig.2). This work was later confirmed by another study showing that Aβ42 could be recovered from cells devoid of PS (**8**). Altogether, our work demonstrates that there exists both PS-dependent and PS-independent Aβ production, at least in fibroblasts.

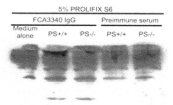

Figure.2: Endogenous Aβ40 production by PS+/+ and PS-/- fibroblasts: 2x150mm dishes of PS+/+ or PS-/- were allowed to secrete putative Aβ in 5% Prolifix S6 medium then media were precipitated with either preimmune serum or IgG-purified fraction of FCA3340 (**9**) and analyzed as described (**8**).

JLK INHIBITORS PREVENT Aβ PRODUCTION WITHOUT AFFECTING NOTCH PROCESSING

The blockade of PS-dependent Aβ production by several inhibitors that physically interact with PS has rapidly demonstrated its theoretical limits as potential therapeutic mean. Thus, these inhibitors trigger severe side-effects, *in vivo*, and are likely not appropriate as therapeutic means in AD treatment. The hope in a strategy aimed at blocking Aβ would stand if selective inhibitors preventing Aβ without affecting other γ-secretase-mediated cleavages could be designed. Indeed, we described a series of isocoumarin derivatives, referred to as JLK inhibitors (**9**), that inhibit Aβ40 and Aβ42 production by cells over-expressing wild-type and Swedish mutated βAPP (Fig.3A) in a dose-dependent manner (IC$_{50}$ about 30μM). As expected, these inhibitors concomitantly increased C99, the β-secretase-derived βAPP fragment targeted by γ-secretase (**9**).

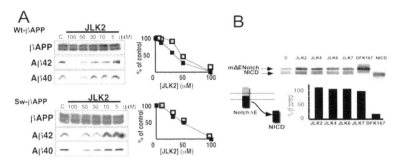

Figure.3: JLK inhibitors inhibit Aβ production but do not affect Notch cleavage: A) HEK293 cells overexpressing wild-type (Wt)- or Swedish mutated (Sw)-βAPP were incubated with the indicated concentrations of JLK2 then Aβ42 and Aβ40 were sequentially immunoprecipitated with FCA3542 and FCA3340, respectively as described (**9**). Controls correspond to cells incubated without inhibitor. B) Cells over-expressing myc-tagged mΔENotch were incubated with several JLK inhibitors and with DFK167, a PS-directed γ-secretase inhibitor. DFK167 but not JLK inhibitors prevent NICD production (detected as in **9**)

We established that JLK did not display toxicity in our experimental conditions and did not interfere with secretory process and did not affect the β -secretase BACE, the proteasome, calpains and caspases, demonstrating a rather encouraging specificity for the γ-secretase pathway (**10**). Conversely, NICD production by cells over-expressing mΔENotch, a constitutively processed Notch construct, was not affected by JLK inhibitors while PS-directed inhibitor DFK167 fully abolishes its formation (Fig.3B). JLK inhibitors did not affect Notch-related phenotype, *in vivo*(**11**). Altogether, our data demonstrate the theoretical possibility of designing selective blockers of the Aβ-producing pathway without triggering unexpected deleterious side-effects.

REFERENCES

1. Checler, F. (1995) Processing of the β-amyloid precursor protein and its regulation in Alzheimer's disease. *J. Neurochem.* 65, 1431-1444.
2. Citron, M. (2004) β-secretase inhibition for the treatment of Alzheimer's disease-promise and challenge. TIPS. 25, 92-97
3. Suh, Y-H. and Checler, F. (2002) Amyloid precursor protein, presenilins and α-synuclein: molecular pathogenesis and pharmacological applications in Alzheimer's disease. *Pharm. Rev.* 54, 469-525
4. Herreman, A., Serneels, L., Annaert, W;, Collen, D., Schoonjans, L. and De Strooper, B Total inactivation of γ-secretase activity in presenilin-deficient embryonic stem cells. *Nat.Cell.Biol.*.2, 461-462
5. Wolfe, M.S., Xia, W., Ostaszewski, B.L., Diehl, T.S., Kimberly, W.T. and Selkoe, D.J. (1999) Two transmembrane aspartates in presenilin-1 required for presenilin endoproteolysis and γ-secretase activity. *Nature.*398, 513-517
6. Kopan, R., Schroeter, E.H., Weintraub, H.and Nye, J.S. (1996) Signal transduction by activated mNotch: importance of proteolytic processing and its regulation by the extracellular domain. *Proc. Natl. Acad. Sci. USA.*
7. Armogida, M., Petit, A., Vincent, B., Scarzello, S., Alves da Costa, C. and Checler, F. (2001) Endogenous β-amyloid production in presenilin-deficient embryonic mice fibroblasts. *Nat. Cell. Biol.* 3, 1030-1033
8. Wilson, C.A., Doms, R.W., Zheng, H. and Lee, V.M-Y. (2002) Presenilins are not required for Aβ42 production in the early secretory pathway. *Nat. Neurosci.* 5, 849-855.
9. Petit, A., Bihel, F., Alves da Costa, C., Pourquié, O., Kraus, J.L. and Checler, F (2001) New protease inhibitors prevent γ-secretase-mediated Aβ40/42 production without affecting Notch cleavage. *Nat. Cell. Biol.* 3, 507-511
10. Petit, A., Bihel, F., Alves da Costa, C., Pourquié, O., Suh, Y-H., Kraus, J.L. and Checler, F. (2002) New non peptidic inhibitors of γ-secretase abolish Aβ production without modifying Notch cleavage. In: *Research and Perspectives in Alzheimer's disease*, Eds: A.Israël, B. De Strooper, F. Checler, Y. Christen, pp. 63-70
11. Petit, A., Pasini, A., Alves da Costa, C., Ayral, E., Hernandez, J.F., Dumanchin-Njock, C., Phiel, C., Marambaud, P., Wilk, S., Farzan, M., Fulcrand, P., Martinez, J., Andrau, D. and Checler, F. (2003) JLK isocoumarin inhibitors: selective γ-secretase inhibitors that do not interfere with notch pathway, *in vitro* and *in vivo*. *J. Neurosci. Res.* 74, 370-377

PRION DISEASES AND THE PRION PROTEIN

Pierre Aucouturier, Etienne Levavasseur, Clara Ballerini, Claude Carnaud.

Inserm E-0209, Université Pierre et Marie Curie, Hôpital St Antoine, F-75012 Paris, France

Transmissible spongiform encephalopathies (TSE) are fatal neurodegenerative disorders that are well defined at clinical, pathological and biochemical levels (1). A most original aspect is the implication of the host cellular 'prion' protein (PrPc), which adopts abnormal conformations (PrPsc) that seem to participate in the transmissibility of TSE. PrPsc aggregates and precipitates in the brain where it may form amyloid plaques.

TSE are transmitted by unconventional infectious agents termed prions, which include PrPsc as the main (or possibly only) component. TSE display strikingly long incubation periods when prions accumulate in peripheral lymphoid organs, without raising an immune response (2). We will review some aspects of prion diseases, including the peculiar nature of PrP, and the role of immune cells in propagating prion infection.

NATURAL AND EXPERIMENTAL TRANSMISSIBLE SPONGIFORM ENCEPHALOPATHIES

Natural prion diseases may have distinct aetiologies. Hereditary forms of Creutzfeldt-Jakob disease (CJD) and related diseases are described, with dominant autosomal transmission; in all cases there is a mutation of the gene that encodes PrP. Sporadic cases are the most frequent in humans; it is not clear whether they occur in animals also. They are defined by their unknown aetiology, likely a somatic event in the central nervous system (CNS), but a possible infectious origin was suspected in some of them (3). Finally, many TSE are transmitted. Most are weakly contagious and certain epidemics relate to rare or unusual human behaviours (kuru, iatrogenic CJD, bovine spongiform encephalopathy). In all cases –including familial and sporadic- TSEs can be transmitted by inoculation of infected nervous tissue. This transmission may cross certain species barriers. Thus, natural TSE isolates have been adapted to experimental animals, especially mice and hamster, which are helpful models for studying the pathogenesis. In spite of strain-related differences, TSE are a well defined pathological entity. Typical features include brain 'spongiform' changes, proliferation of astrocytes and microglial cells, and neuronal loss. Above all, a specific characteristic is the accumulation of proteinase K (PK)-resistant PrPsc in tissue lesions.

THE PRION PROTEIN

Recombinant PrP (supposedly similar to PrPc) includes a flexible, unstructured glycine- and alanine-rich N-terminal half, a central hydrophobic segment and a globular C-terminal domain with 3 α-helices and a small anti-parallel β-sheet. Two asparagines in the structured region are irregularly glycosylated; different glycoform patterns of PrPc exist in distinct cell types. Interestingly, strain-specific patterns of N-glycosylation and sizes of PK-resistant fragments relate to a variety of PrPsc conformations (4). Spectral studies and molecular modelling suggest that PrPsc structure is variable between strains and quite different from PrPc (5).

PrPc is a GPI-anchored membrane protein expressed in many tissues. Although its functions are not clearly understood, low inter-species divergence of the gene (6) and the fine tuning of PrPc expression during differentiation and activation of certain cell subsets suggest important physiological roles. This is particularly evident in the immune system, where CD4+ T lymphocytes and antigen presenting cells such as dendritic cells (DC) display strong PrP staining; this PrPc level correlates with DC maturation and T lymphocyte activation, suggesting a role in the 'immunological synapse' that occurs during antigen presentation. Indeed, T-cell proliferation in response to allogenic DC deprived of PrPc is reduced as compared to that elicited with PrP+ DC (figure 1).

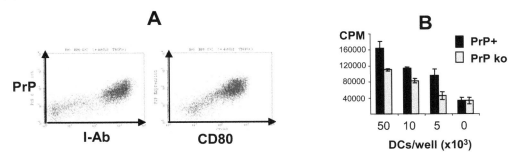

Figure 1. The prion protein (PrPc) is involved in T-cell activation by DCs. **A** : Correlated PrPc expression with MHC class II (I-Ab) and co-stimulatory (CD80) molecules, two markers of mature DC. **B** : Reduced T cell responses to allogenic stimulation by PrP° DC as compared with PrP+ DC (Ballerini et al, in preparation).

THE IMMUNE SYSTEM IN TSE PATHOGENESIS

The strikingly long incubation period of TSE is featured by prion accumulation in secondary lymphoid organs, where infectivity titers increase before they become detectable in the CNS (2). This stage proves determinant for disease progression, since mice that lack lymphoid follicles resist peripheral inoculation of scrapie agent. The most evident sites of prion storing are follicular dendritic cells (FDC) that differentiate in germinal centers and are maintained by sustained interaction with membrane lymphotoxins on B lymphocytes. Suppression of FDC networks early in the pathogenesis impairs scrapie neuro-invasion, suggesting that accumulation of the agent in lymphoid structures close to site of penetration is important (7). Other studies, with different prion strains, show that lymphoid accumulation of prions and disease may occur when FDC are absent (8;9). As a matter of fact, several aspects of early TSE pathogenesis remain obscure. First, eventhough FDC are important sites of accumulation, cells that actually replicate prions are not strictly identified. Second, it is unclear how prions propagate from epithelia to lymphoid structures and sites of neuro-invasion. After peripheral inoculation, PrPsc is soon detected in immobile structures such as FDC and peripheral nervous system (10), but also in migrating DC (11). DC display unique properties such as being recruited in peripheral tissues, phagocyte foreign agents and migrate to lymphoid tissues. A possible role of DCs in TSE pathogenesis is supported by their high infectivity titers in scrapie infected mice, and their ability to infect the brain in the absence of lymphoid follicles (12). Experiments from our laboratory indicate that peripheral inoculation of scrapie into 'plt' mutant mice, which manifest impaired homing of naive T-cells and DC to lymphoid tissues due to CCL19 and CCL21 chemokine deficiency, results in delayed disease (figure 2).

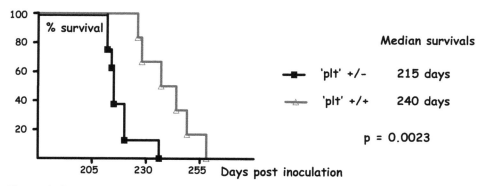

Figure 2. Survival times after subcutaneous inoculation of 'plt' mice and heterozygous controls with 139A scrapie agent (Levavasseur et al, in preparation).

These data suggest that, in addition to a likely role of peripheral nervous tissue in the progression of prions to the CNS, spreading of these agents by sessile lymphoid elements contributes to the pathogenesis. Identifying all the actors involved in the transport of prions and studying their actual implication in neuro-invasion, will help designing preventive strategies applicable at the pre-clinical stages of TSE.

Supported by European Community RTD action number QLK5-CT-2002-01044.

REFERENCES

1. Wisniewski,T., Aucouturier,P., Soto,C., and Frangione,B. 1998. The prionoses and other conformational disorders. *Amyloid: Int.J.Exp.Clin.Invest.* 5:212-224.
2. Aucouturier,P. and Carnaud,C. 2002. The immune system and prion diseases: a relationship of complicity and blindness. *J.Leukoc.Biol.* 72:1075-1083.
3. Glatzel,M., Abela,E., Maissen,M., and Aguzzi,A. 2003. Extraneural pathologic prion protein in sporadic Creutzfeldt-Jakob disease. *N.Engl.J. Med.* 349:1812-1820.
4. Aucouturier,P., Kascsak,R.J., Frangione,B., and Wisniewski,T. 1999. Biochemical and conformational variability of human prion strains in sporadic Creutzfeldt-Jakob disease. *Neurosci.Lett.* 274:33-36.
5. Mornon,J.P., Prat,K., Dupuis,F., Boisset,N., and Callebaut,I. 2002. Structural features of prions explored by sequence analysis. II. A PrP(Sc) model. *Cell Mol.Life Sci.* 59:2144-2154.
6. Van Rheede,T., Smolenaars,M.M., Madsen,O., and De Jong,W.W. 2003. Molecular evolution of the Mammalian prion protein. *Mol.Biol.Evol.* 20:111-121.
7. Mabbott,N.A., Young,J., McConnell,I., and Bruce,M.E. 2003. Follicular dendritic cell dedifferentiation by treatment with an inhibitor of the lymphotoxin pathway dramatically reduces scrapie susceptibility. *J.Virol.* 77:6845-6854.
8. Shlomchik,M.J., Radebold,K., Duclos,N., and Manuelidis,L. 2001. Neuroinvasion by a Creutzfeldt-Jakob disease agent in the absence of B cells and follicular dendritic cells. *Proc.Natl.Acad.Sci.Usa* 98:9289-9294.
9. Prinz,M., Montrasio,F., Klein,M.A., Schwarz,P., Priller,J., Odermatt,B., Pfeffer,K., and Aguzzi,A. 2002. Lymph nodal prion replication and neuroinvasion in mice devoid of follicular dendritic cells. *Proc.Natl.Acad.Sci.Usa* 99:919-924.
10. Andreoletti,O., Berthon,P., Marc,D., Sarradin,P., Grosclaude,J., vanKeulen,L., Schelcher,F., Elsen,J.M., and Lantier,F. 2000. Early accumulation of PrPSc in gut-associated lymphoid and nervous tissues of susceptible sheep from a Romanov flock with natural scrapie. *J.Gen.Virol.* 81:3115-3126.
11. Huang,F.P., Farquhar,C.F., Mabbott,N.A., Bruce,M.E., and MacPherson,G.G. 2002. Migrating intestinal dendritic cells transport PrPsc from the gut. *J.Gen.Virol.* 83:267-271.
12. Aucouturier,P., Geissmann,F., Damotte,D., Saborio,G.P., Meeker,H.C., Kascsak,R., Kascsak,R., Carp,R.I., and Wisniewski,T. 2001. Infected splenic dendritic cells are sufficient for prion transmission to the CNS in mouse scrapie. *J.Clin.Invest* 108:703-708.

EXPLORING THE FOLDING AND AGGREGATION MECHANISMS OF AMYLOID-FORMING PEPTIDES BY COMPUTER SIMULATIONS

Sébastien. Santini,[1] Guang-Hong Wei,[2] Normand Mousseau,[2] and Philippe Derreumaux[3]

1 Laboratoire Information Génomique et Structurale, UPR 2589 CNRS, Marseille, France
2 Département de physique and RQMP, Université de Montréal, Québec, Canada
3 Laboratoire de Biochimie Théorique, UPR 9080 CNRS, et Université Paris 7 Denis-Diderot, Paris, France

An important feature of Alzheimer's disease is that the toxic species of the amyloid-β protein (Aß) might be both soluble oligomers and insoluble fibrils. This finding makes Aβ oligomers attractive therapeutic targets. A second feature is that small peptides as short as 4, 7, 8 and 15 amino acids can also form amyloid fibrils, rendering the analysis of the early steps of Aβ aggregation more tractable by molecular modelling studies. In this work, we simulate the folding of a dimer and a trimer of the fragment Aβ16-22 using the activation-relaxation technique and a simplified energy model. Starting from randomly chosen configurations, our sampling procedure, which is not biased towards a specific intra- and inter-molecular arrangement, leads to an in register, antiparallel β-sheet organization in agreement with solid state NMR. Analysis of the trajectories allows to characterize the oligomeric intermediates in atomic detail. We find that occurrence of an intermediate containing 30% α-helix is not an obligatory step for folding Aβ16-22 in its dimeric and trimeric forms. This indicates that there are alternative folding pathways for fibril formation of the Aβ1-42 and Aβ1-40 peptides.

INTRODUCTION

A detailed experimental characterization of the Aβ oligomeric intermediates is thus far very difficult and only limited data are available because the intermediates are typically short-lived and are present in a wide range of conformations and degrees of aggregation. Because rapid equilibrium between monomers, dimers and larger units has been found for Aβ1-40 (1), we have attempted to understand the folding mechanisms of the dimers and trimers of the Aβ16-22 peptide using the Activation Relaxation technique (ART) (2) coupled with OPEP, which provides for a detailed protein and unbiased energy model (3). The Aβ16-22 fragment of sequence KLVFFAE was chosen because it forms fibrils at pH 7.4 with antiparallel beta-sheets (4), comprises the central hydrophobic core that is thought to be important in full length Aβ assembly and contains four occurring Alzheimer's disease-causing mutations, A21G, E22Q, E22K and E22G, which accelerate fibril formation.

MATERIALS AND METHODS

ART can be used to optimize any cost function in a high-dimensional space through a series of activated steps. Here we apply its most recent version which uses the Lanczos algorithm to extract a limited spectrum of eigenvectors and eigenvalues without requiring the evaluation and diagonalization of the full Hessian matrix (1,5). A basic event in ART consists of four steps. Starting from a minimum, the system is first pushed in a random direction outside the harmonic well until a negative eigenvalue appears in the Hessian matrix. The system is then pushed along the corresponding eigenvector until the total force approaches zero, indicating a saddle point. Subsequently, the configuration is then relaxed into a new local energy minimum, using standard minimization technique. Finallly, the newly-generated conformation is accepted/rejected using the Metropolis criterion at 300 K.

We use a flexible-geometry model where each amino acid is represented by six particles, i.e. N, H, Calpha, C, O and one bead for the side chains. All coordinates are free to vary. The OPEP potential form was optimized on the structures of six monomeric peptides with 10-38 residues adopting various secondary structures in solution (5,6). OPEP, which treats solvent effects implicitly, includes three types of interactions: pairwise 6-12 contact energies between the side chains considering the hydrophobic and hydrophilic character of each amino acid, potentials to maintain stereochemistry (excluded volume, bond lengths, bond angles) and two-body and four-body terms for the hydrogen bonding interactions.

RESULTS

In this work, we have performed 20 ART simulations on both the dimer (7) and trimer of the Aβ 16-22 peptide for 11,000 events using OPEP starting from various conformations and orientations of the chains: fully unfolded, and mixing of α-helices and parallel or antiparallel beta-strands. The results presented here are independent of the starting points. We emphasize that the ART-OPEP procedure is not biased towards the formation of any particular structures. For instance, simulations on the tetramer of the KPGE peptide does not form any regular structures (unpublished results).

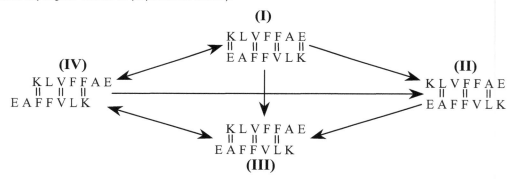

Figure 1. The four predicted β-sheet registries of the dimer of the Aβ16-22 peptide. Vertical double lines indicate hydrogen bonds. The patterns I and II are detected by NMR solid state of the fibrils (4). One-way and two-way arrows indicate transitions between the patterns observed during the ART-OPEP simulations.

Analysis of the lowest-energy structures generated by all ART simulations shows several interesting features. Firstly, although the antiparallel beta-sheet arrangement is the most stable structure for the dimer and the trimer , in agreement with the NMR solid state of the fibril Aβ16-22 at neutral pH (4), several hydrogen-bond patterns with similar free energies exist (see the results on the dimer in Figure 1). This indicates that full structural order in the fibrils requires larger aggregates. Secondly, a parallel beta-sheet structure with low free energy is possible for the dimer, and mixed antiparallel-parallel organizations are also possible for the trimer. This result is significant because it helps clarify the dependency of beta-sheet registries on pH conditions (8). Thirdly, the folding pathways do not require obligatory α-helical intermediates. This finding indicates that destabilization of α-helical inter-mediates is unlikely to abolish oligomerization of full-length Aβ peptides. Finally, our simulations also emphasize the crucial role of the reptation move of one strand of the beta-sheet with respect to the others during the assembly process. Again, this is consistent with recent isotope-edited infrared spectroscopy analysis on the prion peptide spanning residues 109-122 (9).

REFERENCES

1. G. Bitan, S. S. Vollers and D.B. Teplow, Elucidation of primary structure elements controlling early amyloid-beta protein oligomerisation, J. Biol. Chem. (2003) 278: 34882-34889.

2. G.T. Barkema and N. Mousseau, Event-based relaxation of continuous disordered systems, Phys. Rev. Lett. (1996) 77: 4358-4361.

3. P. Derreumaux, Generating ensemble averages for small proteins from extended conformations by Monte Carlo simulations, Phys. Rev. Lett. (2000) 85: 206-209.

4. J.J. Balbach, Y. Ishii, O.N. Antzutkin, R.D. Leapman, N.W. Rizzo, F. Dyda, J. Reed and R. Tycko, Amyloid fibril formation by Abeta 16-22, a seven-residue fragment of the Alzheimer's beta-amyloid peptide, and structural characterization by solid state NMR, Biochemistry (2000) 39: 13748-59.

5. N. Mousseau, P. Derreumaux, G. T. Barkema, and R. Malek, Sampling activated mechanisms in proteins with the activation relaxation technique, J. Mol. Graph. Model. (2001) 19: 78-86.

6. G. Wei, N. Mousseau, P. Derreumaux, Complex folding pathways in a β-hairpin, Proteins, in press (2004).

7. S. Santini, G-H. Wei, N. Mousseau and P. Derreumaux, Pathway complexity of Alzheimer's beta-amyloid Abeta(16-22) peptide assembly, Structure, in press (2004).

8. A.T. Petkova, G. Buntkowsky, F. Dyda, R.D. Leapman, V. Yau and R. Tycko, Solid State NMR Reveals a pH-dependent Antiparallel beta-Sheet Registry in Fibrils Formed by a beta-Amyloid Peptide., J. Mol. Biol. (2004) 335: 247-60.

9. R.A. Silva, W. Barber-Armstrong and S. Decatur, The organization and assembly of a β-sheet formed by a prion peptide in solution: an isotope-edited FTIR study, J. Am. Chem. Soc. (2003) 125: 13674-5.

CAMP-SPECIFIC Pde4B MEDIATES Aβ- INDUCED MICROGLIAL ACTIVATION

G. Sebastiani[1], C. Morissette[2], C. Lagacé[2], M. Boulé[2], M.J. Ouellette[2], R.W. McLaughlin[2], D. Lacombe[2], F. Gervais[1], P. Tremblay[2]

[1]Neurochem (International) Limited, 275 Armand-Frappier Blvd, Laval, Quebec, Canada H7V 4A7 and Neuhausstrasse 5, 6318 Walchwil, Canton Zug, Switzerland.
[2]Neurochem Inc., 275 Armand-Frappier Blvd, Laval, Quebec, Canada H7V 4A7

Alzheimer's disease (AD) is a neurodegenerative condition characterized by the presence of neurofibrillary tangles together with the accumulation and extracellular deposition of amyloid-β (Aβ) peptide in the form of senile plaques. In addition to these hallmarks, profound neuron loss and infiltration of inflammatory cells are also observed (1).

The precise role of inflammation in AD has been widely debated. Early on, microglia and astrocytes were shown not only to increase the production of Aβ peptide but also to induce an inflammatory autotoxic state. In addition, retrospective epidemiological studies seem to support the use of non-steroidal anti-inflammatory drugs as a therapeutic approach for the treatment of AD (2-4). More recently, new evidence has revealed that microglial cells may have a beneficial role, as the recruitment and activation of microglial cells appear essential for the clearance of both Aβ complexes and senile plaques (5,6). We have used cDNA array technology in order to characterize better the activation response of microglial cells to Aβ peptide at the gene expression level.

Aβ'S STATE OF AGGREGATION AFFECTS ITS CAPACITY TO ACTIVATE PRIMARY RAT MICROGLIA

Aβ peptide, produced from the processing of the amyloid precursor protein, exists in various oligomeric forms and ultimately assembles into fibers. In our studies, we used two forms of Aβ: soluble and fibrillar. Soluble $A\beta_{1-42}$ ($sA\beta_{1-42}$) is composed not only of monomers of $A\beta_{1-42}$ but also of some oligomeric forms (<200 kDa) that are intermediates in the assembly process. The fibrillar preparation ($fA\beta_{1-42}$) possesses large amounts of $A\beta_{1-42}$ fibrils readily detectable by Thioflavin T staining and electron microscopy. We showed that $sA\beta_{1-42}$ has a greater ability to activate primary microglia than its fibrillar counterpart. Exposure to 5 µM of $sA\beta_{1-42}$ induced the release of approximately 200 pg/ml of TNF-α as measured by ELISA, whereas with an equal amount of $fA\beta_{1-42}$ less than 100 pg/ml can be measured at 5 hours post-stimulation. These differences are more pronounced at higher concentrations (7.5, 10 and 15 µM). Since different $A\beta_{1-42}$ preparations elicited responses of varying intensities, we hypothesized that a comparative gene expression analysis of cells stimulated with soluble versus fibrillar preparations might elucidate the molecular mechanisms underlying microglial activation.

GENE EXPRESSION PROFILE OF Aβ-STIMULATED MICROGLIAL CELLS

After 4 hours of stimulation with either soluble or fibrillar $A\beta_{1-42}$, RNA was isolated from the microglial cells, radioactively labeled and hybridized to Atlas™ 1.2 cDNA arrays purchased from BD biosciences (Mississauga, ON). After quantification of the signal intensities, data were normalized and fold stimulation ratios were calculated. Of the 1176 genes on the array, 38 were found to be upregulated ($P < 0.01$): 20 by $sA\beta_{1-42}$, 7 by $fA\beta_{1-42}$ and 11 by both preparations. Subsets of these are depicted in Figure 1. Only 6 transcripts were found to be down-regulated: 4 by $sA\beta_{1-42}$ and 2 by $fA\beta_{1-42}$. No genes were found to be significantly down-regulated by both $A\beta_{1-42}$ preparations. These findings further support the notion that even though some common cellular pathways are activated in response to both soluble and fibrillar $A\beta_{1-42}$, other microglial responses may be preferentially modulated by one of the two forms. As cDNA array analysis is a semi-quantitative approach, the modulation of some of the identified transcripts was confirmed by real-time reverse-transcription PCR.

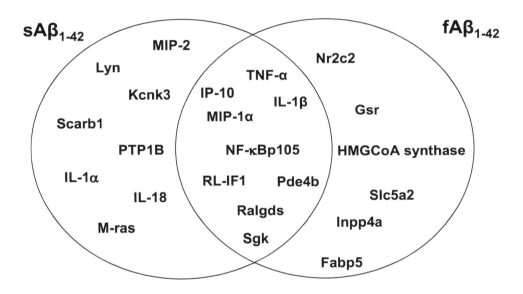

Figure 1. Genes up-regulated by either $sA\beta_{1-42}$, $fA\beta_{1-42}$ or both preparations

cAMP-SPECIFIC PHOSPHODIESTERASE 4B (Pde4B)

One particularly interesting gene identified in our analysis was Pde4B. Pde4B is part of a multigene family of evolutionarily highly-conserved phosphodiesterases. Its main role is to regulate the availability of intracellular cAMP, which itself has been implicated in the control of the activation of various signaling pathways including those of microglial cells (7). Using TNF-α as a robust marker of $A\beta_{1-42}$-induced activation in our primary rat microglial cultures, we examined whether inhibition of Pde4B would block activation. In fact, treatment of microglial cells for a period of 15 minutes with Rolipram, a specific Pde4 inhibitor, prior to incubation with $sA\beta_{1-42}$, results in a 2-fold decrease in the amount of TNF-α mRNA expressed (measured by real-time RT-PCR). This decrease in TNF-α mRNA correlates well with the amount of TNF-α secreted by the microglial cells. When stimulated with $sA\beta_{1-42}$ for 4 hours microglial cells produce 689 ± 22 pg/ml of TNF-α. As little as 0.1 μM Rolipram is sufficient to significantly reduce TNF-α to 572 ± 25 pg/ml ($P < 0.001$). A dose-dependence is seen up to 5 μM Rolipram where there is a maximal inhibtion of 70%. Even relatively higher doses of Rolipram (20-

50 μM) do not augment the inhibition achieved as depicted in Figure 2. This indicates that other pathways, in addition to the Pde4b/cAMP pathway, are involved in the activation of microglia by sAβ$_{1-42}$.

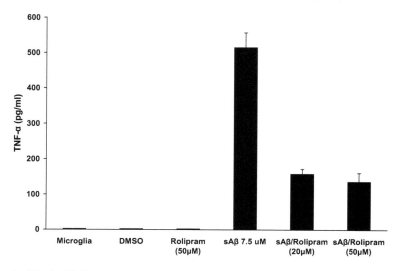

Figure 2. Effect of Rolipram, a specific Pde4 Inhibitor on sAβ$_{1-42}$-stimulated TNF-α production.

CONCLUSIONS

Our results indicate that different forms of Aβ$_{1-42}$ (soluble versus fibrillar) induce some common but also distinct mediators leading to microglial cell activation. In particular, we show that Pde4B is upregulated in Aβ-stimulated cells implicating cAMP as an important modulator of Aβ-induced responses. As such, Pde4B may be considered a potential therapeutic target in AD. A major task would be to develop an inhibitor specific to this cerebral isoform of Pde4B thus avoiding the serious side effects that have been described for some Pde4B inhibitors. Further characterization of the other genes identified in our array analysis may lead to novel candidates for drug discovery.

REFERENCES

1. Akiyama H, Barger S, Barnum S et al.(2000) Inflammation and Alzheimer's disease. *Neurobiol Aging* **21**: 383-421.
2. Bas A, Ruitenberg A, Hofman A et al.(2001) Nonsteroidal antiinflammatory drugs and the risk of Alzheimer's disease. *N Engl J Med* **345**: 1515-21.
3. McGeer PL, McGeer EG(1996) Anti-inflammatory drugs in the fight against Alzheimer's disease. *Ann NY Acad Sci* **777**: 213-20.
4. McGeer PL, Schulzer M, McGeer EG(1996) Arthritis and anti-inflammatory agents as possible protective factors for Alzheimer's disease: a review of 17 epidemiologic studies. *Neurology* **47**: 425-32.
5. DiCarlo G, Wilcock D, Henderson D, Gordon M, Morgan D(2001) Intrahippocampal LPS injections reduce Abeta load in APP+PS1 transgenic mice. *Neurobiol Aging* **22**: 1007-12.
6. Jantzen PT, Connor KE, DiCarlo G et al.(2002) Microglial activation and beta -amyloid deposit reduction caused by a nitric oxide-releasing nonsteroidal anti-inflammatory drug in amyloid precursor protein plus presenilin-1 transgenic mice. *J Neurosci* **22**: 2246-54.
7. Zuckerman SH, Gustin J, Evans GF(1998) Expression of CD54 (intercellular adhesion molecule-1) and the beta 1 integrin CD29 is modulated by a cyclic AMP dependent pathway in activated primary rat microglial cell cultures. *Inflammation* **22**: 95-106.

PERSPECTIVES OF IMMUNOTHERAPY AGAINST TSE

C. Carnaud[1], S. Gregoire[1], F. Chatelet[2], M. Bruley-Rosset[1], P. Aucouturier[1]

INSERM E209[1]- University Paris VI, and Service of Pathology[2], Hospital Saint-Antoine, Paris, FRANCE

Accumulating evidence suggests that neurodegenerative conditions such as Alzheimer disease (AD) or transmissible spongiform encephalopathies (TSE) might be tractable by immunointervention. Transgenic mice developing AD show improvement following passive or active immunization against the Aβ peptide (1,2). Similarly, antibodies against PrP, the prion protein, delay disease onset in mice experimentally inoculated with a scrapie agent (3,4). Several obstacles need however to be overcome before efficient and safe immunotherapies can be realistically proposed. The first of these difficulties concerns the ways to overcome immunological tolerance. TSE patients fail to develop immune responses against PrP^C or PrP^{Sc} as a consequence of tolerance (5). A second major concern is the risk of eliciting autoimmune complications following immunization against self-PrP. The inflammatory manifestations reported in AD patients actively vaccinated against the Aβ peptide constitute a severe warning (6). Last, but not least, it is essential to identify the immune effectors (antibodies, helper T cells, cytotoxic lymphocytes, macrophages etc..) which antagonize prion spread in vivo. This is important for developing efficient vaccines and for avoiding generating autoimmune effectors that would be potentially hazardous. We shall review in this chapter our most recent data in relation with these issues.

HOW TO ELICIT ANTI-PrP RESPONSES IN WILD TYPE MICE?

We have designed protocols combining vaccine DNA and PrP synthetic peptides, first in PrP-knockout (PrP°/°) mice. PrP°/° mice respond to vaccine DNA and to 3 peptides of 30 residues. Those 3 peptides elicit specific T cell proliferation and anti-PrP antibodies (Table 1). Some antibodies might be particularly relevant in a vaccine context as they recognize native cell-bound PrP^C (Table 1) and cure infected cell lines.

Table 1. Features of anti-PrP responses elicited by vaccine DNA or peptides

	CD4+ T cell response[1]	Antibodies against recombinant PrP[4]	Antibodies against native PrP[5]
vaccine DNA[2]	3.6 ± 0.2	+++	+++
peptide 98-127[3]	2.1 ± 0.3	+++	++
peptide 143-172[3]	4.5 ± 0.4	++	+
peptide 158-187[3]	5.3 ± 0.6	+++	-

[1] Expressed as the mean (± SE) of the proliferation indexes from 4 to 7 independent experiments.
[2] T cells primed in vivo with DNA and challenged in vitro with peptide 158-187 (mean of 7 experiments)
[3] T cells primed in vivo and recalled in vitro with the same peptide (mean of 4 to 7 experiments)
[4] Measured by ELISA against plastic-bound recombinant mouse PrP from baculovirus
[5] Revealed by indirect immunofluorescence on EL4 cells over-expressing cell-bound PrP^C

The same protocols were ineffective in PrP+ wild type mice, but tolerance could be overcome when unmethylated CpG nucleotides were associated to the immunogenic peptides. Wild type mice immunized under such conditions, develop specific T and B cell responses characterized by interferon-γ secreting T cells and by circulating antibodies against recombinant PrP. Interestingly, antibodies against native PrP are not detected. Thus, a fraction of the T and B repertoires against PrP^C escapes deletion in wild type mice and remains functional provided that stimulation is accompanied by appropriate adjuvants. Experiments are in progress in order to find out whether active immunization with peptides plus oligo-CpG can protect against TSE. Results from a first cohort indicate that immunized mice accumulate less PrP^{Sc} in the spleen and survive significantly longer than controls receiving oligo-CpG only.

HOW DO IMMUNE PrP°/° T CELLS SURVIVE IN A PRP+ ENVIRONMENT?

In order to gain insights into the protective immunological mechanisms and the risk of adverse autoimmunity, we have developed adoptive transfer models whereby wild type mice receive T or B lymphocytes from immune PrP-knockout mice. Such models are ideal for evaluating the protective potential of individual cell subsets and for observing possible autoimmune reactions caused by massive injections of primed cells.

We transferred first immune or naive T cells from knockout mice into PrP+ mice devoid of autotochtonous T cells. The recipients were periodically re-challenged with vaccine PrP-DNA or peptides in order to sustain anti-PrP memory T cells. We asked first whether primed T cells would be deleted or repressed in a PrP+ environment and whether continuous re-challenge would possibly accelerate this process. The results show that this is not the case. Primed T cells survive and even expand especially when supported by sustained antigenic challenge. Vigorous in vitro responses with high stimulation can be detected, as late as 3 months after transfer.

Surprisingly, sera from such recipients contained very few antibodies. This could mean either that the antibodies synthesized by autochthonous B cells were adsorbed by the surrounding PrP-expressing tissues, or alternatively, that the B cell repertoire was severely censored in wild-type mice. In order to test the latter hypothesis, we compared the antibody responses of Prp-knockout mice lethally irradiated and reconstituted with primed T cells and naive B cells of knockout or wild type origin. The recipients were then challenged with vaccine DNA in order to stimulate a response. The results reveal a clear difference in the quantity and the quality of the antibodies produced by the two B cell types. Those which differentiated in a PrP+ environment respond considerably less well and produce a narrower range of antibodies (in particular no antibodies against cell-bound PrP^C) than those originating from knockout mice. A strong censorship seems therefore to imposed upon B cell precursors with high affinity for PrP.

IMMUNE PrP°/° T CELLS INFILTRATE THE BRAIN OF PRP+ RECIPIENTS

T-less mice reconstituted with PrP-primed T cells and challenged every other week with peptides, were sacrificed 3 months after transfer for histological examination of brain and spinal chord together with controls reconstituted with naive T cells. The only samples showing lymphocytic infiltrates located close to the ventricles were from brains of mice transferred with immune T cells (Figure 1). Interestingly, the most florid lesions were generated by T cells which had been boosted by peptide 158-187, which elicits the most potent CD4+ T cell inducer (Table 1). It is still not clear whether such brain infiltrates have a clinical impact. The mice look normal, but more refined functional studies need to be performed. In any event, these observations plead for some caution.

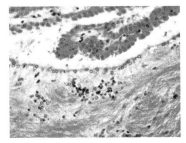

Transferred T cells	Challenge	Score
Naive	none	2/2 no infiltrate
DNA-primed	none	1/3 mild infiltrate
id	p 98-127	2/2 mild infiltrates
id	p 158-187	2/2 infiltrates

Figure 1. scattered lymphocytes close to a brain ventricule (X 400)

CONCLUDING REMARKS

Our results contain both good and bad news. The good news is that wild type mice can be actively immunized against PrP and be partially protected. Other strategies will be attempted in a near future, but it is already acquired that a functional repertoire persists in wild type mice. Another good news is that primed T cells adoptively transferred can survive and probably expand in a tolerogenic environment. The bad news is that the B cell repertoire is evidently modified by endogenous PrP^C, resulting in an antibody response which is both quantitatively and qualitatively altered. Whether the missing antibodies are important for fighting prions and achieving solid remission, will have to be properly evaluated. Another bad news is that anti-PrP lymphocytes infiltrate the brain of recipient PrP+ mice periodically challenged with antigen. The fact that infiltrates are more florid with T cells boosted with peptide 158-187 compared with peptide 98-127, clearly suggests that it will be important to identify the protective peptides (and the efficient immune effectors) and in order to find out whether they are different from those which are hazardous. It may turn out that the most effective ones will also be the most harmful in terms of autoimmunity. Clinicians would then be confronted to a dilemma similar to the one presently posed by AD.

REFERENCES

1. Schenk D, Barbour R, Dunn W, Gordon G, Grajeda H, Guido T, Hu K et al. 1999. Immunization with amyloid-beta attenuates Alzheimer-disease-like pathology in the PDAPP mouse. *Nature* 400:173-177.
2. Monsonego A, Weiner HL.Immunotherapeutic approaches to Alzheimer's disease. 2003. *Science* 302:834-838.
3. Aucouturier P, Carnaud C. 2002. The immune system and prion diseases: a relationship of complicity and blindness. *J Leukoc Biol.* 72:1075-1083.
4. Heppner FL, Arrighi I, Kalinke U, Aguzzi A. 2001. Immunity against prions? *Trends Mol Med.* 7:477-479.
5. Porter DD, Porter HG, Cox NA. 1973. Failure to demonstrate a humoral immune response to scrapie infection in mice. *J Immunol.* 111:1407-1410.
6. Nicoll JA, Wilkinson D, Holmes C, Steart P, Markham H, Weller RO. 2003. Neuropathology of human Alzheimer disease after immunization with amyloid-beta peptide: a case report. *Nat Med.* 9:448-452.

SERPINOPATHIES AND THE CENTRAL NERVOUS SYSTEM

David A Lomas, Didier Belorgey, Meera Mallya, Maki Onda, Kerri J. Kinghorn, Lynda K. Sharp, Russell L. Phillips, Richard Page, Elena Miranda and Damian C. Crowther

Department of Medicine, University of Cambridge, Cambridge Institute for Medical Research, Wellcome Trust/MRC Building, Hills Road, Cambridge. CB2 2XY, UK

The serpinopathies

Alpha-1-antitrypsin is the archetypal member of the serine proteinase inhibitor or serpin superfamily. This family includes members such as α_1-antichymotrypsin, C1 esterase inhibitor, antithrombin and plasminogen activator inhibitor-1 which play an important role in the control of proteinases involved in the inflammatory, complement, coagulation and fibrinolytic cascades. The family is characterised by more than 30% sequence homology with α_1-antitrypsin and conservation of tertiary structure. Consequently physiological and pathological processes that affect one member may be extrapolated to another. The structure of the serpins is based on three β-sheets (A-C) and nine β-helices. This structure supports an exposed mobile reactive loop that presents a peptide sequence as a pseudosubstrate for the target proteinase (Fig. 1, left). After docking the proteinase is inactivated by a mousetrap action that swings it from the top to the bottom of the protein in association with the insertion of an extra strand in β-sheet A.

The structure of the serpins is very much a dual edged sword in that it is central to their role as effective antiproteinases but also renders them liable to undergo conformational change in association with disease. This process is best characterised for the severe Z deficiency variant of α_1-antitrypsin that results in protein retention in hepatocytes in association with cirrhosis (see refs 1,2 for reviews). The Z mutation of α_1-antitrypsin is at residue P_{17} (17 residues proximal to the P_1 reactive centre) at the head of strand 5 of β-sheet A and the base of the mobile reactive loop. The mutation opens β-sheet A thereby favouring the insertion of the reactive loop of a second α_1-antitrypsin molecule to form a dimer (Fig. 1). This can then extend to form polymers that tangle in the endoplasmic reticulum of the liver to form inclusion bodies. Although many α_1-antitrypsin deficiency variants have been described, only two other mutants of α_1-antitrypsin have similarly been associated with plasma deficiency and hepatic inclusions: α_1-antitrypsin Siiyama (Ser53Phe) and α_1-antitrypsin Mmalton (52Phe deleted). Both of these mutants lie within the shutter domain that underlies β-sheet A and both allow the formation of loop-sheet polymers *in vivo*. Polymerisation also underlies the mild plasma deficiency of the S and I variants of α_1-antitrypsin. The point mutations that are responsible for these variants have less effect on β-sheet A than does the Z variant. Thus the rates of polymer formation are much slower than that of Z α_1-antitrypsin which results in less retention of protein within hepatocytes, milder plasma deficiency and the lack of a clinical phenotype. The phenomenon of loop-sheet polymerisation is not restricted to α_1-antitrypsin and has now been reported in other serpin variants to cause disease. Mutants of C1-inhibitor, antithrombin and α_1-antichymotrypsin can also destabilise the serpin architecture to form inactive polymers that are associated with angio-oedema, thrombosis and emphysema respectively.[1,2]   In view of the common underlying disease mechanism we have grouped these conditions together as the serpinopathies

Polymerisation of mutants of neuroserpin cause dementia

The process of disease related polymerisation is most strikingly displayed by the inclusion body dementia, familial encephalopathy with neuroserpin inclusion bodies (FENIB).[3] This is an autosomal dominant dementia characterised by eosinophilic neuronal inclusions of neuroserpin (Collins' bodies) in the deeper layers of the cerebral cortex and the substantia nigra.   The inclusions are PAS positive and diastase resistant and bear a striking resemblance to those of Z α_1-antitrypsin that form within the liver.   The observation that FENIB was associated with a mutation (Ser49Pro) in the neuroserpin gene that was homologous to one in α_1-antitrypsin that causes cirrhosis (Ser53Phe) strongly indicated a common molecular mechanism.   This was confirmed by the finding that the neuronal inclusion bodies of FENIB were formed by entangled polymers of neuroserpin with identical morphology to those present in hepatocytes from a child with α_1-antitrypsin deficiency related cirrhosis.[3]   Moreover we have recently shown that recombinant Ser49Pro neuroserpin has a greatly accelerated rate of polymerisation when compared to the wild type protein.[4]

   The direct relationship between the magnitude of the intracellular accumulation of neuroserpin and the severity of disease is clearly shown by the recent identification of other mutations of neuroserpin in families with FENIB.[5]   In the original family with Ser49Pro neuroserpin (neuroserpin Syracuse) the affected family members had diffuse small intraneuronal inclusions of neuroserpin with an onset of dementia between the ages of 45 and 60 years.   However, in a second family, with a conformationally more severe mutation (neuroserpin Portland; Ser52Arg) and larger inclusions, the onset of dementia was in early adulthood; and in a third family, with yet another mutation (His338Arg) there were more inclusions and the onset of dementia in adolescence.   The most striking example was the family with the most 'polymerogenic' mutation of neuroserpin, Gly392Glu.   This replacement of a consistently conserved residue in the shutter region resulted in large multiple inclusions in every neurone with affected family members dying by age 20 years.[5]   Thus FENIB shows a clear genotype-phenotype correlation, with the severity of disease correlating closely with the propensity of the mutated neuroserpin to form polymers.

Conclusion

The serpinopathies underlie a range of diseases as diverse as cirrhosis, thrombosis, angio-oedema and dementia.   We have characterised the molecular mechanism and have used the serpinopathies as a paradigm for a new category of conditions that we have called the 'conformational diseases'.   These are characterised by protein misfolding and aberrant β-strand linkage and include the amyloidoses, Huntington's disease, Parkinson's disease, Alzheimer's disease and the spongiform encephalopathies

References

1.   Lomas, D.A. and Mahadeva, R., Alpha-1-antitrypsin polymerisation and the serpinopathies: pathobiology and prospects for therapy, *J. Clin. Invest.,* 110, 1585, 2002.

2.   Carrell, R.W. and Lomas, D.A., Alpha$_1$-antitrypsin deficiency: a model for conformational diseases, *New. Engl. J. Med.,* 346, 45, 2002.

3.   Davis, R.L. et al., Familial dementia caused by polymerisation of mutant neuroserpin, *Nature,* 401, 376, 1999.

4.   Belorgey, D. et al., Mutant neuroserpin (Ser49Pro) that causes the familial dementia FENIB is a poor proteinase inhibitor and readily forms polymers *in vitro, J. Biol. Chem.,* 277, 17367, 2002.

5.   Davis, R.L. et al., Association between conformational mutations in neuroserpin and onset and severity of dementia, *Lancet,* 359, 2242, 2002.

6.   Gooptu, B. et al., Inactive conformation of the serpin a$_1$-antichymotrypsin indicates two stage insertion of the reactive loop; implications for inhibitory function and conformational disease, *Proc. Natl. Acad. Sci (USA),* 97, 67, 2000.

7.   Mahadeva, R. et al, Six-mer peptide selectively anneals to a pathogenic serpin conformation and blocks polymerisation: implications for the prevention of Z a$_1$-antitrypsin related cirrhosis, *J. Biol. Chem.,* 277, 6771, 2002.

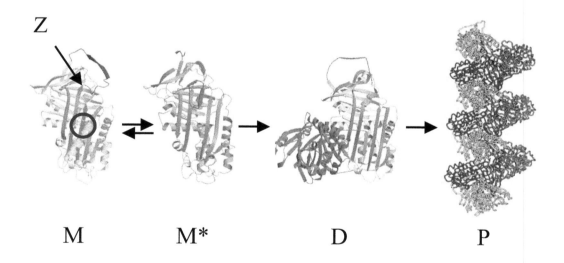

Figure 1.   The structure of α_1-antitrypsin is centred on α-sheet A (green) and the mobile reactive centre loop (red).   Polymer formation results from the Z variant of α_1-antitrypsin (Glu342Lys at P$_{17}$; arrowed) or mutations in the shutter domain (blue circle) that open β-sheet A to favour partial loop insertion and the formation of an unstable intermediate (M*) [6,7].   The patent β-sheet A can either accept the loop of another molecule to form a dimer (D) which then extends into polymers (P).   The individual molecules of α_1-antitrypsin within the polymer are coloured red, yellow and blue.

HEREDITARY CEREBRAL AMYLOID ANGIOPATHY ASSOCIATED WITH ATTR TYR114CYS

T. Yamashita[1], Y. Ando[2], M. Nakamura[2], K. Haraoka[3], S. Xuguo[2], H. Terazaki[3], M. Ueda[1], T. Ikeda[1], S.Saito[2], T. Kawaji[4], T. Hirai[5], Y. Washimi[6], and M. Uchino[1]

[1]Department of Neurology, [2]Department of Diagnostic Medicine, [3]Department of Gastroenterology and Hepatology, [4]Department of Ophthalmology, and [5]Department of Radiology, Graduate School of Medical Sciences, Kumamoto University, 1-1-1 Honjo, Kumamoto 860-0811, Japan. [6]Department of Neurology, Chubu National Hospital, 36-3, Gengo, Morioka, Ofu, Japan

INTRODUCTION

Cerebral amyloid angiopathy (CAA) is characterized by the amyloid deposition in the cerebral blood vessels of the brain and the leptomeninges. Typical clinical manifestations of the central nervous system (CNS) include dementia, cerebral hemorrhage, transient ischemic attacks, cerebral infarction, and convulsion. Although amyloid deposits in the meningo-cerebrovascular system were thought to be the cause of the manifestations of the CNS, precise mechanism remains to be elucidated.

FAP ATTR Tyr114Cys is classified into the ocluloleptomeningeal type of familial amyloidotic polyneuropathy (FAP). Our recent clinical studies revealed that most of FAP ATTR Tyr114Cys patients showed various manifestations of the CNS in addition to vitreous opacity and polyneuropathy during the course of the disease. We examined the manifestations of the CNS and amyloid formation mechanism in FAP ATTR Tyr114Cys patients.

PATIENTS AND METHODS

Eight patients (2 males, 6 females, age 46.0 ± 6.8 years old) living in Nagasaki Japan from a large kindred of FAP ATTR Tyr114Cys were examined neurologically and radiologically. Total protein and IgG levels in the cerebrospinal fluid (CSF) of the patients was also analyzed. We examined the autopsied spinal and cerebral samples of a 51-year-old FAP patient with Congo red staining and immunohistochemistry.

RESULTS

Neurological Findings in FAP ATTR Tyr114Cys Patients.

Cerebral hemorrhage without hypertension was noted in the 3 patients. Moreover, fluctuating consciousness, rapidly progressive dementia, and transient ischemic attacks were noted in 1, 2 and 2 patients, respectively.

Analysis of the CSF Protein and IgG Levels.

Total protein levels were within normal ranges in an asymptomatic carrier of ATTR Tyr114Cys gene. Total protein and IgG levels gradually elevated in correlation with duration of the illness. However, IgG/ total protein ratio did not elevated significantly, suggesting increased vascular permeability in blood vessels on the surface of the brain and intrusion of serum proteins into the CFS.

MRI Findings.

Serial MIR studies were performed with T1 weighted images and FLAIR images after Gd enhancement. In addition to leptomeningeal enhancement in the spinal cord, leakage of Gd into the subarachnoidal space just after enhancement was observed.

Histopathological Findings of the Spine and Brain.

In the autopsied spinal and brain samples, a large amount of amyloid deposits was confirmed by Congo red staining and in polarized light: a large amount of amyloid deposition in the leptomeninges and in the blood vessels of the leptomeninges was observed. Moreover, amyloid deposition was observed in the blood vessels of the brain parenchyma. The amyloid was positive against a polyclonal anti-TTR antibody. In addition to remarkable amyloid deposition in the blood vessels, intrusion of amyloid out of blood vessels into the subarachnoidal space and the brain parenchyma was observed.

Amyloid Deposition in the Parenchyma of the CNS.

It is generally accepted that TTR amyloid can not be detected in the parenchyma of the CNS except for leptomeninges and blood vessels' wall in most of TTR related FAP. However, in the parenchymal edge of the spinal cord, amyloid deposition was observed in the FAP ATTR Tyr114Cys patient.

DISCUSSION

The neurological findings in FAP ATTR Tyr114Cys patients accorded with those of cerebral amyloid angiopathy. Moreover, histopathological findings such as amyloid deposition in the blood vessels of the cerebral parenchyma also accorded with those of cerebral amyloid angiopathy. These results suggest that CAA caused by TTR amyloid deposition contributed to development of the manifestations of the CNS in FAP ATTR Tyr114Cys.

Serial MRI studies were performed with FLAIR images after Gd enhancement. Leakage of Gd into the subarachnoid space after Gd injection was observed, indicating the increased vascular permeability in blood vessels on the surface of the brain.

Pathological findings of cerebral blood vessels in the autopsied FAP ATTR Tyr114Cys patient revealed a large amount of amyloid deposition in the blood vessels, double-barrel, and intrusion of amyloid out of blood vessels into the subarachnoid space and the brain parenchyma indicating the disruption of the cerebral blood vessels' walls. These findings suggest that the blood cerebrospinal fluid barrier and the blood brain barrier might be disrupted by CAA in FAP ATTR Tyr114Cys.

Our data suggest that TTR amyloid deposited in the blood vessels of the CNS might be derived from the serum. Three patients with FAP ATTR Trr114Cys underwent liver transplantation. Effect of liver transplantation on clinical manifestations can not be evaluated because the duration of following up is not so long period. Future serial observations and accumulation of cases are necessary to give conclusions.

In conclusions, CAA is deeply connected with development of manifestations of the CNS in FAP ATTR Tyr114Cys patients. Disruption of the blood brain barrier and the blood cerebrospinal fluid barrier might occur in CAA in FAP ATTR Tyr114Cys.

REFERENCES

1. Greenberg, S.M. et al., The clinical spectrum of cerebral amyloid angiopathy: presentations without lobar hemorrhage, *Neurology*, 43,2073, 1993.
2. Yong, W.H. et al., Cerebral hemorrhage with biopsy-proved amyloid angiopathy, *Arch.Neurol.*, 49:51, 1992.
3. Sakashita, N. et al., Familial amyloidotic polyneuropathy (ATTR Val30Met) with widespread cerebral amyloid angiopathy and lethal cerebral hemorrhage, *Pathol.Inter.*, 51:476, 2001.

DISTRIBUTION OF AMYLOID PEPTIDE BETA 1-40 AND 1-42 IN CEREBRAL AMYLOID ANGIOPATHY AND CORRELATION WITH ALZHEIMER DISEASE PATHOLOGY

J. Attems (1), F. Lintner (1), and K. A. Jellinger (2)

(1) Institute of Pathology, Otto Wagner Hospital, Vienna, Austria; (2) Institute of Clinical Neurobiology, Vienna, Austria.

INTRODUCTION

The pathological hallmarks of sporadic late onset Alzheimer disease (AD), the most common cause of dementia in the elderly, include extracellular deposition of β-amyloid peptide (Aβ) in the neuropil (senile plaques, SPs) and in the cerebral vasculature (cerebral amyloid angiopathy, CAA), and neuritic cytoskeletal lesions with deposition of hyperphosphorylated microtubule-associated tau protein in neurons as neurofibrillary tangles (NFTs) and around amyloid deposits as neuritic plaques (NPs). The neuropathological diagnostic criteria for AD is the (semi) quantitative assessment and topographical staging of NPs and NFTs (1).

CAA is defined as deposition of a congophilic material in meningeal and cerebral arteries, arterioles, capillaries and veins, representing deposition of Aß in the vessel walls. It is a common finding in the brains of elderly demented and non-demented individuals; its incidence and severity increase with age, and are associated with cerebral hemorrhages and infarctions. The prevalence of CAA in AD patients, according to different studies, varies from 70% to 100% (2).

By immunohistochemistry, two forms of Aβ can be distinguished. Aβ terminating at amino acid position 40 (Aβ40) and at position 42 or 43 (Aβ42). In CAA Aβ40 affects vessel walls more frequently and more severely than Aβ42. In SPs and NPs, however, Aβ42 is predominant and Aβ40 is rarely detected.

The relationship between CAA and AD is poorly understood and the origin of Aβ in CAA remains unclear. Basically, three different mechanisms have been proposed (2):

- derivation of Aβ from blood and/or cerebrospinal fluid
- production of Aβ by smooth muscle cells within vessel walls and/or pericytes
- derivation of Aβ from the neuropil (i.e. SP, NP), in the course of its perivascular drainage

The aim of the present study was to investigate the distribution of Aβ40 and Aβ42 in leptomeningeal and intracortical arteries/arterioles, cortical capillaries and plaques, to detect possible differences in the correlation between CAA, CapCAA, and AD plaque pathology, with respect to different patterns of Aβ deposits.

MATERIAL AND METHODS

We investigated 100 human autopsy brains of both genders aged 60 to 100 years. Our cohort consisted of 50 brains each with high and low grade or negative AD pathology. Neuropathological assessment of AD was performed using CERAD, Braak scores and NIA-RI Criteria; details were described previously (3).

The severity of Aβ40/Aβ42 depositions in each leptomenigeal vessels, cortical vessels, capillaries, and plaques was assessed semiquantitatively on a four point scale, leading to an Aβ40/Aβ42 leptomenigeal score (Aβ40LS/Aβ42LS), Aβ40/Aβ42 cortical score (Aβ40CS/Aβ42CS), Aβ40/Aβ42 capillary score (Aβ40CapS/Aβ42CapS), and Aβ40/Aβ42 plaque score (Aβ40PS/Aβ42PS) respectively. Mean values of Aβ40LS+Aβ40CS, Aβ42LS+Aβ42CS, Aβ40LS+Aβ42LS, and Aβ40CS+Aβ42CS lead to Aβ40, Aß42, leptomeningeal, and cortical total scores respectively (Aβ40TS, Aβ42TS, AβLTS, AβCTS). Mean values of Aβ40TS+Aβ42TS and AβLTS+AβCTS were equal and each lead to CAA total score (CAATS; for details see 4).

RESULTS

Aβ40LS and Aβ42LS showed no significant difference (Aβ40LS mean: 0.91± 1.093, Aβ42LS mean: 0.71±0.924, P>0.01) and highly correlated with each other (ρ=0.828, P<0.01), Aβ40CS was significantly higher than Aβ42CS (Aβ40CS mean: 0.85±1.132, Aβ42CS mean: 0.85±0.966, P<0.01) and highly correlated with Aβ42CS (ρ=0.725, P<0.01). No significant difference and high correlations were seen between Aβ40LS and Aβ40CS/Aβ42LS and Aβ42CS (ρ=0.728/ρ=0.712, P<0.01).

Aβ40TS was significantly higher than Aβ42TS (Aβ40TS mean: 0.88±1.035, Aβ42TS mean: 0.65±0.871, P<0.01) and highly correlated with Aβ42TS (ρ=0.819, P<0.01). AβLTS and AβCTS showed no significant differences (AβLTS mean: 0.81±0.942, AβCTS mean: 0.715±0.975) and higly correlated with each other (ρ=0.755, P<0.01).

Aβ42CapS and Aβ42PS were significantly higher than Aβ40CapS and Aβ40PS respectively (Aβ40CapS mean: 0.15±359, Aβ42CapS mean: 2.02±1.279, Aβ40PS mean: 0.19±0.456, Aβ42PS mean: 2.12±1.241; P<0.01). Low correlations were seen between Aβ40CapS and Aβ42CapS/Aβ40PS and Aβ42PS (ρ=0.260/ρ=0.275, P<0.01), a high correlation between Aβ40CapS and Aβ40PS (ρ=0.855, P<0.01), and a very high correlation between Aβ42CapS and Aβ42PS (ρ=0.946, P<0.01).

Aβ40CapS, Aβ42CapS, Aβ40PS, and Aβ42PS showed no to low positive correlations with each Aβ40TS and Aβ42TS respectively (ρ<0.5, P<0.05), with the exception of a medium correlation between Aβ40PS and Aβ40TS (ρ=0.511, P<0.01).

We correlated all CAA/CapCAA scores with CERAD, Braak stages, and NIA-RI criteria; high positive correlations were seen between each Aβ42CapS and Aβ42PS and each each of the morphological AD Criteria (ρ>0.76, P<0.01), low positive correlations between each Aβ40LS, Aβ40CS, Aβ40TS, AβLTS, AβCTS, CAATS, Aβ40CapS, and Aβ40PS and all three AD criteria (0.2<ρ<0.5, P<0.05).

DISCUSSION

The present study clearly indicates, that the severity of Aβ42 deposits in/at capillaries (i.e. Aβ42CapS) significantly correlates with the severity of neuritic AD pathology using CERAD, Braak stages, and NIA-RI criteria, and to a very high degree with the severity of Aβ42 deposits in plaques (i.e. Aβ42PS). The severity of Aβ40 in both leptomeningeal and cortical vessels (i.e. Aβ40LS, Aβ40CS, Aβ40TS) increased significantly with increasing CERAD, Braak stages, and NIA-RI criteria; however, only low positive correlations between non-capillary CAA and AD pathology were observed. The very high correlation between Aβ42CapS and Aβ42PS in the present study, supports the concept of perivascular drainage of Aβ, with Aβ entering the perivascular pathways at the level of capillaries, leading to capillary Aβ deposition and consequently to CapCAA (3-5).

It is widely accepted, however, that CapCAA only occurs in the presence of CAA and thus may be a part of CAA (6). This assumption, however, is contrasted by our findings, since Aβ42CapS showed no correlation with Aβ42TS and only low correlation with Aβ40TS. The low correlation of Aβ42PS with each Aβ40TS and Aβ42LS, and the lack of a correlation with Aβ42CS, indicates that general CAA is not a result of perivascular draining of Aβ.

Our findings of low/no correlations between Aβ40TS/Aβ42TS and neuritic AD pathology argue against the possibility of one common pathomechanism for both AD pathology and non-capillary CAA.

The presence of CAA could, under yet unknown additional influences, have a promoting effect on AD pathology, which is reflected by the significant increase of Aβ40TS with increasing neuritic AD lesions in our cohort and the significant association between CAA and AD pathology in previous studies (6).

In conclusion our results suggest that:

- CAA is characterized by Aβ40/Aβ42 deposits in leptomeningeal and cortical arterial vessels, with Aβ40 being more frequent and more severe. Affection of capillaries is very rare and is considered to represent an indicator of high grade CAA. CAA (together with other influences) possibly promotes AD pathology.

- By contrast, CapCAA is characterized by globular Aβ42 deposits in cortical capillaries and pericapillary compartments, often in conjunction with parenchymal Aβ42 deposits. It is presumably a result of Aβ42 drainage from SPs and NPs along basemnt membranes (i.e. perivascular drainage) and, thus, closely related to both plaques and neurofibrillary AD pathology, but not to general (non-capillary) CAA.

Acknowledgements: The authors thank Mrs. Veronika Rappelsberger and Mrs. Barbara Weidinger for excellent laboratory work.

REFERENCES

1. Duyckaerts, C., Dickson, D.W., *Neurodegenaration. The Molecular Pathology of Dementia and Movement Disorders*, 1st.ed., ISN Neuropath Press, Basel, 2003, pp 47-65.
2. Revesz, T. et al., Cerebral amyloid angiopathies : a pathologic, biochemical, and genetic view, *J Neuropathol Exp Neurol*, 62, 885, 2003.
3. Attems, J., and Jellinger, K.A., Only cerebral capillary amyloid angiopathy correlates with Alzheimer pathology – a pilot study, *Acta Neuropathol (Berl)*, 107, 83, 2004.
4. Attems, J., Lintner, F., and Jellinger, K.A., Amyloid β peptide 1-42 highly correlates with capillary cerebral amyloid angiopathy and Alzheimer disease pathology, *Acta Neuropathol (Berl)*, 107, 283, 2004.
5. Preston, S.D. et al., Capillary and arterial cerebral amyloid angiopathy in Alzheimer´s disease: defining the perivascular route for the elimination of amyloid beta from the human brain, *Neuropathol Appl Neurobiol*, 29, 106, 2003.
6. Thal, D.R. et al., Two types of sporadic cerebral amyloid angiopathy, *J Neuropathol Exp Neurol*, 62, 282, 2002.

HEREDITARY CEREBRAL AMYLOID ANGIOPATHY ASSOCIATED WITH A NOVEL AMYLOID BETA PRECURSOR PROTEIN MUTATION

L. Obici[1], G. De Rosa[3], G. Palladini[1], S. Marciano[1], S. Donadei[1], E. Arbustini[2], L. Verga[2], M. Concardi[2], G. Ferrari[3], G. Merlini[1].

Biotechnology Research Laboratory[1] and Institute of Human Pathology[2], IRCCS Policlinico San Matteo, Pavia; Neurology Unit, Ospedale di Ivrea[3], Ivrea, Italy

The cerebrovascular deposition of amyloid fibrils, particularly in the walls of leptomeningeal and cortical arteries and arterioles, is a well-recognized cause of primary intracerebral hemorrahagic stroke, ischemic lesions and dementia (1). The Aβ peptide is the amyloid subunit in most forms of cerebral amyloid angiopathy (CAA), including age-related CAA and CAA associated with sporadic and familial Alzheimer's disease (1).

Missense mutations within the Aβ sequence of the amyloid-β precursor protein gene (AβPP) cause autosomal dominant, presenile cerebral amyloid angiopathy in families with different ethnic background. Five Aβ mutants have been associated to date with extensive vascular amyloid deposition and fatal intracerebral hemorrhages. These mutations are clustered at residues 21-23 of the Aβ peptide. Three different amino acid substitutions are known to affect glutamic acid at position 22, namely E22Q underlying prototypic hereditary CAA-Dutch type, E22K reported in Italian families and E22G occurring in a Swedish kindred. The clinical phenotype associated with these variants also includes a variable degree of cognitive decline and dementia on a vascular basis (2). Moreover, histopathologic features of Alzheimer's disease, including senile plaques and neurofibrillary tangles, are associated to severe amyloid angiopathy in carriers of either the Flemish A21G variant or the Iowa D23N mutant (3).

Here we report a novel mutation in the AβPP gene causing hereditary CAA and fatal intracerebral hemorrhages in a large Italian kindred.

KINDRED

We studied an Italian family, originating from Puglia, in the Southern part of the country, in which nine members over two generations presented with recurrent haemorragic stroke in the sixth to eighth decade of life. The proband presented at the age of 70 years with confusional state without headache or motor symptoms. A CT scan demonstrated the presence of a large left temporal hemorrhage involving the deep white matter. A younger sister presented aged 50 years with motor impairment and paresthesias on the right arm. She had no history of hypertension. A CT scan showed a single fronto-parietal hemorrhage in the left hemisphere, involving the cortical and subcortical white matter. Seven years later she complained of weakness and paresthesias on her left arm. An MRI scan showed a large poroencephalic cavity in the left fronto-parietal lobe and a recent right frontal hemorrhage. A cerebral vessels angiography was normal and no arteriovenus malformations were found. A third hemorrhagic stroke developed three years later. She died aged 58 from a fourth intracerebral haemorrhage (ICH) and underwent autopsy. One brother also presented with severe headache and confusion aged 75. A CT scan showed an acute left frontal hemorrhage and three poroencephalic cavities affecting the right frontal and both occipital lobes. The disease rapidly progressed thereafter. The patient developed a progressive cognitive decline and three ICH over one year. A 63-year-old cousin presented with a large temporal ICH, first appearing as a subarachnoid hemorrhage on a CT scan. Old haemorrhagic foci were also shown on gradient-echo MRI. He died two months after the first stroke from another ICH.

METHODS

Neuropathological examination was performed on paraffin-embedded brain tissue obtained at autopsy from two family members. Sections from frontal, temporal and parietal lobes were stained with hematoxylin and Congo red according to standard procedures.

For immunoelectron microscopy characterization formalin-fixed, paraffin embedded sections were rehydrated through a series of graded ethanols and fixed in osmium tetroxide 1%. After dehydration in ethanol solutions of increasing concentration, the specimens were embedded in Epon-Araldite resin. Immunohistochemical study was carried out on ultrathin sections mounted on parlodion-coated nickel grids. with either anti-Aβ40 or anti-Aβ1-42 monoclonal antibody (ABETA GmbH, Heidelberg, Germany).

Mutation analysis of exon 17 of the *AβPP* gene was performed by PCR amplification using the following, flanking intronic primers: 5'-ACC TCA TCC AAA TGT CCC CTG C-3' and 5'-TCT CAT AGT CTT AAT TCC CAC-3'. Amplification conditions for AmpliTaq Gold (Applied Biosystems, Foster City, CA, USA) were 95°C for 10 minutes for maximal enzyme activation, followed by 30 cycles at 95°C for 1 minute, 60°C for 1 minute, and 72°C for 1.5 minutes. PCR products were cleaned by gel extraction and both strands were sequenced with the Big Dye terminator cycle sequencing kit (Applied Biosystems, Foster City, CA, USA) and the products were analysed on an ABI PRISM 310 DNA sequencer (Applied Biosystems, Foster City, CA, USA).

To perform restriction analysis in patients, in one healthy relative and in 100 controls, a Tsp45I endonuclease recognition site was artificially introduced in the mutated allele by PCR amplification, using a different sense primer (5'-CTA ATT GCG TTT ATA AAT TG-3') coupled with the following anti-sense, mismatched primer (5'-ATG ACA ACA CCG CCC ACC gTG-3' where the lower case indicated the mismatched nucleotide). Amplification conditions for HotStart Taq (Qiagen, Chatsworth, CA, USA) were 95°C for 15 minutes followed by 30 cycles at 94°C for 1 minute, 55°C for 1 minute and 72 °C 1 minute. Products were digested at 65°C according to manufacturer's instructions, electrophoresed on 3,2 % agarose gel, stained with ethidium bromide and visualised with UV light.

RESULTS

Postmortem examination demonstrated in both patients the presence of vascular amyloid deposits in leptomeningeal and cortical vessels of cerebral lobes. Features of severe amyloid angiopathy were observed, such as the "vessel-within-vessel" configuration (Figure 1) and evidence of hemorrages associated with affected vessels. Sections stained with Congo red showed typical apple-green birefringence in the walls of small and medium-sized arteries after visualization under polarized light.

Immunoelectron microscopy characterization of amyloid fibrils in the walls of affected vessel showed specific immunostaining with both anti-Aβ40 and anti-Aβ42 antibodies.

DNA sequencing of exon 17 of *AβPP* showed that affected family members were carriers of a C to G transversion in the first nucleotide of codon 705, resulting in the novel valine for leucine substitution. All patients were heterozygous for this mutation, as confirmed by RFLP analysis with the enzyme Tsp45I, for which a specific cleavage site was artificially introduced with a PCR-based method in the mutated allele only. The mutation was absent in one healthy relative. Screening of 200 control chromosomes did not reveal any carrier.

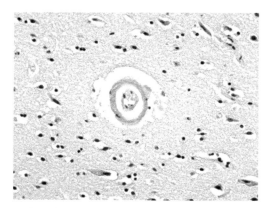

Figure 1. Vessel-within-vessel configuration in a section from the frontal cortex (X20)

DISCUSSION

We identified in an Italian family a novel *AβPP* mutation causing severe cerebral amyloid angiopathy manifesting with recurrent, symptomatic hemorrhagic stroke and death by age 77. The mutation site is located within the part of the AβPP gene which corresponds to the Aβ sequence, resulting in a Val for Leu substitution at residue 34 of the Aβ peptide.

  Although further neuropathological characterization is still in progress, postmortem examination of brain tissue from two members has shown evidence of severe CAA without the presence of parenchimal amyloid plaques. This new variant expands the spectrum of Aβ mutants associated with CAA, strengthening previous observation that mutations within the coding region of Aβ are predominantly associated with hemorragic stroke instead of typical AD clinical picture.

REFERENCES

1.   Revesz T, Holton J, Lashley T, Plant G, Rostagno A, Ghiso J and Frangione B. Sporadic and familial cerebral amyloid angiopathies. Brain Pathol 2002; 12:343-357.
2.   Rensink A, de Waal RMW, Kremer B and Verbeek MM. Pathogenesis of cerebral amyloid angiopathy. Brain Research Reviews 2003; 43:207-223.
3.   Grabowski TJ, Cho HS, Vonsattel JPG, Rebeck GW and Greenberg SM. Novel amyloid precursor protein mutation in an Iowa family with dementia and severe cerebral amyloid angiopathy. Ann Neurol 2001; 49:697-705.

COGNITIVE DISTURBANCE AND BRAIN PATHOLOGY IN OLD DOGS

J. Rofina, M.J.M. Toussaint, A.M. van Ederen, M. Secrève and E. Gruys

Department of Pathobiology, Faculty of Veterinary Medicine, Utrecht University, Utrecht, The Netherlands

INTRODUCTION

Cognitive disturbance occurs frequently in geriatric dogs[1-3]. Some authors have classified this behavior with questionnaires and described cognitive dysfunctions to be related with Alzheimer pathology[4, 5].

The aims of this investigation were:

- Comparing three questionnaires (A, B, C) with each other.
- To correlate various items of changed behavior with brain pathology.
- To compare pathology scores in demented and non-demented dogs without possible influence of age.

MATERIAL AND METHODS

The behavior of 30 pet dogs of various ages was evaluated and scored with three questionnaires (A, B and C) for the owners. In the age class of 13 years and older two comparable groups of nine dogs, with and without behavioral changes, were formed and compared for pathological brain lesions.

Questionnaires A[4] and B[5] were as described; questionnaire C was a combination of a former one[6] based on the human Mini-Mental State test and questions used in A and B. Their scoring results were compared with each other. The results of questionnaire C were used to calculate Spearman rank correlation coefficients (Statistix® software) with pathological lesions in the brains of all 30 dogs.

The lesions studied were:

- Amyloid depositions, semi-quantitatively scored after Congo red staining (congophilic material was detected by fluorescence microscopy[7] and identified as amyloid when it exhibited apple-green birefringence in polarized light) and anti Aβ42 immunohistochemistry.
- Oxidative damage products, semi-quantitatively scored after immunostaining for 4-hydroxynonenal (HNE) and 8-hydroxy-2'-deoxyguanosine (OHDG), Ziehl-Neelsen staining and autofluorescence of lipofuscin in unstained sections studied with multi-photon excitation microscopy.
- Relative cortical atrophy, diameter of the cerebral cortex (mm) in relationship to the total diameter of the brain cross section.

RESULTS

Despite slight differences in outcome (Figure 1a) the three questionnaires were highly correlated (Spearman Rank correlation coefficient 0.9; P < 0.001). Questionnaire C scored better, especially when comparing only old dogs (Figure 1b).

When comparing amyloid deposits (the sum of results of Congo red and Anti-Aβ42 stainings) and oxidative damage (the sum of score results of anti-HNE and anti-OHDG, Ziehl-Neelsen and auto-fluorescence) with behavior changes in the questionnaires A, B and C, oxidative damage had a higher correlation compared to amyloid deposits. This difference was even greater within the group of old dogs (Figure 1b).

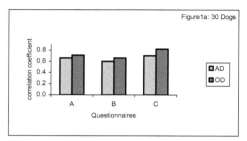 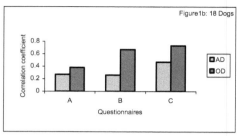

Figure 1a/b. Correlation coefficients for the dementia scores of the questionnaires and the sum of results of pathological lesions concerning amyloid and oxidative damage of a group of 30 dogs (1a) and 18 dogs (1b). AD: Amyloid deposits. OD: Oxidative damage. A: scoring according to Kiatipattanasakul et al[4], B according to Colle et al[5] and C: after modified Rofina et al[6].

Each item used in questionnaire C has a different correlation with the brain lesions (Table 1). However, all questions together in a dementia score, gave acceptable correlation coefficient values with the amyloid depositions, oxidative damage signs and cortical atrophy. The dementia score and sum of lesions correlated significantly better. Age showed high correlation coefficients except for cortical atrophy (Table 1).

In order to minimize the influence of age two groups of old dogs, with and without behavior changes, were compared (Table 2) which revealed that Congo red fluorescence, cortical atrophy and most of the oxidative parameters were significantly correlated with the behavioral changes.

Table 1. Correlation coefficients of the score of each separate item used in questionnaire C and their sum (dementia score) with the results of each pathological lesion and the sum of the results of these lesions of a group of 30 dogs.

	Congo red	Aβ 42	HNE	OHDG	Fluor	Ziehl-N	Cortical atrophy	Sum of lesions
Items								
Appetite	0.5	0.5	0.5	0.6	0.5	0.4	-0.6	0.6
Incontinent	0.6	0.5	0.6	0.7	0.5	0.5	-0.6	0.6
Day/night rhythm	0.6	0.5	0.5	0.6	0.6	0.5	-0.6	0.6
Aimless behavior	0.7	0.6	0.5	0.6	0.6	0.5	-0.6	0.6
Expression	0.7	0.6	0.6	0.6	0.7	0.7	-0.5	0.6
Interaction	0.6	0.6	0.6	0.7	0.6	0.6	-0.6	0.7
Memory	0.7	0.5	0.5	0.6	0.5	0.6	-0.5	0.6
Disorientation	0.7	0.6	0.5	0.5	0.6	0.5	-0.6	0.6
Dementia score	0.7	0.6	0.7	0.7	0.8	0.7	-0.6	0.8
Age	0.8	0.8	0.8	0.7	0.9	0.9	-0.4	0.9

0.4: P < 0.05; 0.5: P < 0.01; ≥ 0.6: P < 0.001; Fluor.: Auto-fluorescence of lipofuscin; Ziehl-N: Ziehl-Neelsen staining; cortical atrophy: width of cerebral cortex divided by total width of coronal brain section (brain cut at the level of the *chiasma opticum*)

Table 2. Correlation coefficients of the score of each pathological lesion and the sum of the results of these lesions with the sum of the item scores of questionnaire C (dementia score) for 18 dogs of 13 years and older.

Pathology	Dementia score
Congo red	0.6^b
Aβ 42	0.3
HNE	0.5^a
OHDG	0.7^c
Autofluorescence.	0.6^b
Ziehl-Neelsen	0.6^b
Cortical atrophy	0.6^b
Sum of lesions	0.6^b

a: $P < 0.05$; b: $P < 0.01$; c: $P < 0.001$

CONCLUSION

The results showed that a questionnaire can be used to diagnose Alzheimer changes in canine practice. Questionnaire C used in this study correlated best with the pathological lesions. Therefore questionnaire C is recommended for further studies.

Age is a major factor in behavior changes but when only old dogs are considered, amyloid deposits and oxidative damage both have an important role.

It appears that the oxidative changes, in addition to amyloid, are markers for the more severe behavior changes called dementia. From our results it can be concluded that therapy should focus on both amyloid as oxidative damage.

REFERENCES

1. Adams, B., Chan, A., Callahan, H., Siwal, C., Tapp, D., Ikeda-Douglas, C., Atkinson, P., Head, E., Cotman, C.W. and Milgram, N.W., Use of a delayed non-matching to position task to model age-dependent cognitive decline in the dog, *Behav. Brain Res.*, 108, 47, 1999.
2. Callahan, H., Ikeda-Douglas, C., Head, E., Cotman, C.W. and Milgram, N.W., Development of a protocol for studying object recognition memory in the dog, *Prog. Neuro-Psychopharmacol. Biol. Psychiat.*, 24, 693, 2000.
3. Head, E., Callahan, H., Muggenburg, B.A., Cotman, C.W. and Milgram, N.W., Visual-discrimination learning ability and β-amyloid accumulation in the dog, *Neurobiol. Aging*, 19, 415, 1998.
4. Kiatipattanasakul, W., Nakamura, S., Hossain, M.M., Nakayama, H., Uchino, T., Shumiya, S., Coto, N. and Doi, K., Apoptosis in the aged dog brain, *Acta Neuropathol.*, 92, 242, 1996.
5. Colle, M.A., Hauw, J.J., Crespeau, T., Uchilhara, T., Akiyama, H., Checler, F., Pageat, P. and Duykaerts, C., Vascular and parenchymal Aβ deposition in the aging dog: correlation with behavior, *Neurobiol. Aging*, 21, 695, 2000.
6. Rofina, J.E., van der Meer, I., Goossens, M., Secrève, M., van Ederen, A.M., Schilder, M. and Gruys, E., Preliminary inquiry to assess behaviour changes in aging pet dogs, in *Proc. Amyloid and Amyloidosis IX*, Bely, M. and Apathy, A., Eds., Budapest, 2001, 464.
7. Linke, R.P., Highly sensitive diagnosis of amyloid and various amyloid syndromes using Congo red fluorescence, *Virchows Arch.*, 436, 439, 2000.

D-AMINO-ACID PEPTIDE SPECIFIC FOR CEREBRAL Aβ-AMYLOID IN MAN CROSS-REACTS WITH CANINE CEREBRAL AMYLOID

R.P. Linke[1], R. Oos[1], J.E. Rofina[2], K. Wiesehan[3], D. Willbold[3], E. Gruys[2]

[1] *Max-Planck-Institut für Biochemie, Martinsried, Germany;*

[2] *Dep. Veterinary Pathology, University of Utrecht, Utrecht, The Netherlands;*

[3] *Inst. f. Physik. Biologie u. BMFZ, Universität Düsseldorf, and IBI-2 Forschungszentrum Jülich, Germany*

ABSTRACT

The aim of this ongoing study is to produce a tracer which can assist in diagnosing the Alzheimer's diseases in vivo by binding specifically to the cerebral Aβ-amyloid. To this end a D-amino acid peptide has been selected (and published) that specifically binds to human Aβ-amyloid in tissue sections. The question in this report was whether this D-peptide would also bind specifically to Aβ-amyloid in tissue sections of aged and demented dogs due to Aβ-amyloid deposits found mainly in leptomeningeal vessels. Here we show specific binding of this D-peptide to canine Aβ–amyloid on four demented dogs while cerebral tissue of two non-demented dogs lacking amyloid did not bind this peptide.

INTRODUCTION

Alzheimer's disease is the most common cause of progressive dementia affecting more than 1 % of the population in more developed states (1). The diagnosis of early and very early Alzheimer's disease in vivo has still not been established. A selected D-amino-acid peptide (DAAP) specifically binding the $Aβ_{1-42}$ synthetic peptide has been selected by mirror image phage display that also binds to human Aβ amyloid deposits in tissue sections, even after tagging with a fluorochrome (FDAAP) (2) and could be attempted with the aid of scintigraphy in vivo. Before this could be done in man this DAAP should be tried in animals. Demented dogs have been shown to succumb to a similar senile dementia with Aβ amyloid (3). Here we show that the FDAAP cross-reacts with canine Aβ-amyloid in tissue sections and thus may be applicable in vivo for diagnostic purposes.

MATERIAL AND METHODS

The principle of the phage display (4) is shown in Fig. 1. The DAAP was fluoresceinated with FITC as described (2). Formalin-fixed paraffin-sections (6 μm) of four demented dogs (dog 1: cross-breed, male, 18

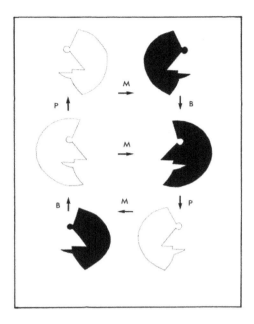

Figure 1. The principle of selecting the DAAP (4) in form of a cartoon.

Explanations: white, L-peptides; *black*, D-peptides; *M*, production of a mirror image via synthesizing peptides by employing either L-amino acids or D-amino acids, *2/3 moon*, target (Aβ1-42, synthetic peptide); *1/3 moon*, selected peptide via phage display (P) which binds (B) to target.

How to read the cartoon: The **L**-target is used for selecting an **L**-peptide (top left). This **L**-peptide can be mirrored to produce a **D**-peptide which will bind the **D**-target (top right). The **D**-target has been used for selecting a further **L**-peptide. Mirroring this **L**-peptide will yield the **D**-peptide which binds the **L**-target.

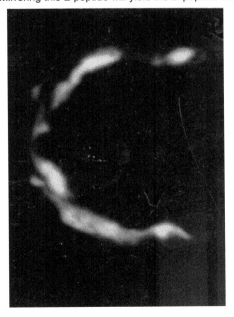

Figure 2. FDAAP binding to leptomeningeal amyloid in cerebral tissue section of a demented dog (240 x). The light arc is fluorescent green in original print. It represents leptomeningeal canine Aβ-amyloid.

years; dog 2: Belgian Sheepdog, m, 18.8; dog 3: cross-bred, m, 12.5; dog 4: Yorkshire Terrier, m, 16 – euthanized for various reasons) and two control dogs (Danish, m, 0.5, euthanized, from another research project) were available from Utrecht/NL. Amyloid was diagnosed on tissue sections after Congo red staining according to Puchtler et al. (ref. in 6), and classified through immunohistochemistry using a polyclonal antibody directed against human ex vivo Aβ protein and the peroxidase-anti-peroxidase procedure as described (5) including appropriate controls (antisera against human cerebral non-Aβ amyloids; sections lacking Aβ-amyloid). The FDAAP was applied to re-hydrated tissue sections at 1 µg/ml in PBS for one hour in the dark (as the entire procedure). After rinsing with PBS and mounted in glycerine jelly the sections were evaluated and photographed using blue light for excitation and a green-orange barrier filter (6).

RESULTS

The tissue sections of the four demented dogs showed the presence of amyloid (Congo red binding and green birefringence) in the leptomeningeal vessels. Immunohistochemistry revealed this amyloid to be of the Aβ class. After the application of FDAAP the entire amyloid which was identified with Congo red was bright green fluorescent (Fig. 2). When the sections were pre-stained with Congo red followed by FDAAP staining the stained areas of both were congruent proving specific binding of the FDAAP to canine Aβ-amyloid.

CONCLUSION

The fluoresceinated D-amino acid peptide that binds specifically to human cerebral Aβ-amyloid in tissue sections also cross-reacts with respective canine Aβ-amyloid in tissue sections and may bind also to canine Aβ-deposits in vivo.

ACKNOWLEGEMENTS

We thank J. Lindermeyer and M. Bandmann for technical assistance, the Deutsche Forschungsgemeinschaft, Bonn-Bad Godesberg (grant Li 247 12-3) and the Max-Planck-Institut (Prof. Dr. R. Huber) for support of this work.

REFERENCES

1. Glenner, G.G., and Murphy, M.A., Amyloidosis of the nervous system. *J. Neurol. Sci.* 894, 1, 1989.
2. Wiesehan, K., et al., Selection of D-amino-acid peptides that bind to Alzheimer's disease amyloid peptide Aß1-42 by mirrow image phage display, *Chem BioChem* 4, 748, 2003.
3. Rofina, J., et al., Canine counterpart of senile dementia of the Alzheimer type: amyloid plaques near capillaries but lack of special relationship with activated microglia and macrophages. *Amyloid* 10, 86, 2003.
4. Schumacher, T.N., et al. *Science* 271, 1854, 1996.
5. Linke, R.P., et al. Antigenic heterogeneity of cerebral amyloid deposits in Alzheimer's disease. In: *"Amyloid and Amyloidosis"*, Natvig, J.,B. et al. (eds.) , pp. 741, Kluwer Acad. Pupl. Dordrecht/NL, 1991.
6. Linke, R.P., Highly sensitive diagnosis of amyloid and various amyloid syndromes using congo red fluorescence. *Virchows Arch. Path. Anat.* 436, 439, 2000.

INHIBITION OF β-AMYLOID PEPTIDE-INDUCED MICROGLIAL CELL ACTIVATION BY GLYCOSAMINOGLYCANS

C. Morissette[1], *F. Gervais*[2], *P. Tremblay*[1]

[1]*Neurochem Inc., 275 Armand-Frappier Blvd, Laval, Quebec, Canada H7V 4A7*
[2]*Neurochem (International) Limited, 275 Armand-Frappier Blvd, Laval, Quebec, Canada H7V 4A7 and Neuhausstrasse 5, 6318 Walchwil, Canton Zug, Switzerland*

Glycosaminoglycans (GAGs) have been shown to contribute to the development of amyloid diseases. In Alzheimer's disease (AD), senile plaques, which are mostly composed of the β-amyloid peptide (Aβ), colocalize with GAGs (1-5). *In vitro*, the anionic carbohydrate moieties of GAGs appear to interact with residues 13 to 16 (HHQK) of the Aβ peptide. These interactions favor the assembly of Aβ into oligomers and fibers thereby promoting their toxicity and increasing their resistance to proteolytic degradation (6-8).

Activated microglial cells are also associated with amyloid plaques and contribute to the neuronal damage seen in AD patients. Aggregated and soluble Aβ have been shown to differentially activate microglia *in vitro* leading to the production of inflammatory mediators such as nitric oxide (NO) and cytokines. By binding to Aβ and modulating its assembly process, or by interfering with the direct binding of Aβ to the microglial cell surface, GAGs can modulate the inflammatory process. We therefore verified the effect of GAGs by studying whether Congo red (CR) and various GAGs can modulate the response of microglia to soluble Aβ.

MATERIALS AND METHODS

Rat Primary Microglial Cell Culture – Cortices from newborn Sprague Dawley-CD rats were mechanically dissociated in cold Dulbecco's Modified Eagle Medium (DMEM)-gentamycin (10 µg/mL). Following a 30-minute enzymatic treatment using dispase solution (Roche Diagnostics, Cat. #165 859) in DMEM, the cell preparation was filtered through a 60-µm nylon membrane, centrifuged, and resuspended in DMEM-10% Fetal Bovine Serum (FBS) without antibiotic. On day 10 or 11 of culture, cell-containing flasks were gently agitated on an orbital shaker for 2 hours and the detached microglial-enriched population was plated and allowed to adhere for 2 hours and subsequently washed to remove non-adherent cells. Microglial cells were acclimated for 18-20 hours before being used for experiments.

Aβ1-42 Peptide – Synthetic Aβ1-42 peptide (American Peptide, Cat. #62-0-80, lot 011113T1, lot Q03121X) was treated with 1,1,1,3,3,3-hexafluoro-2-propanol (HFIP), sonicated, and stored at –80°C. Preparations were thawed, dried under nitrogen gas, and resuspended in 0.04 M Tris to a final concentration of 150 µM. Aβ1-42 peptide was then diluted to 30 µM in 20 mM N-[2-hydroxyethyl]piperazine-N'-[2-ethanesulfonic acid] (HEPES), pH 7.0 to 7.4.

Compounds – CR, MW 696.7 (Cat. # C-6277), heparin (Cat. # H-3149) of high MW mainly around 17,000 - 19,000, low MW heparin (Cat. # H-3400) of approximatively 3,000 MW, and chondroitin sulfate C (Cat. #C4384), were purchased from Sigma, St-Louis.

Nitrite Assay – 5 µM of soluble Aβ and 5 U/mL of IFN-γ was added to the microglial culture in the presence or absence of compounds and the production of nitrite was determined in the supernatant after 48 hours. Levels of NO_2^-, a breakdown product of NO, were measured in the culture supernatant with a fluorescent method using 2,3-diaminonaphthalene (DAN), as described by Misko et al. (9).

RESULTS

We compared the ability of GAGs and the sulfonate-containing dye CR, which binds preferentially to aggregated as opposed to soluble Aβ, to interfere with Aβ-induced microglial activation. The levels of nitrite production were measured in rat primary microglial cultures following activation using soluble Aβ and a priming concentration of IFN-γ, in the presence of GAGs or CR.

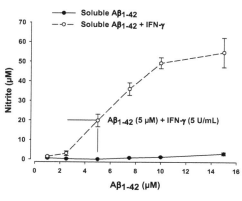

Figure 1. Conditions for the activation of the microglial cell culture were determined by incubating increasing levels of soluble Aβ1-42 with a low concentration of rat IFN-γ (5 U/mL, Gibco, Cat. # 13283-023). In the presence of a priming concentration of IFN-γ, Aβ1-42 induced the activation of the microglial culture as measured by the nitrite levels in the supernatant. Nitrite production increased linearly in a dose-dependent manner with concentrations ranging from 2.5 to 10 µM of soluble Aβ1-42 in the presence of IFN-γ. In the absence of IFN-γ, no nitrite was produced even in the presence of up to 15 µM of soluble Aβ1-42.

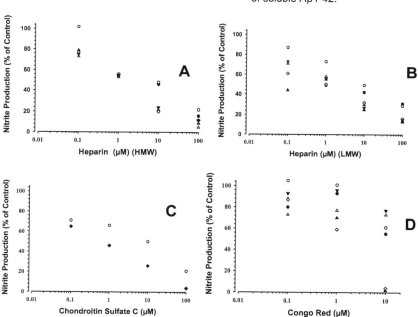

Figure 2. GAGs influence the nitrite levels produced by microglia following stimulation with soluble Aβ1–42 and IFN-γ: **A)** High MW heparin reduces the nitrite production induced by 5 µM of soluble Aβ1-42 and IFN-γ in a dose-dependent fashion. The IC_{50} was estimated to be ~2 µM. **B)** Low MW heparin reduces in dose-dependent

fashion the nitrite production induced by Aβ1-42 and IFN-γ. The IC_{50} was estimated to be ~5 μM. **C)** Chondroitin sulfate C reduces in a dose-dependent fashion nitrite production induced by soluble Aβ1-42 and IFN-γ. The IC_{50} was estimated to be ~2 μM. **D)** Only 10 μM of CR marginally decreases nitrite production following stimulation with soluble Aβ1-42 and IFN-γ. Higher doses of CR interfere directly with the nitrite assay. **A, B, D)** Symbols represent mean for each of 6 independent experiments performed in duplicate. **C)** Symbols represent the mean for each of 2 independent experiments performed in duplicate.

SUMMARY OF RESULTS

In the presence of a priming concentration of IFN-γ, Aβ1-42 induced the activation of the microglial culture as measured by the amount of nitrite in the supernatant. GAG compounds interfered with the production of nitrite by microglial cultures following stimulation with soluble Aβ1-42 and IFN-γ. Heparin (HMW and LMW), and chondroitin sulftate C inhibited in a dose-dependent manner the Aβ1-42–induced microglial activation. In contrast, CR had no effect on microglial activation.

CONCLUSION

Our observations suggest that by binding to soluble Aβ, GAGs can prevent both microglial activation and the ensuing release of toxic inflammatory mediators. Alternatively, GAGs may prevent activation by binding directly to the cell surface and preventing interaction with Aβ. However, a heparin-based approach, using low MW heparin, could present a safety risk as chronic use would be required for treating AD. A therapeutic approach based on the minimal structural motif of the GAGs necessary to target Aβ and other amyloidogenic proteins has been previously proposed and may therefore be preferable (10, 11). By binding to soluble Aβ at the GAG-binding site, compounds that mimick the anionic properties of GAGs could inhibit the formation of oligomeric and fibrillar Aβ states *in vivo* and facilitate their clearance. GAG-mimetic compounds may present a better pharmacokinetic, pharmacodynamic, and safety profile as they can be administered orally without interfering with biological processes.

REFERENCES

1. Kroger S, Schroder JE(2002) Agrin in the developing CNS: new roles for a synapse organizer. *News Physiol Sci* **17**: 207-12.
2. Cotman SL, Halfter W, Cole GJ(2000) Agrin binds to beta-amyloid (Abeta), accelerates abeta fibril formation, and is localized to Abeta deposits in Alzheimer's disease brain. *Mol Cell Neurosci* **15**: 183-98.
3. Snow AD, Mar H, Nochlin D *et al.*(1988) The presence of heparan sulfate proteoglycans in the neuritic plaques and congophilic angiopathy in Alzheimer's disease. *Am J Pathol* **133**: 456-63.
4. Berzin TM, Zipser BD, Rafii MS *et al.*(2000) Agrin and microvascular damage in Alzheimer's disease *Neurobiol Aging* **21**: 349-55.
5. Donahue JE, Berzin TM, Rafii MS *et al.*(1999) Agrin in Alzheimer's disease: altered solubility and abnormal distribution within microvasculature and brain parenchyma. *Proc Natl Acad Sci U S A* **96**: 6468-72.
6. Sadler IIJ, Smith DW, Shearman MS, Ragan CI, Tailor VJ, Pollack SJ (1995) Sulphated compounds attenuate β-amyloid toxicity by inhibiting its association with cells. *NeuroReport* **7**: 49-53.
7. Gupta-Bansal R, Frederickson RCA, Brunden KR (1995) Proteoglycan-mediated inhibition of Aβ proteolysis: A potential cause of senile plaque accumulation. *J Biol Chem* **270**: 18666-18671.
8. Pollack SJ, Sadler II, Hawtin SR, Tailor VJ, Shearman MS(1995) Sulfated glycosaminoglycans and dyes attenuate the neurotoxic effects of beta-amyloid in rat PC12 cells. *Neurosci Lett* **184**: 113-6.
9. Misko TP, Schilling RJ, Salvemini D, Moore WM, Currie MG(1993) A fluorometric assay for the measurement of nitrite in biological samples. *Anal Biochem* **214**: 11-6.
10. Gervais F, Morissette C, Kong X(2003) Proteoglycans and amyloidogenic proteins in peripheral amyloidosis. *Curr Med Chem (IEMA)* **3**: 361-70.
11. Gervais F, Chalifour R, Garceau D *et al.*(2001) Glycosaminoglycan mimetics: a therapeutic approach to cerebral amyloid angiopathy. *Amyloid* **8**: 28-35.

MORPHOLOGY AND GROWTH KINETICS OF ALZHEIMER'S AMYLOID β- PEPTIDES, Aβ AND THE ARCTIC MUTATION Aβ (E22G): IN SITU ATOMIC FORCE MICROSCOPY STUDY

M. Hellberg[1], N. Norlin[2], O. N. Antzutkin[2] and N. Almqvist[1]

[1]*Department of Applied Physics and Mechanical Engineering, Division of Physics,* [2]*Department of Applied Chemistry and Geosciences, Division of Chemistry, Luleå University of Technology, SE-971 87 Luleå, Sweden. Corresponding authors: Oleg.Antzutkin@ltu.se and Nils.Almqvist@sirius.luth.se*

A principal constituent of Alzheimer's disease (AD) amyloid plaques is the human amyloid-β-peptide (Aβ), a 39-43 amino acid peptide of known primary sequence (1). Currently, it is widely believed that Aβ deposits contribute directly to the disease's progressive neurodegeneration (2). Aggregation cascade for Aβ peptides, its relevance to neurotoxicity in the course of AD and experimental methods useful for these studies have been also recently discussed (3). In the present study (4) a commercial atomic force microscope (AFM) with a custom designed tapping mode system (5) was used to *in situ* monitor the 'wild type' and the 'Arctic' mutant (6) of the Alzheimer's Aβ peptides, $A\beta_{(1-40)}$ and arc-$A\beta_{(1-40)}$(E22G). The aggregation behaviour was studied *in vitro* in a physiologically relevant TRIS buffer (pH 7.4) at peptide concentrations ranging from 2.5 to 50 µM. Peptide concentration effects on fibrillization were studied on fully matured samples, whereas the aggregation and fibrillization processes were studied during a few weeks at discrete time points. The Arctic mutant was shown to form aggregates and fibrils at much lower concentrations (<2.5 µM) than the wild type peptide (*ca.* 12.5 µM). Moreover, at the same concentration, 50 µM, the aggregation process occurs more rapidly for the Arctic mutant compared with the wild type peptide: first amyloid fibrils appear after 3 versus 7 days after the onset of sample incubation without additional rocking. Initially, the self-aggregation process is characterized as a slow transition of aggregates to a highly distinct and ordered morphology, putatively, a key kinetic intermediates, so called spherical bodies. Then, the process proceeds with rapid polymerization of the peptides into amyloid fibrils. Interestingly, the Arctic mutant amyloid fibrils exhibit a large variety of polymorphs: both coiled and non-coiled fibril structures of at list five distinct types were recognized and their dimensions were measured precisely by AFM.

Table 1. AFM measured dimensions on aggregates and fibrils

Aggregate:	wt-$A\beta_{(1-40)}$ SB*	arc-$A\beta_{(1-40)}$ SB*	wt-$A\beta_{(1-40)}$ fibrils	arc-$A\beta_{(1-40)}$ fibrils				
				Non-coiled types		Coiled types		
				1	2	3	4	5
Height (nm)	6.6±0.9	3.1±0.8	4.8±0.5	3.4±0.3	5.0±0.5	6.5±0.5	7.1±0.6	5.2±0.7
CoH#(nm)	–	–	8.1±0.8	–	–	8.5±0.5	10.2±0.9	11.1±0.8
CoS## (nm)	–	–	~110, ~170	–	–	31±3.1	40±4.7	88±9.9

*SB, #CoH and ##CoS denote spherical bodies, crossover height and crossover spacing, respectively. The heights on the coiled fibrils are measured in between and on the top of the crossovers, respectively.

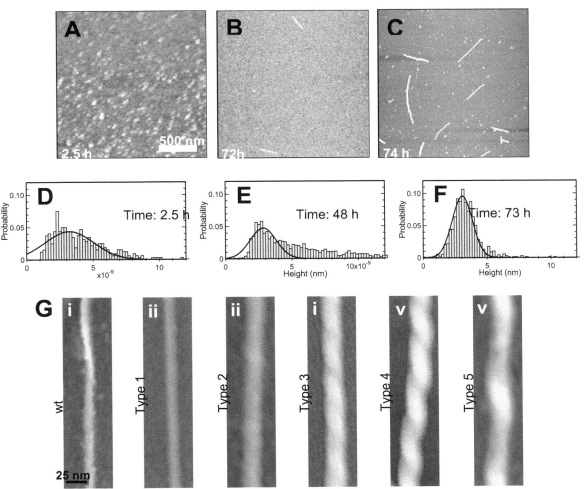

Figure 1. A–C: AFM height images of 'Arctic' arc-Aβ$_{(1-40)}$ aggregates at different time points ((A) 2.5 h, (B) 72 h and (C) 74 h) after the onset of the aggregation reaction, 50 μM in TRIS buffer, pH 7.4. Note an abrupt transition from the uniformly shaped spherical aggregates (B) to amyloid fibrils (C). The image sizes are 2000 nm and the height is greyscale coded). **D, E, F**: Probability histograms of the arc-Aβ$_{(1-40)}$ aggregate heights at different times after the onset of the aggregation reaction; (D) 2.5 h, diffuse aggregates with a large distribution of heights; (E) 48 h, some more structured aggregates have been formed; (F) 72 h, well structured uniformly shaped aggregates are short lived intermediates and putative neurotoxins. **G(i–vi)**: Morphology of Aβ fibrils visualized with AFM height images. The height scale is greyscale coded in every image from 0 nm (dark) to 18 nm (bright). (i) A typical wild type (wt) Aβ$_{(1-40)}$ fibril; (ii–iii) polymorphism of arc-Aβ$_{(1-40)}$ fibrils classified into different types: the *types 1* and *2* are non-coiled fibrils (ii–iv) and the *types 3, 4* and *5* are left-handed coiled fibrils (iv–vi). The types 1 and 2 non-coiled fibrils were classified due to their different heights, while the types 3, 4 and 5 coiled fibrils were classified due to the differences in cross-over heights (CoH) and cross-over spacing (CoS).

MATERIALS AND METHODS

Peptides were synthesized on a Perkin Elmer 433A peptide synthesizer using solid-phase fast F-moc synthesis and HPLC purified as described in Antzutkin *et al.* (7). 50 μM aqueous stock solutions of peptides were prepared by mixing peptides into a buffer solution of 10 mM TRIS, 5 mM EDTA, 10 mM KCl and 0.01 wt% NaN$_3$ with pH adjusted by NaOH(aq) to 7.4. Other concentrations of peptides were achieved by diluting the stock solution by the buffer. All samples were kept in 1 ml plastic tubes without additional agitation. The AFM samples were prepared by disposing a droplet of the peptide solution (*ca.* 15 μl) directly onto a freshly cleaved mica surface and leaving it to incubate for approximately 15 minutes in a small container to avoid contamination and minimize evaporation. The samples were gently rinsed with excess buffer and transferred to the AFM microscope for imaging. Additional buffer, 40-50 µl, was added to fill the AFM fluid cell. AFM images were recorded using a Nanoscope II (Digital Instruments/Veeco, Santa Barbara, CA) with a custom design for tapping mode (amplitude modulation AFM) in fluids. The tapping piezo is driven by an external function generator (Dagatron) or from a multifunction data acquisition board (National Instruments, PCI-6110). The drive amplitude is typically 50-200 mV. Dedicated electronics are used to detect the amplitude of the preamplified deflection sensor signal. The amplitude is compared with the user defined amplitude setpoint and the difference is sent as the error signal to the regular feedback loop of the Nanoscope II. The time constant of this amplitude detection system is 0.25 ms, which approximately corresponds to 2.5 oscillations of the tapping cantilever under normal imaging conditions. This setup has shown to provide gentle imaging. The AFM images (400 x 400 pixels) were recorded using the D or J scanners with maximum lateral scan areas of 14.9 µm and 174 µm, respectively. Standard or oxide-sharpened commercially available, 100 µm long, Si_3N_4 cantilevers, with integrated tips (Digital Instruments) and a nominal spring constant, of 0.32 N/m, were used. These cantilevers were driven close to their resonance frequencies at ~9-11 kHz in the aqueous buffer. Typical scan rates were 1-3 scan lines per second, i.e. a typical acquisition times for a complete AFM image were between 1 and 6 minutes. The Nanoscope III software (version 4.23R2) was used for image visualization and manual measurements. IGOR Pro 4.07 (Wavemetrics, Lake Oswego, OR) was used with procedures developed in our laboratory, to semi-automatically extract volume, statistics and height distributions of Aβ aggregates and fibrils. Typically, 200-600 aggregates in an image are used to form an aggregate height histogram. More details about the morphological particle analysis algorithm are given in ref. 4.

ACKNOWLEDGMENTS

Peptides were synthesized by ONA in the group of R. Tycko, LCP, NIDDK, NIH. We thank Dr. Claire Goldsbury for useful technical discussions regarding time-lapse imaging of fibril. This work was supported by grants from the Swedish Research Council (ONA and NA). Alzheimer's foundation in Sweden is acknowledged for a grant used for a stipend (NN).

REFERENCES

1.   G.G. Glenner, C.W. Wong, *Biochem. Biophys. Res. Commun.* **120** (1984) 885-889.
2.   D.M. Walsh, *et al.*, *Nature* **416** (2002) 535-539; D.J. Selkoe, *J. NIH. Res.* April **7** (1995) 57-64.
3.   O.N. Antzutkin, *Magn. Reson. Chem.* **42** (2004) 231-246.
4.   M. Hellberg, N. Norlin, O.N. Antzutkin, N. Almqvist, 2004, submitted to *Proc. Natl. Acad. Sci. U.S.A.*
5.   M. Hellberg, N. Norlin, *In situ* Studies of Aggregation Dynamics and Structure of Alzheimer's Amyloid Fibrils Using AFM, MSc. thesis, Luleå University of Technology, ISRN: LTU - EX -- 03/193 –SE, 52 pp.
6.   C. Nilsberth, *et al.*, *Nature Neurosc.* **4(9)** (2001) 887-893.
7.   O.N. Antzutkin, *et al.*, *Proc. Nat. Acad. Sci. U.S.A.* **97(24)** (2000) 13045-13050.

REVERSING THE FORMATION OF ALZHEIMER AMYLOID FIBRILS AND BLOCKING TOXIC Ca^{2+} INFLUX

*‡Ingram VM, *Blanchard BJ, *Chen A, *Stafford K & **Stockwell B.

*Dept. Biology, Massachusetts Institute of technology, Cambridge, MA, USA.
** Whitehead Institute for Biomedical Research, Massachusetts Institute of technology, Cambridge, MA, USA.
‡corresponding author: vingram@mit.edu

It has been proposed by several laboratories that the neurotoxicity of fibrils of Abeta-peptides in Alzheimer's Disease (AD) involves a disturbance of calcium homeostasis in neuronal cells. Using neuronal cell cultures, we find that within a very few seconds of contact between aggregated Abeta1-42 and the neuronal cell there is an immediate and rapid influx of external cakcium ions via calcium-permeant AMPA-receptors, which is blocked by the specific antagonist NBQX. We consider that calcium-influx is the initiating event of AD pathology, followed by a whole series of downstream events, including eventual cell death.

Using cultured neuronal cells, we have developed two series of useful compounds: (i) ten short so-called Decoy Peptides of D-amino acids, selected from a large peptide library, that block the Abeta-induced calcium-influx; (ii) the selection in a High Throughput Screen of several small molecules that block and reverse Abeta fibril formation and also eliminate the Abeta-induced calcium-influx. We illustrate the effectiveness of decoy peptides and of the first HTS compound in eliminating calcium-influx. Because this HTS compound also effectively reverses Abeta fibrils, it is a good candidate for Alzheimer therapy, as are the Decoy Peptides.

SECTION 6

LOCALIZED AMYLOIDOSIS

β2 MICROGLOBULIN AMYLOIDOSIS

A. Argilés, A. Charnet, C.F. Schmitt-Bernard, M. García-García, C. Méjean, G. Mourad.

Laboratoire de Génomique Fonctionnelle (LGF) – UPR 2580 CNRS

Assemat *et al* identified β2-microglobulin amyloidosis (β2m-A) in carpal tunnel from 9 dialysis patients (out of 230) (1). The prevalence of β2m-A while progressing the dialysis treatment neighbored 100 % (2). The amyloid material was predominantly observed in tendons and joints, but it was also reported to involve heart, digestive tract, ears and other locations (reviewed in 3). This new form of amyloid is an invalidating disease that both limits the day to day activity and is live threatening in other occasions following the location of the amyloid deposits (3). The progress in the understanding of this newly identified disease came with the identification of the protein components of the amyloid material.

PROTEIN CHARACTERIZATION AND ADVANCED GLYCATION END PRODUCTS

In 1985, Gejyo *et al* purified and sequenced the main protein component of amyloid deposits obtained from the carpal tunnel of a dialysis patient (4). The first 13 aminoacids from the N-terminus had an identical sequence to that known for β2m (4). Very early after Gejyo's work, analytical chemistry of amyloid proteins showed that the β2m contained in amyloid deposits was constituted of several isoforms with lower pI (5,6) than that previously reported in normal serum or urine (7). These more acidic β2m isoforms were considered to be specific for β2m amyloidosis.

The first modifications that were proposed to explain the more acidic pI of the modified β2m were deamidation of the Asn-17 (converting it into Asp-17) (8) and lysine specific N-terminal proteolysis (9). However, both modifications were not invariably observed in the more acidic β2m isoforms purified from serum, urine and amyloid deposits (10).

Miyata *et al* reported the presence of AGEP modified β2m in serum and urine of renal failure patients and were able to reproduce a shift in the pI of β2m by submitting it to AGEP modification (11). Therefore, this paper provided a possible explanation for the appearance of the most acidic β2m isoforms, although did not exclude other causes. More importantly, Miyata's work opened a new path of research in dialysis related amyloidosis: AGEP modification of the protein constituents. Serum levels of AGEP modified proteins are increased in renal failure, are higher in hemodialysis patients than in peritoneal dialysis patients and decrease towards the normal level after a successful renal transplantation (12,13,14). It was proposed that AGEP-β2m would be the main component of the amyloid fibrils and would participate in the recruitment of macrophages, thereby participating in the genesis of the amyloid deposits (15). However, several subsequent findings do not support the AGEP hypothesis. First, we observed that the AGE pentosidine is not in the core of amyloid fibrils and it is one of the components which is more easily removed with a solution of PBS (16). Further, we found that the more acidic isoforms of β2m were not specific of β2m amyloidosis and even not of renal failure: they can be observed in urine of subjects with normal renal function (10). Second, the work from Hou *et al* showed that AGEP-β2m had no more effect than AGEP-albumin in the biological systems they studied (synthesis of MP1 and delay in monocyte apoptosis) (17), and the work of Moe *et al* demonstrated that AGEP modification of β2m abrogates even the effect of non glycated β2m (18). Finally, Dinarello's group has shown that IL-1 and TNFα were increased by AGEP modified proteins (both albumin and β2m modified AGEP), only in the presence of LPS in the medium, suggesting that two triggers are necessary to obtain the cellular effect of AGEP modified proteins and again that these effects are not specific for β2m (19).

THE OTHER COMPOUNDS IN β2 MICROGLOBULIN AMYLOIDOSIS: PROTEINS AND CELLS

The cellular infiltrate in β2m amyloidosis was found to be mainly constituted of macrophages (). However, both ultrastructural and necropsic studies provided evidence showing that macrophages infiltrate is a secondary phenomenon, mainly inducing phagocytosis (20,21). Thus, the hypothesis by which AGEP-β2m would participate in the genesis of amyloid fibrils by its chemotactic effects was challenged by these findings.

Many other constituents of amyloid deposits have been identified and ultrastructural analysis of the amyloid fibrils by Kisilevsky's group has allowed the proposition of a model of fibril consisting on a helically wound 3 nm wide ribbon-like chondroitin sulphate proteoglycan (CSPG) "double tracks" enclosing amyloid P component and an outer layer of CSPG onto which β2m would be attached (22). Among the other protein constituents of amyloid fibrils, we have been interested in α2-macroglobulin, a major antiprotease that we identified in amyloid fibrils, that has specific binding capacities for β2m and that would inhibit amyloid deposit resorption "in vitro" (23).

SUMMARY AND PRESENT STAND POINT: THE INFLAMMATORY HYPOTHESIS

Since the clinical description of β2m amyloidosis, much progress has been achieved. The main protein constituents have been identified, the cellular participation has been documented and some protein-protein and protein-cell interactions have been elucidated. Further, several different "in vitro" models of β2m amyloidosis have been reported and used to analyse the physiopathology of the disease. However, again, the news come from the clinical side: Flöege's group in Germany have reported a decrease of 80% in the prevalence of β2m amyloidosis between 1988 and 1992 in a single centre (24). The factors that these authors have retained that would explain this decrease include the improvement in the quality of the water used for dialysis (24).

The observations reported by Schwalbe *et al* (24) are confirmed by our own clinical experience. Thus, we have to admit that we have progressed in the prevention of β2m amyloidosis without fully understanding its pathogenesis. The decrease in β2m prevalence has been observed in the absence of any major modification in serum levels of β2m, but with a clear improvement in water quality. The hypothesis of a chronic inflammatory stress induced by repeated stimuli with bio-incompatible dialysis systems would bring both the information from the clinics and that from the laboratory together. In keeping with Dinarello's data showing that LPS stimulus could have a central role in the activation of the cells and in enhancing the effect of the other amyloid components, using endotoxin-free dialysates in the clinics could account for the observed decrease in β2m amyloidosis prevalence.

REFERENCES

1.　Assenat H, Calemard E, Charra B, et al. Nouv Press Med 24:1715, 1980
2.　Jadoul M, Garbar C, Noel H, *et al*. Kidney Int. 51:1928-32, 1997

3. Argilés A. Nephrology 2:373-86, 1996

4. Gejyo F, Yamada T, Odani S, *et al*. Biochem Biophys Res Commun, 129:701-6, 1985

5. Gorevic PD, Muñoz PC, Casey TT, *et al*. Proc Nat Acad Science USA, 83:7908-12, 1986

6. Argilés A, Mourad G, Axelrud-Cavadore C, *et al*. Néphrologie, 8:51-4, 1987

7. Hall PW, Ricanati ES, Vacca CV. Clinica Chemica Acta , 77:37-42, 1977

8. Odani H, Oyama R, Titani K, Ogawa H, Saito A. Biochem Biophys Res Commun, 168:1223-29, 1990

9. Linke RP, Hampl H, Lobeck H, *et al* Kidney Int, 36:675-81, 1989

10. Argilés A, García-García M, Derancourt J, Mourad G, Demaille JG. Kidney Int 48:1397-405, 1995

11. Miyata T, Oda O, Inagi R, *et al* J Clin Invest, 92:1243-52, 1993

12. Makita A, Bucala R, Rayfield EJ, *et al*. Lancet 343:1519-22, 1994

13. Friedlander MA, Wu YC, Schulak JA, Monnier VM, Hricik DE. Am J Kidney Dis, 25:445-51, 1995

14. Hricik DE, Schulak JA, Sell DR, Fogarty JF, Monnier VM. Kidney Int 43:398-403, 1993

15. Miyata T. Editorial. Nephrol Dial Transpl 11:934-6, 1996

16. Argilés A, Gouin-Charnet A, Wu YC, Mourad G, Friedlander M. Nephrol Dial Transpl 11:2371-2, 1996

17. Hou FF, Jiang JP, Guo JQ, *et al* J Am Soc Nephrol. 13:1296-306, 2002

18. Moe S, Singh GK, Bailey AM. Kidney Int 57:2023-34, 2000

19. Dinarello C, Reznikov L, Waksman J, Pichetsrider M, Shaldon S. Blood Purif 21:347-9, 2003

20. García-García M, Argilés A, Gouin-Charnet A, *et al* Kidney Int 55:899-906, 1999

21. Garbar C, Jadoul M, Noel H, van Ypersele de Strihou C. Kidney Int. 55:1983-90, 1999

22. Inoue S, Kuroiwa M, Ohashi K, Hara M, Kisilevsky R. Kidney Int. 52:1543-9, 1997

23. Gouin-Charnet A, Mourad G, Argilés A. Biochem Biophys Res Commun 231:48-51, 1997

24. Schwalbe S, Holzhauer M, Schaeffer J, Galanski M, Koch KM, Floege J. Kidney Int. 52:1077-83, 1997

AMYLOID TUMORS OF BREAST, LUNG AND RIB IN NONSECRETORY MULTIPLE MYELOMA

S. Richter[1]. G. Sauter[2]. L. Jonas[1]. E. Gafumbegete[1]. J. Makovitzky[1].

[1]University of Rostock, Dep. Path. [2]University of Basel, Dep. Path.

INTRODUCTION

Nonsecretory multiple myeloma (NSMM) represents a variant of multiple myeloma characterized by unde-tectable monoclonal immunoglobulins in serum and urine, and accounts for one to five percent of all cases of MM. The following case is of interest because of the rare association between NSMM and amyloidosis.

CASE REPORT

A 49-year-old woman was hospitalized in May 1996 because of a 6-month history of painful breathing. On physical examination, there was a localized tenderness on the right seventh rib. A fracture was presumed. X-ray, computed tomography and magnet resonance imaging showed a single lytic lesion on the right seventh rib. Her past history included a carpal tunnel syndrome on both sides with following surgical procedure. The rib tumor has been resected because of a suspected osteolytic metastasis. The histopathologic examination revealed a plasmocytic plasmocytoma with extensive amyloidosis. Bone marrow biopsy showed a massive infiltration of 80% mature small plasma cells. Laboratory values were hemoglobin 6,9 mmol/l, leucocytes 5400/µl and thrombocytes 306000/µl. Electrolyte, creatinin, liver parameters and ß$_2$-microglobulin were normal. Quantitative immunoglobulin studies detected hypogammaglobulinemia with a serum IgG value of 4,75 g/l (normal 8-18 g/l), IgA 0,14 g/l (0,9-4,4 g/l) and IgM 0,18 g/l (0,45-2,5 g/l). Serum and urine protein electrophoresis were both negative for monoclonal gammopathy. Also immunoelectrophoresis failed to show any monoclonal immunoglobulins. Immunofixation of the concentrated urine however detected light chains of κ-type. A diagnosis of nonsecretory multiple myeloma was made. Treatment with melphalan and prednisone was initiated. A bone marrow aspiration performed after 6 month of therapy showed only 25% plasma cells. This therapy was followed by interferon α, idarubicin (Zavedos®) and dexamethasone. Therapy was completed with an allogenic bone marrow transplantation. Four years after diagnosis the patient was admitted to the hospital again because she developed cough and shortness of breath. Computed tomography showed a fibrosis in both upper pulmonary lobes. Mammography indicated a possible mastopathy. A pulmonary and breast biopsy revealed multiple amyloid tumors. Congo red staining showed an anisotropy and green polarization color.

MATERIALS AND METHODS

All specimens were fixed in PBS buffered formalin and embedded in paraffin. Deparaffinized sections (2-4 µm; 80°C for 14 to 16 hours) were stained with hematoxylin and eosin. The presence of amyloid was demonstrated by the appearance of green polarization color due to Congo red staining according to Romhanyi [6] in polarized light. Performic acid pretreatment [7] as well as potassium permanganate oxidation [8] with and without trypsin digestion served to demonstrate the resistance of the amyloid tumors. Glycosaminoglycans of the amyloid deposits were determined with the critical electrolyte concentration (CEC) method using the topooptical staining reaction with toluidine blue at pH 5,2 [5]. Immunostaining was performed with monoclonal antibodies directed against VS38 (1:50), IgA (1:10), IgG (1:100) and IgM (1:100) as well as the antibodies against the light chains kappa (1:500) and lambda (1:3000, all the preceding from Dako, Hamburg). Two different antisera against amyloid of lambda light chain (Aλ [HAR]; dilution 1:10000) and kappa light chain origin (Aκ [SIN]; dilution 1:50000) were used. Immunoreaction was visualized using the avidin-biotin complex.

RESULTS

Histology and Histochemistry

Tissue sections of the lung stained with hematoxylin and eosin showed an eosinophilic mass with an anisotropic effect in polarized light. Staining with Congo red induces congophilia and strong birefringence of amyloid with a green polarization color. The lung tissue was extensively replaced by nodular amyloid deposits. Amyloid was also found as distinct concentric accumulation in the left breast being resistant to induced proteolysis with potassium permanganate (Fig. 1). Distinguish between amyloid (green) with a linear positive sign after Congo red staining and collagen fibers (grey) with a linear negative sign as shown in figure 1b. Applying the CEC method the amyloid deposits exhibit metachromasia in normal light and birefringence with a red polarization color in polarized light indicating the presence of highly sulfated glycosaminoglycans.

Immunohistochemistry

The plasma cell infiltrates were immunoreactive with the antibody VS38, against the 64D proteine as specific for plasma cells. The following antibodies were repeatedly negative concerning the plasma cell infiltrates: IgA, IgG, IgM, antibodies against κ- and λ- light chains. Amyloid deposits in the rib tumor and the lung reacted positively with anti- amyloid kappa (Aκ) and negatively with anti- amyloid lambda (Aλ).

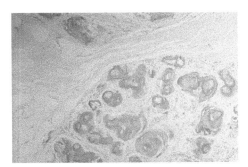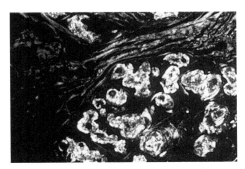

Figure 1. Mammary gland showing nodular amyloid deposits with a green polarization color a) light microscopic b) polarization microscopic. Congo red staining with KMnO$_4$ pretreatment (×40)

CONCLUSION

Nonsecretory multiple myeloma (NSMM) is a rare variant of multiple myeloma without monoclonal immunoglobulins in either serum and urine. It occurs in 1% to 5% of all patients with myeloma. A review of 869 cases of MM seen at the Mayo Clinic revealed that only three patients had no monoclonal protein in the serum or urine and were designated as having nonsecretory myeloma [3]. The lack of M-protein in the serum and urine in nonsecretory myeloma may be due to the inability of plasma cells to excrete the immunoglobulin, the low synthetic capacity of immunoglobulin, degradation of the M-protein within the plasma cells or rapid extracellular degradation of the M-protein after secretion from the plasma cells [2].

In this case serum and urine protein electrophoresis were negative for monoclonal gammopathy. Serum immunoelectrophoresis was normal. Immunofixation of the concentrated urine revealed κ-light chains more likely consistent with a lowsecretory myeloma. Development of amyloid tumors in NSMM has been described only occasionally in the literature. Azar et al. described 123 patients with MM, seven patients of them had a NSMM and only one of whom developed amyloidosis [1]. The patient described above suffered from NSMM with AL amyloidosis affecting the breast and lung. AL amyloidosis is characterized by fibrils composed of the variable portion of a monoclonal light chain. Bladé et Kyle observed that in nonsecretory multiple myeloma the light chain is of κ-type in 75% similar to our case [2]. Amyloid deposits are known to have three major components: the amyloid fibril protein, the amyloid P- component and glycosaminoglycans (GAGs). GAGs are important in the amyloidogenic pathway. They accelerate the formation of fibrils and protect them against proteolytic degradation [4]. The CEC method with constant toluidine blue concentration but variable salt ($MgCl_2$) concentration serves to show various GAGs within the amyloid deposits: hyaluronic acid (0,1M), chondroitin sulphate (0,5M), keratan sulphate (1M) and heparan sulphate (1,8/1,9 M).

It is assumed that the plasmocytic plasma cells possess an excretory mechanism, which allows the pathologic immunoglobulins to be secreted either as amyloid proteins polymerizing into amyloid fibrils or as immunoglobulin fragments falling in degradation as soon as they are excreted out of the tumor cell.

REFERENCES

1.   Azar HA, Zaino EC, Pham TD, Yannopoulos K. "Nonsecretory" plasma cell myeloma: observations on seven cases with electron microscopic studies. Am J Clin Pathol 58:618-29, 1972.

2.   Bladé J, Kyle RA. Nonsecretory myeloma, Immunoglobulin D myeloma, and plasma cell leukemia. Hematol Oncol Clin North Am 13:1259-72, 1999.

3.   Kyle RA. Multiple Myeloma. Mayo Clin Proc 50:29-40, 1975.

4.   McLaurin J, Franklin T, Kuhns WJ, Fraser PE. A sulfated proteoglycan aggregation factor mediates amyloid-β peptide fibril formation and neurotoxicity. Amyloid 6:233-243, 1999.

5.   Módis L. Organization of the extracellular matrix: A polarization microscopic approach. CRC Press 207-227, 1991

6.   Romhányi G. Selective differentiation between amyloid and connective tissue structures based on the collagen specific topo-optical staining reaction with congo red. Virch Arch Pathol Anat 354:209-22, 1971

7.   Romhányi G. Selective demonstration of amyloid deposits and methodical possibilities of analysis of their ultrastructural differences. Zentralbl Allg Pathol 123:9-16, 1979

8.   Wright JR, Calkins E, Humphrey RL. Potassium permanganate reaction in amyloidosis. A histologic method to assist in differentiating forms of this disease. Lab Invest 36:274-81, 1977

THE SPECTRUM OF LOCALIZED AMYLOIDOSIS: A CASE SERIES OF TWENTY PATIENTS

ML Biewend, MD; DM Menke, MD; KT Calamia, MD

Mayo Clinic, Jacksonville, Florida, USA

ABSTRACT

Localized deposition of amyloid may occur in individual organs, in the absence of systemic involvement. The reason for localized deposition is unknown, but it is hypothesized that deposits result from local synthesis of amyloid protein, rather than the deposition of light chains produced elsewhere.

We identified twenty cases of localized amyloidosis at our institution between 1993-2003. There were eleven males and nine females in the group. The mean age at the time of diagnosis was 65.5 years. Organs involved included skin, soft tissues, oropharynx, larynx, lung, bladder, colon, conjunctiva, and lymph node. In six of nine patients typed, the amyloid light chain was lambda. In those patients where follow-up was available (mean 7.6 years), none developed systemic disease.

Localized amyloidosis occurs in a variety of organ systems. Evolution into systemic amyloidosis was not seen in our series of patients, supporting the hypothesis of local production of amyloid protein in these cases.

INTRODUCTION

In AL amyloidosis, circulating immunoglobulin light chains aggregate in fibrillar deposits in diverse organs. Amyloidogenicity is determined by certain structural features of the protein favoring misfolding, and by local tissue conditions favoring protein instability and aggregation into beta pleated sheets. Localized deposition of light chains can also occur.[1] Infiltration of plasma cells has been observed near localized amyloid deposits suggesting local production of the amyloid precursor protein.[2] Immunohistochemical and molecular genetic studies have characterized clonal plasma cell populations associated with local amyloid deposits, and have matched the amyloid protein with DNA sequencing of the plasma cell clone.[3,4] One case report failed to reproduce this finding.[5] Amyloid lesions are distinct from plasmacytomas in which cellular elements dominate with associated background amyloid material; the presence of both elements is extremely rare.[6] Our goal was to review the spectrum of disease presentation of localized amyloidosis and to evaluate the data on progression to systemic disease.

METHODS

Patients with biopsy proven amyloidosis confined to a single organ seen at Mayo Clinic Jacksonville, Florida, between 1993 and 2003 were identified by searching computer generated diagnosis lists and pathology records. The clinical records and histopathologic findings were reviewed.

RESULTS

We identified twenty cases of localized amyloidosis at our institution between 1993-2003. These patients had biopsy-proven amyloid without clinical or laboratory evidence of systemic disease. By comparison, over a six-year interval, 67 patients were diagnosed with systemic amyloidosis at our institution. Systemic disease was excluded in seventeen of the patients by appropriate studies and examinations. Three of the patients underwent only serum protein immunoelectrophoresis and urine protein immunofixation. The age at diagnosis, protein typing, treatment rendered and findings at follow-up of each patient, are described in the Table. There were eleven males and nine females in the group. The mean age at the time of diagnosis was 65.5 years, median 69 years. The distribution of organ involvement in this group was: skin or soft tissue, four cases; ear, nose or throat, five; lung, four; bladder three; colon, two; conjunctiva, one; lymph node, one. Representative photographic and radiographic findings are illustrated in the Figure. In those patients where follow-up was available (mean 7.6 years), no patient developed systemic disease. Seven patients were treated with excision of the amyloid deposits; three developed local recurrence, three were free of disease, and one was lost to follow-up. Interestingly, both patients with amyloidosis of the colon later displayed no evidence of disease despite lack of treatment. Follow-up data were available for twelve patients; none developed systemic amyloidosis.

DISCUSSION

We identified twenty cases of localized amyloidosis occurring in a variety of organ systems, affecting both sexes in middle to late adulthood. Subcutaneous fat aspiration was used to exclude systemic disease. This is a sensitive and specific method for excluding systemic amyloidosis.[7] The patients did not exhibit any clinical or laboratory evidence of systemic disease at presentation, nor did they evolve into systemic disease over time, supporting the hypothesis of local production of amyloid protein in these cases.

Sites of localized amyloidosis have been reported in a variety of organ systems: brain and spinal nerve roots; larynx, trachea, bronchi, and pulmonary parenchyma; oropharynx, esophagus, stomach, and colon; renal pelvis, ureter, bladder, seminal vesicle and vas deferens; skin and soft tissue; spine, scapula and long bones. Development of systemic disease following a diagnosis of localized amyloidosis was rare, occurring in 2% of cases, with a reported follow-up interval of six months to fifteen years. One quarter of patients with primary systemic amyloidosis present with clinical involvement of only one organ, but investigation leads to the recognition of widespread involvement.[1] The data for exclusion of systemic disease at presentation are not always available in the cited articles; it therefore remains a possibility that some cases appearing to develop systemic disease are instead a failure to diagnose systemic disease at presentation. Our experience suggests that without clinical or laboratory evidence of systemic involvement, progression from localized amyloidosis to systemic disease is uncommon. Further study of localized light chain amyloidosis may elucidate the underlying pathophysiology of tissue tropism for amyloidogenesis, providing research direction for treatment of systemic amyloidosis.

BIBLIOGRAPHY

1. Merlini G and Belloti V. Molecular mechanisms of amyloidosis. N Engl J Med 2003;349:583-96.
2. HagariY, Mihara M, Konohana I, Ueki H, Yamamoto O, Koizumi H. Nodular localized cutaneous amyloidosis. Br J Dermatol 1998;135(4):630-633.
3. Setoguchi M, Hoshii Y, Kawano H, Ishihara T. Analysis of plasma cell clonality in localized AL amyloidosis. Amyloid 2000;7:41-45.

4. Asl K, Liepnieks J, Nakamura M, Benson M. Organ specific (localized) synthesis of Ig light chain amyloid. J Immunol 1999;162:5556-60.

5. Livneh A, et al. Light chain amyloidosis of the urinary bladder. A site restricted deposition of an externally produced immunoglobulin. J Clin Path 2001;54:920-23.

6. Yin H. Soft tissue amyloidoma with features of plasmacytoma: a case report and review. Arch Pathol Lab Med 2002;126(8):969-71.

7. Gertz MA, Li CY, Shirahame T, Kyle R. Utility of subcutaneous fat aspiration for the diagnosis of systemic amyloidosis. Arch Int Med 1988;148:929-33.

Table. Clinical and laboratory data of study patients.

Location	Age	Type	Treatment	Years at followup	Findings at followup
Conjunctiva	65	Lambda	No treatment	7	Local progression
Tongue	73	Lambda	‡	‡	‡
Tongue	69	†	‡	‡	‡
Tonsillar pillar	55	†	‡	‡	‡
Parotid and submandibular glands	73	†	Excision	15	No disease
False vocal cord	77	Kappa	Excision	12	Local recurrence
Tracheobronchial	80	Lambda	No treatment	6	Stable disease
Pulmonary amyloidoma	75	Lambda	Excision	6	No disease
Pulmonary nodules, bilateral	73	†	No treatment	2	Stable disease
Pulmonary nodules, bilateral	65	Kappa	‡	‡	‡
Bladder	35	†	‡	‡	‡
Bladder	53	†	DMSO	12	Local recurrence
Ureter	64	Lambda	Excision	8	No disease
Colon	63	†	No treatment	10	No disease
Colon	68	†	No treatment	7	No disease
Subcutaneous tissues of the legs	67	Kappa	‡	‡	‡
Left axillary nodes	82	Lambda	‡	‡	‡
Skin of nose and chin	41	†	Excision	7	Local recurrence
Skin of nose and cheek	62	†	Excision	12	Local recurrence
Skin of foot	69	†	Excision	2	Local recurrence

†Typing not performed.

‡Follow-up data not available.

LOCALIZED AMYLOIDOSIS: ABOUT 36 CASES

M. Paccalin (1), M. Rubi (1), E. Hachulla (2), M. Carreiro (3), L. Tricot (4), B. Wechsler (5),
J. Cabane (6), C. Cazalet (7), P. Roblot (1)

Department of Internal Medicine : (1) Poitiers, (2) Lille, (3) Toulouse, (5) Paris La Pitié,
(6) Paris Saint-Antoine, (7) Rennes - (4) Department of Nephrology Paris-Necker.

Amyloidosis is the deposition of abnormal proteinaceous substance between cells of various tissues and organs. Several cases of localized amyloidosis are reported in the literature but few data are known concerning the risk of evolution towards systemic amyloidosis (1, 2). We conducted a two-year national inquiry to determine the clinical characteristics, the therapeutic management and the prognosis of amyloidosis localised either at the pharyngo-laryngeal tract, the respiratory tract, the genito-urinary tract, or the gut.

Between January 2001 and December 2002, 19 centers contributed to the inquiry, reporting 36 patients with localised amyloidosis (mean age : 55.5 years, [33-85]).

	Pharyngo-laryngeal (n = 15)	Respiratory tract (n = 10)	Genito-urinary tract (n = 10)	Gut (n = 1)
Mean age, sex (F/M)	52.5 [33-71], (7/8)	59.8 [35-73], (5/5)	54.4 [35-85], (7/3)	73, (0/1)
Symptoms	Hoarseness, dyspnea, change in phonation	Hoarseness, dyspnea, cough	Hematuria, irritative signs	
Immuno-histochemistry	AL : 6	AL : 6	AL : 3	AL : 1
Serum paraprotein	0	IgMλ : 1	0	0
Bone marrow plasmocytosis	2/8	1/6	2/4	0
Biopsies : (fat aspirate / labial salivary gland / rectal biopsy)	Normal (2/8/1)	Normal (3/9/2)	Normal (2/4/6)	Normal (0/1/0)

Table 1. Clinical and biological investigations

The amyloid fibrils derived from monoclonal immunoglobulin light chains (AL type) in 16 cases. Too often, because of a fortuitous discovery, no frozen sample was performed.

Four patients had a cancer associated with amyloidosis : prostate (n =3) and lungs (n = 1). One patient was also treated for sarcoidosis.

Mean follow-up was 5.2 years [1-16]. No case of systemic amyloidosis was reported among the patients. Symptomatic treatment was usually prescribed : laser resection (n=7) and stenting (n=1) in pharyngo-laryngeal amyloidosis, endoscopic resection (n=4) and stenting (n=1) when amyloidosis deposition was in the urinary tract. Three patients were treated shortly with corticosteroids and melphalan, without any improvement. One patient died because of lung cancer, another one because of laryngeal perforation after progression of amyloidosis deposit.

CONCLUSION

Localized amyloidosis lesions might be progressive and require long term follow-up, but this disease does not improve towards systemic amyloidosis.

REFERENCES

1.   A O'Reagan, HM Fenlon, JF Beamis et al. Tracheobronchial amyloidosis. The Boston University experience from 1984 to 1999. Medicine 2000;79:69-79
2.   O Tirzaman, DL Wahner-Roedler, RS Malek et al. Primary localized amyloidosis of the urinary bladder: a case series of 31 patients. Mayo Clin Proc 2000;75:1264-8

STUDIES ON UNFOLDED β2-MICROGLOBULIN C-TERMINAL 92 TO 99 IN DIALYSIS RELATED AMYLOIDOSIS

Y. Motomiya[1], K. Haraoka[2], X. Sun[2], H. Morita[3], Y. Ando[2]

[1]Suiyukai Clinic, [2]Department of Laboratory Medicine, Kumamoto University School of Medicine, [3]Department of Hemodialysis, Chukyo Hospital, [1]Suiyukai Clinic, Kashihara, Nara, Japan

β2 microglobulin (β2M) is well known as the precursor protein of dialysis-related amyloidosis and several variants of β2M have been reported for hemodialysis patients since Gejyo's report in 1985 (1). However, no structural variant directly leading to an amyloid β2M has yet been reported.

β2M consists of 99 amino acids and the monoclonal antibody for C-terminal 92-99 residues has been reported by Stoppini et al.(2) to inhibit fibril formation *in vitro*. More recently, Ivanova et al.(3)reported that C-terminal 28 residues, per se, could form an amyloid fibril, too. On the hypothesis that the unfolding of C-terminal may essentially contribute to amyloidogenesis of β2M *in vivo*, we undertook this study by preparing a monoclonal antibody against the C-terminal 92-99 residues of β2M.

METHODS

First, the peptide identical to the C-terminal octa peptide was synthesized by an automated machine and conjugated with KLH. Then, injected into Balb/c mice and finally the monoclonal antibody was prepared according to Galfre/Milstein's method (4). As to preparation of the denatured β2M, we first isolated β2M from plasma ultrafiltrate and then treated with acidification which had been reported to denature the precursor into the amyloidogenic one (5).

RESULTS

1. Reactivity in several clinical specimens by immunoblotting.

β2M sample of neither serum nor plasma ultrafiltrate showed any blotting for mAb92-99, but an acidified β2M sample of ultlafiltrate showed a weak, but distinct, stain at higher titration of mAb92-99, X500, and X1000, respectively. Carpal specimens all of which was positive for a Congo red stain showed a strong blotting (figure 1)

2. Immunohistochemical studies

Seven carpal tissues were positive for Congo red and other 10 tissue specimens were negative for Congo red stain. All of the Congophilic areas showed a distinct staining for mAb92-99 and, conversely, no staining could be found in non-Congophilic areas as shown by arrows in the picture (figure 2).

Whereas, in 10 Congo red negative tissues, immunohistochemistry of mAb92-99 was differed from one among specimen to another such as positive, co-existing of both positive/negative and negative staining, though an accumulation of β₂M was confirmed by polyclonal anti-β₂M serum evenly in all those specimens (Figure 3).

Serum	β₂m pH 2.5	β₂m pH 7.3	Amyloid fibril	Amyloid fibril	Antibody dilution
					X 100
					X 200
					X 500
					X 1000
(a)	(b)	(c)	(d)	(e)	

Figure 1. Immunoblotting for various specimens. a):serum, b): acidified β₂m of ultrafiltrate from HD patient's plasma, c): β₂m of the ultrafiltrate from HD patient's plasma, d) and e): homogenate from carpal tissue.

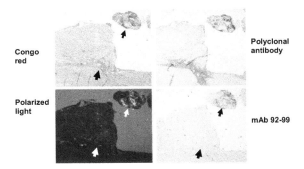

Figure 2 Immunohistochemistry of carpal tissue (small arrows indicate positive stain, large arrows indicate negative)

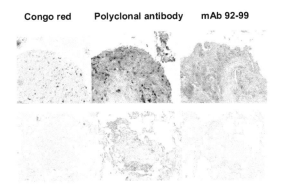

Figure 3. Immunohistochemistry of coracoacromial ligament (Upper:Congo red negative/ mAb92-99 positive, Lower: Congored negative/ mAb92-99 negative)

CONCLUSION

1) The monoclonal antibody specific for the C-terminal of β₂M was prepared.

2) This monoclonal antibody doesn't react to the native β₂M but react to the denatured β₂M by acidification.

3) The immunohistochemical study showed the presence of β_2M with unfolded C-terminal not only in amyloid tissues unexceptionally but also a majority of, although not all, non-amyloid tissues.

4) We consider that the unfolding in the C-terminal of β_2M might precede a fibril formation of β_2M, which suggest a possible molecular structure of intermediate.

5) The existence of β_2M with the unfolded C-terminal in non-Congophilic areas might be considered to provide a new significance of a preemptive accumulation of the native β_2M in β_2-amyloidosis.

REFERENCES

1   Gejyo F, Yamada T, Odani S, Nakagawa Y, Arakawa M, Kunitomo T, et al. (1985). A new form of amyloid protein associated with chronic hemodialysis was identified as $\beta_2$microglobulin. *Biochem .Biophys. Res.Commun.*, **129**, 701-6.

2   Stoppini M, Bellotti V, Mangione P, Merlini G, Ferri G. (1997). Use of anti-($\beta_2$microglobulin) mAb to study formation of amyloid fibrils. *Eur. J. Biochem.*, **249**, 21-6.

3   Galfre G, Milstein C. (1981). Preparation of monoclonal antibodies: strategies and procedures. *Methods. Enzymol.*, **73**, 3-46.

4   Ivanova MI, Gingery M, Whitson LJ, Eisenberg D. (2003). Role of the C-terminal 28 residues of β_2-microglobulin in amyloid fibril formation. *Biochemistry.* **42**,13536-40.

5   Colon W, Kelly JW. (1992). Partial denaturation of transthyretin is sufficient for amyloid fibril formation in vitro. *Biochemistry.*, **31**,8654-60.

SYSTEMIC B₂-MICROGLOBULIN-RELATED AMYLOIDOSIS IN A PATIENT WITH CHRONIC RENAL FAILURE, RECEIVING LONG TERM HEMODIALYSIS FOR 3 YEARS – A LIGHT AND ELECTRON MICROSCOPIC STUDY

Miklós Bély[1] and Tamás Lakatos[2], Ágnes Apáthy[3]

Department of Pathology[2] and Department of Orthopedic Surgery[2], Policlinic of the Hospitaller Brothers of St.John of God in Budapest, and Department of Rheumatology[3] National Institute of Rheumatology and Physiotherapy, H-1027 Budapest, Frankel L Street 17-19.

INRTODUCTION

Gejyo et al (1985) described β₂-Microglobulin related amyloidosis in a patient who had been on regular hemodialysis for 13 years (1). Since then it has become known that long term hemodialysis (HD) may be complicated by systemic, or localized β₂-Microglobulin amyloidosis, frequently involving bones and joints. β₂-Microglobulin amyloid deposits are rare in patients receiving HD treatment for less than 5 years (2).

HD is frequently associated or complicated by disorders of the locomotor system including renal osteodystrophy, pyrophosphate arthritis, avascular necrosis of bone, destructive arthropathy, subchondral bone erosions, cystic bone lesions, periarticular calcification, recurrent hemarthrosis, carpal tunnel syndrome, etc. β₂-Microglobulin amyloid deposits may be present in these disorders, even having a direct causal role, and may cause signs and symptoms. The **aim** of this study was to characterize β2-Microglobulin amyloid depositions in both hip joints of a patient, who was treated with HD for 3 years.

PATIENT'S HISTORY

A **69-year-old woman** was in chronic renal failure (nephrosclerosis) and since May 15, 1996 underwent regular hemodialysis (twice a week) for chronic uraemia.

Clininal Complaints: Since 1984 she had musculoskeletal complaints (pain and restriction of movement). She had constant pain of the ***left hip*** joint since February 1999; at that time X-ray examination revealed destructive arthropathy suspicious for aseptic bone necrosis. In September ***1999***, she underwent total hip replacement.

Since 2001 there was constant pain due to progressive destructive arthropathy of the ***right hip*** joint, suspicious for aseptic bone necrosis. Because of subcapital neck fracture of the right femur [May ***2001***], a total endoprosthesis of right hip joint was implanted.

Clinical Diagnosis: Dialysis associated arthropathy, aseptic bone necrosis of the left femoral head, subcapital neck fracture of the right femur.

Methods: The tissue specimens were fixed in 8% formaldehyde solution and embedded in paraffin. Serial sections were cut and stained with Hematoxylin and Eosin, PAS or Congo red according to Romhányi (1971), without alcoholic differentiation, and sealed with gum arabic. The amyloid deposits were determined and characterized histochemically by Congo red staining after performate pretreatment (85% HCOOH (8 ml), 30% H_2O_2 (3 ml) and 96% H_2SO_4 (220 βl) according to Romhányi (1979) at 20°C for 1, 3, 5, 10, 15, 20 or 25 sec, and after oxidation (0.25% $KMnO_4$ and 0.15% H_2SO_4, 1/1) induced degradation of amyloid deposits according to

Wright et al. (1977) at 20°C for 30 sec, and 1, 2, 3, 4, 5, 6 or 10 min and covered with gum arabic (Bély and Apáthy 2001, Bély 2003).

The classical histological and histochemical observations were confirmed by electron microscopy and immunohistochemical staining using the streptavidin-biotin-complex/horseradish peroxidase method (a-hu beta-2-Microglobulin 1:400; [polyclonal antibody A0072; DAKO, Glostrup, DK).

Microscopic examination of left hip joint showed chronic synovitis with intra- and extracellular, eosinophilic and PAS positive globular deposits of amyloid in the villous synovial membrane (Fig.1.a-b.). With Congo red stain, these globular eosinophilic deposits had a non-birefringent central core and at their periphery there was a ring like birefringence with green polarizations color (Fig.2.a-b.). This peripheral ring showed a positive reaction for anti-human β2-microglobulin immunohistochemically (Fig.3.a-b.). In some areas intracellular amyloid deposits were demonstrable within CD68 positive macrophages. The beta-2-Microglobulin amyloid deposits of the synovial membrane were histochemically **sensitive** to **KMnO4 oxidation** for 30 sec, and were **resistant** to **performate pretreatment** for 1-15 sec, resistant/sensitive for 20 sec. By electron microscop the intra- and extracellular filamentous amyloid deposits showed a dense core-like accumulation and a loose peripheral marginal zone (Figs. 4-5.).

Histologic examination of right hip joint revealed the same changes as previously seen in the contralateral joint: chronic synovitis with β2-microglobulin amyloid deposits, furthermore Congo red positive dystrophic amyloid deposits in the joint capsule, and tumor-like amyloid deposits in the femoral head - negative for anti-human β2-microglobulin, AA, or AL amyloid.

DISCUSSION

The HD associated beta-2-Microglobulin amyloid deposits in the synovial membrane may co-existent with different types of amyloidosis. Dystropic (or other types) of amyloid localized to the articular capsule, may be present in the same patient at the same time.

The presence of congophilic, birefringent amyloid deposits in different joint structures does not necessarily mean the existence of a beta-2-Microglobulin amyloidosis in a patient treated with long term hemodialysis.

The early diagnosis of amyloidosis is important and may indicate a better prognosis in case of adequate treatment.

REFERENCES

1.   Gejyo, F. et al., A new form of amyloid protein associated with chronic hemodialysis was identified as $\square_2$-microglobulin. *Biochem. Biophys. Res. Commun.,* 129, 701-706, 1985.

2.   Maury, C.P.J., $\square_2$-microglobulin amyloidosis. A review. *Rheumatol. Int.,* 10, 1-8, 1990.

3.   Romhányi, G., Selective differentiation between amyloid and connective tissue structures based on the collagen specific topo-optical staining reaction with congo red. *Virchows Arch.,* 354, 209-222, 1971.

4.   Romhányi, G., Selektive Darstellung sowie methodologische Möglichkeiten der Analyse ultrastruktureller Unterschiede von Amyloidablagerungen. Zbl. Allg. Pathol. Pathol. Anat., 123, 9-16, 1979.

5.   Wright, J.R., Calkins, E., and Humphrey, R.L., Potassium permanganate reaction in amyloidosis. Laboratory Investigation, 36, 274-281, 1977.

6.   Bély, M., and Apáthy, Á., Identification of Amyloid Deposits by Histochemical Methods of Romhányi: Applied Histochemistry. Systemic secondary (AA) amyloidosis in Rheumatoid Arthritis. Amyloid: J. Protein Folding Disord., 8, Suppl.2. (Guest Editor: M Bély) 177-182, 2001.

7.   Bély, M., Differential diagnosis of amyloid and amyloidosis by the histochemical methods of Romhányi and Wright. Acta Histochemica, 105, 361-365, 2003.

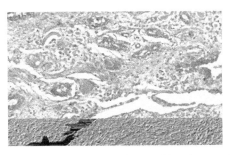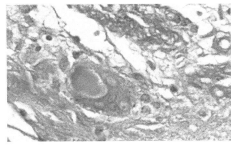

Fig.1.a-b. Synovial membrane. Chronic synovitis. Homogenous - eosinophilic and PAS positive - globular deposits, localized to the villous stroma of the synovial membrane, partly incorporated by macrophages. **(a)** PAS, x125 **(b)** Same as Fig. 1a x400

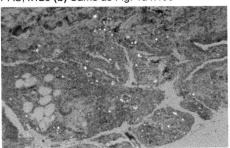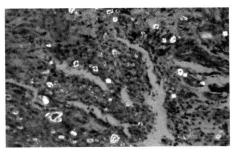

Fig.2.a-b. Synovial membrane. Eosinophilic globular deposits are congophilic and birefringent under polarised light. (a) Congo red staining, without alcoholic differentiation, covered with gum arabic. Viewed under polarised light, x50 **(b)** Same as Fig. 2a x125

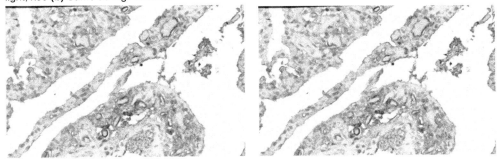

Fig.3.a-b. Synovial membrane. Chronic synovitis. The homogenous eosinophilic globular deposits show diffuse, or superficial positivity for anti-human (a-hu) beta-2-Microglobulin. Negative for a-hu AA, or AL amyloid. **(a)** StrABC/HRP: anti-human (a-hu) beta-2-Microglobulin 1:400, x50 **(b)** Same as Fig. 3a x125

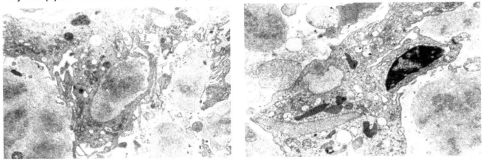

Figs. 4-5. Synovial membrane. Macrophage. Intra- and extracellular filamentous amyloid deposits, with a dense core-like accumulation and a loose peripheral marginal zone. The intracellular amyloid is bordered by membrane. (O: 5000x⇒ A: 11000x)

AN UNUSUAL TUMOR OF AMYLOID IN THE CORNER OF THE EYE: A CASE REPORT WITH POLARIZATION OPTICAL AND IMMUNOHISTOCHEMICAL ANALYSIS

J. Makovitzky and A. Chott

Department of Obstetrics and Gynaecology, University of Rostock, Germany and Department of Pathology, Vienna General Hospital, Vienna, Austria

INTRODUCTION

Tumour-like masses of amyloid, also referred to as amyloid tumour or amyloidoma, may occur as part of systemic amyloidosis or as a primary lesion in which they may coexist with a B-cell lymphoproliferative disorder. Amyloid tumour involving the ocular adnexa/orbit is extremely rare. We report the case of a 48-year old female patient who presented with a 2 cm large tumour mass in the medial corner of the left eye. An initial biopsy in May 1997 showed fatty and fibrous tissue with amyloid deposits. One year later, in June 1998, the amyloid-tumour was partially exstirpated, and the histological and immunohistochemical examination revealed an extranodal lymphoma of mature B-cell type, most consistent with MALT lymphoma with extensive plasmacytic differentiation. Six months later the tumour recurred in a retrobulbar localization, measuring 12.4 x 16 mm and was extirpated in toto. After local radiation therapy (2 courses of 40 Gy) the patient is well as of March 2002, without any evidence of disease.

MATERIALS AND METHODS

The biopsies and the tumour were fixed in 4% formalin in PBS at pH 7.2. The deparaffinized sections (2-4µm) were stained with HE, Giemsa, Gömöri silver impregnation and PAS staining reaction. Immunohistochemical staining reactions for VS38c, IgG, kappa-/ lambda-light chains, CD3, CD20, CD79a and Ki-67 was performed. Congo-red staining was done as described before [1], and potassium-permanganate and/or potassium permanganate-trypsin digestion was done as described by Romhányi [2].

For the polarization-histochemically analysis of the amyloid type we used various topo-optical staining reactions: toluidine blue (tb) and 1.9 dimethyl methylene blue (1.9 dmmb) staining reactions with postprecipitation at various pH values. For the detection of oligosaccharide chains in the amyloid we used the aldehyde bisulfite-toluidine blue reaction (ABT-r) [3], for the sialic acid component the sialic acid specific topo-optical staining reaction [4], and for the detection of RNA residue in amyloid-deposits we used the potassium-permanganate-bisulfite/tb-reaction (PBT-r) [5]. Immunohistochemical examinations were performed with reactions against amyloid (AA, AL kappa and lambda).

RESULTS

Histologically the specimens showed fibrous connective tissue and massive amyloid deposit, surrounded by plasma cells and few giant cells of the foreign-body type. Additionally, foci of lymphoid cells with reactive germinal centres surrounded by plasma-cells could be seen. The plasma cells were positive for

VS38c and IgG and expressed monotypic kappa light chains. The plasma cells were also positive for CD79a, but not for CD20. In these cells the activity of Ki-67, a marker for proliferation, was below 10%.

A resistance to Congo-red anisotropy after $KMnO_4$ and/or $KMnO_4$-trypsin digestion was seen, detecting amyloid of the AL type. The amyloid deposits were positive for AL kappa by immunohistochemistry. (Fig. 1, Fig. 2 a, b)

Fig. 1. Giemsa staining X40

The oriented GAG-components (heparan, keratane, chrondroitine sulfate and hyaluronic acid) could be demonstrated with the tb- topo-optical staining reaction combined with the CEC method at pH 5.2 [6,7]. Using the ABT-reaction, the sialic acid specific topo-optical staining reaction and the PBT-reaction we could selectively demonstrate the oligosaccharide chains, the sialic acid component and residues of RNA in the amyloid deposits.

 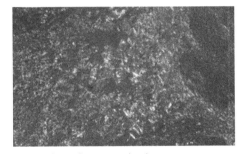

Fig. 2. Congo-red staining, X40   b. under crossed polars

a. lightmicroscopic picture   anisotropy with green polarization color

All non-fibrillar components are oriented sterically in a highly ordered fashion, as in various amyloid types: tumour like-amyloid-deposits, localized amyloid (menisci, Os labrum, breast, synovial membrane), in association with medullary carcinoma of the thyroid etc. [8].

CONCLUSION

MALT lymphoma associated with massive amyloid deposits is very rare and usually occurs in the lung [9]. Case reports describe involvement the breast, stomach, and thyroid [10,11,12,13,14]. Differential diagnosis includes extra-medullary plasmacytoma/ non-secretory myeloma with amyloidosis which we cannot exclude with certainty, however, regard as unlikely, largely because of the presence of reactive germinal centers which usually are not present in plasmacytoma. The criteria for the diagnosis of the non-secretory myeloma have previously been reviewed [15].

This amyloid type has fibrillar (protein skeleton) and nonfibrillar components: glycosaminoglycane (heparan, keratane, chondroitin sulfate and hyaluronic acid) P component, sialic acid and RNA-residues, all these components are oriented in a highly ordered fashion, based on our polarization optical analysis.

REFERENCES

1.   Romhányi, G. Selective differentation between amyloid and connective tissue structures based on the collagen specific topo-optical staining reaction with congo red. Virch Arch Pathol Anat 354:209-222, 1971.

2.   Romhányi, G. Differences in ultrastructural organization of amyloid as revealed by sensitivity or resistance to induced proteolysis. Virch Arch Pathol Anat 357:29-52, 1972.

3.   Romhányi, Gy., Deák, Gy. and J. Fischer. Aldehyde-bisulfite-toluidine blue (ABT) staining as a topo-optical reaction for demonstration of linear order of vicinal OH groups in biological strucures. Histochemistry 43:333-345, 1975.

4.   Makovitzky, J. Ein topo-optischer Nachweis von C_8-C_9- unsubstituierten Neuraminsäureresten in der Glykokalyx von Erythrocyten. Acta histochem 66:192-196, 1980.

5.   Fischer, J. Optical studies on the molecular arrangement of RNA in tissue with a selective topo-optical reaction of RNA. Histochemistry 59:325-333, 1979.

6.   Módis, L. Polarization microscopic of amyloid. In: Organization of the extracellular matrix: a polarization microscopic approach. CRC Press, Boca Raton, FL,USA pp 208-225. 1991.

7.   Appel, R.T. and J. Makovitzky. Romhányi´s staining methods applied to tissue-isolated amyloid fibrils. Acta histochem 105:371-372, 2003.

8.   Makovitzky, J. Polarization optical analysis of amyloid deposits with various topo-optical reactions. Acta histochem 105:369-370, 2003.

9.   Lim, L.K., Lacy, M.Q., Kurtin P.J., Kyle, R.A. and M.A. Gertz. Pulmonary marginal zone lymphome of MALT type as a cause of localised pulmonary amyloidosis. J Clin Pathol 54:642-646, 2001.

10. Wieker, K., Röcken, C., Koenigsmann, M., Roessner, A. and A. Franke. Pulmonary low-grade MALT-lymphoma associated with localized pulmonary amyloidosis. A case report. Amyloid 9:190-193, 2002.

11. Matsumoto, H., Koga, H., Iida, M., Suekane, H., Tarumi, K., Hoshika, K., Mikami, Y. and K. Haruma. Helicobacter-independent, chemotherapy-resistant, radiosensitiv gastric MALT lymphoma with massive deposits of amyloidlike sunstrance. Dig Dis Sci 48:2018-2222, 2003.

12. Caulet, S., Robert, I., Bardaxoglou, E., Noret, P., Tas, P., Le Prise, Y., Launois, B. and M.P. Ramee. Malignant lymphoma of mucosa associated lymphoid tissue: a new etiology of amyloidosis. Pathol Res Pract 191:1203-1207, 1995.

13. Gupta, D., Shidham, V., Zemba-Palko, V. and A. Keshgegian. Primary bilateral mucosa-associated lymphoid tissue lymphoma of the breast with atypical ductal hyperplasia and localized amyloidosis. A case report and reviews of the literature. Arch Pathol Lab Med 124:1233-1236, 2000.

14. Wilk, W. and B. Papla. Amyloidoma and marginal zone malignant lymphoma (MALT type) in gastric antrum. A case report. Pol J Pathol 49:183-186, 1998.

15. Blade, J. and R. Kyle. Non-secretory myeloma, Immunglobulin D myeloma, and plasma cell leukemia. Hematol Oncol Clin North A 13:1259-1972, 1999.

KAPPA III IMMUNOGLOBULIN LIGHT CHAIN ORIGIN OF LOCALIZED ORBITAL AMYLOIDOSIS

K. Hamidi Asl(1), J.J. Liepnieks(1), W.R. Nunery(2), M. Yazaki(1), M.D. Benson(1,3)

(1) Department of Pathology and Laboratory Medicine, (2) Department of Ophthalmology, Indiana University School of Medicine (3) Richard L. Roudebush Veterans Affairs Medical Center, Indianapolis, Indiana USA

INTRODUCTION

Localized orbital amyloidosis is a rare but well recognized entity in which fibril proteins are deposited within the extraocular muscles or adnexae (1). Orbital amyloidosis usually presents as unilateral proptosis with restricted eye movement but occasional bilateral involvement has been reported. Affected individuals seek medical attention because of diplopia or exophthalmos. The condition is usually slowly progressive without visual acuity impairment but with increasing disability due to restricted eye movement. Several studies have suggested an immunoglobulin origin for the amyloid (2-4). We report the biochemical characterization of amyloid from the lateral rectus muscle of the right eye of a patient whose disease progressed from unilateral to bilateral orbital involvement over ten years.

PATIENT

A 27 year old female was first evaluated by an ophthalmologist with the complaint of diplopia of one year duration. Her medical history was significant for trauma to the right eye from an orange thrown during a "pillow fight" in college two years prior. Examination at age 28 revealed limitation of right eye movement with proptosis (5). A CT scan of the orbits revealed thickened right medial rectus and right lateral rectus muscles with proptosis. An orbital muscle biopsy gave the diagnosis of amyloid, and evaluation for systemic amyloid disease, including bone marrow biopsy, immunoelectrophoresis and rectal biopsy, was negative. The patient first presented to Indiana University Medical Center at age 32 when she had progression of right eye proptosis and restricted movement of both eyes. A CT scan showed muscle enlargement of the left orbit as well as the right. A thorough evaluation for systemic amyloidosis was non-revealing and she was treated by surgical decompression of the right orbit. The patient was evaluated again at age 40 at which time all movement in the right eye was lacking and the left eye had severe limitation. Again, a thorough evaluation for systemic amyloidosis was non-revealing. A right lateral orbitotomy with bone flap for removal of amyloid tumor from the orbit was performed and resected tissue was obtained for biochemical analysis. Histologic sections showed massive infiltration of muscle with amyloid.

METHODS

Lateral rectus muscle (1g) was placed in liquid nitrogen and pulverized in a mortar and pestle. The sample was then homogenized with a hand homogenizer in 0.1M sodium citrate, 0.15M sodium chloride and centrifuged at 12,000 RPM for 30 minutes. The pellet was homogenized as above three more times. The final pellet was dialyzed against water and lyophilized. The lyophilized fibrils stained with Congo red and exhibited green birefringence under polarized light. Fibrils were solubilzed in 6M guanidine hydrochloride containing 10mg dithiothreitol/ml, alkylated with iodoacetic acid, and the supernatant after centrifugation chromatographed on a Sepharose CL6B column eluted with 4M guanidine hydrochloride. Pooled fractions were exhaustively dialyzed against water and lyophilized. The Sepharose CL6B pool was digested with trypsin, lyophilized, solubilized with 50% acetic acid, centrifuged, and the supernatant was fractioned by reverse phase HPLC on a Beckman Ultrasphere ODS column eluted with an acetonitrile gradient. Peaks were analyzed by Edman degradation. The Sepharose CL6B pool was also digested with pyroglutamate aminopeptidase, applied to PVDF membrane in a ProSorb cartridge, and analyzed by Edman degradation.

RESULTS

Chromatography of guanidine solubilized fibrils on Sepharose CL6B yielded two retarded peaks eluting in the approximately 10kD and ≤ 5kD areas. SDS-PAGE analysis of the higher molecular weight pool showed a band at approximately 10kD, and of the lower molecular weight pool a broad smear in the 3-5kD region. Sequence analysis of the 10kD pool showed a major sequence starting with residue 150 of the constant region of an immunoglobulin kappa light chain. Analysis of the lower molecular weight pool indicated the presence of numerous peptides, some of which started with residues 127 and 150 of kappa constant region. Analysis of tryptic peptides from the 10kD pool identified residues 3-61 of kappa light chain variable region as the major constituents (Figure 1). Peptides from residues 150-207 of kappa constant region were present, but in approximately 20-25% the molar amounts of variable region. The 50% acetic acid insoluble pellet from the trypsin digest contained a peptide starting with residue 62 and continuing to residue 99 but in less than 5% the molar amount of other variable region tryptic peptides. Analysis of the 10kD pool after pyroglutamate aminopeptidase digestion showed a new sequence starting with Arg2 of the variable region and identified the N-terminus as pyroglutamic acid.

```
              10                  20                  30
<Q-R-V-M-T-Q-S-P-A-T-L-S-V-S-A-G-E-R-A-T-L-S-C-R-A-S-Q-S-V-G-N-N-L-A-W-
              T2                  T3                  T4
  |─────────────────────────>  |────>  |────────────────────
                 <Q-AP
  |──────────────────────────>
```

```
             40                  50                  60                  70
Y-Q-Q-K-P-G-Q-A-P-R-L-L-I-Y-G-A-S-T-R-A-T-G-V-P-A-R-F-S-A-S-G-S-G-T-E-
   T4                       T5                  T6                  T7
  ──────────────────>  |──────────────>  |──────────────>  |──────────────
```

```
             80                  90          95A      99
F-T-L-T-I-S-S-L-Q-S-E-D-F-A-V-Y-Y-C-Q-Q-Y-H-N-W-P-P-Y-T-F-G
                 T7
  ──────────────────────────────────────────────────────>
```

Figure 1. Amino acid sequence of the variable region of kappa III amyloid protein RIV. N-terminal <Q denotes pyroglutamic acid. The arrows indicate the peptides used to determine the sequence. Peptides identified with T were isolated after digestion of the 10kD pool with trypsin. <Q-AP denotes the sequence obtained from the 10kD pool after pyroglutamate aminopeptidase treatment. Residues are numbered according to Kabat, *et al.* (6).

DISCUSSION

Analysis of fibrils isolated from the lateral rectus muscle showed that the amyloid protein was derived from a monoclonal kappa light chain whose variable region is most homologous to the kappa III subgroup. These results demonstrate that isolated orbital amyloidosis is a monoclonal plasma cell disease. As with many reported cases, it is restricted to the orbital muscles. In this case, several evaluations over 15 years failed to show any evidence for systemic amyloidosis. The amyloid process is progressive and may be bilateral. The possibility of chronic inflammatory reaction with subsequent local proliferation of a plasma cell clone is raised.

REFERENCES

1. Nehen JH (1979). Primary localized orbital amyloidosis. *Acta Ophthalmol* **57**: 287-295.
2. Conlon MR, Chapman WB, Burt WL, Larocque BJ and Hearn SA (1991). Primary localized amyloidosis of the lacrimal glands. *Ophthalmology* **98**: 1556-1559.
3. Tan SY, Murdoch IE, Sullivan TJ, Wright JE, Truong O, Hsuan JJ, Hawkins PN and Pepys MB (1994). Primary localized orbital amyloidosis composed of the immunoglobulin ϒ heavy chain CH3 domain. *Clin Sci* **87**: 487-491.
4. Pasternak S, White VA, Gascoyne RD, Perry SR, Johnson RLC and Rootman J (1996). Monoclonal origin of the localized orbital amyloidosis detected by molecular analysis. *Br J Ophthalmol* **80**: 1013-1017.
5. Erie JC, Garrity JA and Norman ME (1989). Orbital amyloidosis involving the extraocular muscles. *Arch Ophthalmol* **107**: 1428-1429.
6. Kabat EA, Wu TT, Perry HM, Gottesman KS and Foeller C (1991). *Sequences of Proteins of Immunological Interest, 5th Edition*, Vol. 1; pp. 103-150. U.S. Department of Health and Human Services, National Institutes of Health, Bethesda, MD.

CORNEAL DYSTROPHY CAUSED BY A NOVEL MUTATION OF THE *TGFBI* GENE: A CASE REPORT

B. Stix[1], J. Rüschoff[2], A. Roessner[1], and C. Röcken[1]

Institutes of Pathology of the [1]Otto-von-Guericke-University Magdeburg and the [2]Klinikum Kassel, Germany

INTRODUCTION

Corneal dystrophy causes visual impairment by interfering with corneal transparency and is subdivided into the granular and the lattice type of corneal dystrophy. The term lattice corneal dystrophy (LCD) contains several forms of disorders, all of which are characterised by the accumulation of amyloid within the cornea. LCD I is the classical form of the disease with an onset of symptoms (epithelial erosions and visual difficulties) in the first decade and is inherited in an autosomal dominant manner. Lattice corneal dystrophy types II, III, and IIIA represent additional variations of LCD with a distinct mode of inheritance and/or clinical features. In all cases, a keratoplasty is frequently required for visual rehabilitation.

Studies of the molecular basis of the corneal dystrophies have revealed that most reported cases of corneal dystrophies are caused by amino acid substitutions within the *TGFBI* gene, located on the long arm of chromosome 5 (5q31) [1-6]. LCD I, for example, has been almost exclusively associated with a substitution of cysteine for arginine at codon 124 in the fourth exon of the *TGFBI* gene. Some additional mutational hotspots are located in the exons 11, 12, and 14 [1-6].

AIMS

We report on a patient who, at 56-years of age, was found to suffer from amyloid deposits within the cornea probably caused by an unknown mutation in the transforming growth factor-beta-induced (*TGFBI*) gene on chromosome 5q31. These are thought to cause abnormal folding and precipitation of the encoded protein, keratoepithelin. Our initial investigations have focused on the exons 4, 11, 12, and 14, the exons with the most mutational hotspots.

MATERIALS AND METHDOS

All biopsy specimens were re-examined. Specimens had been fixed in formalin and embedded in paraffin. Deparaffinized sections were stained with haematoxylin and eosin. The presence of amyloid was demonstrated by the appearance of green birefringence from alkaline alcoholic Congo Red staining under polarized light.

DNA isolation

After full informed consent was obtained, genomic DNA was extracted from peripheral blood leukocytes from the patient using the NucleoSpin Blood L Kit from Macherey & Nagel (Düren, Germany). Exon 4, 11, 12, and 14 of the TGFBI gene, which encoded for kerato-epithelin were amplified by polymerase chain reaction (PCR) using 1 μg of genomic DNA and the primers listed in Table 1. Initial denaturation and activation of Taq polymerase at 95°C for 5 min was followed by 35 cycles (30 for exon 4) with denaturation at 95°C for 30 sec, annealing at 59°C (exon 4), 60°C (exon 11 and 12), and 61°C for exon 14, respectively, for 30 sec, and elongation at 72°C for 60 sec. Final elongation was performed at 72°C for 15 min. The PCR products were isolated (Nucleo Spin Extract, Macherey & Nagel, Düren, Germany) and sequenced using the ABI PRISM Big Dye - Terminator v.1.1 Ready Reaction kit (Applied Biosystems, Darmstadt, Germany) and an ABI PRISM Model 310 capillary sequencer.

RESULTS AND DISCUSSION

The biopsies obtained from the cornea showed large patchy deposits of homogenous eosinophilic material with the green birefringence in poalrized light after Congo red staining characteristic for amyloid. Immunhistochemical staining procedures showed a strong immunoreactivity for amyloid P component, but no reaction for antibodies against amyloid A, apolipoprotein A1, fibrinogen, gelsolin, lambda and kappa light chain, lysozyme and transthyretin. No mutation was found in the gelsolin-gene. DNA analyses of the *TGFBI* gene revealed a heterozygotous T to C substitution at the second position of codon 540. In the *TGFBI* gene, there are 8 different mutations described so far (R124C, R124L, L518P, A546D, P551Q, H626R, P551T, and L527R), which result in the lattice corneal dystrophy with the development of amyloid. In our continuing studies, we have analysed the amino acid sequences of the exons 7, 9, and 10 without finding any mutations. Here, we describe a novel single nucleotide substitution (T540C), which would result in a replacement of phenylalanine by serine at this position. This mutation is thought to cause abnormal folding and precipitation of the encoded protein, keratoepithelin.

To the best of our knowledge this is the first report of a SNP at position 540 of the keratoepithelin gene *TGFBI* [1-6]. However, at this time point we are unable to classify this new mutation within the group of different forms of corneal dystrophy.

REFERENCES

1. Dighiero, P., et al. A new mutation (A546T) of the betaig-h3 gene responsible for a French lattice corneal dystrophy type IIIA. *Am J Ophthalmol.*, 129, 248-251, 2000.

2. Hirano, K., et al. Corneal amyloidosis caused by Leu518Pro mutation of beta ig-h3 gene. *Brit. J. Ophthalmol.,* 84, 583-585, 2000.

3. Hirano, K., et al. Late-onset form of lattice corneal dystrophy caused by leu527Arg mutation of the TGFBI gene. *Cornea*, 20, 525-529, 2001.

4. Hotta, Y. et al. Arg124Cys mutation of the betaig-h3 bene in a Japanese family with lattice corneal dystrophy type I. *Jpn.J Ophthalmol.*, 42, 450-455, 1998.

5. Klintworth, G.K., Bao, W., and Afshari,N.A. Two mutations in the TGFBI (BIGH3) gene associated with lattice corneal dystrophy in an extensively studied family. *Invest Ophthalmol. Vis. Sci.*, 45, 1382-1388, 2004.

6. Munier, F.L., et al. Kerato-epithelin mutations in four 5q31-linked corneal dystrophies. *Nature Genet.* 15, 247-251, 1997.

FREE FATTY ACIDS PROMOTE AMYLOID FORMATION WITHOUT PARTICIPATING IN THE AMYLOID LIKE FIBRIL

Gunilla T Westermark

Division of Cell Biology, Faculty of Health Sciences, Linköping University, Linköping, Sweden

Introduction

Islet amyloid is the most common pathological finding in the islets of Langerhans and depositions are present to various degrees in the majority of individuals with type 2 diabetes. Islet amyloid polypeptide (IAPP) is the main constituent of the islet amyloid and its deposition is accompanied by a 50-60 % reduction of beta-cell mass (1, 2). IAPP is produced by the beta-cells in which it is stored and released together with insulin after stimulation. The estimated molar ratio of IAPP vs. insulin is one to ten in the secretory granules.

Studies on transgenic mice over-expressing human IAPP indicate that solely over-expression of the precursor protein is not sufficient for initiation of amyloid formation. Instead, an interesting, but occasional finding was that islet amyloid occurred in transgenic animals after a prolonged intake of a diet high in fat. A majority of diabetic patients are obese with increased circulating levels of free fatty acids (FFAs). FFAs affect the beta cells directly and stimulate acute insulin secretion from isolated islets. Both the length of the chain and the number of double bounds are of importance. We have shown that culturing isolated islets from a transgenic mouse expressing the gene for human IAPP but deficient of the endogenous IAPP gene, in the presence of FFAs results in the formation of an intra-granular deposition. These deposits reveal a fibrilar appearance and immunolabels with antibodies specific for human IAPP. *In in vitro* studies on amyloid formation with the sensitive Thioflavin T assay we have shown that the presence of FFAs can promote fibril formation. This work describes the morphological appearance of IAPP amyloid-like fibrils formed in the presence of FFAs and the results of biochemical analyses of IAPP fibrils assembled in the presence of FFAs.

Material and methods

Human IAPP 1-37 was synthesized by Keck Laboratories, Yale University, USA. Prior to the study, the IAPP was dissolved in hexa fluoro isopropanol (HFIP), filtered and lyophilized. Stock solutions of the sodium salts of FFAs (myristic acid, oleic acid, palmitic acid, stearatic acid and linoleic acid) were prepared in 100 % ethanol. Amyloid-like fibrils were formed at an IAPP concentration of 2.5 mM and 10 mM of FFA over 24 hours. The FFAs were tested individually. The formed fibrils were pelleted by centrifugation at 15,000xg for 60 minutes in an Eppendorf bench centrifuge. The supernatant was collected and lyophilized and the pellet was washed with water and lyophilized. The supernatant and pellet fraction were analyzed with matrix assisted laser desorption ionization time of flight (MALDI-TOF) (Voyager-DE Pro, Applied Biosystems, USA) and electro spray ionization mass spectrometry at positive and negative mode (Protana, Odense, Denmark).

Samples analyzed with MALDI-TOF were re-dissolved in 90% acetonitrile and heated to 90°C for 10 minutes and samples analyzed with ion spray mass spectrometry were dissolved in 50% acetonitrile in water.

The IAPP fibrils, assembled in the presence or absence of FFAs were negatively contrasted with 5% uranyl acetate in 50 % ethanol and studied in a Jeol-2100 electron microscope (Jeol, Tokyo, Japan).

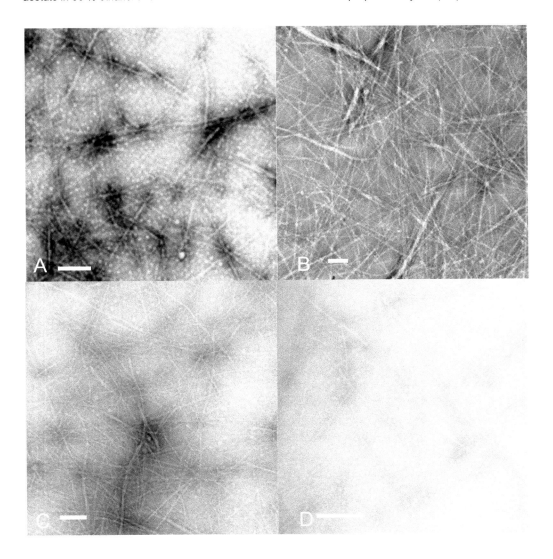

Figure 1 IAPP (1-37) assembled in 4 % ethanol (A), IAPP (1-37) assembled in the presence NA-oleate and 4 % ethanol (B), IAPP (1-37) assembled in the presence of Na-stearat and 4 % ethanol (C) and IAPP (1-37) assembled in the presence of Na-linoleic and 4 % ethanol (D). Bar = 200 nm

Results

IAPP fibrils were straight and unbranched and about 8 nm in width. Fibrils assembled in the presence of FFAs did not differ morphologically from those without FFAs (Figure 1 A-D). Mass spectrometry analysis of the pellet material of the samples assembled in the presence of FFAs revealed in all samples IAPP with the expected molecular mass, identically with the results in material without FFAs. Fatty acids were not identified here. In examination of the supernatants with ion spray mass spectrometry in negative mode, FFAs were detected. The pellet material with fibrils contained only IAPP and no FFAs were identified.

Discussion

The reason why islet amyloid develops in type 2 diabetes is unknown but may be multifactorial. FFAs occur in an increased concentration in plasma in diabetes and may affect fibril formation directly. In an *in vitro* system, where we used the Thioflavin T assay, we showed previously that FFAs enhance formation of amyloid like fibrils from IAPP. This finding raised the question whether FFAs are incorporated in the fibrils. The results of the present study indicate that no FFAs occur in the amyloid like fibrils themselves. Instead, FFAs may affect fibril formation from IAPP by other mechanisms. One possibility is that FFAs induce beta sheet structure of the normally non-structured peptide IAPP and thereby makes it more prone to aggregate into amyloid like fibrils. It may also be possible that FFAs participate more directly at an early stage of IAPP aggregation, e.g. during the formation of the nucleus for fibril formation. Studies of the secondary structure of IAPP in the presence and absence of FFAs are obviously of interest and are currently performed. It would also be of interest to study whether FFAs only affects fibril formation from IAPP or whether the effects are the same with other amyloid fibril proteins, e.g. Abeta or ATTR (5).

Acknowledgments

Supported by the Swedish Research Council (project no 14040-03A), the Swedish Diabetes Association, The Novo Nordic Insulin Foundation, Ollie and Elof Ericsson Foundation and Magnus Bergvall Foundation

References

1.   Westermark, P., Wernstedt, C., Wilander, E. and Sletten, K. (1986). A novel peptide in the calcitonin gene related peptide family as an amyloid fibril protein in the endocrine pancreas. *Biochem. Biophys. Res. Commun.,* **140**, 827-831.

2.   Clark, A., Wells, C. A., Buley, I. D., Cruickshank, J. K., Vanhegan R.I., Matthews, D. R., Cooper G. J. S., Holman R. R. and Turner, R.C. (1988). Islet amyloid, increased A-cells, reduced B-cells and exocrine fibrosis: quantitative changes in the pancreas in type 2 diabetes. *Diab. Res.,* **9**, 151-159.

3.   Verchere, C. B., D'Alessio, D. A., Palmiter, R. D., Weir, G. C., Bonner-Weir, S., Baskin, D. G. and Kahn, S. E. (1996). Islet amyloid formation associated with hyperglycemia in transgenic mice with pancreatic beta cell expression of human islet amyloid polypeptide. *Proc. Natl. Acad. Sci. U.S.A.,* **93**, 3492-3496.

4.   Zhi, M. and Westermark, G. T. (2002) Effects of free fatty acid on polymerization of islet amyloidpolypeptide (IAPP) in vitro and on amyloid fibril formation in cultivated isolated islets oftransgenic mice overexpressing human IAPP. Mol Med., **8**, 863-868.

5.   Wilson, D. M. and Binder, L. I. (1997) Free fatty acids stimulate the polymerization of tau and amyloid beta peptides. In vitro evidence for a common effector of pathogenesis in Alzheimer's disease. Am J Pathol. , **150**, 2181-95.

MEDULLARY THYROID CARCINOMA-ASSOCIATED AMYLOID IS COMPRISED OF CALCITONIN GENE-RELATED PEPTIDE II

C. L. Murphy, S. Wang, B. Crombie, S.D. Macy, T.K. Williams, D.T. Weiss, and A. Solomon

Human Immunology and Cancer Program, Department of Medicine, University of Tennessee Graduate School of Medicine, Knoxville, TN, USA

I. INTRODUCTION

Localized deposits of amyloid are commonly found in human medullary thyroid carcinoma (MTC). Heretofore, sequence analyses of this material had revealed the presence of calcitonin (CT) and/or its precursor, procalcitonin (PCT).[1,2] We now report that this type of amyloid is composed of another closely related and, perhaps more relevant molecule; namely, calcitonin gene-related peptide II (CGRP II). This component (in addition to CT and PCT) was identified through chemical analyses of fibrils extracted from formalin-fixed, paraffin-embedded sections of amyloid-containing tumors from two patients with MTC. Immunohistochemical analyses using antisera specific for each of the 3 proteins, revealed diffuse staining of the thyroid tissues by the anti-CT and –PCT reagents, whereas the anti-CGRP II antibody reacted solely with the green birefringent congophilic deposits. The results of our chemical and immunologic studies indicate that the amyloid found in MTC is formed, not from CT or PCT, but rather, from CGRP II.

II. MATERIALS AND METHODS

Fibrils were extracted from 4 μm-thick sections of formalin-fixed, paraffin-embedded amyloid-laden tissue from 2 patients with MTC. For control purposes, the same procedures were used to obtain material from a specimen of normal thyroid. The extracts were reduced and alkylated, passaged through a reverse phase-HPLC column, and the relevant peaks subjected to direct (automated) Edman degradation, as well as trypsin digestion. The tryptic peptides were isolated by HPLC for sequence and MS/MS spectrometric analyses.[3]

III. RESULTS

The reverse phase-HPLC profiles of the 2 amyloid extracts are shown in Figure 1. In Cases #1 and #2, Peak 2 contained peptides encompassing the entire CGRP II molecule (Figure 2). A fragment lacking the N-terminal 11 residues of this protein was present in Peak 1. Additionally, in Case #1, CT and a peptide corresponding to the N-terminal portion of PCT were detected in Peaks 1 and 2, respectively (histone-related peptides were found in Peak 3). In Case #2, defensin-related peptides were found in Peaks 1 and 2, and in Peak 3, those corresponding to PCT precursor I, as well as hemoglobin. In contrast, only PCT, but not CGRP II (or CT) was found in the normal thyroid extract.

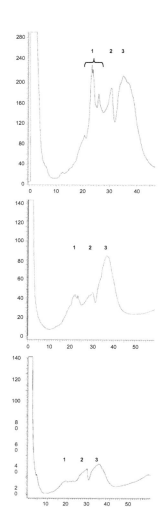

MTC Case #1 (S96-30348): *Peak 1*; C-terminal CGRP II fragment, CT. *Peak 2*; CGRP II, PCT fragment. *Peak 3*; Histones

MTC Case #2 (01-7637): *Peak 1*; C-terminal CGRP II fragment, defensin. *Peak 2*; CGRP II, defensin. *Peak 3*; PCT precursor I, hemoglobin.

Normal thyroid Case #3: *Peak 1*; HSP 10 kD, crystalline, prostatic binding protein, chromogranin a. *Peak 2*; hemoglobin, ribosomal protein, glyceraldehyde 3P dehydrogenase, chromogranin b. *Peak 3*; PCT, ribosomal proteins, hemoglobin, HSP 27 kD.

Figure1. Reverse phase-HPLC profiles of water soluble, protein extracted from 2 cases of MTC (#1 and #2) and one of normal thyroid gland (Case #3).

```
CGRP II      ACNTATCVTH RLAGLLSRSG GMVKSNFVPT NVGSKAF
S96-3048     ---------- ---------- ---------- -------
01-7637      ---------- --------[       ]------ -------
```

Figure 2. Comparison of the amino acid sequences of CGRP II and MTC-amyloid associated proteins. The dashes indicate residue identity. [ ], residues not identified.

In immunohistochemical analyses, consecutively cut sections from each of the MTC cases were stained with Congo red, anti-CGRP II, anti-CT, and anti-PCT antibodies. In both, the apple green birefringent congophilic deposits were recognized only by the anti-CGRP II antibody. In contrast, there was diffuse tissue reactivity with the anti-CT and –PCT antisera and the amyloid was not immunostained by these reagents.

IV. DISCUSSION

Although CT (or PCT) has been implicated as the amyloidogenic precursor in MTC, our chemical studies of fibrillar material extracted from tissue obtained from 2 patients with this disorder have revealed the presence of yet another protein that is encoded by a CT-related gene, namely, CGRP II. That this molecule was the fibrillar constituent was evidenced through immunohistochemical analyses that showed the deposits to be immunostained by the anti-CGRP II antibody, whereas the reactivity of the anti-CT and -PCT reagents was limited to the tumor cells and surrounding thyroid tissue. Based on these results, we posit that the detection of CT and/or PCT in water soluble thyroid tumor extracts may have resulted from their entrapment within the fibril structure and that, in fact, CGRP II is the amyloidogenic protein associated with MTC.

VI. CONCLUSION

We have demonstrated that MTC-associated amyloid is comprised of CGRP II and provisionally designate this entity as ACGRP II.

VII. ACKNOWLEDGEMENTS

The authors thank Dr. John Neff for providing the MTC and normal thyroid specimens. This study was supported, in part, by USPHS Research Grant CA10056 from the National Cancer Institute and the Aslan Foundation. A.S. is an American Cancer Society Clinical Research Professor.

VIII. REFERENCES

1. Sletten, K., Westermark, P., and Natvig, J.B., Characterization of amyloid fibril proteins from medullary carcinoma of the thyroid, *J. Exp. Med.* 143, 993, 1976.
2. Sletten, K., Westermark, P., and Natvig, J.B., Characterization of molecular forms of calcitonin in amyloid fibrils from medullary carcinoma of the thyroid, in *Amyloid and Amyloidosis*, Natvig, J.B. et al., Eds., Kluwer Academic Publishers, Boston, 1990, p.477.
3. Murphy, C.L., et al., Chemical typing of amyloid protein contained in formalin-fixed paraffin-embedded biopsy specimens, *Am. J. Clin. Pathol.*, 116, 135, 2001.

STUDIES OF THE AMYLOIDOGENIC PROPERTIES OF proIAPP

J F Paulsson and G T Westermark

Division of Cell Biology, Faculty of Health Sciences, Linköping University, Sweden

Introduction

Islet amyloid is associated with type 2 diabetes and is found in more than 90% of diabetic individuals post mortem. Islet amyloid consists of islet amyloid polypeptide (IAPP) which is produced by pancreatic beta-cells and is co-stored with insulin in the secretory granula (1).

Both insulin and IAPP are expressed as prohormones and undergo enzymatic cleavage at dibasic amino acids performed by the prohormone convertases PC 2 and PC 1/3. On proIAPP processing PC 2 is active at the N-terminus while PC 1/3 has its activity at the C-terminus. In studies of PC 1/3 null mice it was shown that PC 2 has the ability to fully process proIAPP into IAPP (2) (Figure 1). Processing is a prerequisite for biological activity. It takes place in the maturing secretory granules and the enzymes require a low pH for their activity.

Islet amyloid is described as an extra cellular deposit but intracellular amyloid has been found in our hIAPP+/+ mIAPP-/- mouse strain and in beta-cells from human transplanted islets (3). Diabetic patients have elevated levels of circulating proinsulin and most probably also high levels of proIAPP, perhaps depending on chronically elevated levels of free fatty acids which reduce the activation of the prohormone convertases (4, 5). Absent or aberrant processing of proIAPP has been postulated as a possible mechanism for the initiation of islet amyloid. Here we have set up a model for studying amyloidogenic properties of proIAPP.

Figure 1. Amino acid sequence of proIAPP with the cleavage sites for PC 2 and PC 1/3 indicated with arrows. Epitope for antiserum A 133 is shown below proIAPP.

Material and Methods

Transfection: Beta-TC-6 cells and GH4C1 cells were used for transfection.

PreproIAPP was ligated into pcDNA3 eukaryotic expression vector and cells were transfected with polyethylenimine (low molecular weight). Cells were cultured overnight on cover slips in cell culture dish and transfected with 40µg DNA in 10mM polyethylenimine in 5% sucrose solution in serum free RPMI medium. After 6 hours of incubation, fetal bovine serum was added to the medium. Cells were fixed with 2% paraformaldehyde in PBS 48 hours after transfection.

Immunohistochemistry: Monoclonal antibodies against PC 2 and PC 1/3 were produced with regular hybridoma technique using sp2/0 myeloma cells as fusion partners. The monoclonal antibodies against PC 2 were IgM and the antibodies against PC 1/3 were IgG. The presence of PC 2 and PC 1/3 expression was studied in untransfected Beta-TC-6 cells and GH4C1 cells cultured on cover slips. The PC 2 and PC 1/3 monoclonal antibodies were used directly from the culture medium without purification, and their reactivity was detected with secondary anti mouse antibodies conjugated with Alexa 488 (anti mouse IgG) and 595 (anti mouse IgM).

Expression of human IAPP was shown with antiserum A 133 (Figure 1), specific for human IAPP without cross reactivity with mouse IAPP. The reaction was detected by Alexa 488 conjugated secondary anti rabbit antibody and visualized in a Nikon C1 confocal microscope. Cells were also stained with Congo red and investigated by confocal microscopy using the He/Ne laser 543 where the positively stained amyloid reveals a red fluorescence (6).

Recombinant expression of proIAPP and IAPP: cDNA for proIAPP and IAPP was ligated into the pGEX 2TK vector and transformed in Y1090 bacteria. Protein expression was induced by 3mM IPTG. Protein expressed in this vector is produced as fusion protein with a GST-tag N-terminally of the expressed peptide. After sonication of the bacteria the GST-tag-peptide were bound to Sepharose 4B beds followed by several washing steps to remove bacterial proteins. The peptides were then removed from the GST-tag by enzymatic cleavage. For electron microscopy, small aliquots were applied to fomvar coated grids, contrasted with uranyl acetate and studied in a Jeol-2100 (Jeol, Tokyo, Japan) electron microscope.

Results

Transfection: Beta-TC-6 cells and GH4C1 cells were transfected with the human proIAPP and the transfection efficiency was determined to 5% and 10% respectively.

Immunohistochemistry: Beta-TC-6 cells immunolabeled with antibodies against PC 2 and PC 1/3 while reactivity was absent in the GH4C1 cells.

When transfected with human preproIAPP, cells from both cell lines show immunolabeling with antiserum A 133, specific for human IAPP (Table 2). In Beta-TC-6 cells a granular staining pattern occurred different from the larger intracellular aggregates present in the GH4C1 cells expressing human proIAPP. Amyloid specific red fluorescence after Congo red staining was detected in GH4C1 but this was absent in the Beta-TC-6 cells (Table 2).

Recombinant expression of proIAPP and IAPP: GST-proIAPP and GST-IAPP peptides were expressed in Y1090 bacteria without the formation of insoluble inclusion bodies. ProIAPP and IAPP were cleaved of from the GST- tag after the addition of thrombin protease. After 6 hours, aggregates were present in the solution and electron microscopical analysis revealed the presence of amyloid-like fibrils (Figure 2).

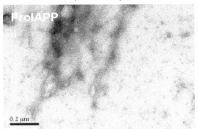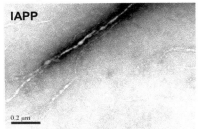

Figure 2. Aliquots from aggregate containing solution, taken 6 hours after the addition of thrombin protease, show that amyloid like fibrils appears in both proIAPP and IAPP solutions.

Table 2 The results of immunohistochemistry with antiserum A133, anti-PC 2 and anti-PC 1/3 and Congo red staining of human ProIAPP transfected Beta-TC-6 and GH4C1 cells

Cell line	PC	A 133	Congo red
Beta-TC-6	PC 1/3 and P C2	+	-
GH4C1	-	+	+

Discussion

We have used two different cell lines which both contain the regulatory pathway for secretion for transfection of human preproIAPP. The Beta-TC-6 cells express PC 2 and PC 1/3 and in these cells proIAPP is expressed into IAPP while GH4C1 cells do not express any of the prohormone convertases and are therefore unable of process proIAPP.

In transfected GH4C1 cells large intracellular aggregates of IAPP immunoreactive material could be detected. When these cells were stained with Congo red, amyloid specific red fluorescence was detected. This amyloid is made up by proIAPP because the GH4C1 cells lack the processing enzymes necessary for proIAPP processing. Amyloid was not detected in the beta-TC-6 cells. Instead, a granular labelling pattern appeared after IAPP immunolabeling.

Expressing IAPP and proIAPP as GST-fusion proteins prevents formation of inclusion bodies in Y1090 bacteria and keeps the recombinant IAPP peptides in solution during the extraction steps. Interestingly, when the GST-tag was removed from the expressed peptides both proIAPP and IAPP aggregated and formed amyloid within a few hours.

Transfection studies with human IAPP have been reported previously. These studies were performed in COS-1 fibroblast cells and amyloid was present intracellularly (7). It is not clear if preproIAPP or IAPP was used for these studies. In our model we used endocrine cell lines and found that when the amyloid was accumulated it consisted of proIAPP. Recombinantly expressed ProIAPP and IAPP showed similar amyloidogenic properties and both formed amyloid like fibrils. In conclusion absent processing of proIAPP can be a possible mechanism for formation of intracellular amyloid.

Supported by the Swedish Research Council (project no 14040-03A), the Swedish Diabetes Association, The Novo Nordic Insulin Foundation, Ollie and Elof Ericsson Foundation, Magnus Bergvall Foundation and Östergötland County Council medical research foundation.

References

1. Westermark, P., Wernstedt, C., Wilander, E. and Sletten, K. (1986). A novel peptide in the calcitonin gene related peptide family as an amyloid fibril protein in the endocrine pancreas. *Biochem. Biophys. Res. Commun.*, **140**, 827-31.
2. Badman, M. K., Shennan, K. I., Jermany, J. L., Docherty, K. and Clark, A. (1996). Processing of pro-islet amyloid polypeptide (proIAPP) by the prohormone convertase PC2. *FEBS Lett.*, **378**, 227-31.
3. Westermark, G. T., Gebre-Medhin, S., Steiner, D. F. and Westermark, P. (2000). Islet amyloid development in a mouse strain lacking endogenous islet amyloid polypeptide (IAPP) but expressing human IAPP. *Mol. Med.*, **6**, 998-1007.
4. Yoshioka, N., Kuzuya, T., Matsuda, A., Taniguchi, M. and Iwamoto, Y. (1988). Serum proinsulin levels at fasting and after oral glucose load in patients with type 2 (non-insulin-dependent) diabetes mellitus. *Diabetologia*, **31**, 355-60.
5. Furukawa, H., Carroll, R.J., Swift, H. H. and Steiner, D. F. (1999). Long-term elevation of free fatty acids leads to delayed processing of proinsulin and prohormone convertases 2 and 3 in the pancreatic beta-cell line MIN6. *Diabetes*, **48**, 1395-401.
6. Puchtle, H. and Sweat, F. (1965). Congo red as a stain for fluorescence microscopy of amyloid. *J. Histochem. Cytochem.*, **13**, 693-4.
7. Hiddinga, H. J. and Eberhardt, N. L. (1999). Intracellular amyloidogenesis by human islet amyloid polypeptide induces apoptosis in COS-1 cells. *Am. J. Pathol.*, **154**, 1077-88.

LARYNGEAL AMYLOIDOSIS: LOCALIZED VERSUS SYSTEMIC DISEASE

H. Bartels [1], F.G. Dikkers [1], J.E. van der Wal [2], H.M. Lokhorst [3] and B.P.C. Hazenberg [4]

Departments of Otorhinolaryngology (1), Pathology (2), and Rheumatology (4), University Hospital Groningen, Groningen, The Netherlands and Department of Hematology, University Medical Center Utrecht (3), Utrecht, The Netherlands

OBJECTIVE

To study clinical and pathological characteristics, possibility of systemic disease, and effect of local therapy in patients with laryngeal amyloidosis.

PATIENTS AND METHODS

Records of all patients with localized laryngeal amyloidosis in a single tertiary referral center were examined retrospectively at diagnosis and after local therapy. Out of 188 new patients with amyloidosis between 1990 and 2003, five patients had localized laryngeal amyloidosis. One of these patients (*) already had ocular amyloidosis without systemic involvement. A sixth patient (#) known elsewhere with localized laryngeal amyloid turned out to have systemic AL amyloidosis 8 years later. This patient was added tot the study group. Patient characteristics are listed in table 1.

Table 1. Characteristics at initial presentation of the patients with localized laryngeal amyloidosis

	sex	anatomic characteristics		hoarseness	medical history of amyloidosis	
		location	site	G R B A S	age at first laryngeal treatment	follow-up since (years)
	male	supraglottis	unilateral right	3 3 3 0 2	24	5
*	female	supraglottis and glottis	unilateral left	2 0 2 2 1	41	1
	female	glottis	bilateral	3 3 2 0 3	42	7
	male	glottis	bilateral	3 3 3 0 3	50	5
	male	supraglottis	unilateral right	3 3 3 0 0	57	3
#	male	Supraglottis	bilateral	3 2 3 0 3	46	13

* patient with also ocular amyloidosis; # patient developing systemic amyloidosis after eight years (case six); G = overall grade; R = roughness; B = breathiness; A = asthenicity; S = strained quality

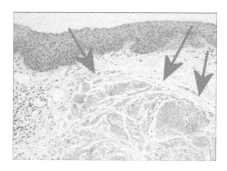

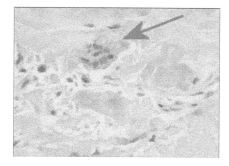

Figure 1. Congo red stained amyloid deposit.   **Figure 2.** Giant cells at the border of a deposit.

RESULTS

The ratio of free light chains in serum was abnormal in two patients: the sixth patient with systemic amyloidosis (#) and the patient already known with ocular amyloidosis (*). Histological examination showed giant cells at the peripheral margins of amyloid in two patients (fig. 1 and 2). Immunohistology helped to exclude AA amyloid, confirmed the presence of SAP, but failed to detect specific reactivity for either kappa or lambda light chain. The most important indications for treatment were dramatic voice changes, objectified by phonetograms, and decreased tolerance of exertion. Amyloid interfering with laryngeal or airway function was removed during microlaryngoscopy with CO2-laser or cold endoscopic excision. The best results were seen when glottic deposits (fig. 3) were removed by cold endoscopic excision and supraglottic deposits (fig. 4) were removed by CO2-laser. Four patients showed recurrent disease, requiring surgery in only one patient after 7 months.

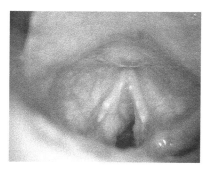

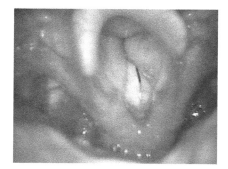

Figure 3. Glottic amyloidosis   **Figure 4.** Supraglottic amyloidosis

CONCLUSIONS

Laryngeal amyloidosis is a slowly progressive disease, which should be treated depending on the complaints of the patient. Stabilization often occurs after a number of years. A systematic work-up, including measurement of free light chains, helps to detect systemic disease. Follow-up should be yearly for at least 10 years.

FORMATION OF THE VASCULAR AMYLOID FIBRIL PROTEIN MEDIN FROM ITS PRECURSOR LACTADHERIN

S. Peng[1], A. Persson[1], E. Wassberg[2], P. Westermark[1]

[1]Department of Genetics and Pathology and [2]Department of Surgical Sciences, Uppsala University, Uppsala, Sweden

INTRODUCTION

Localized amyloid in the media of the artery wall was described decades ago (1), but the protein making up the amyloid was not characterized until 1999 by Häggqvist et.al. (2), as being an internal cleavage product of the protein lactadherin. The amyloid protein was, due to its localization to the media termed medin. Medin amyloid, one of the most common amyloids, affects the population above 50 years old (3, 4). It is mostly found in the aorta, but other arteries are also affected especially in the upper part of the body (Peng et al.,submitted). Medin amyloid is found predominantly in the inner part of the media layer. The mechanism by which medin is formed from lactadherin and if this event is normal or pathological are not known, nor are the consequences of the deposition on the surrounding tissue. No vascular disease is yet found to be associated with medin amyloid.

Medin is a 50 amino acid residue peptide intercleaved from the C2-like domain of the precursor protein lactadherin (2). Lactadherin was first reported as a milk fat globule protein in human milk (5). Later studies showed that it is expressed by various kinds of mammalian cells such as breast epithelial cells, smooth muscle cells, and macrophages (6-8). Many functions have been proposed for lactadherin. It may be involved in cell adhesion via the N-terminal RGD motif in the EGF-like domain or mediate macrophage engulfment of apoptotic cells (8).

The aim of this study was to further elucidate the exact location of AMed amyloid in the aortic wall. The aim was also to find out whether medin occurs in aortic tissue devoid of amyloid.

MATERIAL AND METHODS

Biopsies from the thoracic aorta were obtained at surgery. Part of the material was fixed in 4% formaldehyde, prior to paraffin embedding and immunohistochemistry and the second part of the material was stored at -20 °C. The frozen material was used for protein extraction followed by western blotting and or was sectioned for Congo red fluorescence studies. Antibodies A-179 (against medin) and A-180 (against lactadherin, but outside of medin) were used in both immunohistochemistry and Western blotting. The study was approved by the ethical committee at the University Hospital.

RESULTS

In immunohistochemistry, antibody A-179 recognized medin amyloid but not lactadherin. Medin amyloid was often found close to the elastic lamina of the media of the aorta, and sometimes around the nuclei of smooth muscle cells. A-180 immunoreactivity was usually seen in close association to the elastin of arteries.

Confocal microscopy of Congo red stained frozen tissue sections clearly showed amyloid deposits within smooth muscle cells in the media layer of aorta, as well as extracellular amyloid. The intracellular amyloid seemed to wrap around the nuclei, sometimes even intruding into it.

Western blotting showed various bands with antibody A-179, but only aortic tissue with amyloid showed two medin specific bands, one at 6.5 Kd and one at 3.5 kD. As expected, lactadherin was found in both amyloid- and non-amyloid aorta.

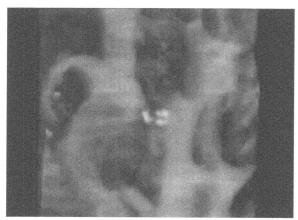

Figure 1. Image of human amyloid-containing aorta stained with Congo-red and studied in a fluorescence microscope. Light spots are partially intracellular amyloid deposits.
Gray areas correspond to elastic lamina.

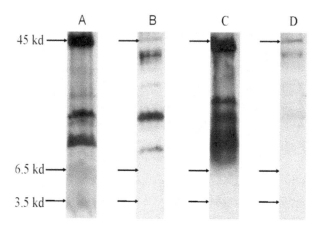

Figure 2. Western-blotting of amyloid (A,C) and non-amyloid (B,D) cases with antibody against medin (A,B), and antibody against lactadherin outside medin (C,D)

CONCLUSIONS

In immunohistochemistry antibody A-179 labeled medin amyloid, but did not seem to label lactadherin. The A-179 epitope is probably hidden in the native protein and is only exposed when medin has formed. Antibody A-180 labeled lactadherin but not medin. The labeling was often in close association to elastin, which might indicate that lactadherin has an elastin-binding role in arteries (Persson et al., manuscript in preparation).

Intracellular medin amyloid was detected by confocal microscopy. Intracellular fibrillar deposits might be the start point of extracellular medin amyloid. If the cleavage of lactadherin to medin occurs within the smooth muscle cell, this may give rise to protofibrillar formation, toxic to the cell. We believe that this may lead to cell death and that the intracellular amyloid thereby becomes extracellular.

In western blot analyses of amyloid extracted from aortic tissue, antibody A-179 showed reaction with several protein species with higher molecular masses in addition to medin. It is yet unclear whether these bands represent medin oligomers or different cleavage products of lactadherin. Elucidation of the nature of these protein species is important and mass spectrometry analyses are presently performed.

Further experiments such as electron microscopy should be done to confirm the intracellular amyloid formation. An *in vitro* study is under progress in order to further understand the mechanism of intracellular amyloid formation.

ACKNOWLEDGMENTS

Supported by the Swedish Research Council and the Swedish Heart Lung Foundation.

REFERENCES

1. Störkel, S., Bohl, J., Schneider, H. M. (1983). Senile amyloidosis: principles of localization in a heterogeneous form of amyloidosis. *Virchows Arch. B Cell Patho.l Inc.l Mol. Pathol.*, **44**, 145-161.

2. Häggqvist, B., Näslund, J., Sletten, K., Westermark, G. T., Mucchiano, G., Tjernberg, L. O., Nordstedt, C., Engström, U., Westermark, P. (1999). Medin: an integral fragment of aortic smooth muscle cell-produced lactadherin forms the most common human amyloid. *Proc. Natl. Acad. Sci. U S A.*, **96**, 8669-8674.

3. Cornwell, G.G. III., Murdoch, W. L., Kyle, R.A., Westermark, P., Pitkänen, P. (1983). Frequency and distribution of senile cardiovascular amyloid. A clinicopathologic correlation. *Am. J. Med.*, **75**, 618-623.

4. Mucchiano, G., Cornwell, G. G. III., Westermark, P. (1992). Senile aortic amyloid. Evidence for two distinct forms of localized deposits. *Am. J. Pathol.*, **140**, 871-877.

5. Black, M.E., Armstrong, D. (1998). Human-milk lactadherin in protection against rotavirus. *Lancet.*, **351**, 1815-1816.

6. Taylor, M. R., Couto, J. R., Scallan, C. D., Ceriani, R. L., Peterson, J. A. (1997). Lactadherin (formerly BA46), a membrane-associated glycoprotein expressed in human milk and breast carcinomas, promotes Arg-Gly-Asp (RGD)-dependent cell adhesion. *DNA. Cell Biol.*, **16**, 861-869.

7. Carmon, L., Bobilev-Priel, I.,Brenner, B., Bobilev, D., Paz, A., Bar-Haim, E.,Tirosh, B., Klein, T., Fridkin, M., Lemonnier, F.,Tzehoval, E., Eisenbach, L. (2002). Characterization of novel breast carcinoma-associated BA46-derived peptides in HLA- A2.1/D(b)-beta2m transgenic mice. *J .Clin. Invest.*, **110**, 453-462.

8. Hanayama, R., Tanaka, M., Miwa, K., Shinohara, A., Iwamatsu, A., Nagata, S. (2002). Identification of a factor that links apoptotic cells to phagocytes. Nature, **417**, 182-187.

EFFECTS OF MEDIN ON SMOOTH MUSCLE CELLS

A. Persson[1], S. Peng[1], P. Gerwins[1], X. Fu[2] and P. Westermark[1]

[1]Department of Genetics and Pathology, [2]Department of Women's and Children's Health, Uppsala University, SE-751 85 Uppsala, Sweden

INTRODUCTION

AMed or medin amyloid is the most common form of senile amyloid found with an occurrence of almost 100 % in the population above fifty years of age (1). The name medin is derived from its localization to the media of arteries. A recent study by Peng and co-workers (2) shows that it is mainly found in arteries of the upper part of the body. Whether medin has any effect or is involved in any disease still needs to be studied. But given the fact that other amyloid proteins are causing toxicity and disease we believe that medin might have a similar effect in vessels and thus might be involved in several pathological conditions of arteries. Medin is a 50-amino acid long internal fragment of the precursor protein lactadherin (3). Lactadherin is a 46 kD large protein composed of 364 amino acids. It contains three different domains; an EGF-like domain in the N-terminus and coagulation factor V- and VIII-like domains in the C-terminus. The EGF-like domain of lactadherin contains the tripeptide RGD, which binds integrins, whereas the C-terminus binds phosphatidyl serine (4,5). Lactadherin was first discovered in breast milk but it is expressed by various cell types, such as smooth muscle cells of arteries (3). Many functions have been ascribed lactadherin. Given the similarity to coagulation factor V and VIII Shi and Gilbert (6) demonstrated that lactadherin, in vitro, functions as an anticoagulant by competing for the membrane binding sites of factor V and VIII. Recent studies show that lactadherin might act as a linker between macrophages and dying cells and thus be involved in apoptosis. Lactadherin binds to phosphatidyl serine on dying cells and to integrins on macrophages, thereby facilitating engulfment (7, 8). How and why medin is formed from lactadherin is not known. The purpose of this study was to examine the effects of medin amyloid in vitro.

MATERIALS AND METHODS

Cell culture

Uteri smooth muscle cells were cultured in 48-well-plates (Falcon, Becton Dickinson Labware, Franklin Lakes, NJ) in 200 µl of Dulbecco's modified eagle's medium supplemented with 10% FBS, L-glutamine, and penicillin/streptomycin (Sigma, St. Louis, MO) and incubated at 37°C in a 5% CO_2 atmosphere. Synthetically produced medin (Keck Biotechnology Resource Laboratory, New Haven, CT) was dissolved and incubated for various time periods and was then added to the uteri smooth muscle cells in a 20 µM final concentration.

Gelatin zymography

At various time points cell medium was taken from the cell culture and was electrophoresed on 10% SDS-PAGE gels containing gelatin. After electrophoresis gels were agitated in 2.5% Triton X-100 to remove the SDS and to restore enzyme activity. Gels were then incubated for 1-2 days in 1% Triton X-100 at 37°C, during which present gelatinases cleave the gelatin in the gel. The gel was stained in Coomassie Blue solution until proteolytic activity was observed as white bands against a blue background.

<u>Western blot</u>

In order to identify the enzymes responsible for the upregulated gelatinolytic activity western blot was performed with different antibodies directed against matrix metalloproteinases (Oncogene, Boston, MA). Cell media was electrophoresed on 10% SDS-PAGE gels. The separated proteins were blotted onto a 0.3 μm nitrocellulose membrane (Amersham Biosciences, Uppsala, Sweden). After blocking the membrane was exposed to the primary antibody followed by the secondary antibody (horse radish peroxidase-conjugated swine anti-rabbit immunoglobulin (Dako, Glostrup, Denmark)). Finally, by using an enhanced chemiluminescence system (Amersham Biosciences, Uppsala, Sweden) the reaction was visualised.

RESULTS

The morphology of uteri smooth muscle cells was greatly affected by medin. Cells in the control looked more healthy and were more homogenous (Figure 1). An increased gelatin degradation was observed from cell media from medin-treated cells (Figure 2a). Matrix metalloproteinase-2 (MMP-2) was partly responsible for this increased gelatinolytic activity (Figure 2b).

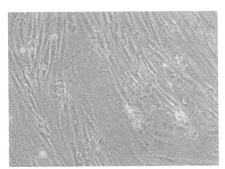

Figure 1. Medin-treated cells to the left and control cells to the right. The morphology is changed when cells are coincubated with medin.

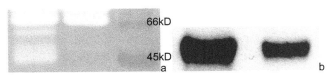

66kD

45kD

a   b

Figure 2. A gelatin zymography (a) showing an increased gelatinolytic activity of medium from cells cultured with medin (medin- left lane; control - middle lane). A Western blot (b) of cell medium shows that MMP-2 is upregulated when medin is present (left lane) compared to the control (right lane).

DISCUSSION

Degradation of the extracellular matrix components elastin and collagen has been implicated in many vascular diseases, such as aortic aneurysm and dissection. Several studies demonstrate the upregulation of matrix metalloproteinases in these disease processes (9-11). Preliminary data from this study indicate that medin might affect the normal function of cells *in vitro*, by inducing changed morphology and increased degradation of extracellular matrix proteins. Medin amyloid might thus be involved in several pathological conditions of arteries where vascular remodelling occurs, including aortic aneurysm and dissection. An *in vivo* study, with surgically removed aortic material from patients with aortic aneurysms and dissections, is currently being performed in our laboratory.

ACKNOWLEDGMENTS

Supported by the Swedish Research Council and the Swedish Heart Lung Foundation.

REFERENCES

1. Mucchiano, G., Cornwell, G.III, and Westermark, P. (1992). Senile aortic amyloid; evidence for two distinct forms of localized deposits. *Am. J. Path.*,**140**, 871-77.

2. Peng, S., Glennert, J., and Westermark P. Medin-amyloid: a recently characterized age-associated arterial amyloid form affects mainly arteries in the upper part of the body. Submitted.

3. Häggqvist, B., Näslund, J., Sletten, K., Westermark, G.T., Mucchiano, G., Tjernberg, L.O., Nordstedt, C., Engström, U., and Westermark, P. (1999). Medin: an integral fragment of aortic smooth muscle cell-produced lactadherin forms the most common human amyloid. *Proc. Natl. Acad. Sci. USA* , **96**, 8669-74.

4. Couto, J.R., Taylor, M.R., Godwin, S.G., Ceriani, R.L., and Peterson, J.A. (1996). Cloning and sequence analysis of human breast epithelial antigen BA46 reveals an RGD cell adhesion sequence presented on an epidermal growth factor-like domain. *DNA and Cell Biol.*, **15**, 281-86.

5. Andersen, M.H., Berglund, L., Rasmussen, J.T., and Petersen, T.E. (1997). Bovine PAS-6/7 binds $\alpha_v\beta_5$ integrin and anionic phospholipids through two domains. *Biochemistry*, **36**, 5441-46.

6. Shi, J. and Gilbert, G.E. (2003). Lactadherin inhibits enzyme complexes of blood coagulation by competing for phospholipid-binding sites. *Blood*, **101**, 2628-36.

7. Hanayama, R., Tanaka, M., Miwa, K., Shinohara, A., Iwamatsu, A., and Nagata, S. (2002). Identification of a factor that links apoptotic cells to phagocytes. *Nature*, **417**, 182-87.

8. Akakura, S. Singh, S., Spataro, M., Akakura, R., Kim, J.I., Albert, M.L. and Birge, R.B. (2004). The opsonin MFG-E8 is a ligand for the alphavbeta5 integrin and triggers DOCK-dependant Rac1 activation for the phagocytosis of apoptotic cells. *Exp. Cell. Res.*, **292**, 403-16.

9. Davis, V., Persidskaia, R., Baca-Regen, L., Itoh, Y., Nagase, H., Persidsky, Y., Ghorpade, A., and Baxter, B.T. (1998). Matrix metalloproteinase-2 production and its binding to the matrix are increased in abdominal aortic aneurysms. *Arterioscler. Thromb. Vasc. Biol.*, **18**, 1625-33.

10. Palombo, D., Maione, M., Cifiello, B.I., Udini, M., Maggio, D., and Lupo, M. (1999). Matrix metalloproteinases. Their role in degenerative chronic diseases of abdominal aorta. *J. Cardiovasc. Surg.*, **40**, 257-60.

11. Crowther, M., Goodall, S., Jones, J.L., Bell, P.R.F., and Thompson, M.M. (2000). Increased matrix metalloproteinase 2 expression in vascular smooth muscle cells cultured from abdominal aortic aneurysms. *J. Vasc. Surg.*, **32**, 575-83.

TRANSMISSION OF MOUSE AApoAII AMYLOIDOSIS BY THE AMYLOID FIBRILS; INHIBITORY EFFECTS ON TRANSMISSION BY THE DENATURATION/ DEGRADATION OF AMYLOID FIBRILS

Jinko Sawashita, Huanyu Zhang, Xiaoying Fu, Tatsumi Korenaga, Masayuki Mori, Keiichi Higuchi

Department of Aging Biology, Institute on Aging and Adaptation, Shinshu University Graduate School of Medicine, Matsumoto, Japan.

Data is accumulating to suggest that a certain amyloidosis including AA and mouse AApoAII amyloidosis are transmissible (1 - 3). It is postulated that invasion of preexisted amyloid seed into the hosts plays a critical role in the transmission. In order to search for a measure to eliminate or attenuate amyloid-seeding (and transmission) activity, we evaluated inhibitory effects of several treatments and medical reagents on amyloid fibril in the mouse AApoAII transmission model system. AApoAII fibril was isolated as a water suspension from the liver of an R1.P1-*Apoa2^c* mouse, and denatured or degradated by the following treatments or reagents; autoclaving, freeze-thaw, guanidine hydrochloride (Gdn-HCl), alkaline solution, formic acid, antibiotics (tetracycline, rifampicin, etc) or polyphenols (resveratrol, piceatannol, etc). The degree of denaturation was evaluated by thioflavine T fluorescence assay of aliquot. Two-month-old female R1.P1-*Apoa2^c* mice were administered intravenously with native or denatured fibril. After two months, the mice were sacrificed and the organs were collected. The degree of AApoAII amyloid deposition was evaluated in sections of these organs by Congo red staining and polarizing microscopy. 6M Gdn-HCl and formic acid completely inhibited amyloid induction. Generally amyloid deposition was decreased in mice administered with denatured fibrils as compared with that in mice administered native fibril. Tetracycline suppressed the amyloid deposition well. Polyphenols, which is known to effectively denature prion and Aß fibrils, failed to denature AApoAII fibrils.

This data should shed a new light on the development of a measure to prevent amyloid transmission.

MATERIAL AND METHODS

Denaturation / degradation of AApoAII amyloid fibrils: The AApoAII fibrils were isolated as a water suspension from the liver of 20-month-old R1.P1-*Apoa2^c* mice (4). AApoAII fibrils (1mg / ml) were sonicated, and were denatured by the following treatments or reagents; autoclaving (at 121 °C, in distilled water for1-3 h or in 1 M NaOH for 0.5 h), delipidization in the soluble state by ethanol-diethyl ether, freeze-thaw (at -80°C for 1 h+ at room temperature (RT) for 1 h, 5 times interval), 10% formalin at RT for 24 h, formic acid at RT for 8 h, 6M Urea at RT for 72 or 120 h, 6M Guanidine hydrochloride (Gdn-HCl) at RT for 24 h, or 2M NaOH at RT for 1 h. After then, each sample was washed or dialyzed for removing of these reagents, and sonicated on ice. On the other side, sonicated AApoAII fibrils were degradated by the following medicines or reagents; nordihydroguaiaretic acid (NDGA), antibiotics (the typical medicines of main 9 groups of antibiotics in Japan; benzylpenicillin (PCG), cephalexin, chloramphenicol, erythromycin, lincomycin, polymyxin B, rifampicin (RFP), streptomycin, or tetracycline (TC)), or polyphenols (resveratrol or piceatannol). And after then, each sample was sonicated on ice.

Detection of denaturation / degradation of AApoAII fibrils: The degree of denaturation / degradation of amyloid fibrils were evaluated by thioflavine T fluorescence assay and electron microscopy of aliquot.

Inhibitory effects on transmission by the denaturation / degradation of AApoAII fibrils: Two-month-old female R1.P1-*Apoa2^c* mice, have amyloidogenic *Apoa2^c* allele, were administered intravenously with native (only sonicated) or treated fibrils. After two months, the mice were sacrificed and the seven major organs (heart, intestine, liver, skin, spleen, stomach and tongue) were collected. The degree of AApoAII amyloid deposition was evaluated in sections of these organs by Congo red staining and polarizing microscopy.

RESULTS AND DISCUSSION

Generally AApoAII amyloid deposition was decreased in mice administered with denatured fibrils as compared with that in mice administered native fibrils. In particular, we succeeded in the completely inhibition of the AApoAII amyloid transmission by formic acid, Gdn-HCl, and autoclaving in 1 M NaOH (Figure 1). In addition, these chemical reagents and treatment were well known to the effective methods on denaturation of prion, so these treatments might be useful to prevent the potential transmission of amyloid fibrils in laboratories. Further more, PCG, RFP and TC of the antibiotics suppressed the amyloid deposition well (Figure 2). RFP and TC were known to the strong inhibitors of Aß fibrils extension / neurotoxicity *in vitro* (5, 6), thus, further investigation might warrant these medicines, including PCG, as therapeutic agents for amyloidosis in the recent future.

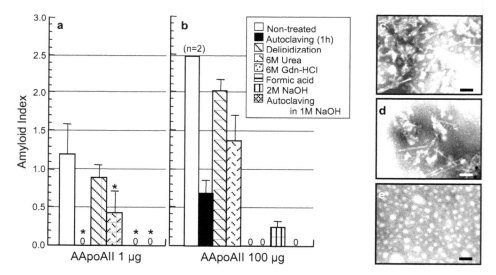

Figure 1. Inhibitory effects of the denaturation / degradation methods on AApoAII amyloid transmission. AApoAII fibrils were denatured or degradated by the physical treatments or chemical reagents described in the text. *a* and *b*, the AI of mice administered intravenously with native (Non-treated) or treated fibrils (1 or 100 µg / mouse). Each bar and line represent the mean + SD (n=3 - 5). Asterisks indicate significant difference (*p*<0.05) against the Non-treated (white bar) with Mann-Whitney's U test. c - e, the images of AApoAII fibrils observed with a JEOL 1200EX electron microscope (JEOL, Japan). Each scale bar represents 50nm. *c*, native; *d*, autoclaved for 3 h; *e*, autoclaved for 0.5 h in 1 M NaOH.

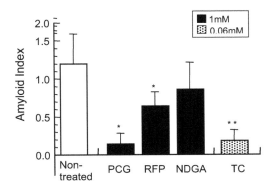

Figure 2. Inhibitory effects of the medical reagents on AApoAII amyloid transmission. AApoAII fibrils were denatured by the antibiotics or other medical reagents described in the text. Figure shows the AI of mice administered intravenously with native (Non-treated) or treated fibrils (1 μg / mouse). Each bar and line represent the mean + SD (n=4 - 5). Asterisks indicate significant difference (*, $p<0.05$; **, $p<0.01$) against the Non-treated (white bar) with Mann-Whitney's U test.

On the other side, NDGA and polyphenols were being watched with keenest interest as special reagents on denature Aß fibrils *in vitro* (7 - 9), but these reagents failed to denature AApoAII fibrils (Figure 2, data of polyphenols not shown). So these data suggested that AApoAII might have any differential mechanisms on denaturation, and further studies remained yet.

REFERENCES

1. Higuchi, K. et al., Fibrilization in mouse senile amyloidosis is fibril conformation-dependent, *Lab. Invest.,* 78, 1535, 1998.

2. Xing, Y. et al., Induction of protein conformational change in mouse senile amyloidosis, *J. Biol. Chem.,* 277, 33164, 2002.

3. Korenaga, T. et al., Tissue distribution, biochemical properties and transmission of mouse type AApoAII amyloid fibrils, *Am. J. Pathol.,* 164, 1597, 2004.

4. Pras, M. et al., The characterization of soluble amyloid prepared in water, *J. Clin. Invest.,* 47, 924, 1968.

5. Tomiyama, T. et al., Inhibition of amyloid ß protein aggregation and neurotoxicity by rifampicin, *J. Biol. Chem.,* 271, 6839, 1996.

6. Forloni, G. et al., Anti-amyloidogenic activity of tetracyclines: studies in vitro, *FEBS Lett.,* 487, 404, 2001.

7. Goodman, Y. et al., Nordihydroguaiaretic acid protects hippocampal neurons against amyloid ß-peptide toxicity, and attenuates free radical and calcium accumulation, *Brain Res.,* 654, 171, 1994.

8. Ono, K. et al., Nordihydroguaiaretic acid potently breaks down pre-formed Alzheimer's ß -amyloid fibrils *in vitro,* *J. Neurochem.,* 81, 434, 2002.

9. Savaskan, E. et al., Red wine ingredient resveratrol protects from beta- amyloid neurotoxicity, *Gerontology,* 49, 380, 2003.

TRANSMISSION OF MOUSE AApoAII AMYLOIDOSIS FROM MOTHER TO PUPS

T. Korenaga[1], X. Fu[1], M. Mori[1], J. Sawashita[1], H. Naiki[2], T. Matsushita[3] and K. Higuchi[1]

[1]Department of Aging Biology, Institute on Aging and Adaptation, Shinshu University Graduate School of Medicine, Matsumoto, Japan; [2]Department of Pathology, Fukui Medical University, Matsuoka, Japan; [3]Field of Regeneration Control, Institute for Frontier Medical Science, Kyoto University, Kyoto, Japan.

In mouse strains with the amyloidogenic apolipoprotein A-II (apoA-II) gene ($Apoa2^C$ allele), which codes the type C apoA-II protein (Pro5Gln), apoA-II polymerize to form amyloid fibrils; AApoAII (1). This formation then leads to the development of early onset and systemic amyloidosis. Prion-like transmission is induced by the injection of preformed AApoAII fibrils in R1.P1-$Apoa2^C$ mice with the $Apoa2^C$ allele (2,3). In the study reported here, we examined the transmission of AApoAII amyloidosis from mother to pups. We induced amyloidosis in female R1.P1-$Apoa2^C$ mice by injection of AApoAII fibrils and mated them with male mice. After pregnancy was confirmed, the male mice were removed, and the pups were suckled for three weeks after birth and then weaned. Amyloid deposition 4, 6, 8 months after birth was significantly higher in the mice born from amyloidosis-induced mothers than from those without amyloidosis. Four months after birth, AApoAII deposition was first detected in the small intestine, after which it extended to the whole body. The pups were changed from amyloidosis-induced to control mothers and vice versa soon after birth. Only in the mice born from control mothers and lactated by the amyloidosis-induced mothers, AApoAII amyloidosis increased significantly. The intraperitoneal injection of milk obtained from amyloidosis-induced mother mice induced amyloidosis in young R1.P1-$Apoa2^C$ mice.

Our findings substantiate the transmissibility of mouse AApoAII amyloidosis from mother to pups. These findings may help to shed light on the pathogenesis and prevention of amyloidosis.

MATERIAL AND METHODS

Animals: The R1.P1-$Apoa2^C$, a congenic strain of mice that has amyloidogenic $Apoa2^C$ allele of the SAMP1 strains on the genetic background of the SAMR1 strain were raised under specific pathogen free (SPF) condition.

Isolation of AApoAII amyloid fibrils: The AApoAII amyloid fibril fraction was from the liver of 20-month-old R1.P1-$Apoa2^C$ mice. Sonicated amyloid samples (100 μg) were injected immediately into tail vain to induce AApoAII amyloidosis.

Detection of amyloid deposition: The intensity of AApoAII amyloid deposition was determined semiquantatively using amyloid index (AI). AI was determined to represent the mean value of the scores of amyloid deposition graded 0 to 4 in the seven major tissues (liver, spleen, tongue, heart, intestine, stomach and skin) stained with Congo red.

Evaluation of maternal transmission of AApoAll amyloidosis: In our first experiment, female 2-month old *R1.P1-Apoa2^C* mice were induced amyloidosis by intravenous injections with AApoAll amyloid fibrils. Three months after injection, mice were mated with male mice. The offspring born from amyloidosis-induced mother were nursed for 21 days and then weaned. Offspring were killed at 4, 6, and 8 months of age for determination of amyloidosis. In our second experiment, the offspring born from amyloidosis-induced and control mothers were exchanged just after their birth. They were separated from foster mothers after 21 days lactation.

Induction of amyloidosis by the milk: Female 2-month-old *R1.P1-Apoa2^C* mice were induced amyloidosis and mated with male mice, bore babies and lactated them. Milk was collected by sucking with a vacuum pump after subcutaneous injection of 0.18 U oxytocin and 8 hours separation from their babies. 10 µl of milk was injected into one-month old *R1.P1-Apoa2^C* mice intraperitonealy. After 4 months from injections, the mice were killed.

Statistical analysis: Significant differences in the values of AI among the various groups of mice were examined with Mann-Whitney' s U-test.

RESULTS

At 4 months of age, slight amyloid deposits were observed in the intestine of offspring of amyloidosis-induced mothers. The intensity of amyloid deposition in these mice increased with age. In contrast, no deposition was observed in offspring born from control mothers at 4 and 6 months of age. Until 8 months of age, the degrees of amyloidosis (AI) in offspring of amyloidosis-induced mothers were significantly higher than that in offspring of control mothers. (Figure. 1). This amyloid deposit was stained positively with anti-AApoAll antiserum and negatively with the anti-AA antiserum. Amyloid was deposited in the intestine at first, and then extended to the tongue, liver, spleen, heart, stomach and skin.

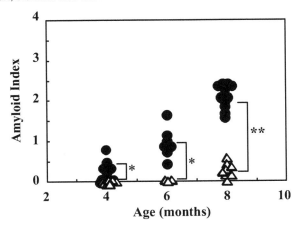

Figure 1. Intensity of amyloid deposition in R1.P1-*Apoa2^C* mice born from amyloidosis-induced mother (●) or control mother mice (O). 10, 7, 11 offspring of amyloidosis-induced mothers and 7, 3, 9 offspring of control mothers were killed at the age of 4, 6 and 8 months respectively. * P<0.05, **p<0.01 with Mann-Whitney's U-test.

In the second experiment, significant acceleration of amyloidosis was observed in offspring of control mothers which were nursed by amyloidosis-induced foster mothers. The first deposition of amyloid fibrils were detected in the intestine of offspring at 4 months of age and the degree of deposition became more severe as age increased. On the other hand, only one of nine offspring of an amyloidosis induced mother which were exchanged to a control foster mother, had amyloid deposit at the age of 6 months.

Four months after injection of milk of amyloidosis-induced mother mice, amyloid deposition was observed in the tongue, lungs and intestine in 9 of 11 mice. In contrast, 5 control mice injected with milk of mothers without amyloidosis had no amyloid deposition.

DISCUSSION

We induced amyloidosis in female mice by injection of AApoAII amyloid fibrils to evaluate the possibility of the vertical transmission of amyloidosis. Amyloid deposition was significantly accelerated in offspring of amyloidosis-induced mothers. Initial amyloid deposition was found in the intestine of offspring. Acceleration of amyloidosis can be seen only in the offspring that were nursed by amyloidosis-induced mice. These findings suggest oral transmission of amyloid fibrils rather than *in utero* transmission. Only the injection of the milk collected from amyloidosis-induced mothers to young R1.P1-*Apoa2^C* mice induced amyloidosis. We speculate that a very small amount of AApoAII amyloid fibrils in milk might induce amyloidosis. The origin of these fibrils should be elucidated in the future.

Concerning human CJD, transmission from mother to child never been observed until now. However offspring from scrapie-affected ewes show a higher risk of developing scrapie than offspring from asymptomatic ewes. Recent studies suggest an early lateral post-natal event for the pathway of scrapie transmission from ewes to lambs. Anticipation, the tendency for earlier ages of onset and increased severity of clinical symptoms among younger than older generations has been recognized in type I FAP. But, we recently revealed that intake of amyloid fibrils ATTR may not explain differences in the mean age at onset and the progression of amyloidosis in FAP.

Mouse AApoAII amyloidosis has the following characteristics, 1) precursor apoA-II protein is constitutively circulating in the blood and forms amyloid fibrils without degradation or modification. 2) amyloid fibrils deposit in the whole body, 3) Addition of preexisting amyloid fibrils into the solution of amyloid protein monomer accelerates fibril formation in neutral pH. Thus, mouse AApoAII amyloidosis will be the most interesting and suitable model for amyloidosis transmission. We believe that mouse AApoAII amyloidosis may propose a general mechanism which explains the onset, progress and prevention of various amyloidosis.

REFERENCES

1. Xing, Y., Nakamura, A., Chiba, T., Kogishi, K., Matsushita, T., Li, F., Guo, Z., Hosokawa, M., Mori, M., and Higuchi, K. (2001). Transmission of mouse senile amyloidosis. *Lab. Invest.*, 81, 493-499
2. Xing, Y., Nakamura, A., Korenaga, T., Guo, Z., Yao, J., Fu, X., Matsushita, T., Kogishi, K., Hosokawa, M., Kametani, F., Mori, M. and Higuchi, K. (2002). Induction of protein conformational change in mouse senile amyloidosis. *J. Biol. Chem.* 277, 33164-33169
3. Korenaga, T., Fu, X., Xing, Y., Matsusita, T., Kuramoto, K., Syumiya, S., Hasegawa, Z., Naiki, H., Ueno, M., Ishihara, T., Hosokawa, M., Mori, M. and Higuchi, K. (2004) Tissue distribution, biochemical properties and transmission of mouse type A AApoAII amyloid fibrils. Am. J. Pathol. 164, 1597-1606.

EXTERNAL BEAM RADIATION FOR TRACHEOBRONCHIAL AMYLOIDOSIS: THE BOSTON UNIVERSITY EXPERIENCE

J.L. Berk, G.A. Grillone, M. Skinner, and A.C. Hartford

Departments of Medicine, Otolaryngology, and Radiation Medicine, Boston University School of Medicine, Boston MA USA

INTRODUCTION

Localized amyloidosis refers to the deposition of β-sheets of insoluble protein into single organs. The bladder, eye, breast, skin or brain may be involved. In the lung, localized amyloid may occur as parenchymal nodules, diffuse alveolar disease, or submucosal deposits involving the larynx, subglottis, or tracheobronchial network. Tracheobronchial amyloidosis (TBA) is a rare disorder, representing 1.1% of patients referred to Boston University [1]. To date, approximately 100 cases have been published [2]. Patients die of respiratory insufficiency, pneumonia, or massive hemoptysis due to progressive airway narrowing. Surgical debulking by manual techniques or laser excision offers immediate relief but does not prevent disease recurrence. Low dose radiation induces plasma cell apoptosis and has been used successfully to treat conjunctival amyloidosis and plasma cell-related neuropathy [3,4]. Over the past 3 years, we have treated TBA patients with low dose external beam radiation (EBRT).

METHODS

Biopsy-proven TBA patients with progressive airflow obstruction by spirometry and airway narrowing by chest CT or bronchoscopy were treated with EBRT at Boston Medical Center. Low dose radiation (20 Gy) was delivered in 10 fractions to diseased airways identified by bronchoscopy and chest CT imaging. Spirometry, lung volumes, DLCO, ABG, chest CT, and standardized stair climbing with oximetry monitoring were performed 12 months after radiation therapy. Performance status was assessed at the initial and follow-up evaluations. The Institutional Review Board at Boston Medical Center approved data review and manuscript preparation.

RESULTS

Seven patients aged 41±15 years (mean ± 1 S.D.) with TBA complicated by progressive respiratory compromise or recurrent hemoptysis were treated with low dose (18-20 Gy) EBRT. Two patients initially expressed a faint light chain monoclonal band on urine immunofixation electrophoresis (IFE) that was not detected on follow-up testing. Similarly, fat pad aspirates in these two patients revealed borderline congophilia at the initial evaluation but not subsequently. All other IFE and fat pad studies were unremarkable (Table 1). Quantitative free light chain assays were normal in all patients.

Table 1. Demographics

CASE	SEX	AGE Dx	Congo Red	SIFE	UIFE	Fat Pad	κ FLC	λ FLC	κ/λ FLC	ESR
1	F	37	+	--	--	--				15
2	M	62	+	--	--	--	11.6	11.8	0.98	23
3	F	34	+	--	--	--	10.2	11.6	0.88	39
4	F	32	+	--	--	+/-	12.5	13.1	0.95	16
5	F	37	+	--	+/-	+/-	13.4	11.4	1.18	11
6	F	23	+	--	--	--	10.2	10.3	0.99	10
7	F	62	+	--	--	--	11.7	12.5	0.94	19
Mean	6:1	41	100%				11.60	11.78	0.99	19.00
S.D.		15.1					1.26	0.96	0.10	9.88

Patients had longstanding disease, with 3.4 ± 3.4 years (mean ± 1 S.D.) between symptoms onset and tissue diagnosis. Delay from diagnosis to EBRT averaged 1 ± 0.75 years (range 106-855 days). Follow up-to-date is 0.94 ± 0.60 years (range 0.48-2.24 years) (Table 2).

Table 2. Pre- and post-radiation data collection

CASE	SXS => Dx (Yrs)	Dx => XRT (Yrs)	XRT	Dose (Gy)	F/U (mos)
1	3.45	1.13	R lung	18	11.3
2	1.23	0.32	L lung	20	5.6
3	7.05	0.42	R lung	30	24.6
4	1.36	2.34	Bilateral	20	12.6
5	0.34	1.05	Neck	20	27.2
6	9.16	1.50	Bilateral	19.8	12.7
7	1.23	0.29	L lung	19.8	18.3
Mean	3.40	1.01		21.1	16.0
S.D.	3.40	0.75		4.0	7.7

Chest CT imaging did not reveal uniform changes in airway wall thickening or lumen caliber following treatment. Fiberoptic bronchoscopy documented decreased mucosal edema and stable endobronchial pathology occasional case-specific significant disease regression at 12 months post-EBRT. Airflow obstruction increased at six months follow-up spirometry, recovering to baseline or improving 12 months post-EBRT. Heart rate response and oximetry did not change but stair climbing capabilities improved on a case-specific basis. At 1-year follow-up, total lung capacity, residual volume, and functional residual capacity were not significantly changed from baseline. Diffusing capacity declined in all cases after EBRT, decreasing 12.7 ± 7.0% at 6 months and 15.7 ± 8.1% at 12 months follow-up. All patients reported significantly improved ability to perform activities of daily life (ADL). Performance status improvement approached statistical significance (p=0.0698). Treatment complications were limited to esophagitis (43%). One patient was treated for community-acquired pneumonia 2 months after completing EBRT.

CONCLUSIONS

Low dose EBRT is well tolerated in patients with tracheobronchial amyloidosis, with mild and transient declines in gas diffusion being the most significant complication. EBRT appeared to prevent progressive amyloid deposition in localized airway disease for up to 44 months follow-up. Although disease regression following EBRT was not a consistent finding, all patients reported significantly improved functional capacities at 1-year follow-up. Pulmonary function tests documented airflow preservation. Whether low dose EBRT induces subclinical lung injury remains unclear. EBRT should be considered when standard laser therapy cannot a) control disease progression or b) properly address the distribution of disease.

REFERENCES

1. O'Regan, A., Fenlon, H.M., Beamis, Jr., J.F., Steele, M.P., Skinner, M. and J.L. Berk. Tracheobronchial amyloidosis. The Boston University experience from 1984 to 1999. Medicine 79:69-79, 2000.

2. Cordier, J.F., Loire, R. and J. Brune. Amyloidosis of the lower respiratory tract: Clinical and pathologic features in a series of 21 patients. Chest 90:827-31, 1986.

3. Kelly, J.J. Jr, Kyle, R.A., Miles, J.M., O'Brien, P.C. and P.J. Dyck. The spectrum of pheripheral neuropathy in myeloma. Neurology 31:24-31, 1981.

4. Shinoi, K., Shirai-Shi, U. and J. Yahata. Amyloid tumor of the trachea and lung resembling bronchial asthma. Dis. Chest 42:442-45, 1962.

Supported by the Amyloid Research Fund at Boston Univsersity

WILD TYPE TRANSTHYRETIN-DERIVED SENILE SYSTEMIC AMYLOIDOSIS WITH LIMITED PROTEOLYTIC DIGESTION OF THE MONOMER: CLINICO-PATHOLOGIC STUDIES WITH DNA AND AMYLOID PROTEIN CHARACTERIZATION

Maria M. Picken[1], Fred Leya[1], Roger N. Picken[2]

[1]*Loyola University Medical Center1 and* [2]*Hines VA Hospital, Maywood, IL, USA.*

Senile systemic amyloidosis (SSA) may be derived from the wild type of transthyretin (wtTTR) and it is postulated that the monomer frequently undergoes proteolytic digestion (2).

The most frequent type of cardiac amyloid is derived from the immunoglobulin light chain; however, it is increasingly recognized that among older patients amyloid derived from transthyretin (ATTR) is within the main differential diagnosis (1, 2).

We report clinico-pathologic studies, with DNA sequencing and amyloid protein characterization, in a patient of German origin. His father had systemic amyloidosis (not further characterized); no other family members have thus far been diagnosed with amyloidosis

MATERIALS AND METHODS

Experimental design: we report clinico-pathologic studies, with DNA sequencing and amyloid protein characterization, in a patient of German origin. Materials and methods: Multiple biopsies, autopsy tissue, and peripheral blood were available for studies. Ethical permission was obtained. Endomyocardial tissue samples, obtained at autopsy, were used in this study. As a negative control, a sample of myocardium from an age-matched patient was used. This negative control sample was negative for amyloid by Congo red stain and electron microscopy.

Tissues used in this study were derived from Myocardium and consisted of both amyloid negative and amyloid positive material. Tissue was prepared by mechanical disruption in lysis buffer (NP40 1%, glycerol 10%, NaCl 100mM, Tris 20mM pH7.5, EDTA 1mM, DTT 1mM, EDTA free protease inhibitor) using a polytron (Powergen 125) homogenizer. The lysate was spun at 13,000 rpm for 15 min and the supernatant was separated from the pellets, aliquoted and frozen at −80°C until use. At the time of analysis, supernatants were thawed, and 100ul supernatant incubated with 7ul (packed volume) bead affinity reagent in 100ul of 50mM Tris (pH 7.4). Reagents were incubated for 90 min with constant agitation at 4°C. After incubation, affinity beads were washed three times with PBS (20mM, pH 7) containing 0.3% tween-20, twice with PBS (20mM, pH 7) , and once with deionized water. Bound proteins were extracted with 50% acetonitrile containing 0.3% TFA, then re-concentrated on an H50 ProteinChip® Array. After a final wash to remove any contaminating salts, 1ul of sinapinic acid (50% saturated solution in 50% acetonitrile and 0.5% trifluoroacetic acid) was applied and the array analyzed using a mass reader (model PBSIIc; Ciphergen).

RESULTS AND DISCUSSION

Examination of the amyloid tissue deposits (by SELDI-mass spectrometry) demonstrated two closely related peaks corresponding to TTR monomer.   One peak, which was shorter by 2 amino acids, was twice as abundant.   The negative control sample did not demonstrate a dominant peak in the corresponding region.

DNA extracted from the buffy coat showed no evidence of the TTR gene mutation.

The patient's father had systemic amyloidosis (not further characterized); no other family members have thus far been diagnosed with amyloidosis.   This male patient was diagnosed with restrictive cardiomyopathy and ATTR at the age of 77.   Subsequently, he required a pacemaker, developed renal and multiple endocrine failure, and died 5 years after the diagnosis of ATTR. Endomyocardial biopsy showed abundant interstitial deposits of ATTR; deposits were also detectable in several gastro-intestinal biopsies.   At autopsy, very abundant interstitial deposits of amyloid were present in the myocardium (1100 gm) and both lungs (1040 and 740 gm right & left respectively).   Less abundant interstitial deposits were also seen in the skeletal muscles and adjacent adipose tissue.   Other tissues/organs predominantly showed involvement of the arteries of the gastrointestinal tract, endocrine glands, and kidneys (cortex and medulla); many small peripheral nerves were also involved. No amyloid deposits were seen in the brain and spinal cord.

CONCLUSIONS

1. This patient, with a family history positive for an unknown type of amyloid, had SSA derived from wtTTR with only minimal proteolysis of the monomer
2. Myocardium was the most severely affected organ, followed by the lungs; systemic arteries also showed deposits of amyloid

ACKNOWLEDGEMENT

This work is supported by a research grant from the Amyloidosis Foundation, Inc.

REFERENCES

1.   Picken MM., The changing concepts of amyloid. Arch Pathol Lab Med 2001, 125:25-37.
2.   Westermark P., Bergstrom J., Solomon A., Murphy C., Sletten K. Transthyretin-derived senile systemic amyloidosis: clinicopathologic and structural considerations. Amyloid: J Protein Folding Disord 2003, 10, Suppl. 1, 48-54.

PINDBORG TUMOR-ASSOCIATED AMYLOID PROTEINS

C. L. Murphy,[1] S. Wang,[1] D.P. Kestler,[1] S.D. Macy,[1] T.K.. Williams,[1] E.R. Carlson,[2] D.T. Weiss,[1] and A. Solomon[1]

[1]Human Immunology and Cancer Program, Department of Medicine, University of Tennessee Graduate School of Medicine, Knoxville, TN, USA and [2]Department of Oral and Maxillofacial Surgery, University of Tennessee Graduate School of Medicine, Knoxville, TN, USA

I. INTRODUCTION

Amyloid deposits are commonly associated with Pindborg tumors, *i.e.*, calcifying epithelial odontogenic tumors (CEOT). Reports regarding the nature of the congophilic protein have been conflicting; however, in recent analyses of fibrillar extracts from 3 such cases, we demonstrated that this material was comprised of N-terminal fragments of a hitherto unknown protein product derived from the FLJ20513 gene.[1] We provisionally designated this type of amyloid as APin. Our studies now have been extended to include chemical and immunologic examination of 3 additional cases where the presence of this molecule, as well as another novel, but related protein encoded by FLJ40850 mRNA was detected. Both of these components are encompassed within the 943-bp ORF of an expressed sequence tag (EST) scan obtained by the Institute for Genomic Research (TIGR) and designated THC 1981099.

II. MATERIALS AND METHODS

A. AMYLOID EXTRACTION AND CHARACTERIZATION

Amyloid fibrils were extracted from sections of formalin-fixed, paraffin-embedded blocks obtained from 3 patients with CEOTs. The proteins were purified by reverse phase-HPLC, the material contained in the peak eluates treated with cyanogen bromide, and the resultant peptides subjected to automated amino acid sequence analyses as previously described.[1]

B. IMMUNOHISTOCHEMISTRY

Tissue was immunostained using the ABC technique. The primary reagents were rabbit antisera made against synthetic peptides encompassing the first 14 residues located within either the FLJ40850 or FLJ20513 sequence; the secondary was a biotinylated goat anti-rabbit antibody.

III. RESULTS

The HPLC elution profiles of the 3 CEOT amyloid extracts are shown in Figure 1. Sequence analyses of the cyanogen bromide-derived peptides revealed the presence, in each case, of 37 and 10 to 39 N-terminal residues of FLJ20513 and FLJ40850, respectively. The locations of these two elements within the predicted amino acid sequence of the ORF contained within the THC 1981099 EST are given in Figure 2 and their first amino acid is designated #4 and #1, respectively. Additionally, in cases 973462 and 922839A, two other components were detected. Although each (#2 and #3) had different N-termini, both were located within the ORF upstream from FLJ20513 and were not part of the FLJ40850-encoded sequence.

Case #1 (973462): *Peak 1*; N-terminal 37 residues of #4 (FLJ20513) (90%), N-terminal 17 residues of #2 (10%). *Peak 2*; N-terminal 37 residues of #4 (FLJ20513) (90%) and 11 residues of #1 (FLJ40850) (10%).

Case #2 (922839A): *Peak 1*; N-terminal 37 residues of #4 (FLJ20513) (90%). *Peaks 2 and 3*; N-terminal 39 residues of #1 (FLJ40850) (75%), N-terminal 37 residues of #4 (FLJ20513) (20%), and trace amounts of the first 18 residues of #3.

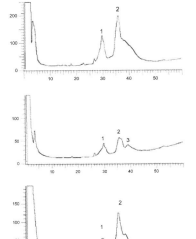

Case #3 (3011899): *Peak 1*; N-terminal 37 residues of #4 (FLJ20513) (90%). *Peak 2*; Mixture, at a ratio of 1.5:1, of N-terminal 10 residues of #1 (FLJ40850) and 37 residues of #4 (FLJ20513), respectively.

Figure 1. Reverse-phase HPLC profiles of amyloid extracted from formalin-fixed, paraffin-embedded tissue sections from 3 CEOT cases.

#1 ▼ #2 ▼ #3 ▼ #4 ▼

SWIPPFSGILQQQQQAQIPGLSQFSLSALDQFAGLLPNQIPLTGEASFAQGAQAGQVDPLQLQTPPQTQPGPSHVM

PYVFSFKMPQEQGQMFQYYPVYMVLPWEQPQQTVPRSPQQTRQQQYEEQIPFYAQFGYIPQLAEPAISGGQQQLAF

DPQLGTAPEIAVMSTGEEIPYLQKEAINFRHDSAGVFMPSTSPKPSTTNVFTSAVDQTITPELPEEKDKTDSLREP

Figure 2. Predicted amino acid sequence encoded by ORF THC 1981099. **#1, 2, 3,** and **4** designate the N-terminal residue of CEOT amyloid proteins and, in the case of **#1** and **#4**, represent the N-terminal products of FJ40850 and FLJ20513 genes, respectively. (═══) and (────) indicate regions encoded by the FLJ40850 and FLJ20513 genes, respectively.

In immunohistochemical analyses, the intensity of immunostained amyloid was consonant with the amount of FLJ40850- or FLJ20513-related amyloid present in the specimen (see Figure 1).

DISCUSSION

We have previously reported that the amyloid extracted from surgical specimens resected from 3 patients with CEOTs was derived from a unique and previously undescribed protein encoded by the FLJ20513 gene.[1] This element was cloned from a signet-ring gastric carcinoma cell line[2,3] and, subsequently, was found to be part of a 943-bp ORF contained in an EST scan designated THC 1981099.

From our studies of 3 additional CEOT amyloid-associated cases, we have confirmed that the FLJ20513-related protein was present and, further, that all the samples contained a heretofore unrecognized product of the FLJ40850 gene. In two cases, other components derived from different mRNA transcripts arising from this ORF also were detected. Notably, the N-termini of #3 and #2, located 8 and 29 residues proximal to the translated portion of cDNA FLJ20513, respectively, seemingly were not derived from either the FLJ20513 or FLJ40850 gene. Conceivably, these molecules could have resulted from degradation of the hypothetical protein arising from the entire ORF transcript contained in THC 1981099. Although the function of this gene (or products) is as yet unknown, it is of interest that it is located on chromosome 4 adjacent to those specifying salivary proteins.

V. CONCLUSION

Based on the results of our studies to date, we posit that these unusual CEOT amyloid-associated proteins are the products of a single gene with alternative splice forms.

VII. ACKNOWLEDGEMENTS

This study was supported, in part, by USPHS Research Grant CA10056 from the National Cancer Institute and the Aslan Foundation. A.S. is an American Cancer Society Clinical Research Professor.

VIII. REFERENCES

1. Solomon, A., et al., Calcifying epithelial odontogenic (Pindborg) tumor-associated amyloid consists of a novel human protein, *J. Lab. Clin. Med.,* 142, 348, 2003.
2. Sugano, S., et al., Homo sapiens hypothetical protein FLJ20513 (FLJ20513 mRNA), 2001:Internet, Oct 24, Accession NP_060325, Available from: http://www.ncbi.nlm.nih.gov/.
3. Sekiguchi, M., Sakakibara, K., and Fujii, G. Establishment of cultured cell lines derived from a human gastric carcinoma, *Jpn. J. Exp. Med.*, 1978, 48, 61,1978.

SEMENOGELIN I IS THE AMYLOIDOGENIC PROTEIN IN SENILE SEMINAL VESICLES

R.P. Linke[1], R. Joswig[1], C.L. Murphy[2], S. Wang[3], H. Zhou[4], U. Gross[5], C.B. Röcken[6], W. B.J. Nathrath[7], P. Westermark[8], D. Weiss[2], A. Solomon[2]

[1] *Max Planck-Institute of Biochemistry, Am Klopferspitz 18a, D-82152 Martinsried, Germany*

[2] *Human Immunology & Cancer Program, University of Tennessee. Knoxville, TN 37920, USA*

[3] *Institute of Pathology, Friedrich-Wilhelm-University, D-53127 Bonn, Germany*

[4] *Institute of Pathology, Clinicum Free University Steglitz, D-12203 Berlin, Germany*

[5] *Institute of Pathology, Otto-Guericke-University, D-39120 Magdeburg, Germany*

[6] *Institute of Pathology, City Hospital Harlaching, D-81545 München, Germany*

[7] *Department of Genetics and Pathology, Rudbeck Laboratory, S-751 85 Uppsala, Sweden*

ABSTRACT

The protein of origin in senile amyloid of the seminal vesicles has been identified in this study. Amyloid fibrils were isolated from frozen seminal vesicles in a pure form. Immunochemical and chemical studies revealed that a major polypeptide of 13 KDa represents an N-terminal fragment of semenogelin I (SgI). Similar polypeptides of SgI were isolated from formalin-fixed seminal vesicles containing amyloid. Although it had been reported previously that this material contained lactoferrin and it was suggested that lactoferrin was the protein of origin, the results of our studies revealed that lactoferrin was not present in significant amounts in pure amyloid fibrils. Since semenogelin I was identified as the protein of origin of senile seminal vesicle amyloid, we provisionally designate this entity as ASgI.

INTRODUCTION

One of the most common forms of amyloid affecting virtually all men over age 80 is that confined to the seminal vesicles. The first analysis was reported by Cornwell et al. (1). A 14 KDa polypeptide was identified which was blocked at the N-terminus. An antiserum against it stained the amyloid, as well as the epithelial cells of the seminal vesicles, thus indicating that these cells produce the protein of origin (1). Here, we idientified the amyloid protein of senile seminal vesicle amyloid.

MATERIAL AND METHODS

Tissues and diagnosis: Frozen seminal vesicles were available from two patients and formalin-fixed material from another patient. Diagnosis was performed on paraffin sections that were stained with Congo red according to the method of Puchtler and then microscopically evaluated.

Extraction of amyloid proteins: Amyloid proteins were extracted from frozen (2) and fixed tissues (3) and from paraffin embedded tissue sections (4).

Western Blotting: The amyloid proteins from various extractions were separated on SDS-PAGE and after electrotransfer were immunochemically characterized using various antibodies that included anti-ASgl. Anti-ASgl, an antiserum against the major low molecular amyloid protein, an antibody against the Sgl, 22-41 polypeptide (both prepared by us) and anti-lactoferrin (Sigma). As controls, a monoclonal anti-Sgl antibody (5) and pure Sgl (6) were made available by courtesy of Dr. Lilja and Dr. Malm, Göteborg/Sweden.

Amino acid sequence analysis was performed on extracted native and extracted formalin-fixed amyloid proteins as published (3, 4).

Immunohistochemistry: Various amyloids in tissue sections including ASgl were probed with the antibodies mentioned above and antibodies directed against the major amyloid classes using the peroxidase-anti-peroxidase method of Sternberger as published previously (7, 8, 9).

RESULTS AND CONCLUSIONS

Immunohistochemical studies on senile seminal amyloid did not reveal any decisive reactions with any of the antibodies prepared against the major amyloid classes (9). Therefore, an antibody against the low molecular weight amyloid protein was developed (anti-ASgl) which recognized only senile seminal amyloid and not the major known amyloids, thus suggesting the presence of a novel protein. This type of amyloid was immunochemically (Western blotting, immunohistochemistry) and chemically (amino acid sequence) analyzed. The amino acid sequence analysis showed Sgl fragments that encompassed positions 1-108 and a minor component of Sgl, 1-148, demonstrating limited proteolysis of the 439-residue Sgl molecule. Sgl is a major protein in semen, the function of which is to coagulate the semen after ejaculation. It liquefies after approximately 10 min due to limited proteolysis by the prostate specific antigen (PSA) and thereby releases and activates the sperm cells (5, 6).

These results on seminal vesicle amyloid were corroborated by Western blotting and by immunohistochemistry where it was shown that the anti-ASgl antibody stained only senile seminal amyloid and no other amyloid tested. In addition, control antibodies, which were directed against the C-terminal part of Sgl, did not react with the amyloid protein but recognized Sgl. Furthermore, the anti-ASgl antibody could be absorbed with pure amyloid fibrils, with pure Sgl but not with lactoferrin. Similar results were obtained with the Sgl-peptide antiserum. Anti-lactoferrin did not react with pure amyloid proteins separated by Western blotting but with amyloid deposits in tissue sections. This strong reaction could be absorbed with lactoferrin but not with pure Sgl and pure amyloid fibrils of Sgl origin. This strong reaction with anti-lactoferrin of senile seminal vesicle amyloid was confirmed by us. However, lactoferrin was removed during isolation of pure amyloid fibrils.

Based on these results, which unequivocally demonstrate that seminal vesicle amyloid is formed from semenogelin I, we provisionally designate this entity as ASgl.

ACKNOWLEDGEMENTS

Supported by the Deutsche Forschungsgemeinschaft, Bonn-Bad Godesberg, Germany (Grand Li 247, 12-3), and by the Swedish Research Council. For technical assistance we thank R. Oos, J. Lindermeyer and M. Bandmann.

REFERENCES

1. Cornwell III., G.G., et al., Seminal vesicle amyloid: the first example of exocrine cell origin of an amyloid fibril precursor. *J. Pathol.* 167, 297, 1992.

2. Pras, M., et al., The characterization of soluble amyloid prepared in water. *J. Clin. Invest.* 47, 927, 1968.

3. Linke, R.P., et al., Identification of amyloid A protein in a sporadic Muckle-Wells syndrome. N-terminal amino acid sequence analysis after isolation from formalin-fixed tissue. *Lab. Invest.* 48, 698, 1983.

4. Murphy, C.L., et al., Chemical typing of amyloid protein contained in formalin-fixed paraffin-embedded biopsy specimens. *Amer. J. Clin. Pathol.* 116, 135, 2001.

5. Lilja, H., Abrahamsson, P.A., and Lundwall, A., Semenogelin, the predominant protein in human semen. *J. Biol. Chem.* 264, 1894, 1989.

6. Malm, J., et al., Isolation and characterization of the major gel proteins in human semen, semenogelin I and semenogelin II. *J. Biochem.* 238, 48, 1996.

7. Linke, R.P., Monoclonal antibodies against amyloid fibril protein AA. Production, specificity and use for immunohistochemical localization and classification of AA-type amyloidosis. *J. Histochem. Cytochem.* 32, 322, 1984.

8. Linke, R.P., Highly sensitive diagnosis of amyloid and various amyloid syndromes using congo red fluorescence. *Virchows Arch. Path. Anat.* 436, 439, 2000.

9. Schröder, R., Nennesmo, I., and Linke, R.P., Amyloid in multiple sclerosis lesions is clearly of Aλ-type. *Acta Neuropath.* 100, 709, 2000.

CARDIAC DISEASE AND HEART VALVE AMYLOID: WHAT IS THE CONNECTION?

Stina Enqvist[1], Erik Wassberg[2], Malin Wilhelmsson[1], Ulf Hellman[3] and Per Westermark[1]

[1]*Department of Genetics and Pathology, Uppsala University, Sweden*
[2]*Department of Surgical Sciences, Uppsala University, Sweden*
[3]*Ludwig Institute for Cancer Research, Sweden*

INTRODUCTION

Cardiac amyloidosis is a well-known condition, which has been reported with varying incidence rates. Systemic amyloidosis of AL and TTR type is often associated with amyloid deposition in heart valves, in addition to blood vessels and myocardium (1, 2). In this classical type of valvular amyloidosis are the deposits occur in previously unaltered valves.

There is also another type of cardiac amyloidosis restricted entirely to the heart valves and can be found in valves, surgically removed for chronic valvular disease. The deposits are small and restricted to the areas of scarring and calcification (3). The amyloid protein has not been identified but it has been demonstrated not to be AL or AA (4). The purpose of this study was to characterize the amyloidogenic protein and investigate the prevalence of this amyloid in surgically removed heart valves.

MATERIAL AND METHODS

Tissues

42 heart valves surgically removed at Uppsala Academic Hospital were used in this study. Each heart valve was divided into 2 pieces and these were processed differently. One was kept at -20° C and the other piece was formalin fixed and paraffin embedded.

Light microscopy

5 μm sections were stained with alkaline Congo red for identification of amyloid.

Purification, electrophoresis and mass spectrometry analysis of amyloid

Frozen valvular tissue was homogenized 3 times in 0.15 M NaCl/ 0.05 M sodium citrate followed by 3 homogenizations in distilled water. A drop of the homogenate was put on a microscopic slide and stained with Congo red. The lyophilized pellet was extracted with 6M guanidine HCl, centrifuged, dialyzed against saturated ammonium sulfate and water, followed by lyophilization. Material was then subjected to sodium dodecylsulfate polyacrylamide gel electrophoresis. The gel was stained with silver, dried and bands of interest were cut out of the gel, digested in situ with trypsin (porcine, modified, sequence grade Promega, Madison WI) and analyzed by matrix assisted laser desorption/ionization time-of-flight mass spectrometry (MALDI-TOF MS) (Ultraflex TOF/TOF, Bruker Daltonics, Bremen, Germany).

Western blot analysis

Proteins electrophoresed on SDS-PAGE gels were blotted onto a nitrocellulose membrane (Amersham Biosciences, Uppsala, Sweden) and exposed to primary antibodies followed by horse radish peroxidase-conjugated swine anti-rabbit immunoglobulin (Dako, Glostrup, Denmark) diluted 1:10 000. The reaction was visualized by using an enhanced chemilumniscence system (Amersham Biosciences, Uppsala, Sweden). The primary antisera had been raised in rabbits against keyhole limpet hemocyanine-linked synthetic peptides corresponding to AA, AL, transthyretin, medin and apolipoprotein A-I. Commercially available antibodies against C3 and SAP were also used (Dako, Glostrup, Denmark).

RESULTS

Light microscopy

The combined result of Congo red stained paraffin embedded material and homogenates showed presence of amyloid in 22 out of 42 (52 %) valves.

Mass spectrometrical analysis of amyloid

Dissolved amyloid fibrillar material, analyzed by SDS-PAGE and stained with silver nitrate, revealed a low-molecular but fairly diffuse protein band in 3 cases. This band was cut out of the gel, digested with trypsin and the resulting peptides were analyzed by MALDI-TOF MS. Fragments corresponding to complement factor C3 were identified.

Western blot analysis

Western blot studies of the heart valve material was performed in order to exclude some of the most common types of amyloidosis, including AA, AL and ATTR. As seen in Table 1, all heart valves examined contained SAP while only 5/10 had a positive reaction with the antibody against C3. Notably, this positive C3 reaction did not correlate with the amount of amyloid in the heart valves. Antibodies against AA, AL and medin gave completely negative results whereas with anti TTR, two faint bands were detected. These probably represented plasma TTR and not the amyloid since in immunohistochemistry no TTR reactivity was seen with the amyloid. Seven of 10 materials showed a band corresponding to full length apoA-1.

Table 1. Western blot result of 10 homogenized heart valves. Antibodies used were directed against apoA-I, SAP, medin, AA, AL, TTR and complement factor C3.

Valve	AapoA I	SAP	AMed	AA	AL	ATTR	C3	Amyloid content*
1	+	+	-	-	-	-	-	1
2	+	+	-	-	-	-	+	1
3	+	+	-	-	-	-	+	2.5
4	+	+	-	-	-	-	+	2
5	-	+	-	-	-	-	+	1
6	+	+	-	-	-	-	+	2
7	+	+	-	-	-	-	-	2
8	-	+	-	-	-	+	-	1
9	+	+	-	-	-	-	-	1
10	-	+	-	-	-	+	-	1.5

* Amyloid content scored 0-3, where 0 contains no amyloid and 3 has extensive deposition.

DISCUSSION

In this study we demonstrated that amyloid was present in 22 out of 42 (52%) surgically removed heart valves. The nature of the amyloidogenic protein was not elucidated but complement factor C3 was shown to be present in three cases. Since tryptic fragments covered different parts of this large protein, it is highly unlikely that it is the precursor protein but complement factor C3 might be involved in the pathogenesis of the amyloid in calcified heart valves. Its role in the disease process has to be further investigated. Western blot results supports earlier findings by Goffin et al. (3, 4) that AA, AL and TTR are not involved in this type of cardiac amyloidosis. All apoA-I amyloids known today are built up by N-terminal 70-100 amino acid residue fragments of the precursor protein. It is therefore not likely that apoA-I is the amyloidogenic protein in localized heart valve amyloidosis. The high occurrence of apoA-I in the fibrils suggests a role of this protein in the progress to mature amyloid fibrils. Our new studies, and some studies of other groups indicate that apolipoproteins, including apoA-I ,are attached to amyloid fibrils of different biochemical nature and that these proteins may have a more general role in the amyloidogenesis. This role is yet not known.

ACKNOWLEDGMENTS

Supported by the Swedish Research Council and the Swedish Heart Lung Foundation.

REFERENCES

1. Pomerance, A. (1966). The pathology of senile cardiac amyloidosis. *J. Pathol. Bacteriol.*, **91**, 357.
2. Buja, L. M., Khoi, N. B., and Roberts W. C. (1970). Clinically significant cardiac amyloidosis. Clinicopathologic findings in 15 patients. *Am. J. Cardiol.*, **26**, 394.
3. Goffin, Y. A. (1980). Microscopic amyloid deposits in the heart valves: a common local complication of chronic damage and scarring. *J. Clin. Pathol.*, **33**, 262.
4. Goffin, Y. A., Murdoch, W., Cornwell, G. G. III, and Sorenson, G. D. (1983). Microdeposits of amyloid in sclerocalcific heart valves: a histochemical and immunoflourescence study. *J. Clin. Pathol.*, **36**, 1342.

SECTION 7

THERAPEUTICS

IMMUNOTHERAPY OF AL AMYLOIDOSIS

Alan Solomon, Deborah T. Weiss, and Jonathan S. Wall

Human Immunology & Cancer Program, Department of Medicine, University of Tennessee Graduate School of Medicine, Knoxville, TN, USA

I. INTRODUCTION

Presently, therapeutic options for patients with primary systemic (AL) amyloidosis are limited to reducing light chain synthesis using anti-plasma cell chemotherapy given in conventional amounts or high doses combined with autologous stem cell transplantation.[1,2] This approach, particularly dose-intensive therapy, has extended survival and, in some cases, resulted in improved organ function over time.[3] However, such treatment can be associated with numerous complications[4] and an exceedingly high mortality rate, especially in the elderly or those with multi-system or extensive cardiac involvement.[5]

Importantly, amyloid deposition is not necessarily irreversible, as evidenced experimentally[6] and clinically in cases of both primary (AL) and secondary (AA) amyloidosis.[3,7,8] Compelling evidence of *in situ* resolution of AL deposits has come from chemical analyses of fibrillar extracts where it has been determined that, most often, the protein constituents are not composed of intact light chains, but rather, consist of fragments that seemingly result from proteolytic degradation.[9] The failure of the body to remove such material totally may be attributed to its non-foreign nature or the presence of other co-depositing components, *e.g.*, glycosaminoglycans or P-component that have been alleged to interfere with amyloidolysis.[10,11]

II. AMYLOID-REACTIVE ANTIBODIES

To investigate humoral or cellular factors that could facilitate AL resolution, we developed an *in vivo* animal model in which amyloidomas were produced in mice by subcutaneous injection between the scapula of up to 200 mg of water-soluble human AL extracts. We found that this substance was removed by an immune mechanism associated with the generation of amyloid-reactive antibodies that recognized, as evidenced using *in vitro* assays, the light chain constituent of the amyloid protein injected, as well as heterologous ALκ or ALλ extracts.[12] Additionally, when the fibrillar preparation was re-administered to immunized animals, the rate of disappearance increased approximately two-fold. Elimination of amyloid tumors also was expedited when the extracts were first incubated overnight with mouse immune serum. Further, the resolving amyloidomas were extensively infiltrated by activated neutrophils. Under similar conditions, amyloidolysis was not accelerated in immunodeficient (SCID) mice or in animals where the neutrophils were suppressed by administration of an anti-neutrophil mAb or were functionally impaired, *i.e.*, in CD-18 knockout mice.

These results led us to hypothesize that the persistence of amyloid results from the patient's inability to elicit an immune response directed towards the fibrillar protein. We reasoned that, if available, passive administration of fibril-reactive antibodies could initiate resolution of the pathologic deposits. To test this theory, we immunized mice with light chain variable region (V$_L$) fragments obtained by proteolytic cleavage of human

Bence Jones proteins that, upon thermal denaturation under acidic conditions, formed fibrils that possessed the characteristic tinctorial and ultrastructural features of amyloid. The monoclonal antibodies (mAbs) that were generated reacted specifically with AL fibrils, regardless of their κ or λ constant region (C_L) isotype or V_L subgroup, but did not recognize native (soluble) light chains. When one such prototypic IgG1κ antibody, designated 11-1F4, was labeled with fluorescein and injected at a contra-lateral site into an amyloidoma-bearing mouse, it localized only in the tumor. Based on these results, experiments were performed to determine if the 11-1F4 reagent would be effective therapeutically. Single or multiple (every 48 hrs x 6) 100-μg doses were given to mice with human ALκ or ALλ amyloidomas (4 κ, 8 λ) and resulted in an up to 4-fold more rapid elimination of the tumors, as compared to control animals.[12]

We posit that amyloid resolution results from a three-step process that includes: (1) binding or opsonization of fibrils by the amyloid-reactive mAb; (2) attraction and activation of neutrophils via their interaction with specific regions on the Fcγ portion of the antibody molecule; and (3) enzymatic and/or chemical proteolysis of the amyloid by neutrophil-derived endopeptidases or free radicals, respectively. (Presumably, complement is not involved in this process since it is not activated by mouse IgG1 molecules.)

To determine if this form of passive immunotherapy could be effective clinically, the murine IgG1κ 11-1F4 amyloid-reactive mAb was chimerized (the mouse C_H and C_L domains were replaced by their human counterparts). Comparative immunologic analyses, including fibril binding, dot blot, and immunohistochemical assays, indicated that the specificity of the chimeric 11-1F4 mAb was identical to that of the original murine antibody. Moreover, as demonstrated in studies involving mice with human ALκ or ALλ amyloidomas, the modified reagent also was found to accelerate amyloidolysis, was well tolerated, and did not react with normal human tissue.[13]

The demonstration of the effective performance of the chimeric 11-1F4 mAb has formed the basis for the National Cancer Institute's Drug Development Group to proceed with full-scale GMP production of mAb 11-1F4 and to perform the pharmokinetic and toxicology studies required for an eventual Phase I clinical trial.

III. CONCLUSIONS

Given the generally poor prognosis of patients with primary (AL) amyloidosis, who often are not diagnosed until extensive fibril deposition and compromised organ function have occurred, and the fact that only limited therapeutic options are now available for such individuals, the use of passive immunotherapy to effect amyloid resolution would be an adjunct to anti-plasma cell chemotherapy and would represent an important new approach in the treatment of this disorder.

IV. ACKNOWLEDGEMENTS

Supported in part by USPHS Research Grant CA10056 from the National Cancer Institute, Contract 21X5034A from SAIC-Frederick, and the Aslan Foundation. AS is an American Cancer Society Clinical Research Professor.

V. REFERENCES

1. Kyle, R.A. et al., A trial of three regimens for primary amyloidosis: colchicine alone, melphalan and prednisone, and melphalan, prednisone, and colchicine, *N. Engl. J. Med.,* 336, 1202, 1997.
2. Comenzo, R.L. and Gertz, M.A., Autologous stem cell transplantation for primary systemic amyloidosis, *Blood,* 99, 4276, 2002.

3. Skinner, M. et al., High-dose melphalan and autologous stem-cell transplantation in patients with AL amyloidosis: an 8-year study, *Ann. Int. Med.*, 140, 85, 2004.

4. Saba, N. et al., High treatment-related mortality in cardiac amyloid patients undergoing autologous stem cell transplant, *Bone Marrow Transplant,* 24, 853, 1999.

5. Kumar, S. et al., High incidence of gastrointestinal tract bleeding after autologous stem cell transplant for primary systemic amyloidosis, *Bone Marrow Transplant,* 28, 381, 2001.

6. Richter, G., The resorption of amyloid under experimental conditions, *Am. J. Pathol.*, 30, 239, 1954.

7. Hawkins, P.N. et al., Scintigraphic quantification and serial monitoring of human visceral amyloid deposits provide for turnover and regression, *Q. J. Med.,* 86, 365, 1993.

8. Gianni, L. et al., New drug therapy of amyloidoses: resorption of AL-type deposits with 4'-iodo-4'-deoxydoxorubicin, *Blood,* 86, 855, 1995.

9. Solomon, A. and Weiss, D.T., Protein and host factors implicated in the pathogenesis of light chain amyloidosis (AL amyloidosis). *Amyloid: Int. J. Exp. Clin. Invest.,* 2, 269, 1995.

10. Kisilevsky, R. et al., Arresting amyloidosis in vivo using small-molecule anionic sulphonates or sulphates: implications for Alzheimer's disease. *Nat. Med.,* 1, 143, 1995.

11. Tennent, G.A., Lovat, L.B., and Pepys, M.B., Serum amyloid P component prevents proteolysis of the amyloid fibrils of Alzheimer disease and systemic amyloidosis, *Proc. Natl. Acad. Sci. USA,* 92, 4299, 1995.

12. Hrncic, R. et al., Antibody-mediated resolution of light chain-associated amyloid deposits, *Am. J. Pathol.*, 157, 1239, 2000.

13. Solomon, A., Weiss, D.T., and Wall, J.S., Therapeutic potential of chimeric amyloid-reactive monoclonal antibody 11-1F4, *Clin. Cancer Res.,* 9, 3831s, 2003.

THE AMYLOID-REACTIVE MONOCLONAL ANITBODY 11-1F4 BINDS A CRYPTIC EPITOPE ON FIBRILS AND PARTIALLY DENATURED IMMUNOGLOBULIN LIGHT CHAINS AND INHIBITS FIBRILLOGENESIS

O'Nuallain B.[1], Murphy C.L.[1], Wolfenbarger D.A.[1], Kennel S.[2], Solomon A.[1], and Wall J.S.[1].

[1]Human Immunology & Cancer Program, Department of Medicine, University of Tennessee Graduate School of Medicine, Knoxville, TN; [2] Life Sciences Division, Oak Ridge National Laboratory, Oak Ridge, TN, USA.

ABSTRACT

We have previously reported [1] that a murine mAb, designated 11-1F4, prepared against a human Vκ4 fibrils (LEN), reacted specifically with AL amyloid, regardless of the V_L subgroup of the protein. In order to determine the molecular bases of this interaction, we have utilized a europium-linked immunoabsorbent assay (EuLISA) to measure the interaction of mAb 11-1F4 with light chain-related proteins. These studies show that this antibody does not recognize native κ and λ chains but, rather, binds with nanomolar affinity to proteins that have been partially denatured, i.e., when coated onto plastic wells used in the assay. EuLISA of peptides derived from proteolytic cleavage of the Vκ4 immunogen revealed that the cryptic epitope was located in a region encompassing the first 59 amino acids. Through epitope mapping of synthetic peptides, this site was placed more precisely within the first (N-terminal) 22 residues that comprise the first framework region (FRI) of Ig light chains, and especially involves those located between positions 1 and 4. Additionally, our finding that mAb11-1F4 inhibited at sub-equimolar concentrations *de novo* V_L fibrillogenesis suggests that interaction of this antibody with a partially unfolded amyloidogenic intermediate may act to stabilize the molecule and/or sterically inhibit fibril formation. The discovery that mAb 11-1F4 can prevent fibrillogenesis, as well as accelerate amyloidolysis, has provided new information on its immunoreactivity and therapeutic potential for patients with primary (AL) amyloidosis.

RESULTS AND DISCUSSION

Using a sample size of sixteen V_L and BJP proteins of both the κ and λ isotypes we showed that 11-1F4, a murine mAb prepared against human Vκ4 fibrils, bound with nanomolar affinities to light chain proteins in an isotype-independent manner (data not shown). To characterize this "cross-reactive" epitope we used as our model system 11-1F4's reactivity with the Vκ4 immunogen (LEN), a heat denatured preparation that consisted of fibrils and amorphous-like material. The 11-1F4 mAb binds to the soluble and fibrillar forms of the LEN V_L when adsorbed to dryness onto high binding (COSTAR) microtiter plates with the same affinity (EC_{50}'s of about 0.1 nM: Fig. 1). However, there is a 2-fold lower amplitude signal for the binding of 11-1F4 to LEN fibrils relative to the non-fibrillar form that might be due to the burial of mAb binding sites within the structure of each fibril and/or through higher order fibril-fibril interactions (Fig. 1). The LEN immunogen when incubated with 11-1F4 in solution resulted in efficient inhibition of the mAb binding to adsorbed LEN V_L in the plate (Fig. 1)

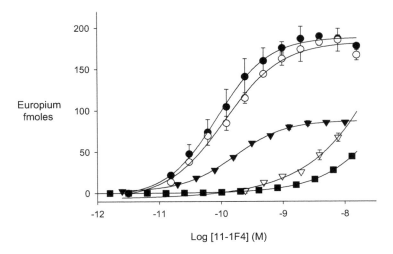

Figure 1. 11-1F4 binding to adsorbed fibrillar and monomeric LEN Vκ4 protein in the presence or absence of soluble monomeric or heat denatured LEN or the LEN 1-30 peptide. LEN fibrils (▼), and monomeric LEN V_L alone (●), or in the presence of a 50-fold molar excess of monomeric LEN V_L (○), heat denatured LEN V_L (■) or LEN 1-30 peptide (∇). All Ab interactions were carried out at 37°C in PBS containing 1% BSA and 0.05% Tween 20 (assay buffer). Europium time-resolved fluorescence was measured with excitation and emission wavelengths of 396 nm and 636 nm, respectively.

In contrast, solution-phase LEN V_L had no effect on the interaction of 11-1F4 with LEN V_L (Fig. 1). These results suggest that the mAb recognizes a non-native epitope that is exposed on surface adsorbed LEN and LEN fibrils [both plate adsorbed and in solution (data not shown)]. In the literature, the denaturation of proteins on microtiter plate surfaces and the resultant exposure of antibody neo-epitopes as well as the existence of amyloid fibril-specific conformational mAb binding sites are well documented (for examples see [2, 3]).

To further characterize the 11-1F4 epitope on the LEN protein we generated 11-1F4 binding curves for synthetic LEN peptides made to overlapping regions within the N-terminal 59 amino acids. These studies revealed the epitope to be within the first ~22 residues (data not shown). Furthermore, the 11-1F4 binding analyses indicated that the interaction with the N-terminal 4 amino acids contributed ~30% of the total binding energy - the remainder being distributed throughout the other 18 residues (data not shown). The LEN 1-30 peptide was shown to be as good a competitor as the LEN immunogen at inhibiting 11-1F4 binding to plate-adsorbed LEN V_L. These data suggest that 11-1F4 binds to a discontinuous, cryptic epitope expressed by plate-adsorbed and fibrillar LEN but not monomeric LEN in solution.

The interaction of 11-1F4 with the Wil $V_\lambda 6$ protein inhibited fibrillogenesis by increasing the lag time in the kinetics of fibril growth by 8-fold (Fig. 2). In addition, a 25% decrease in the ThT fluorescence emission at equilibrium was observed in the presence of 11-1F4, consistent with destabilization of the fibrils or a reduction in the number of available ThT binding sites on the fibrils. In contrast to 11-1F4, two other isotype-matched mAbs had little or no effect on the fibrillogenesis of the Wil $V_\lambda 6$ protein (Fig. 2). Interestingly, the maximum rate of fibril growth, depicted by the dashed lines in Fig. 2, was unaffected by the presence 11-1F4.

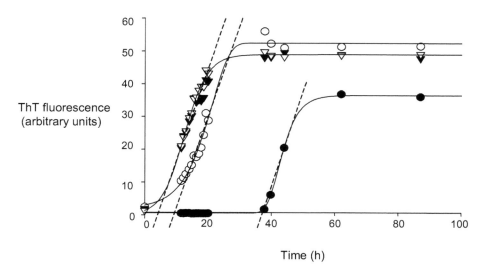

Figure 2 Inhibition of *de novo* WIL (V$_\lambda$6) fibrillogenesis by 11-1F4.. (∇) Progress curve for Wil fibrils grown without a mAb, Wil fibril growth in the presence of subequimolar 11-1F4 (●), a V$_\lambda$8-specific mAb (▼) or MOPC-31c mAb (○). The lag time was calculated from the X-axis intersection of a hypothetical line drawn parallel to the maximum rate of fibril growth. The experiment was carried out in an unblocked low binding microtiter plate (COSTAR) at 37°C in PBS with 5 µM WIL and 2.5 µM mAb.

CONCLUSION

The 11-1F4 mAb has been shown to accelerate the rate of amyloidolysis in a murine model of AL amyloidosis [1]. The studies presented herein demonstrate that the reactivity of the 11-1F4 is dependent upon the expression of a cryptic antigen expressed on non-native light chain proteins and synthetic fibrils but not on proteins within the native structural ensemble. The specificity with which 11-1F4 has been shown to react with light chain fibril renders this reagent an invaluable addition to the armamentarium of the physician treating patients with this invariably fatal and devastating from of amyloid disease.

REFERENCES

1.   Hrncic R. *et al.*, Antibody-mediated resolution of light chain-associated amyloid deposits. *Am. J. Pathol.* 157, 1239-1246, 2000.
2.   Butler JE., Solid supports in enzyme-linked immunosorbent assay and other solid-phase immunoassays. Methods Mol. Med., 94, 333-372, 2004.
3.   O'Nuallain B. and Wetzel R., Conformational Abs recognizing a generic amyloid fibril epitope. *Proc. Natl. Acad. Sci.*, 99 (3),1485-90, 2002.

ACKNOWLEDGEMENTS

AS is an American Cancer Society Professor. This work was supported in part by NCI grant # CA10056 and a donation from the Aslan Foundation.

MOLECULAR RECOGNITION AND SELF-ASSEMBLY OF AMYLOID FIBRILS: THE ROLE OF AROMATIC INTERACTIONS

E. Gazit

Department of Molecular Microbiology and Biotechnology, Tel Aviv University, Tel Aviv 69978, Israel

The formation of amyloid fibril plaques is associated with major diseases of unrelated origin. A partial list includes Alzheimer's diseases, Parkinson's disease, Type II diabetes, Prion diseases, and various familial amyloidosis diseases. The mechanism of amyloid fibrils formation is assumed to be a nucleation-dependent process. According to the common models, unfolding events are followed by a series of equilibrium steps to form a prefibrillar nucleus of a critical size. This is followed by thermodynamically-favorable growth steps of addition of monomers to the growing nucleus. In all these cases normal proteins undergo a self-organization process that result in the formation of well-ordered assemblies with fibrilar structure and a diameter of 7-10 nm, as observed by electron microscopy and atomic force microcopy, and a clear X-ray fiber diffraction with a 4.6-4.8 Å reflection on the meridian. However, in spite of the key medical importance of the process of amyloid self-assembly, the molecular mechanism by which amyloid fibrils are being formed is not fully understood.

ANALYSIS OF SHORT AMYLOIDOGENIC MOTIFS

Amyloid fibrils formation is notably distinct from a simple process of non-specific aggregation. This is reflected by the fact that amyloid fibrils are well-ordered structures as compared to amorphous aggregates. Therefore, we assume that a specific pattern of molecular recognition, rather than non-specific hydrophobic interactions, is the basis for the formation of such an ordered assembly. The determination of such interactions, which underlay the basis for the molecular recognition and self-assembly, is therefore crucial for profound understanding of the amyloid formation process. Furthermore, such knowledge is valuable for the future design of drugs that will be able to block these interactions and thus have a key clinical importance as described in the previous section. The study of very short peptide fragments that retain all the molecular information needed to mediate the process of amyloid fibrils formation is therefore a very useful reductionist model to understand the specific interactions that mediate this process.

THE USE OF PEPTIDE MODELS TO STUDY AMYLOID FORMATION MECHANISM

A very significant advantage of peptide analysis is their simple chemical synthesis using solid-phase techniques. The well-known peptide synthesis methods facilitate the production of many site-directed analogues of amyloidogenic peptides to be studied. This allows pinpointing determinants in the peptides that mediate the process of amyloid formation. Solid-phase peptide synthesis also allows the site-specific labeling of peptides with fluorescent probes during. Indeed, a very siginifacnt amount of work in recent years was directed toward the identification of minimal amyloid-forming motifs. Table 1 summarizes some of the recent identification of very short amyloidogenic motifs that form fibrils that are highly similar to those formed by the full length proteins and polypeptides. It is quite striking that such short elements have all the information to mediate a well-ordered structural self-assembly process as amyloid formation

IMPLICATIONS FOR THE DEVELOPMENT OF AMYLOID FORMATION INHIBITORS

We suggest, based on experimental and bioinformatical analysis, that aromatic stacking interactions may provide energetic contribution as well as order and directionality in the self-assembly of ordered amyloid structures (2-9). This is in line with the well-known central role of aromatic stacking interactions in self-assembly of many supramolecular structures. In the path of our reductionist approach toward the identification of the shortest motifs that mediate the self-assembly of the fibrils, we recently demonstrated that the dipeptide core-recognition motif of Alzheimer's β-amyloid polypeptide contains all the molecular information needed for efficient self-assembly into well-ordered, stiff, and elongated nanotubes (10). This observation provided an experimental link to the suggestion made by the late Max Perutz that amyloid fibrils are water-filled nanotubes. Furthermore, we demonstrated that the peptide nanotubes could serve as a degradable mold for the casting of nano-scale objects. Similar peptides can alternatively assemble to closed-cage nanostructures (11). Taken together, our hypothesis provides a new approach to understand the self-assembly mechanism that governs amyloid formation, indicates possible ways to control this process, and demonstrate the ability of amyloidogenic recognition elements to self-assemble into well-ordered nanostructures.

Table 1. Summary of short amyloidogenic motifs

Parent Polypeptide name	Associated disease	Amino acid sequence	Reference
Islet Amyloid Polypeptide	Type II diabetes	NFGAIL	
		NFLVH	
Medin	Aortic medial amyloid	NFGSVQ	
Amyloid β Polypeptide	Alzheimer's disease	KLVFFAE	
Calcitonin	Medullary carcinoma	DFNKF	

CONCLUSIONS

Amyloid fibrils formation represents a major medical concern. In the last few years, it was shown in several studies that short aromatic peptides can form amyloid fibrils that are very similar to those form by the full-length polypeptides. We speculate that the stacking interactions rather than mere hydrophobicity may provide energetic contribution as well as order and directionality in the self-assembly of amyloid structures. The stacking hypothesis suggests a new approach to understand the self-assembly mechanism that governs amyloid formation, ables the identification of novel motifs, and indicates possible ways to control this process.

REFERENCES

1.   Harper, J.D. and Lansbury, P.T.Jr., Models of amyloid seeding in Alzheimer disease and scrapie: mechanistic truths and physiological consequences of the time-dependent solubility of amyloid proteins. *Annu. Rev. Biochem.* 66, 385, 1997.

2.   Dobson, C.M., Protein misfolding, evolution and disease. *Trends Biochem. Sci.* 24, 329, 1999.

3. Sipe, J. D. and Cohen, A. S., Review: history of the amyloid fibril. *J. Struct. Biol.* 130, 88, 2000.

4. Gazit, E., The "correctly-folded" state of proteins: Is it a metastable state? *Angew. Chem.* 41, 257, 2002.

5. Jarrett, J.T. and Lansbury, P.T.Jr., Seeding "one-dimensional crystallization" of amyloid: a pathogenic mechanism in Alzheimer's disease and scrapie? *Cell* 73, 1055, 1993.

6. Tenidis, K., Waldner, M., Bernhagen, J., Fischle, W., Bergmann. M., Weber, M., Merkle, M.L., Voelter, W., Brunner, H. and Kapurniotu, A., Identification of a penta- and hexapeptide of islet amyloid polypeptide (IAPP) with amyloidogenic and cytotoxic properties. *J. Mol. Biol.* 295, 1055, 2000.

7. Balbach, J.J., Ishii, Y., Antzutkin, O.N., Leapman, R.D., Rizzo, N. W., Dyda, F., Reed, J., and Tycko, R., Amyloid fibril formation by Aβ 16-22, a seven-residue fragment of the Alzheimer's beta-amyloid peptide, and structural characterization by solid state NMR. *Biochemistry,* 39, 13748, 2000.

8. Gazit, E., A Possible Role for π-Stacking in the Self-Assembly of Amyloid Fibrils. *FASEB J.* 16, 77, 2002.

9. Azriel, R. and Gazit, E., Analysis of the Structural and Functional Elements of the Minimal Active Fragment of Islet Amyloid Polypeptide (IAPP): An Experimental Support for the Key Role for the Phenylalanine Residue in Amyloid Formation. *J. Biol. Chem.* 276, 34156, 2001.

10. Gazit, E., Global Analysis of Tandem Aromatic Octapeptide Repeats: The Significance of Aromatic-Glycine Motif. *Bioinformatics* 18, 880, 2002.

11. Gazit, E., Mechanistic Studies of the Process of Amyloid Fibrils Formation by the Use of Peptide Fragments and Analogues: Implications for the Design of Fibrillization Inhibitors. *Curr. Med. Chem.* 9, 1725, 2002.

12. Mazor, Y., Gilead, S., Benhar, I., and Gazit, E., Identification and Characterization of a Novel Molecular-Recognition and Self-Assembly Domain in the Islet Amyloid Polypeptide. *J. Mol. Biol.* 322, 1013, 2002.

13. Reches, M., Porat, Y., and Gazit, E., Amyloid Fibrils Formation by Pentapeptide and Tetrapeptide Fragments of Human Calcitonin. *J. Biol. Chem.* 277, 35475, 2002.

14. Porat, Y., Kolusheva, S., Jelinek, R., and Gazit, E., The Human Islet Amyloid Polypeptide Forms Transient Membrane-Active Prefibrillar Assemblies. *Biochemistry* 42, 10971, 2003.

15. Reches, M., and Gazit, E., Casting Metal Nanowires within Discrete Self-Assembled Peptide Nanotubes, *Science,* 300, 625, 2003.

16. Reches, M. and Gazit, E., Formation of Closed-Cage Nanostructures by Self-Assembly of Aromatic Dipeptides. *Nano Letters* , 4, 581, 2004.

17. Reches, M. and Gazit, E., Amyloidogenic Hexapeptide Fragment of Medin: Homology to Functional Islet Amyloid Polypeptide Fragments. *Amyloid* , in press, 2004.

SERUM AMYLOID P COMPONENT AS A THERAPEUTIC TARGET IN AMYLOIDOSIS

M. B. Pepys

Centre for Amyloidosis and Acute Phase Proteins, Department of Medicine, Royal Free and University College Medical School, University College London, Rowland Hill Street, London NW3 2PF.

WHAT IS AMYLOID?

Amyloid is a pathological extracellular deposit composed predominantly of amyloid fibrils with characteristic morphology and pathognomonic tinctorial properties, specifically the capacity to bind Congo Red dye and then display red-green dichroism when viewed in strong cross polarised light. Amyloid fibrils are formed from normally soluble autologous precursor proteins that have undergone misfolding and aggregated into fibrils with a common cross β-sheet core structure. The deposits are rich in proteoglycans and always contain the non-fibrillar normal plasma glycoprotein, serum amyloid P component (SAP), which binds specifically to all amyloid fibrils.

WHAT IS AMYLOIDOSIS?

Amyloidosis is disease caused by amyloid deposits in the tissues. The deposits may be local or systemic in distribution and acquired or hereditary in aetiology. All patients with amyloidosis have amyloid deposits in their tissues when they first present clinically. In each individual, disease is always worse when there is more amyloid present, and arrest or regression of amyloid deposition is associated with clinical benefit. Only organs or tissues in which amyloid deposits are present are directly functionally compromised, even when the amyloid fibril precursor protein is ubiquitous.

WHAT IS NOT AMYLOIDOSIS?

Several important diseases are associated with amyloid deposits but are not amyloidosis. The neuropathology of Alzheimer's disease by definition includes intra-cerebral Aβ amyloid deposits, but it is not known whether these deposits are responsible for the neuronal cell death that causes cognitive decline; non-fibrillar Aβ aggregates may be more directly involved. In contrast, by analogy with the notable friability of systemic blood vessels containing AL amyloid deposits, it seems likely that the cerebrovascular Aβ amyloid deposits in cerebral amyloid angiopathy are responsible for cerebral haemorrhage. Most patients with type 2 diabetes have amyloid deposits in the islets of Langerhans, which may exacerbate islet dysfunction, but amyloid is not the original cause of their diabetes. Cerebral amyloid deposits of the prion protein are present in many forms of transmissible spongiform encephalopathy but are absent in others, including bovine spongiform encephalopathy in cows and fatal familial insomnia in humans, and amyloid is thus not essential for pathogenesis of TSE. Many other diseases are associated with, and possibly caused by, protein misfolding and aggregation. Some of these aggregates share the amyloid cross β-sheet fold, for example in Huntington's and related poly-glutamine repeat hereditary neurodegenerative diseases, and in Parkinson's disease. However intracellular and intranuclear protein aggregates have very different pathogenetic effects than extracellular amyloid deposits, and are in radically different locations for drug intervention. They are also not associated with the non-fibrillar components of amyloid deposits, the proteoglycans and SAP. These intracellular protein aggregates are thus not amyloid and the diseases with which they are associated are not amyloidosis. Although there may be informative similarities and parallels, especially in protein misfolding and aggregation studied *in vitro*, to conflate such widely different processes is misleading and potentially dangerous when extrapolated to development of therapeutic interventions. Other protein misfolding diseases, such as the serpinopathies, are even more remote from amyloidosis, with intracellular deposition of protein aggregates that do not share the typical amyloid cross-β fold and where much of the pathogenesis of disease is related to loss of normal serpin function rather than adverse effects of the aggregates.

HOW DOES AMYLOID CAUSE DISEASE?

In amyloidosis the link between physical presence of amyloid deposits and disease is irrefutably strong. Pre-fibrillar protein aggregates are cytotoxic for cultured cells exposed under artefactual *in vitro* conditions but there is no evidence that this is pathophysiologically relevant to amyloidosis *in vivo*. No clinical or pathological effects are detectable in any form of acquired or hereditary systemic amyloidosis until amyloid deposits are demonstrable, regardless of the length of time that the predisposition exists before amyloid develops. Clinical impairment progresses as amyloid deposits increase in size, and regression of amyloid is associated with clinical improvement and survival. If damaged organs are replaced there is no organ dysfunction, even in the face of unchanged abundance of amyloid fibril precursor proteins fully capable of aggregation and fibril formation, until new amyloid deposits form and accumulate to a critical mass. This usually takes years. There is usually no histological evidence of inflammation, or cell death or dysfunction, even in massively amyloid laden organs, and the organs continue to function until the amyloid deposits critically compromise structure and/or function. For example, cardiac amyloid deposition impairs diastolic filling by stiffening the ventricle. There is no evidence of cytotoxicity in the myocardium and the heart muscle continues to contract excellently and propel the progressively reduced stroke volume until the end. Similarly in hereditary transthyretin amyloidosis, amyloid fibrils may accumulate over years in the vitreous humour of the eye until fibril density impairs optical clarity and causes blindness. The amyloidogenic variant transthyretin in the eye is synthesised in the choroid plexus and by the ciliary epithelium directly overlying the retina, and continuously bathes the retina. However replacement of the amyloid laden vitreous with optically clear medium restores perfect vision in every case! The putative toxic pre-fibrillar aggregates evidently have no deleterious effect on retinal neurones, whereas accumulation of amyloid fibrils causes blindness and their removal cures it. Finally there are many clinical situations in which the local physical presence of substantial amyloid deposits alone is the only possible cause of disease and death, and where reduction or removal of the deposits relieves symptoms and signs and is life saving.

OUR GOAL IN TREATMENT OF SYSTEMIC AMYLOIDOSIS

Clinical assessment and SAP scintigraphy in thousands of patients with systemic amyloidosis over the past 20 years, have established that sufficient reduction of the abundance of amyloid fibril precursor proteins, in both acquired and hereditary amyloidosis, is associated with arrest of amyloid deposition, regression of deposits, and clinical benefit. Further measures to reduce amyloid deposition and/or promote regression will provide incremental benefits, hence our longstanding interest in SAP.

WHAT IS SAP?

SAP is a highly conserved constitutive normal trace plasma protein with specific avid calcium dependent binding to all amyloid fibrils that causes its remarkable selective concentration and persistence in amyloid deposits of all types. SAP is intrinsically resistant to proteolysis and is further stabilised by its tight binding to amyloid fibrils, which in turn stabilises the fibrils and protects them from proteolysis. SAP knockout mice show delayed and reduced deposition of experimentally induced reactive systemic amyloid. SAP is therefore a valid therapeutic target, and *in vivo* inhibition of SAP binding and/or depletion of circulating SAP should reduce amyloid burden.

TARGETED PHARMACOLOGICAL DEPLETION OF SAP FOR TREATMENT OF AMYLOIDOSIS

In collaboration with F Hoffmann-la Roche & Co Ltd we have developed a new chemical entity, carboxy pyrrolidone hexanoyl pyrrolidone carboxylic acid (CPHPC), that inhibits binding of SAP to amyloid fibrils *in vitro* with approximately micromolar IC_{50}. Administration of a sufficient dose to amyloidotic mice *in vivo* completely removes mouse SAP from the deposits (1). Human SAP binds CPHPC much more avidly than does mouse SAP and the stable complex that forms *in vivo* between 2 human SAP molecules and 5 CPHPC molecules is immediately cleared by the liver, leading to swift depletion of all circulating SAP for as long as the drug is administered (2). This is the first example of a novel pharmacological mechanism in which a pathological human protein is depleted by a low molecular weight drug. However *ex vivo* dissociation of human SAP that has already bound to amyloid deposits *in vivo*, requires millimolar CPHPC concentrations. Nevertheless, since circulating SAP is made exclusively in the liver, is the sole source of SAP in the amyloid deposits, and is in dynamic equilibrium with the amyloid SAP pool, sustained plasma depletion should eventually clear all the SAP from the deposits. CPHPC for treatment of amyloidosis is licensed exclusively to a UCL spinout company and we now have about 30 patient years of experience of CPHPC administration to patients with various forms of systemic amyloidosis. There have been no drug related adverse effects, nor any toxicity or abnormal investigational findings. CPHPC is not metabolised and is very rapidly cleared, mostly by the kidney, with a plasma half life of about 1.5 h. Thus although the modest CPHPC doses deployed so far maintain complete depletion of plasma SAP and cause measurable depletion of SAP from the amyloid deposits (2), significant amounts of SAP remain in the deposits even after months of treatment. Consequently we have not yet effectively tested the hypothesis that complete removal of SAP from amyloid deposits would reduce new amyloid deposition and/or promote amyloid regression with clinical benefit. We are therefore currently exploring alternative dose regimes and routes of CPHPC administration in order to achieve the high blood and tissue concentrations of the drug required for clearance of all SAP from the tissue amyloid deposits. We are also conducting a preliminary study in Alzheimer's disease, in which the amyloid deposits are many orders of magnitude smaller than in systemic amyloidosis, and where depletion of circulating SAP should be sufficient to deplete SAP from the cerebrospinal fluid and the deposits.

REFERENCES

1. Pepys, M.B. (1999) The Lumleian Lecture. C-reactive protein and amyloidosis: from proteins to drugs? *In: Horizons in Medicine, Vol. 10* (Williams, G., eds.), Royal College of Physicians of London, London, pp. 397-414.
2. Pepys, M.B., *et al.* (2002) Targeted pharmacological depletion of serum amyloid P component for treatment of human amyloidosis. *Nature,* **417**: 254-259.

Glycosaminoglycans as a therapeutic target of amyloid disorders

B.P.C. Hazenberg

Department of Rheumatology, University Hospital, Groningen, The Netherlands

HISTORY

Rudolph Virchow introduced in 1854 the term amyloid (starch-like) for the material found in lardaceous, waxy livers at autopsy. He observed that this material reacted similarly to starch after staining with iodine followed by sulphuric acid (change of colour from brown to blue). However, soon thereafter (in 1859) Friedreich and Kékulé demonstrated the protein nature of this waxy material. Boyd described in 1932 the glycosaminoglycan chondroitin sulphate to be a constituent of amyloid and Bitter and Muir did so in 1966 for heparan sulphate.

GLYCOSAMINOGLYCANS (GAGS)

Detailed information about the role of glycosaminoglycans (GAGs, formerly called acid mucopolysaccharides) in amyloidosis can be found in literature (1-4). Heparan sulphate is firmly involved in AA amyloidosis and Alzheimer's disease, as both human and animal studies have shown. GAGs are synthesized in the Golgi apparatus and are composed of 20 or more disaccharide repeats of hexuronic acid, and either glucosamine or galactosamine. For heparin and heparan sulphate a specific transferase sequentially adds N-acetylglucosamine and D-glucuronic acid in an alternating pattern to a tetrasaccharide, which is attached to sulphate residues of a core protein, followed by further enzymatic modifications by epimerase and sulphotransferases (3). Heparan sulphate is also present in many other types of amyloid and appears to be a universal component of amyloid.

Snow and Kisilevsky have demonstrated in a series of articles the intimate spatial and temporal relationship of heparan sulphate and AA amyloid in experimentally induced AA amyloidosis in the mouse. GAGS were detected by Alcian blue stain and heparan sulphate by antibodies. In two mouse models, in which amyloid can be induced slowly (1 week) or quickly (36-48 hours), was shown that accumulation of heparan sulphate in splenic perifollicular regions preceded deposition of AA amyloid (2).

PROTEOGLYCANS

Proteoglycans are a heterogeneous family of GAGs covalently attached to a core protein. Proteoglycans are important constituents of the extracellular matrix and basement membranes (5). Of the five different heparan sulphate proteoglycans two appear to be important in amyloidosis, *i.e.* perlecan and agrin (3). Ailles *et al* showed in the fast amyloid induction mouse model that messenger RNA of perlecan increased after 24 hours in the perifollicular region, whereas AA amyloid was not detected in the same region before 36 hours. Therefore local synthesis of perlecan precedes deposition of AA amyloid in the mouse model (2, 4). Heparan sulphate (in contrast to the other GAGs) can significantly increase the beta-sheet content of amyloid precursor proteins, such as serum amyloid A (SAA) and Aβ. Some investigators reasoned that heparan sulphate must physically associate with amyloid precursors to cause these beta-sheet changes. Heparin binding sites are indeed present in human SAA (in the mouse even two, *i.e.* a 33-mer and a 27-mer) and Aβ, as well in some other amyloid precursors. Snow and Castillo postulated some roles for perlecan in amyloidogenesis in Alzheimer's disease, such as precursor processing, binding, enhancing aggregation, accelerating fibril formation, maintaining fibril stability, protection from protease degradation, and hindering clearance (6).

AMYLOID FIBRIL MODEL

These GAG-precursor data are difficult to understand from the perspective of the classic model of the amyloid fibril. Therefore, Inoue and co-workers looked more into detail with high-resolution electron microscopy to amyloid fibrils *in situ* and proposed a new model based upon their observations. They propose that the amyloid core is made up of serum amyloid P component (SAP) around which 3 nm double tracked filaments of chondroitin sulphate (also a GAG) proteoglycan (CSPG) are helically wound enclosing the core. This core complex is, in turn, wrapped in a layer of heparan sulphate proteoglycan 4.5-5.0 nm filaments to which 3 nm amyloid polypeptide filaments (helical rods) are loosely attached parallel to the axis of the fibril (7). This model elegantly integrates the close relation between heparan sulphate proteoglycan and amyloid precursor, although it raises new questions about the genesis of the fibril core.

IN VIVO MODIFICATION OF GAG – AMYLOID PRECURSOR COMPLEX

The universal presence of heparan sulphate proteoglycans, such as perlecan and agrin, and their important role in amyloidogenesis suggests possibilities to modify the relation with precursor proteins. In theory there are three options: modification of precursor, of heparan sulphate, and of the binding of both.

Precursor modification has not been described *in vivo*. Some observations underline the relevance of small structural differences of the precursor protein. Mouse SAA proteins (SAA1.1 and SAA2.1) differ in a small number of amino acids; only SAA1.1 is amyloidogenic, whereas SAA2.1 is not at all. In Alzheimer the precursor protein APP becomes amyloidogenic after cleavage into smaller fragments, *i.e.* Aβ1-42 and Aβ1-40.

Heparan sulphate modification has been described by Kisilevsky (8). He incorporated 4-deoxy analogues of glucosamine in heparan sulphate and showed that two of these anomers seemed to be effective in an AA amyloid mouse model in vivo as well as in a tissue culture model of AA amyloidogenesis. He also studied the effects of presence and up-regulation of heparanase in different tissues on amyloidogenesis.

Modification of the GAG-precursor binding has been a promising approach. Small-molecule anionic sulphonates or sulphates reduced splenic AA amyloid progression in mice and interfered with heparan sulphate-stimulated beta-peptide fibril aggregation *in vitro*. Polyvinylsulphonate appeared to be most effective of all (9). Inoue *et al* used polyvinylsulphonate in murine AA amyloidogenesis and showed that HSPG and AA network were lost from the fibril and the central portion disintegrates into CSPG and SAP. These findings support the concept that polyvinylsulphonate interferes with GAG-precursor binding (10). Snow *et al* demonstrated in posters at this (poster 136) and the former symposium the use of derivatives from *Uncaria Tomentosa*. They used these in a plaque-producing mouse model of Alzheimer's disease and these drugs appeared to be relatively non-toxic, with good passage of the blood-brain barrier. The drugs reduced amyloid plaque load, number, and size by more than 50% following short-term treatment. Walzer *et al* showed in a rat model of β-amyloid induced neuropathy the effect of chronic injections of small GAGs. These small GAGs blocked abnormal intracellular tau changes, inhibited reactive astrocytosis, but had no effect on β-amyloid deposits itself (11).

FIBRILLEX, A NEW ANTI-AMYLOID DRUG

Fibrillex is a small, highly charged sulphonated molecule which was one of the drugs with a beneficial effect on AA amyloidosis in the mouse (9). Although the effect was not as strong as that of polyvinylsulphonate, the drug appeared to be orally bioavailable with a good safety and tolerability profile. Recently a phase II-III multi-centre clinical trial has started in 27 sites in 13 countries to investigate efficacy and safety in patients with clinical AA amyloidosis. Results are expected in the middle of 2005.

CONCLUSIONS

GAGs (mainly heparan sulphate) are essential for genesis and deposition of amyloid fibrils. Interference of the binding between GAG and amyloid precursor seems to be a promising approach for amyloid therapy.

REFERENCES

1. Magnus JH, Stenstad T, Husby G. Proteoglycans, glycosaminoglycans and amyloid deposition. In: Husby G, editor. Reactive amyloidosis and the acute phase response. Bailliere's Clinical Rheumatology. London: Bailliere Tindall, 1994: 575-97.
2. Ancsin JB. Amyloidogenesis: historical and modern observations point to heparan sulfate proteoglycans as a major culprit. Amyloid: J Protein Folding Disord 2003; 10: 67-79.
3. van Horssen J, Wesseling P, van den Heuvel LPWJ, de Waal RMW, Verbeek MM. Heparan sulphate proteoglycans in Alzheimer's disease and amyloid-related disorders. Lancet Neurology 2003; 2: 482-92.
4. Kisilevsky R. Review: amyloidogenesis-unquestioned answers and unanswered questions. J Struct Biol 2000; 130: 99-108.
5. Groffen AJA, Veerkamp JH, Monnens LAH, van de Heuvel LPWJ. Recent insights into the structure and functions of heparan sulfate proteoglycans in the human glomerular basement membrane. Nephrol Dial Transplant 1999; 14:2119-29.
6. Snow AD, Castillo GM. Specific proteoglycans as potential causative agents and relevant targets for therapeutic intervention in Alzheimer's disease and other amyloidoses. Amyloid: Int J Exp Clin Invest 1997; 4: 135-41.
7. Inoue S, Kuroiwa M, Kisilevsky R. AA protein in experimental murine AA amyloid fibrils: a high resolution ultrastructural and immunohistochemical study comparing aldehyde-fixed and cryofixed tissues. Amyloid: J Protein Folding Disord 2002; 9: 115-25.
8. Kisilevsky R, Szarek WA. Novel glycosaminoglycan precursors as anti-amyloid agents part II. J Mol Neurosci 2002; 19: 45-50.
9. Kisilevsky R, Lemieux LJ, Fraser PE, Kong X, Hultin PG, Szarek WA. Arresting amyloidosis in vivo using small-molecule anionic sulphonates or sulphates: implications for Alzheimer's disease. Nature Medicine 1995; 1: 143-8.
10. Inoue S, Hultin PG, Szarek WA, Kisilevsky R. Effect of polyvinylsulphonate on murine AA amyloid: a high-resolution ultrastructural study. Lab Invest 1996; 74: 1081-90.
11. Walzer M, Lorens S, Hejna M, et al. Low molecular weight glycosaminoglycan blockade of β-amyloid induced neuropathology. Eur J Pharmacol 2002; 445: 211-20.

INHIBITION OF IAPP TOXICITY AND AGGREGATION INTO β– SHEETS AND AMYLOID BY A RATIONALLY N-METHYLATED, IAPP AMYLOID CORE-CONTAINING HEXAPEPTIDE

A. Kapurniotu, A. Schmauder, M. Tatarek-Nossol, K. Tenidis

Laboratory of Medicinal and Bioorganic Chemistry, Institute of Biochemistry, University Hospital of the RWTH Aachen, Pauwelstr. 30, D-52074 Aachen, Germany.

Pancreatic amyloid is present in more than 95% of type II diabetes patients and is formed by the aggregation of the 37-residue polypeptide islet amyloid polypeptide (IAPP) (Figure 1) (1,2). Insoluble IAPP amyloid aggregates colocalize with areas of cellular degeneration and have been, therefore, linked to the progressive deterioration of β-cell function and the pathogenesis of type II diabetes. Both the self-assembly state and the morphology of the cytotoxicity-mediating IAPP species are yet not clarified. Several lines of evidence suggest, however, that the cytotoxic species form within the pathway of amyloid formation (2,3). Therefore, the development of strategies or compounds that may interfere with and/or inhibit the amyloid formation process is an important task to understand the molecular basis of IAPP cytotoxicity and to, possibly, slow down or inhibit disease pathogenesis.

Figure 1. Primary structure of IAPP. The amyloid core region IAPP(22-27) or NFGAIL is shown in bold.

IAPP amyloid forms *via* a conformational transition of soluble, mainly disordered IAPP into aggregated β-sheets (4). The hexapeptide IAPP(22-27) or sequence NFGAIL (Figure 1) has been previously identified as a short "IAPP amyloid core" sequence, i.e. a sequence that contains a β-sheet and self-recognition domain of IAPP and is sufficient to self-associate into β-sheet-containing, cytotoxic amyloid fibrils (5). We have recently developed a minimalistic, structure-based chemical strategy to "transform" amyloid core-containing sequences into non-amyloidogenic ones that are able to interact with the native sequences and inhibit amyloid formation and cytotoxicity (6). This has been achieved via the rational N-methylation of a minimum of two amide bonds on the same side of the strand of the β-sheet of the amyloid core region. We have demonstrated the applicability of the strategy by using various amyloid core-containing sequences of IAPP that have been N-methylated based on a NMR-derived structural model of pancreatic amyloid (7). We have shown that N-methylation at G24 and I26 in short IAPP amyloidogenic and cytotoxic sequences, including the NFGAIL one, is able to convert these sequences into non-amyloidogenic and non-cytotoxic ones (6). Moreover, the rationally designed analogue NF(N-Me)GA(N-Me)IL has been shown to be able to inhibit cytotoxic amyloid formation and cytotoxicity of its precursor sequence IAPP(22-27). Here we present biophysical studies on the interaction of NF(N-Me)GA(N-Me)IL with full sequence IAPP and its effect on IAPP aggregation and cytotoxicity.

RESULTS

The interaction of NF(N-Me)GA(N-Me)IL with IAPP and its effect on IAPP aggregation into β-sheets were studied by far-UV circular dichroism spectroscopy (CD) in combination with electron microscopy (EM). Under the applied conditions IAPP alone (5 μM in phosphate buffer, 1% HFIP) aggregated into β-sheets and insoluble amyloid fibrils after a lag time of 5-7 days (2, 4). In the presence of NF(N-Me)GA(N-Me)IL (25 or 50 μM), however, the solution remained clear and at a stable overall conformation for up to one month. Importantly, no aggregation into β-sheets could be observed in the solutions containing the IAPP / NF(N-Me)GA(N-Me)IL mixture by CD and no amyloid fibrils were detected by EM. In addition, the CD spectrum of mixture strongly differed from the CD spectrum of the arithmetic sum of the CD spectra of IAPP and NF(N-Me)GA(N-Me)IL. These results suggested that NF(N-Me)GA(N-Me)IL interacted with IAPP and that this interaction resulted in an inhibition of IAPP aggregation into amyloid fibrils.

Next, the effect of the NF(N-Me)GA(N-Me)IL precursor sequence NFGAIL on IAPP amyloid formation was tested. The CD spectrum of the mixture IAPP (5 μM) with NFGAIL (50 μM) differed from the one of IAPP alone suggesting that the two peptides interacted. However, 14 days after the begin of incubation insoluble IAPP amyloid fibrils formed as confirmed by EM. These results suggested that although NFGAIL could in fact interact with IAPP, as was expected for an amyloid core-containing peptide, it was not capable of inhibiting IAPP aggregation into amyloid fibrils. This result strongly suggested that while the ability of the designed inhibitor NF(N-Me)GA(N-Me)IL to recognize and interact with IAPP is related to the amyloid core-containing sequence NFGAIL, its ability to inhibit amyloid fibril formation by IAPP is due to the presence of the two N-methyl rests. In fact, the presence of these two rests has been previously shown to strongly affect the conformation of NFGAIL (6): While the CD spectrum of NFGAIL was indicative of predominantely random coil structure, the spectrum of NF(N-Me)GA(N-Me)IL had a pronounced minimum at 225-228 nm that suggested that this peptide was significantly conformationally constrained in either β-turn- or β-sheet (or β-strand)-containing conformeric state(s).

We next examined the viability of the rat insulinoma cell line RIN 5fm in the presence of IAPP alone *versus* a mixture of IAPP with NF(N-Me)GA(N-Me)IL. IAPP solutions (5 μM, in phosphate buffer, pH 7.4) without or with NF(N-Me)GA(N-Me)IL (0.5 μM) were incubated for about 20 h with the cells at final concentrations between 500 nM and 100 pM. The solutions containing NF(N-Me)GA(N-Me)IL were about 2-3 fold less cytotoxic than the IAPP alone solution according to the MTT reduction assay (2). These results suggested that NF(N-Me)GA(N-Me)IL was capable of inhibiting both amyloid formation and cytotoxicity of aggregated IAPP.

CONCLUSIONS

NF(N-Me)GA(N-Me)IL is a rationally N-methylated, IAPP amyloid core-containing synthetic hexapeptide that was designed based on a recently developed amyloid inhibitor design strategy (6). The strategy aims at converting amyloidogenic sequences into non-amyloidogenic ones via conformational restriction that is achieved by selective N-methylation of a minimum number of amide bonds on the same side of the strand of the β-sheet of the amyloid core region. The rationally designed, N-methylated peptides were expected to be able to interact with the amyloidogenic precursor sequences and inhibit the lateral extension of the β-sheet structure, intervening, thus, with the amyloid formation pathway and inhibiting amyloid-related cytotoxicity.

Our studies show that NF(N-Me)GA(N-Me)IL fulfils the requirements of the design approach: It is able to interact with full length IAPP and it inhibits IAPP amyloid formation. Importantly, NF(N-Me)GA(N-Me)IL was found to be able to reduce the cytotoxic effect of IAPP aggregates on rat insulinoma cells. Our studies also suggest that these effects are due to the presence of the N-methyl conformational constraints within the amyloid core region NFGAIL. The results presented here provided the first proof-of-principle of the applicability of our amyloid inhibitor design approach to the inhibition of amyloid formation and cytotoxicity of full length IAPP and suggest that this approach may be a promising strategy towards the development of compounds for the treatment of typ II diabetes related β-cell-death and disease pathology. Most importantly, together with the results obtained with N-methylated Aβ peptides our data suggest that N-methylated amyloid core-containing peptide sequences may find therapeutic applications to other protein aggregation diseases too (8).

REFERENCES

1.   Westermark, P., Wernstedt, C., Wilander, E., Hayden, D. W., O'Brien, T. D. and Johnson, K. H. 1987. Amyloid fibrils in human insulinoma and islets of Langerhans of the diabetic cat are derived from a neuropeptide-like protein also present in normal islet cells. *Proc. Natl. Acad. Sci. USA* 84: 3881-3885.

2.   Kapurniotu, A. 2001. Amyloidogenicity and cytotoxicity of islet amyloid polypeptide (IAPP). *Biopolymers* 60: 438-459.

3.   Dobson, C. M. 2003. Protein folding and misfolding. *Nature* 426: 884-890.

4.   Kayed, R., Bernhagen, J., Greenfield, N., Sweimeh, K., Brunner, H., Voelter, W. and Kapurniotu, A. 1999. Conformational transitions of islet amyloid polypeptide (IAPP) in amyloid formation in vitro. *J. Mol. Biol.* 287: 781-796.

5.   Tenidis, K., Waldner, M., Bernhagen, J., Fischle, W., Bergmann, M., Weber, M., Merkle, M.-L., Voelter, W., Brunner, H. and Kapurniotu, A. 2000. Identification of a penta- and hexapeptide of islet amyloid polypeptide (IAPP) with amyloidogenic and cytotoxic properties. *J. Mol. Biol.* 295: 1055-1071.

6.   Kapurniotu, A., Schmauder, A. and Tenidis, K. 2002. Structure-based design and study of non-amyloidogenic, double N-methylated IAPP amyloid core sequences as inhibitors of IAPP amyloid formation and cytotoxicity. *J. Mol. Biol.* 315: 339-350.

7.   Griffiths, J. M., Ashburn, T. T., Auger, M., Costa, P., Griffin, R. G. and Lansbury, P. T. J. 1995. Rotational resonance solid-state NMR elucidates a structural model of pancreatic amyloid. *J. Am. Chem. Soc.* 117: 3539-3546.

8.   Gordon, D. J., Sciarretta, K. L. and Meredith, S. C. 2001. Inhibition of beta-amyloid(40) fibrillogenesis and disassembly of beta-amyloid(40) fibrils by short beta-amyloid congeners containing N-methyl amino acids at alternate residues. *Biochemistry*. 40: 8237-8245.

EFFECTS OF A NEW DIFLUNISAL DERIVATIVE ON TRANSTHYRETIN BINDING AND STABILIZATION IN SERUM FROM FAP PATIENTS

M. R. Almeida [1,2]; B. Macedo [1,2]; I. Cardoso [1]; G.Valencia [3]; G. Arsequell[3]; A. Planas [4], M. J. Saraiva [1,2]

[1] Unidade de Neurobiologia Molecular, Instituto de Biologia Molecular e Celular (IBMC); [2] Departamento de Biologia Molecular, ICBAS -Instituto de Ciências Biomédicas Abel Salazar, Universidade do Porto, Porto, Portugal
[3] Instituto de Investigaciones Químicas y Ambientales de Barcelona, Consejo Superior de Investigaciones Científicas, Barcelona, España
[4] Laboratory of Biochemistry, Institut Quimic de Sarrià, Universitat Ramon Llull, Barcelona, España

INTRODUCTION

Transthyretin (TTR) is a human plasma protein associated with different forms of hereditary amyloidosis. In these diseases TTR variants precipitate as amyloid fibrils, extracellularly. Recent studies indicate that the TTR potency for amyloid formation is related to stability of the tetrameric form of the protein (1). Therefore, a potential therapeutic approach for TTR related amyloidosis is the use of drugs that stabilise TTR inhibiting fibril formation. One of the possibilities are stabilisers binding to the TTR thyroxine (T_4) binding sites. TTR T_4 binding sites are located in a central channel allowing ligands to interact with the different monomers of the molecule and possibly stabilizing it.

In the present work, we tested some drugs resulting from chemical modification of others that have been proposed before as TTR stabilizers namely diflunisal, flufenamic acid and diclofenac. In particular, we tested two diflunisal derivatives, Iododiflunisal and Bromodiflunisal, and Iodoflufenamic acid. The parent compounds and their derivatives were investigated concerning T_4 competition, tetrameric stabilization and aggregation inhibition.

MATERIALS AND METHODS

Whole serum was available from hetero and homozygotic carriers of TTR V30M, from heterozygotic carriers of TTR T119M, compound heterozygotic carriers of both TTR V30M and TTR T119M and from control individuals. Recombinant wild type TTR and TTR variants namely, V30M, T119M and TTR Y78F were produced in a bacterial expression system. The proteins were isolated according to a protocol previously described.

Diflunisal, diclofenac and flufenamic acid are comercially available (Sigma). Iododiflunisal (IDIF), Bromodiflunisal (BrDIF) and Iodoflufenamic acid were produced by chemical modification of the parent compounds. The compounds were dissolved at approximately 10 mg/ml in dimethylsulphoxide (DMSO).

Binding of different compounds to plasma proteins was assessed by ^{125}I - T_4 displacement. Whole plasma (5µL) was incubated with ^{125}I - T_4 in the presence of different compounds (1µ L of 10 mM solution) and subjected to electrophoresis in a glycine-acetate buffer system. The protein band pattern was analysed. Competition with T_4 for the binding to isolated TTR was assayed by gel filtration. TTR 30 nM was incubated with a tracer amount of ^{125}I - T_4 and with increasing concentrations of inhibitor (0-10 µM), overnight at 4°C. Bound T_4 was separated from free T_4 by gel filtration of the sample through a BioGel P6 DG (BIO-RAD) column.

TTR tetramer stability was assayed by Isoelectric Focusing (IEF) in semi-denaturing conditions (4 M urea) (2). 30 µl of plasma were incubated with 5 µl of a 10 mM solution of each compound to be tested, at 37°C for 1 h. This solution was subjected to PAGE and the gel band containing TTR was excised and applied to an IEF gel. Proteins were stained with Coomassie blue.

TTR aggregation was studied by Transmission Electron Microscopy (TEM) (3). Soluble TTR Y78F at 2 mg/ml in PBS was incubated at 37°C with or without, diflunisal and iododiflunisal, at a molar ratio 1:100. After 48 hours, samples were analyzed by TEM.

RESULTS

In order to investigate selectivity of TTR stabilizers we started by determining their capacity to compete with T_4 for protein binding in plasma. We tested some reference compounds such as diflunisal, flufenamic acid and diclofenac and assayed also some derivatives synthetically produced, in particular Iododflunisal (IDIF) and Iodoflufenamic acid (IFLU) and Bromodiflunisal (BrDIF). The results obtained indicated that all the compounds tested presented some inhibition of T_4 binding to TTR however the most potent competitor was found to be IDIF. A similar pattern of competition was obtained when plasma from different carriers was analyzed, namely from carriers of TTR V30M and TTR T119M.

Competition binding studies were then performed for the determination of the relative inhibition potency of binding defined as the EC_{50} T_4 / EC_{50} Competitor ratio. The results demonstrated that IDIF and BrDIF presented relative inhibition potencies of 0.85 and 0.53 being the most potent inhibitors for T_4 binding to TTR. Flufenamic acid and its derivative displayed similar binding potency to TTR albeit lower than IDIF. Diclofenac and diflunisal were the less potent competitors studied. Similar studies performed with the proteins both from serum of TTR WT, TTR V30M, and TTR T119M and the recombinants demonstrated that the affinity of IDIF for TTR binding is very similar to that of T_4 independently of the mutation in TTR.

The effect of the compounds on TTR stability was investigated by IEF analysis in the presence of 4M urea. Plasma from carriers of different TTR mutations and controls was incubated with each of the compounds referred to and, after IEF, the band patterns and tetramer/monomer ratios were compared.

When plasma from carriers of TTR V30M was incubated with IDIF the monomeric band disappears completely and only the tetrameric form of the protein was detected indicating that IDIF, had significative more pronounced effect on the stabilization of the TTR tetramer (Fig. 1A).

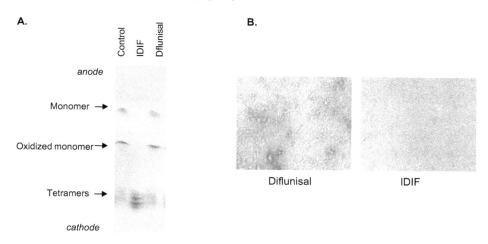

Figure 1 - **A** - Evaluation of TTR stabilization by IEF. **B** - TTR aggregation as observed by TEM.

The stabilization effect of IDIF was also evident when tested in the other plasma samples referred to previously. A similar effect was observed in plasma from TTR knockout mice carrying the TTR V30M human gene.

The inhibitory effect of IDIF on TTR aggregation was tested using TTR Y78F and compared with the effect with diflunisal. When the protein was pre-incubated with IDIF, under physiological conditions (PBS pH 7.4, 37°C) no aggregates were visible and only round particles resembling native TTR were observed. On the contrary, pre-incubation with diflunisal did not avoid the formation of very large aggregates after 48 hours similarly to what was observed in the control situation, indicating that this drug did not inhibit TTR Y78F aggregation (Fig. 1B).

DISCUSSION

It is of most importance to characterise the interaction of the different amyloid fibril formation inhibitors with TTR concerning binding specificity and selectivity to assure their preferential binding to TTR in plasma over other T_4 binding proteins, in particular albumin and TBG, avoiding decreased drug availability in plasma and compromising their use as therapeutic agents for TTR amyloidosis; this fact led us to test drugs already known and novel derivatives in «*ex vivo*» assays for T_4 and tetrameric stability.

Our studies revealed that unlike diclofenac, diflunisal and flufenamic acid IDIF and BrDIF, had high specificity and high binding affinity to TTR as shown in the T_4 binding protein profile obtained after electrophoresis of plasma. The affinity of the IDIF and BrDIF to TTR determined by «*in vitro*» competition with T_4 for binding to isolated recombinant TTR, confirmed the «ex vivo» results. In addition, IDIF and BrDIF efficiently stabilized TTR from dissociation into monomers in plasma from heterozygotic TTR V30M carriers and from controls as demonstrated by the IEF experiments. In this "*ex vivo*" assay diclofenac and flufenamic acid did not seem to stabilize the TTR tetramer.

IDIF inhibited TTR aggregation contrary to what was found for the protein incubated with diflunisal (no inhibition). In the assay conditions, IDIF prevented the formation of TTR oligomers. Being aggregation a multi-step process it cannot be excluded that other compounds may act on different stages of fibril formation.

Taken together, our "*ex vivo*" and "*in vitro*" studies present evidence for the selectivity and efficiency of novel diflunisal derivatives as TTR stabilizers and inhibitors of fibril formation in comparison with other TTR fibril inhibitors previously reported. The evaluation of IDIF as TTR stabilizer will proceed with «in vivo» animal studies.

ACKNOWLEDGEMENTS: The work has been supported by grants of Fundação para a Ciência e Tecnologia (FCT) and Fundação Calouste Gulbenkian from Portugal and a grant of Fundacion LA CAIXA from Spain.

REFERENCES

1. Sebastião M.P., Lamzim V., Saraiva M.J., Damas A.M. (2001) Transthyretin stability as key factor in amyloidogenesis: X-ray analysis at atomic resolution. *J. Mol. Biol.* **306**: 733-744.

2. Almeida, M.R., Alves I.L. Terazaki H, Ando Y., Saraiva M.J. (2000) Comparative studies of two transthyretin variants with protective effects on familial amyloidotic polyneuropathy: TTR R104H and TTR T119M. *Biochem. Biophys. Res. Commun.* **270**, 1024-1028.

3. Cardoso I, Goldsbury, S. A., Muller, V., Olivieri, S., Wirtz, S., Damas, A. M., Aebi, U., Saraiva, M. J. (2002). Transthyretin fibrillogenesis entails the assembly of monomers: a molecular model for the in vitro assembled transthyretin amyloid-like fibrils. *J. Mol. Biol.* **317**: 683-695.

SUPPRESSION OF TRANSTHYRETIN SYNTHESIS BY ANTISENSE OLIGONUCLEOTIDES

M.D. Benson (1,2), B. Kluve-Beckerman(1), K.W. Sloop(3), D.M. Bodenmiller(3)

(1) Department of Pathology and Laboratory Medicine, Indiana University School of Medicine (2) Richard L. Roudebush Veterans Affairs Medical Center (3) Endocrine Discovery Eli Lilly, Indianapolis, Indiana USA

INTRODUCTION

The only specific therapy for transthyretin (TTR) amyloidosis is liver transplantation (1). Essentially all plasma transthyretin is synthesized by the liver and transplantation results in disappearance of the variant TTR from the circulation. This has proven to be an effective therapy for many individuals with TTR amyloidosis, especially those with the Val30Met mutation. A significant number of patients, however, have progression of systemic amyloidosis after liver transplantation, and it has been hypothesized that this is the result of deposition of amyloid fibrils which are a product of normal TTR (2). In particular, patients with TTR Thr60Ala, Leu58His, and Cys10Arg have been observed to have progression of cardiac amyloidosis after liver transplantation. The increased ratio of normal to variant transthyretin in cardiac tissues obtained at autopsy from these patients is in agreement with this hypothesis. As an alternative therapy we have explored the use of antisense oligonucleotides (ASO) to TTR, modified to target the liver to decrease expression of TTR. To test this therapeutic option we have created a transgenic mouse model carrying the Ile84Ser variant gene of human TTR. TTR specific ASOs were then administered to mice, and TTR expression was measured in plasma and TTR mRNA levels in hepatic tissues.

METHODS

A genomic construct including the TTR Ile84Ser gene sequence was used to create transgenic mice which express the variant human TTR. Founder mice were bred to homozygosity and then bred into the C57/Bl strain for a total of seven generations. ASOs were synthesized as 20 base phosphorothioate chimeric oligonucleotides with 2N MOE groups on bases 1-5 and 16-20. TTR specific and random oligonucleotide controls were administered to mice either subcutaneously or intraperitoneally twice per week. TTR levels were determined by SDS-polyacrylamide gel (SDS-PAGE) electrophoresis and densitometry on individual mouse sera. TTR mRNA levels were determined by TaqMan Q-RT PCR.

RESULTS

TTR Ile84Ser transgenic mice expressed human TTR with serum levels of 80-120 mg/ DL. In preliminary experiments it had been shown that the TTR specific ASOs suppressed TTR synthesis by HepG2 cells in a dose dependant fashion (Figure 1). When administered to transgenic mice, TTR specific ASOs were found

Reduction of human transthyretin in HepG2 Cells

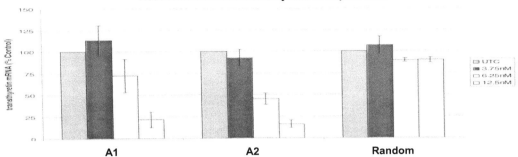

Figure 1. Cultures of HepG2 cell were treated with TTR specific ASO A1, A2, or random oligonucleotide. UTC = untreated control

to suppress plasma levels of human TTR by $\geq$ 80% at a dose of 25 mg/kg (Figure 2). A lesser degree of suppression was seen with random oligonucleotide. TTR mRNA levels from selected mice showed suppression of human TTR mRNA (Figure 3). Mouse TTR mRNA measurements on the same hepatic tissues indicated an increase in mouse TTR mRNA.

Human TTR in Serum

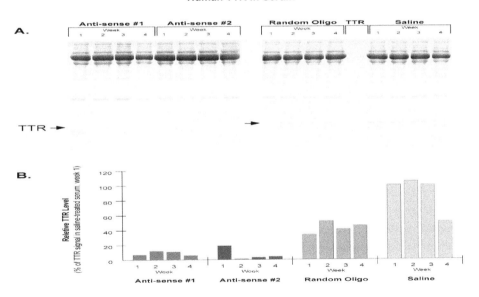

Figure 2. Reduced human TTR serum levels in mice treated with anti-sense oligonucleotides:

A. Tris-Tricine SDS-PAGE fractionation of serum proteins; gels were stained with Coomassie blue.

B. Relative quantification of TTR signals in gels.

Mice were injected with a TTR specific anti-sense oligonucleotide, random oligonucleotide, or saline twice weekly for four weeks. Blood samples were collected once each week. TTR was separated from other serum proteins by Tris-Tricine SDS-PAGE and then quantified by scanning densitometry. TTR values were expressed relative to the TTR signal in week 1 serum of saline-treated mouse. The four values whithin each condition were determined from weekly bleeds of the same mouse.

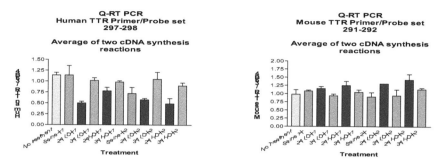

Figure 3. A one week experiment showing suppression of TTR in mRNA by both low dose (LD = 12.5 mg/Kg) and high dose (HD = 25 mg/Kg) TTR specific ASO 2A. 3A = random oligonucleotide. 1X = one injection/week. 2X = two injections/week.

DISCUSSION

These studies indicate that hepatic synthesis of human TTR is suppressed by specific TTR ASOs. This conclusion is supported by both decreased serum levels of human TTR and quantification of hepatic human TTR mRNA. Suppression of human TTR mRNA is associated with an increase in mouse TTR mRNA indicating a possible inhibition of mouse TTR expression by the human transgene expression. These studies indicate the utility of the transgenic mouse model for designing new therapeutic modalities and suggest that antisense oligonucleotide therapy may be an option for altering the course of transthyretin amyloidosis.

REFERENCES

1. Holmgren G, Ericzon BG, Groth CG, Steen L, Suhr O, Andersen O, Wallin BG, Seymour A, Richardson S, Hawkins PN and Pepys MB (1993). Clinical improvement and amyloid regression after liver transplantation in hereditary transthyretin amyloidosis. *Lancet* **341**: 1113-1116.
2. Dubrey SW, Davidoff R, Skinner M, Bergethon P, Lewis WD and Falk RH (1997). Progression of wall thickening after liver transplantation for familial amyloidosis. *Transplantation* **64**: 74.

TARGETED CONVERSION OF THE TRANSTHYRETIN GENE IN VITRO AND IN VIVO

M. Nakamura[1], Y. Ando[1], M. Ueda[2], T. Kawaji[3], T. Yamashita[2], J.M. Kim[1], K. Haraoka[4], H. Terazaki[4], S. Nagahara[5], A. Sano[5], S. Saito[1], and M. Uchino[2].

[1]Department of Diagnostic Medicine, [2]Department of Neurology, [3]Department of Ophthalmology, and [4]Department of Gastroenterology, Graduate School of Medical Sciences, Kumamoto University, 1-1-1 Honjo, Kumamoto 860-0811, Japan [5]Formulation Research Laboratories, Research Division, Sumitomo Pharmaceuticals Co., Ltd., Osaka 567-0878, Japan

INTRODUCTION

Familial amyloidotic polyneuropathy (FAP) is the common form of hereditary generalized amyloidosis and is characterized by the accumulation of amyloid fibrils in the peripheral nerves and other organs. Liver transplantation has been utilized as therapy for FAP, because the variant transthyretin (TTR) is predominantly synthesized by the liver, but this therapy is associated with several problems. Since we need to develop a new treatment, we examined the feasibility of gene therapy in FAP to prevent the production of variant TTR in the liver by using single-stranded oligonucleotides (SSOs).

MATERIALS AND METHODS

We determined optimal conditions for TTR gene conversion by SSOs in cultured HepG2 cells. Fugene6, ExGen 500, and atelocollagen were used for the assessment of transfection efficiency. The length of SSOs (24mer, 45mer, and 74mer) and atelocollagen concentration (0.1% and 0.5%) were also estimated. Finally, we used liver from transgenic mice whose intrinsic wild-type TTR gene was replaced by the murine TTR Val30Met gene for the in vivo study. The level of gene conversion was determined by real-time RCR combined with mutant-allele-specific amplification. As external standards, serial samples of genomic DNA from an FAP homozygous for the ATTR Val30Met gene and a normal mouse were diluted 10-fold with that from a normal subject possessing the wild-type TTR gene and a transgenic mouse expressing murine TTR Val30Met, respectively. The fit-point method was used for gene conversion analysis. To detect the de novo production of murine wild type TTR in serum of the transgenic mice, mass-spectrometric analysis was performed.

RESULTS

1) Evaluation of transfection efficiency in HepG2 cells. Fluorescent-labeled oligonucleotides were used to measure the cellular uptake and localization of the oligonucleotide in individual cells. Approximately 50% of the cells were fluorescent in nucleus when the cells were transfected with atelocollagen. In contrast, approximately 20% of the cells exhibited weak nuclear fluorescence when the cells were transfected with other nonviral vectors.

2) Determination of optimal conditions for SSOs-induced TTR gene conversion in cultured HepG2 cells. In administration of 0.1% atelocollagen, 24mer SSOs were 0%, 45mer SSOs were 0%, and 74mer SSOs were 0.7%. In administration of 0.5% atelocollagen, 24mer SSOs were 0.1%, 45mer SSOs were 2.7%, and 74mer SSOs were 11.1%.

3) TTR gene repair in transgenic mice possessing the murine TTR Val30Met gene. Direct injection of 400 μl of the 74mer SSOs/0.5% atelocollagen complex and the 74mer SSOs/0.05% atelocollagen complex resulted in levels of gene repair of 0.3% and 8.7%, respectively. In MALDI/TOF-mass spectrometry analysis, although only one main peak with the molecular sizes of 13,672 Da which corresponded to the reduced forms of murine TTR Val30Met was observed before 74mer SSOs/0.05% atelocollagen complex injection, the peak with the molecular size of 13,640 Da which corresponded to the reduced forms of murine wild-type TTR was detected in addition to that with the molecular size of 13,672 Da after 74mer SSOs/0.05% atelocollagen complex injection. The ratio of wild-type TTR to total TTR after type I injection was calculated to be 7.7% in comparison of the peak intensities of wild-type TTR and TTR Val30Met.

DISCUSSION

Recently, targeted gene repair by SSOs has attracted considerable attention[1,2]. Kren et al. reported that genomic and phenotypic changes that were induced by targeted gene repair were stable during an 18-month period[3], which suggests that the effect of targeted gene repair could be permanent and that this method might be effective for genomic TTR therapy. Because the gene conversion by SSOs was recognized in vitro and in vivo in our study, gene therapy with SSOs and atelocollagen is thought to be a promising method for gene repair, even though the level of gene conversion was not sufficient for suppression of the production of variant TTR in clinical terms. This method and strategy may be useful for autosomal recessive disorders. Further studies should be performed to determine the optimal delivery system consisting of SSOs with atelocollagen or times of administration.

CONCLUSION

Gene therapy via this method may therefore be a promising alternative to liver transplantation for treatment of FAP.

REFERENCES

1) Gamper, H.B. et al., A plausible mechanism for gene correction by chimeric oligonucleotides, *Biochemistry* 39, 5808, 2000

2) Gamper, H.B. et al., The DNA strand of chimeric RNA/DNA oligonucleotides can direct gene repair/conversion activity in mammalian and plant cell-free extracts, *Nucleic Acids Res.* 28, 4332, 2000.

3) Kren, B.T. et al., Correction of the UDP-glucuronosyltransferase gene defect in the Gunn rat model of Crigler-Najjar syndrome type I with a chimeric oligonucleotide, *Proc.Natl.Acad.Sci.USA*, 96, 10349, 1999.

IDENTIFICATION OF NEW DIFLUNISAL DERIVATIVES AS POTENT *IN VITRO* TRANSTHYRETIN FIBRIL INHIBITORS

T. Mairal[1], G. Arsequell[1], G. Valencia[1], I. Dolado[2], J. Nieto[2], A. Planas[2], J. Barluenga[3], A. Ballesteros[3], R. Almeida[4], M.J. Saraiva[4].

[1]*Unit of Glycoconjugate Chemistry, Institut d'Investigacions Químiques i Ambientals de Barcelona. Consejo Superior de Investigaciones Científicas (IIQAB-CSIC). Barcelona (Spain).* [2]*Laboratory of Biochemistry. Institut Químic de Sarrià. Universitat Ramón Llull. Barcelona (Spain).* [3]*Instituto de Química Organometálica "Enrique Moles". Facultad de Química. Universidad de Oviedo. Oviedo (Spain).* [4]*Molecular Neurobiology Unit. Institute for Molecular and Cell Biology. University of Porto (Portugal).*

Fibril formation and accumulation in peripheral organs is a common feature of both senile systemic amyloidosis (SSA) and familial amyloid polyneuropathy (FAP). The causative agent of such fibrils appears to be the protein transthyretin (TTR). Under physiological conditions, TTR is a stable tetramer, therefore, unit dissociation and subsequent conformational changes of the monomers are necessary events in producing TTR fibrils. In accordance to this hypothesis, therapeutic strategies for FAP and SSA have focused in finding molecules that could bind to TTR and stabilize its tetrameric form (1).

It is known that certain classes of compounds such as salicylates, steroids, antibiotics (penicillin, triiodophenol), flavonoids, inotropic agents (milrinone) and PCBs bind with high affinity to transthyretin (TTR). More recently, some members of a number of these and other families of already known therapeutic compounds such as flavones, tetrahydroquinolines, dihydropyridines, benzodiazepines, NSAIDs, phenoxazines, stilbenes, benzoxazoles and natural products (resveratrol) have also been identified and characterized as *in vitro* TTR fibril inhibitors by limited screening studies (2).

Among the reported compounds salicylates look particularly interesting because more than a hundred years therapeutic tradition and wide clinical applications make them good candidates for chronic treatments such the ones envisaged for amyloid diseases. One of them, diflunisal, has the additional advantage of being a registered drug (3) that is being shown to display suitable TTR amyloid inhibition properties (4).

To further exploit the potential of diflunisal as TTR amyloid inhibitor, a structure-activity relationship analysis of the influence of small changes on its salicylic acid moiety was undertaken.

SYNTHESIS OF DIFLUNISAL ANALOGUES

The three alternatives depicted in Figure 1 were followed: 1) Simple modifications, i.e. acetylation, amidation and methylation, of the functional groups of the molecule; 2) Conjugation with representative polar and non-polar amino acids; 3) Aromatic iodination of diflunisal and derivatives corresponding to alternatives 1 and 2.

Standard chemical procedures for altering salicylic functions, and solution and solid phase peptide synthesis protocols for the conjugation with amino acids were used. The direct electrophilic aromatic iodination of diflunisal analogues was achieved by the use of the iodonium reagent, *bis*(pyridine) iodonium (I) tetrafluoroborate the (IPy$_2$BF$_4$) (5). In this way, a family of 50 analogues were prepared which consisted of a subset of 25 diflunisal derivatives and their corresponding new iodinated counterparts.

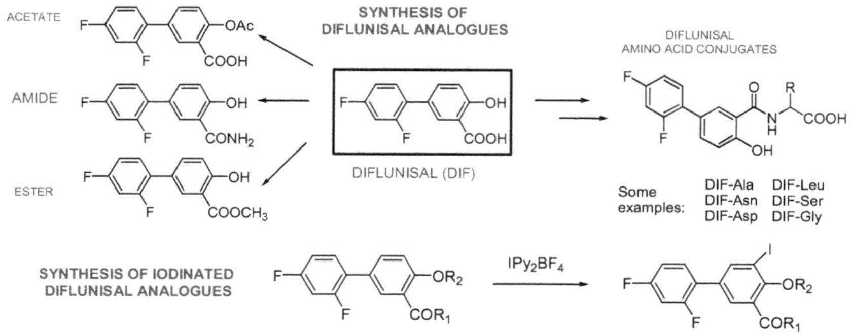

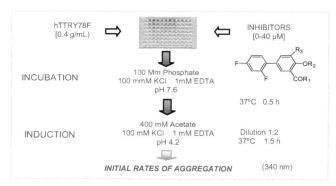

Figure 1. Chemical formulas of analogues.

SCREENING TEST FOR BIOLOGICAL ACTIVITY

For a high throughput screening of these analogues a turbidimetric test based on traditional methods (6) but adapted to microplate technology was devised. The test is performed as depicted in Figure 2 and allows the analysis of up to 13 compounds in a 96-well microplate at 7 different concentrations in 1.5 h rather than the common 72 h time.

Figure 2. Screening procedure.

Time savings and automation of kinetic monitoring of fibril formation are the key features of the method. Such a performance was achieved using the highly amyloidogenic hTTRY78F variant instead of wild type hTTR and acid induction of aggregation at pH:4.2.

RESULTS

Both series of iodinated and non-iodinated compounds were screened as TTR amyloid inhibitors at different concentrations to gain time course curves from which initial rates of fibril formation were calculated. The corresponding plots of initial rates versus inhibitor concentration were then fitted to exponential equations from which parameters such as IC_{50} and RA(%) were derived. The inhibitory potency of the compounds is here compared in terms of RA(%). This parameter is defined as the percentage of reduction of fibril formation rate at a given inhibitor concentration relative to the rate of experiments devoid of inhibitor. A maximum value of RA(%) of 100% indicate that the compound is able to completely prevent fibril formation. In the enclosed diagram (Figure 3), the reported RA(%) values refer to the highest concentrations for a selected set among the 50 inhibitors assayed. As a reference, RA(%) values for: T4, T3, thyronine and triodophenol are also given.

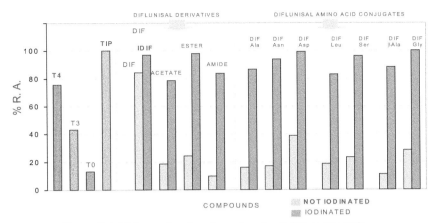

Figure 3. Inhibitory potencies of a series of selected compounds.

The data provided clearly shows that the introduction of a iodine atom in the adjacent position of the phenol group in any of the diflunisal analogues investigated greatly enhaces the inhibitory activity. Among the compounds of the series, iododiflunisal (IDIF) or 2',4'-difluoro-4-hydroxy-5-iodo-[1,1'-biphenyl]-3-carboxylic acid is the simplest compound that shows a inhibitory potency close to the highly potent triodophenol. This and other *ex vivo* properties of IDIF such as high TTR affinity and selectivity (7) makes it a potential drug candidate for TTR related amyloidosis.

ACKNOWLEDGEMENTS.

This study was supported by a grant from Fundación "La Caixa" (Neurodegenerative Disease Program).

REFERENCES

1. a) Saraiva, M.J., *Expert Rev Mol Med.*, 1, **2002**; b) Sacchettini, J. C. and Kelly, J. W. *Nat. Rev. Drug Discov.*, *1*, 267, 2002.
2. a) Miller, S.R., Sekijima, Y., Kelly, J.W. *Lab Invest.* **2004**, 1-8; b) Razavi, H. et al, *Angew Chem Int Ed Engl.*, 42, 2758, **2003** c) Oza, V.B. et al., *J Med Chem.* 45, 321, **2002**; d) Baures, P.W. et al. *Bioorg Med Chem. 7*, 1339, 1999; e) Oza, V.B. et al., *Bioorg Med Chem Lett.* 9, 1, 1999; f) Baures, P.W., Peterson, S.A, and Kelly, J.W. *Bioorg Med Chem.* 6, 1389, 1998.
3. Steelman, S.L., Cirillo, V.J., and Tempero, K.F. *Curr. Med. Res. Opin. 5*, 506, **1978**.
4. a) Adamski-Werner, S.L., et al. , *J. Med. Chem.* 47, 355, 2004; b) Klabunde, T. et al., *Nat Struct Biol. 7*, 312, 2000.
5. a) Barluenga, J. and González, J.M. *Curr. Trends in Org. Synth.* 145, 1999; b) Barluenga, J., *Pure Appl. Chem. 171*,431, 1999.
6. a) Mulkerrin, M. G., and Wetzel, R., *Biochemistry 28*, 6556, 1989; b) Klunk, W. E., Pettegrew, J.W., and Abraham, D. J. *J. Histochem. Cytochem. 37* , 1273, 1989.
7. Almeida, M. R. Et al., *Biochem. J.* (7 April) **2004**, *in press.*

THE FUTURE OF AMYLOIDOSIS

Merrill D. Benson, M.D.

Department of Pathology and Laboratory Medicine, Indiana University School of Medicine and Richard L. Roudebush Veterans Affairs Medical Center, Indianapolis, Indiana USA

Tremendous advances have been made in the study of amyloid and amyloidosis since the beginning of this series of international symposia. In particular, since the IVth International Symposium in 1984, when molecular biology techniques were introduced to this field of study, research has proceeded at an accelerated pace with many new discoveries in protein structure, pathogenesis, identification of new amyloid proteins, and the introduction of new therapeutic strategies. Studies of protein structure have generated hypotheses on the role of protein misfolding and its affect on protein stability which may play an important part in the initiation of fibrillogenesis. This has been particularly true for studies of proteins involved in hereditary amyloidosis (transthyretin, apolipoprotein AI, gelsolin), but also has given some inkling of how the primary structure of proteins such as immunoglobulin light chains may lead to β-fibril formation.

While studies on protein structure have strong implications for the understanding of pathogenesis the development of tissue culture models to study pathogenesis of amyloid fibril formation have also added tremendously to our knowledge of amyloidosis. A prime example is the tissue culture model for studying AA amyloid formation by macrophages/ monocytes. This methodology was an obvious next step in the refinement of SAA/AA studies started in the casein-induced amyloid model of mice. Coupled with molecular biology techniques for altering structure of amyloid precursor proteins it is now possible to test the importance of specific amino acid residues on a protein's ability to enter the amyloid fibril forming pathway.

New amyloid proteins that have been discovered through the techniques of primary protein structure coupled with DNA cloning and sequencing include the many variants of transthyretin, fibrinogen Aα-chain, lysozyme, apolipoprotein AI, apolipoprotein AII, and gelsolin. While the standard scientific approach to identifying a new amyloid protein has been to first isolate amyloid fibrils and determine primary structure of the subunit protein, the ability to extend the knowledge found from classic protein chemistry by the use of molecular biology techniques has proven very powerful.

It is most exciting to see from this present symposium that new ideas on therapy for systemic amyloidosis have come forth. The use of high dose intravenous melphalan with autologous stem cell rescue for the treatment of light chain amyloidosis, while essentially an extension of a chemotherapy approach that has been in existence for several decades, has offered new hope to many patients with this devastating disease. The growing data on liver transplantation for transthyretin amyloidosis are encouraging, and the reports of liver transplantation for treatment of fibrinogen Aα-chain amyloidosis suggest a curative therapy. The introduction of therapies for AA amyloidosis with small molecules, which inhibit proteoglycan components of amyloid fibrils, also is promising and offers a truly new approach to treating this disease.

In addition to the advances in research on classic amyloid and amyloidosis it is gratifying to realize that advances in the study of amyloidosis have been extended and applied to other diseases which are involved in disorders of protein folding. These include Huntington disease, Parkinson disease, and the neuroserpin-opathies. This has enlarged the sphere of amyloid and protein folding diseases in the world of science, and has brought many minds to bear upon the problems of finding effective treatments for many of these diseases.

In summary, research on amyloid and amyloidosis has never been at a stronger position. It continually benefits from the introduction of new research technologies and is poised to continue to advance at the geometric rate exemplified by our series of international symposia.

WE EAT ONLY DISPERSE SYSTEMS: THE PREPARATION OF DISHES IS LARGELY BASED ON THE CONTROL OF THE MICROSTRUCTURE OF FOOD, I.E. CONVALENT AND NON CONVALENT FORCES BETWEEN FOOD MOLECULES

H. J.-M. This

INRA Group of Molecular Gastronomy, Laboratory for chemistry of molecular interactions, Collège de France. 11 place Marcellin Berthelot, 75005 Paris (France) Phone: + 33 (0)1 44 27 13 10; Fax: + 33 (0)1 44 27 13 56; Email: herve.this@college-de-france.fr

Abstract:

We eat only what physical chemists call disperse systems, formerly named colloids. Hence the scientific study of culinary processes, done by the discipline named Molecular Gastronomy, had to devise a way of describing transformations of disperse systems. Such modelization leads to the invention of an infinite number of new dishes.

Keywords:

Disperse systems, colloids, formalism, cooking, culinary transformations, molecular gastronomy.

TABLE OF CONTENTS

1. MOLECULAR GASTRONOMY

What do we eat? Certainly food, but if culinary art developed through centuries, it means that there is something more than just energy of nutriments, contrary to what thought the French chemist Marcellin Berthelot, when he proposed that people would eat nutritive tables in the year 2000 [1].

Hence the question, again: what do we eat? The study of non human primates is giving a lot of information on the kind of food selected, as well as how food cultures could emerge, but analysis of cooking can say more, because human beings are particular animals [2].

1.1 COOKING, BETWEEN LOVE, ART AND CRAFT

As food is prepared through technical processes, the study of culinary techniques can be useful to understand what dishes are. However the taste of food can change considerably depending on the chosen ingredients. Cooks choose carefully which ingredients they mix and this choice has nothing to do with technique; the main question is art. Hence cooking should also be investigated from an aesthetical point of view.

Finally a poor meal shared with friends can be good if the friends are very good friends, and some dishes prepared by someone whose idea is to give love are more appreciated than very good food shared with enemies or prepared by people that we hate. Hence love in culinary investigations should also be explored.

1.2   DEFINITIONS AND PRECISIONS

Molecular Gastronomy is the scientific study of gastronomy, i.e. cooking and eating [3, 4]. It was recently recognized that culinary recipes are composed of two parts: definitions and precisions (old wives tales, proverbs…). Hence the scientific strategy of the discipline, concerning the technical part of cooking: modelization of definitions, and explorations of precisions.

2. WE EAT ONLY DISPERSE SYSTEMS

For both studies, description of food, before and after culinary transformations, is needed. The physical structure of dishes is intermediate between solids (too hard to swallow) and liquids (they are drunk, not eaten); food is made of disperse systems, formerly called colloids [4]. In food systems, both covalent and non covalent forces are important [5-10].

2.1 SOME TRADITIONAL EXAMPLES

2.1.1 Meat, fish, vegetables… and custard

At first order, all plant or animal tissues can be described as gels, as water is dispersed within the tissues, made of cells: there is a continuous phase that can be considered as solid, and the liquid phase is dispersed into it.

 When these tissues are transformed ("cooked", or processed), the dispersion state can change.

The case of custard is interesting to consider, as it is made from egg yolk (a suspension of granules into plasma, S/W), sugar and milk. When this mixture is first whipped, its color turns from yellow to white, because of the introduction of air bubbles. Then, when cooked, the proteins aggregate, forming micro solids dispersed into the liquid phase. Curdling is a flocculation of the sauce, which can be dispersed again, as said the old wives, by shaking with milk.

2.1.2 What does cooking mean? Changes in aggregation state

As was shown with custard, cooking is an important part of culinary activities. What does it mean?

The same word is used for many different processes, as heat treatment generates very different modifications in vegetables or meats, depending on parameters such as time, temperature, pH…

Moreover, cooking is generally considered as the application of a thermal treatment, but there are many other possibilities. For example, "fish cooked à la tahitienne" (*ceviche*, in Spanish speaking countries) are obtained through maceration of fish in lime juice. Is the fish really "cooked"? When no heat treatment is applied, we proposed to introduce a new word, "coction" based on the same Indo-European root "kok" as "cooking".

2.2   A FORMALISM TO DESCRIBE DISHES

In order to describe the physical transformations associated with cooking, we proposed to consider that the phases present in food are gas, liquids or solids. The liquids, hydrophobic or hydrophilic, are named "water" or "oil", depending on their chemical composition. The solids are many, and they generally do not mix, so that different names should be given: solid 1, solid 2...All these different phases can be dispersed, or mixed, or included into one another, or superposed...

Hence the proposal of using letters to envision rapidly all the possible systems. The phases can be written: G (for gas), O (for oil), W (for water), S1 (for solid 1), S2 ... The main processes can also be described by a few symbols: / (dispersed into), + (mixed with), $\supset$ (included into), σ (superposed)... In order to consider all possibilities, a number of k phases A_1 $A2$ ... A_k (k is a natural number) is chosen in the set {G, O, W, S_1, S_2, S_3...}. Then symbols from the set {/, +, $\supset$, σ} are introduced between successive letters. And finally parentheses are added.

Some rules apply, and other details can be included in the formalism, such as the distribution of sizes of dispersed structures, that can be indicated in brackets. Finally, the level of spatial description can be indicated. This "complex disperse systems formalism" (CDS formalism) could be applied successfully to any food considered until now.

It is interesting to note that this formalism make a bridge between macroscopic and microscopic description of systems. It will be shown how it was successfully applied to the description of the hundreds of classical sauces.

3. TECHNOLOGICAL APPLICATIONS OF THE CDS FORMALISM

Is this global description of complex systems useful? The French chemist Antoine Laurent de Lavoisier (Paris, 1743 - id., 1794) introduced the now classical formalism of chemistry because he wanted to make it easier the description of molecules and chemical processes: "In order to better show the state of affairs, and to give directly, in one sight, the result of what goes on in metal dissolutions, I have constructed a special kind of formulas that look like algebra but that do not have the same purpose and that to not derive from the same principles: we are far away the time when chemistry will have the precision of mathematics, and I invite to consider that these formulas are notations whose object is to ease the operations of the mind" [11]. The same usefulness applies to this new CDS formalism.

Also the CDS formalism leads to innovation. Let us consider any formula, with letters A, B, C...K and symbols chosen as described. If connectors and parentheses are added, new systems are predicted. In particular, a dish named "Faraday of lobster", corresponding to the formula ((S1+O+G)/W)/S2 was served by the French cook Pierre Gagnaire (Restaurant Pierre Gagnaire, Paris).

The number of possibilities is innumerable, and art should now choose which one it wants to make.

References

[1] Berthelot, M., Discourse given to the Banquet of the Chamber of chemicals, April 5 1894, in Berthelot, M., *Science et morale*, Calmann-Lévy, Paris, 1897.
[2] Hladik, C. M., Le comportement alimentaire des primates : de la socio-écologie au régime éclectique des hominidés, *Primatologie*, 5, 421-66, 2002.
[3] This, H., Molecular gastronomy, *Angew. Chem. Int. Ed. Engl.,* 41 (1), 83-88, 2002.
[4] This, H., La gastronomie moléculaire, *Sciences des aliments,* 23(2), 187-198, 2003.
[5] Dickinson, E., Emulsions stability, in *Food Hydrocolloids, Structure, properties and functions*, Nishinari, K. & Doi, E., Eds., Plenum Press, New York, 1994.
[6] Everett, D. H., Basic *Principles in Colloid Science*, Royal Society of Chemistry, London, 1988.
[7] Djabourov, M., Architecture of gelatine gels, *Contemp. Phys.,* 29 (3), 273-297, 1988.
[8] Hiemnez, P. C., *Principles of colloid and surface chemistry*, Marcel Dekker Inc., New York, 1986.
[9] Hunter, R. J., *Foundations of Colloid Science*, Oxford University Press, Oxford, 1986.
[10] Lyklema, J., *Fundamentals of Interface and Colloid Science*, Academic Press, London, 1991.
[11] Lavoisier, A. L., *Traité élémentaire de chimie* (2nde ed), Cuchet, Paris, 1793.

SECTION 8

SATELLITE SYMPOSIUM
NEUROCHEM INC.

SATELLITE SYMPOSIUM:
EMERGING CLINICAL PRACTICES IN AA AMYLOIDOSIS

Howard Fillit, M.D.

Institute for the Study of Aging, Inc. and The Mount Sinai Medical Center, NY, NY

It is with great pleasure that I introduce the report on a symposium titled "Emerging Clinical Practices in AA amyloidosis," that was the opening session of the Xth International Symposium on Amyloid and Amyloidosis. The goals of this report are to disseminate knowledge of the disease, help clinicians to achieve early diagnosis, and to employ better therapeutic approaches. The participants in this symposium are all international leaders in the field.

Dr. B Hazenberg is a clinical rheumatologist who has spent 25 years of his career dedicated to the clinical care and research in AA amyloidosis. He is currently the Director of Amyloid Studies at Groningen University Hospital in The Netherlands. He discussed the demographics of AA, including differences and changes in the underlying causes of the disease in different parts of the world.

Dr. Jeffrey Kelly is Dean and VP for academic affairs at Scripps Research Institute. His primary research interests are understanding the mechanism of transthyretin amyloid fibril formation and development of small molecule inhibitors that prevent misfolding. Dr. Kelly is a chemist by training and discussed recent work from his laboratory on novel mechanisms of amyloid formation that may offer new approaches to development of new therapeutics.

Dr. G Merlini is the Director of the Center for Systemic Amyloidosis and the Biotechnology Research Laboratory at the University Hospital in Pavia, Italy. He is also the founding leader of the National Study Group on Amyloidosis that includes 58 clinical centers throughout Italy, all dedicated to the care and research of patients with amyloid diseases. He discussed uncommon conditions that are the underlying cause of AA amyloidosis, and in particular focused on what we can learn from genetic disorders associated with inflammation that lead to amyloid formation.

Dr. Phillip Hawkins trained in the laboratory of Dr. Mark Pepys and is currently the Clinical Director of the National Amyloidosis Centre at the Royal Free and University College Medical School in London. His interests include the diagnosis and monitoring of treatment for amyloidosis. He is particularly expert in the measurement of plasma SAA as a biomarker in AA amyloid and in techniques of SAP scanning. He discussed the value of biomarkers and scanning as diagnostic methods for the early identification of patients with amyloid disease.

Finally, Dr. Martha Skinner discussed new methods for the treatment of AA amyloidosis. Dr. Skinner has devoted 35 years to the care of patients and research on amyloid diseases. She is the Director of the Amyloid Treatment and Research Program at Boston University School of Medicine. Dr. Skinner discussed the treatment available to patients with AA amyloidosis over the years and the importance the clinical trial currently underway employing Fibrillex, a novel therapy for amyloid diseases that may represent the first orally bioavailable disease modifying agent for the treatment of AA amyloid. This clinical trial, sponsored by Neurochem, is being conducted in centers around the world. The results of the trial are expected sometime in 2005.

AA amyloidosis.is a relatively rare systemic form of amyloidosis in the United States, but is more prevalent in Europe, the Middle East and Far Eastern countries. The disease is likely underdiagnosed among patients who suffer from chronic inflammatory conditions. However, the significance of AA amyloid extends well beyond the relatively few sufferers of this disease. For one, AA amyloid is a model of amyloid formation that has had, and will continue to have, important implications for all forms of amyloid, including relatively common diseases that involve amyloid formation as a primary or secondary mechanism of cellular death and clinical morbidity, such as diabetes and Alzheimer's disease. It is fitting and exciting that, after decades of research, we are entering an era in which therapies specific for amyloid diseases that are based on known underlying mechanisms of amyloid formation may soon be available to patients. We are grateful to Neurochem for its leadership and work in discovering and developing novel therapies for amyloid diseases, and for sponsoring this symposium. A portion of the Satellite Symposium follows.

Supported by Neurochem Co., Montreal, CANADA

THE CHANGING FACE OF AA AMYLOIDOSIS

B.P.C. Hazenberg

Department of Rheumatology, University Hospital, Groningen, The Netherlands

INTRODUCTION

Nicolous Fontanus probably described AA amyloidosis for the first time in 1639. In the next centuries pathologists such as Portal, Merat, and Rokitansky recognised the sago spleen and lardaceous liver secondary to tuberculosis and syphilis. Waldenström introduced direct biopsy of tissue during life in the 1920s. Apple-green birefringence of a Congo red stained tissue specimen and positive immunohistology with anti-AA antibodies are currently both required to prove the presence of amyloid of the AA-type.

INCIDENCE AND PREVALENCE OF AA AMYLOIDOSIS

Incidence and prevalence of AA amyloidosis in the general population are unknown. Estimated prevalence from large autopsy studies (with inherent selection bias) between 1955 and 1985 ranged between 0.5% and 0.85% (1). In living patients with juvenile arthritis (JIA) the prevalence of AA amyloidosis varies between 0.1% and 10% (2, 3). In selected populations of patients with rheumatoid arthritis (RA) 5-11% of patients with severe and longstanding disease may suffer from AA amyloidosis, but these are old figures (4, 5). However, recent Finnish studies show a decreasing incidence of amyloid in patients with rheumatic diseases (6, 7). Therefore a realistic estimate of the current prevalence in RA probably does not exceed 1%.

Table 1. Autopsy/histology studies of patients outside the West

Author (ref)	Cooke (8)	James (9)	Chugh (10)	Mody (11)
Period	62-67	51-73	57-79	69-82
Country	N Guinea	Uganda	India	S Africa
Infectious*	47 (96%)	47 (90%)	182 (90%)	23 (79%)
Rheumatol.	1 (2%)	0 (0%)	13 (6%)	4 (14%)
IBD	0 (0%)	0 (0%)	0 (0%)	0 (0%)
Malignant	1 (2%)	0 (0%)	0 (0%)	2 (7%)
Other	0 (0%)	5 (10%)	8 (4%)	0 (0%)
Total	49	52	203	29

* Especially tuberculosis and leprosy

CAUSES OF AA AMYLOID

Autopsy and histology series between 1970 and 1985 of patients outside Europe and North America (table 1) show that infectious diseases (especially tuberculosis and leprosy) cause 80-95% of all cases with AA amyloidosis. Similar series between 1950 and 1990 in Europe and North America (table 2) also show a high frequency of infectious diseases (60-70%) followed by rheumatic diseases (10-30%).

Table 2. Autopsy/histology studies of patients in the West

Author (ref)	Dahlin (12)	Kuhlback (13)	Brownstein (14)	Boussema (15)	Hoshii (16)
Period	21-46	54-64	45-68	56-89	87-89
Country	USA	Finland	USA	France	Japan
Infectious	19 (63%)	22 (69%)	63 (63%)	76 (66%)	23 (27%)
Rheumat	2 (7%)	7 (22%)	15 (15%)	33 (29%)	46 (54%)
IBD	2 (7%)	0 (0%)	4 (4%)	2 (2%)	0 (0%)
Malignant	7 (23%)	3 (9%)	18 (18%)	4 (3%)	14 (17%)
Other	0 (0%)	0 (0%)	0 (0%)	0 (0%)	2 (2%)
Total	30	32	100	115	85

However, series in live patients with AA amyloidosis in the West from 1960-1996 (table 3a) and from 1977-2003 (table 3b) show a shift of underlying diseases: infectious causes fall in this period from about 40% to 10-20%, malignancies fall from 10-15% to 0-5%, and rheumatic diseases rise from about 35% to 65-80%. FMF is an important cause in the Mediterranean area.

Table 3a. Studies of live patients in the West 1960-1996

Author (ref)	Brandt (17)	Gertz (18)	Browning (19)	Janssen (20)	Hazenberg (21)
Period	60-66	56-89	73-82	65-84	85-96
Country	USA	USA	UK	Netherl	Netherl
Infectious	10 (43%)	11 (17%)	13 (17%)	18 (20%)	11 (13%)
Rheumatol.	8 (35%)	42 (66%)	56 (74%)	57 (63%)	68 (82%)
IBD	1 (4%)	6 (9%)	3 (4%)	5 (5%)	0 (0%)
Malignant	3 (13%)	2 (3%)	3 (4%)	4 (4%)	2 (2%)
Other	1 (4%)	3 (5%)	1 (1%)	7 (8%)	2 (2%)
Total	23	64	76	91	83

Table 3b. Recent studies of live patients in the West from 1977-2003

Author (ref)	Joss (22)	Chevrel (23)	Knutar (24)	Garceau (25)	Lachmann (26)
Period	85-99	77-99	89-98	01-03	88-03
Country	UK	France	Finland	Internat	UK
Infectious	6 (14%)	14 (26%)	24 (18%)	21 (12%)	50 (15%)
Rheumatol.	30 (70%)	18 (34%)	96 (74%)	119 (65%)	219 (64%)
IBD	1 (2%)	1 (2%)	0 (0%)	9 (5%)	12 (4%)
FMF	0 (0%)	9 (17%)	0 (0%)	35 (19%)	24 (7%) #
Malignant	1 (2%)	2 (4%)	4 (3%)	0 (0%)	
Other	5 (12%)	9 (17%)	6 (5%)	1 (1%)	23 (7%)
Total	43	53	130	183	340

#, FMF and other hereditary fevers

GRONINGEN REGISTRY 1960-2000

The Groningen registry of patients with amyloidosis shows some fluctuations from 1960 to 2000, but the number of new AA patients secondary to RA remains almost stable at a level of 1-2 patients per year. Although the range is very wide, the mean duration of RA before detection of AA amyloidosis almost doubled from 13 years in 1960 to 22 years in 2000 (figure 1A). Proteinuria at presentation decreased from median 3.0 in the period 1960-80 to 1.8 g/day in the period 1981-2000 (figure 1B), but the median creatinine clearance did not change (45 and 50 ml/min respectively).

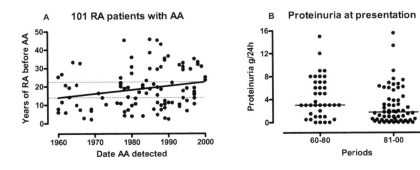

Figure 1. Groningen Registry from 1960-2000. A. Duration of RA before detection of AA amyloidosis (R = 0.22, p < 0.05). B. Proteinuria at presentation (p < 0.05).

SUBCLINICAL AA AMYLOIDOSIS

Three recent studies looked for amyloid in subcutaneous fat tissue of patients with longstanding RA. The prevalence of amyloid ranged from 7% to an amazingly high figure of 27% (table 4). The proportion of patients without characteristic symptoms of AA amyloidosis (so-called subclinical AA amyloidosis) was about two-third (63-74%) of all patients with amyloid detected.

Table 4. Amyloid detected by screening fat tissue in patients with more than 5 years RA.

Author (ref)	El Mansoury (27)	Gómez-Casanovas (28)	Wakhlu (29)
Period	1999	1983-98	~ 2000
Country	Egypt	Spain	India
Number RA	112	313	113
Clinical AA	3	16	8
Subclinical AA	5	45	22
Total AA	8 (7%)	61 (19%)	30 (27%)
Subclinical as % of total AA	63 %	74 %	73 %

CONCLUSIONS

Outside the West infectious diseases are still the main cause of AA amyloidosis, but in the West rheumatic diseases have taken their place. In the Mediterranean area FMF is an important cause. Effective therapy of infectious diseases appears an effective way to prevent clinical AA amyloidosis.

Autopsy studies were the base of our knowledge until 1920, clinical AA amyloidosis in live patients thereafter, and currently it is possible to screen for the presence of subclinical AA amyloid. Therefore the face of AA amyloidosis may look younger then in the past. It generally takes more time in RA to develop clinical AA amyloidosis than 40 years ago. Proteinuria seems to be less prominent today than in the past. Some studies even suggest that subclinical AA amyloidosis is present in a number of patients with longstanding inflammation such as RA.

However, some questions still wait for an answer. What is the current incidence and prevalence of AA amyloidosis in various parts of the world and in different disease populations? Which patient characteristics (genetic as well as acquired) determine the risk of developing AA amyloid? Does subclinical AA amyloidosis really exist and, if so, how common is it and what is its natural course? The answers to these questions may help to reveal the current face of AA amyloidosis.

REFERENCES

1. Simms RW, Prout MN, Cohen AS. The epidemiology of AL and AA amyloidosis. In: Husby G, editor. Reactive amyloidosis and the acute phase response. Bailliere's Clinical Rheumatology. London: Bailliere Tindall, 1994:627-34
2. Baum J, Gutowska G. Death in juvenile rheumatoid arthritis. Arthritis Rheum 1977; 20 (Suppl):253-5
3. Filipowics-Sosnowska AM, Rostropowicz-Denisiewics K, Rosenthal CJ, Baum J. The amyloidosis of juvenile rheumatoid arthritis. Compararive studies in Polish and American children. Arthritis Rheum 1978; 21: 699-703
4. Missen GA, Taylor JD. Amyloidosis in rheumatoid arthritis. J Pathol 1956; 71:179-84
5. Bland JH. Clinical incidence of renal amyloidosis in rheumatoid arthritis. J Maine Med Assoc 1965; 56:251-5
6. Laiho K, Tiitinen S, Kaarela K, Helin H, Isomäki H. Secondary amyloidosis has decreased in patients with inflammatory joint diseases in Finland. Clin Rheum 1999;18:122-3
7. Myllykangas-Luosujärvi R, Aho K, Kautiainen H, Hakala M. Amyloidosis in a nationwide series of 1666 subjects with rheumatoid arthritis who died during 1989 in Finland. Rheumatology 1999;38:499-503
8. Cooke RA, Champness LT. Amyloidosis in Papua-New Guinea. Med J Aust 1970; 2:1177-84
9. James PD, Owor R. Systemic amyloidosis in Uganda: an autopsy study. Trans R Soc Trop Med Hyg 1975 ; 69:480-3
10. Chugh KS, Datta BN, Singhal PC, et al. Pattern of renal amyloidosis in Indian patients. Postgrad Med J 1981; 57:31-5
11. Mody G, Bowen R, Meyers OL. Amyloidosis at Groote Schuur Hospital, Cape Town. S Afr Med J 1984; 66:47-9
12. Dahlin DC. Secondary amyloidosis. Ann Intern Med 1949; 31:105-19
13. Kuhlbäck B, Wegelius O. Secondary amyloidosis: a study of clinical and pathological findings. Acta Med Scand 1966; 180:737-45
14. Brownstein MH, Helwig EB. Secondary systemic amyloidosis: analysis of underlying disorders. South Med J 1971; 64:491-6
15. Boussema E, Labeeuw M, Colon S, Caillette A, Zech P. Maladies associées à l'amylose rénale: À propos de 216 cas. Ann Med Interne 1991; 142:331-4
16. Hoshii Y, Takahashi M, Ishihara T, Uchino M. Immunohistochemical classification of 140 autopsy cases with systemic amyloidosis. Pathol Int 1994; 44:352-8
17. Brandt K, Cathcart ES, Cohen AS. A clinical analysis of the course and prognosis of forty-two patients with amyloidosis. Am J Med 1968; 44:955-69
18. Gertz MA, Kyle RA. Secondary systemic amyloidosis: response and survival in 64 patients. Medicine 1991; 70:246-56
19. Browning MJ, Banks RA, Tribe CR, et al. Ten years' experience of an amyloid clinic - a clinicopathological survey. Q J Med 1985; 54:213-27
20. Janssen S, Van Rijswijk MH, Meijer S, et al. Systemic amyloidosis: a clinical survey of 144 cases. Neth J Med 1986; 29:376-85
21. Hazenberg BP, Van Rijswijk MH. Aspects cliniques de l'amylose AA. In: Grateau G, Benson MD, Delpech M, editors. Les Amyloses. Paris: Flammarion, 2000:377-427
22. Joss N, McLaughlin K, Simpson K, Boulton-Jones JM. Presentation, survival and prognostic markers in AA amyloidosis. Q J Med 2000; 93:535-42
23. Chevrel G, Jenvrin C, McGregor B, Miossec P. Renal type AA amyloidosis associated with rheumatoid arthritis: a cohort study showing improved survival on treatment with pulse cyclophosphamide. Rheumatology 2001; 40:821-5
24. Knutar O, Pettersson T, Tornroth T, Maury CPJ. AA amyloidosis without definable underlying disease. In: Bely M, Apathy A, editors. Amyloid and Amyloidosis. Hungary, 2001:168-70
25. Garceau D, Briand R, Hughes L, Gurbindo C, et al. A prospective analysis of demography, etiology, and clinical findings of AA amyloidosis patients enrolled in the international clinical Phase II/III Fibrillex study. Presented at Xth International Symposium on Amyloid and Amyloidosis, Tours, April 18-22, 2004: O18
26. Lachmann HJ, Goodman HJB, Gallimore J, et al. Characteristics and clinical outcome of 340 patients with systemic AA amyloidosis. Presented at Xth International Symposium on Amyloid and Amyloidosis, Tours, April 18-22, 2004: O16
27. El Mansoury TM, Hazenberg BP, El Badawy SA, et al. Screening for amyloid in subcutaneous fat tissue of Egyptian patients with rheumatoid arthritis: clinical and laboratory characteristics. Ann Rheum Dis 2002;61:42-7
28. Gomez-Casanovas E, Sanmarti R, Sole M, Canete JD, Munoz-Gomez J. The clinical significance of amyloid fat deposits in rheumatoid arthritis: a systematic long-term follow-up study using abdominal fat aspiration. Arthritis Rheum 2001; 44:66-72
29. Wakhlu A, Krisnani N, Hissaria P, Aggarwal A, Misra R. Prevalence of secondary amyloidosis in Asian North Indian patients with rheumatoid arthritis. J Rheumatol 2003; 30:948-51

UNCOMMON CONDITIONS UNDERLYING AA AMYLOIDOSIS

G. Merlini

Amyloidosis Centre, Biotechnology Research Laboratories, University Hospital IRCCS Policlinico San Matteo, Department of Biochemistry, University of Pavia, Italy

Secondary or reactive (AA) amyloidosis represents a substantial proportion of systemic amyloidoses. Its pathogenesis is multifactorial and involves many variables such as the primary structure of the precursor protein, the presence of non-fibrillar proteins (e.g. amyloid P component, apolipoprotein E, proteoglycans and basement membrane proteins), receptors, lipid metabolism and proteases. The persistent and substantial elevation of the serum concentration of the amyloidogenic precursor, serum amyloid A protein (SAA), is necessary, but not sufficient, for the development of AA amyloidosis (1). SAA is an apolipoprotein of high-density lipoprotein, and is synthesized by hepatocytes under the transcriptional control of a number of cytokines, principally IL-1, IL-6 and TNFα. The increased/deregulated production of cytokines can be caused by the conditions listed in Table 1.

Table 1. Conditions associated with AA amyloidosis

Chronic inflammatory dis.	Chronic infections	Neoplasia
Rheumatoid arthritis	Tuberculosis	Hepatoma
Psoriatic arthritis	Osteomyelitis	Renal carcinoma
Chronic juvenile arthritis	Bronchiectasias	Castleman's disease
Ankylosing spondylitis	Leprosy	Hodgkin's disease
Behcet's syndrome	Pyelonephritis	Adult hairy cell leukemia
Reiter's syndrome	Decubitus ulcers	Waldenström's disease
Adult Still's disease	Whipple's disease	
	Acne conglobata	
Inflammatory bowel diseases		
	Common variable immuno-deficiency	
Hereditary periodic fevers		
	Cystic fibrosis	

The detection of the underlying condition may represent a real clinical challenge, and in approximately 7 to 10% of patients with AA amyloidosis this search is vain. This is of paramount clinical relevance, since the timely detection of the underlying disease and its prompt effective treatment could reverse the amyloid process and restore the function of the target organ, which is most frequently the kidney.

LESSONS FROM UNCOMMON CONDITIONS

The hereditary periodic fever syndromes form a unique group of diseases with intermittent clinical manifestations sustained by genetic defects (2). Four syndromes characterized by intermittent bouts of inflammatory symptoms have been clinically and genetically defined: familial Mediterranean fever (FMF), tumor necrosis factor receptor-associated periodic fever syndrome (TRAPS), hyperimmunoglobulinemia D and periodic fever syndrome (HIDS), and cryopyrin-associated periodic syndromes which include Muckle-Wells syndrome, familial cold autoinflammatory syndrome (FCAS), and neonatal-onset multisystem inflammatory disease (NOMID). These conditions have been of great importance in unraveling the genes and mechanisms involved in the inflammatory response. Although the diagnosis of hereditary periodic fever syndromes has been greatly facilitated by genetic testing, their protean clinical manifestations hinder prompt recognition. With the exception of HIDS, all of them were known to be associated, at various frequencies, with AA amyloidosis. HIDS is characterized by a moderate deficiency of mevalonate kinase caused by mutations in the *MVK* gene. It has been reported that lack of isoprenoid products (the end product of the mevalonate enzyme pathway) raises *ex vivo* interleukin-1 secretion in HIDS. However, mechanisms less directly related to reduced isoprenoid output may contribute to IL-1 hypersecretion (2). We report on a 29-year old patient with recurrent fever attacks since the age of six months who recently developed nephrotic syndrome and renal amyloidosis, typed as AA. The fever attacks were associated with clinical manifestations typical of HIDS. Laboratory investigations documented very high values of SAA during attacks as well as in attack-free periods and high concentrations of serum IgD. Genetic analysis identified two missense mutations in the *MVK* gene thus fulfilling the diagnostic criteria for HIDS. Searches for mutations known to be pathogenic in other periodic fever syndromes were negative. This case highlights the importance of maintaining a high level of suspicion for amyloidosis even in conditions which are traditionally considered to be at no risk of developing AA amyloidosis.

The following case is also instructive. A male, born in 1948, presented with intermittent (almost weekly) fever attacks associated with vomiting, diarrhea, arthralgia and urticaria since the age of 25 years. Symptoms attenuated after the age of 35 yrs. A small IgMκ monoclonal component was documented at 25 years of age. In 1999 mild renal insufficiency was documented (s. creatinine 1.5 mg/dL (u.r.l. 1.2) together with proteinuria in the nephrotic range (u. protein 4.5 g/d). Renal function slowly deteriorated, reaching end stage renal failure requiring hemodialysis in October 2003. The patient was evaluated at our Center in November 2003; the serum IgMκ was 9.3 g/L with κ Bence Jones proteinuria. An abdominal fat aspiration showed abundant amyloid deposits. The long history of the presence of a monoclonal protein and the development of amyloidosis could have supported the diagnosis of AL amyloidosis. However, immunoelectron microscopy unequivocally typed the amyloid deposits as AA: high serum concentrations of SAA and IL-6 were then documented. The patient's clinical presentation fulfills the diagnostic criteria for Schnitzler's syndrome, which is characterized by an urticarial skin rash and monoclonal IgM component associated with at least two of the following: fever, arthralgias, palpable lymph nodes, liver or spleen enlargement (3). The etiology of this syndrome is unknown and the role of the monoclonal IgM(κ) in the pathogenesis of urticaria and other features is not yet established. Several authors reported increased serum concentrations of IL-6, while no increased concentrations of known endogenous pyrogens such as TNF-α and IL-1β have been reported (4). Therefore, an improper secretion of IL-6 may be at the basis of this condition. Considering the long history of chronic inflammation associated with Schnitzler's syndrome, it is surprising that reactive amyloidosis has not been previously reported in these patients (3).

Another uncommon condition supported by increased production of IL-6 is Castleman's disease. Several cases of Castleman's disease, predominantly unicentric, plasma cell type, have been reported to be associated with AA (5,6). Almost 50% of patients show IL-6-related symptoms (fever, anemia, high serum concentrations of

CRP and SAA, polyclonal hypergammaglobulinemia and thrombocytosis). However, in many patients the clinical presentation may be very insidious, such as in the case of an asymptomatic woman observed at our Center who presented with nephrotic range proteinuria and renal failure. A previous kidney biopsy had shown amyloidosis, not typed. The patient had no known inflammatory diseases, although her serum concentrations of SAA and IL-6 were high. DNA analysis for hereditary periodic fever syndromes was negative. A CT scan showed a single parasplenic enlarged lymph node which was surgically excised and revealed a characteristic plasma cell type Castleman's disease. After lymph node removal the patient's conditions rapidly improved with normalization of SAA and IL-6 serum concentrations; her renal function, however, remains impaired. This case highlights the paramount importance of early diagnosis: since this is a potentially curable disease, early diagnosis can save the function of the target organ, the kidney.

As the spectrum of diseases underlying reactive amyloidosis is progressively extending, greater expertise and skills are required by clinicians. Nevertheless, the first and most critical step is to suspect reactive amyloidosis. Although genetic testing is necessary for an appropriate work-up of patients with AA amyloidosis, unequivocal characterization of amyloid deposits remains essential because it dictates both prognosis and treatment.

REFERENCES

1   Rocken, C. Shakespeare, A. (2002). Pathology, diagnosis and pathogenesis of AA amyloidosis. *Virchows Arch* **440**, 111-122

2   Drenth, J. P. van der Meer, J. W. (2001). Hereditary periodic fever. *N Engl J Med* **345**, 1748-1757

3   Lipsker, D., Veran, Y., Grunenberger, F., Cribier, B., Heid, E., and Grosshans, E. (2001). The Schnitzler syndrome. Four new cases and review of the literature. *Medicine (Baltimore)* **80**, 37-44

4   Almerigogna, F., Giudizi, M. G., Cappelli, F., and Romagnani, S. (2002). Schnitzler's syndrome: what's new? *J Eur Acad Dermatol Venereol* **16**, 214-219

5   Perfetti, V., Bellotti, V., Maggi, A., Arbustini, E., De Benedetti, F., Paulli, M., Marinone, M. G., and Merlini, G. (1994). Reversal of nephrotic syndrome due to reactive amyloidosis (AA-type) after excision of localized Castleman's disease. *Am J Hematol* **46**, 189-193

6   Lachmann, H. J., Gilbertson, J. A., Gillmore, J. D., Hawkins, P. N., and Pepys, M. B. (2002). Unicentric Castleman's disease complicated by systemic AA amyloidosis: a curable disease. *QJM* **95**, 211-218

THERAPEUTIC MANAGEMENT OF AA AMYLOIDOSIS: FROM BENCH TO BEDSIDE

Martha Skinner, M.D.

Amyloid Treatment and Research Program, Boston University School of Medicine, Boston MA

This topic gives me the opportunity to look back at how far we've come since the discovery of amyloid deposits by Virchow and the development of a method of histologic identification using Congo red stain by Bennhold. And importantly for us how far we've come since the 1st International Symposium on Amyloidosis that took place in Groningen, The Netherlands in 1967. At that first meeting attention was given to the newly discovered P-component and what important role it might play in fibrillogenesis.

In 1972 just before the 2nd International Symposium on Amyloidosis Dr. Earl Benditt identified what he termed Amyloid protein A, now known as AA, as the protein in amyloid deposits associated with the clinical syndrome called secondary amyloidosis. A major cause of secondary or AA amyloidosis in 1972 was infection. In the United States and elsewhere 20-40% of patients with lepromatous leprosy died from amyloidosis of the kidney. AA amyloidosis was a major cause of death among paraplegics. Chronic tuberculosis was a prominent cause of AA amyloidosis. Today effective antibiotics have eradicated or markedly decreased the incidence of chronic infection from any organism, and thus have markedly reduced chronic infection as a cause of AA amyloidosis.

Also in 1972 Familial Mediterranean Fever (FMF) was a chronic inflammatory disorder that occurred in up to 1 in 100 persons in eastern Mediterranean populations. AA amyloidosis was the major serious complication of FMF. In 1974 colchicine was found to decrease the attacks of inflammation associated with FMF and daily lifelong treatment with colchicine has dramatically decreased the complication of AA amyloidosis. Chronic inflammatory rheumatologic diseases and inflammatory bowel disorders are still major causes of AA amyloidosis. However with the discovery of more effective treatments, including methotrexate and tumor necrosis inhibitors, it is expected that fewer patients with these diseases will develop AA amyloidosis.

AA AMYLOIDOSIS: DISAPPEARING OR UNDERDIAGNOSED?

Today AA amyloidosis is rarely seen at the major clinical amyloid referral centers and one wonders if AA amyloidosis is disappearing? Yes, in the past 30 years effective antibiotics have decreased the incidence of AA amyloidosis caused by an underlying infectious disease. However, with increasing awareness of amyloidosis and advances in new precise diagnostic testing, there is reason to believe that AA amyloidosis is also underdiagnosed.

In examining our patient database for the underlying causes of AA amyloidosis in the last 7 years and comparing them with data published on all cases seen at our institution during a 7-year period in the 1960s, the resulting figures are seen in Table 1 (1). The number of cases has not changed, but the underlying causes are different. There are fewer cases associated with infection as expected, but surprisingly more cases associated with inflammatory bowel disease and more cases with proven AA amyloidosis where the underlying disease cause is unknown.

Cause of AA Amyloidosis	1960-1966[1]	1997-2003
Infections	45%	13%
Familial Mediterranean Fever	5%	9%
Rheumatologic Disease	32%	39%
Inflammatory Bowel Disease	5%	17%
Malignancy	13%	9%
Unknown	-	13%
Total N	22	23

MAKING THE CORRECT DIAGNOSIS

The challenge today for amyloid clinicians is to make the correct diagnosis of amyloid type. There is no blood test or genetic marker for AA amyloidosis. Certain steps must be taken to make the correct diagnosis and an AA diagnosis must fit the clinical picture. First, a tissue biopsy must be positive for amyloidosis. Then, more common types of systemic amyloidosis, such as AL and ATTR that have definitive diagnostic tests, must be ruled out. The diagnosis of AA deposits in a tissue biopsy must be confirmed by immunohistochemistry using an antibody to AA protein. Importantly, in the end the clinical evaluation of the patient must fit the diagnosis. If necessary, more rare forms of familial amyloidosis may need to be ruled out by genetic testing. You, here at this Xth International Symposium, are the experts and will be referred the difficult cases for diagnosis.

TREATMENT GUIDELINES

Major advances in the past 10-12 years allow us to discuss definitive treatment for some patients with systemic amyloidosis. Treatments are determined according to amyloid type and within each type further variations occur according to extent of disease. Very aggressive treatments have been developed for the AL and ATTR types of systemic amyloidosis. For patients with AL amyloidosis, treatment with high-dose chemotherapy and autologous stem cell transplantation may be curative, but because it is aggressive, it carries risk and can be given only to selected patients (2). For patients with ATTR amyloidosis, orthotopic liver transplantation can be curative, but is aggressive with a risk associated with major surgery. In addition, it requires early disease status and an available donor (3).

For patients with AA amyloidosis there has been no major therapy. Treatment, until now, has been to recommend treatment for the underlying inflammatory disease. The Neurochem Company has developed a treatment for AA amyloidosis based on a new concept for the treatment target. It does not target the underlying disease factory, but targets the site of fibril deposition with a small molecule that prevents AA fibril formation (Figure 1). It can be classified as a "smart treatment". Most importantly, it carries low risk and is appropriate for all patients.

AA AMYLOIDOSIS

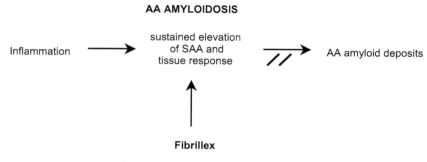

New concept for treatment target

Rating: **Low risk**
          Available to all

Fig. 1 Fibrillex™ is designed to block the formation of AA amyloid deposits in patients with high levels of SAA caused by inflammation

The Fibrillex™ clinical trial is in the final year. If effective, it will be the first drug treatment for a systemic amyloidosis to employ a small molecule inhibitor of amyloid fibril formation.   It will be a major breakthrough. Neurochem, we wish you success. And we thank you for putting enormous resources towards the treatment of a rare disease.

REFERENCES

1.   Brandt K, Cathcart ES, Cohen AS: A clinical analysis of the course and prognosis of forty-two patients with amyloidosis. The American Journal of Medicine 44: 955-969, 1968.
2.   Seldin DS, Sanchorawala V, Malek K, Wright DG, Quillen K, Finn KT, Falk RH, Berk JL, Dember LM, Wiesman J, Anderson J, Skinner M: High dose melphalan and autologous stem cell transplantation (HDM/SCT) for treatment of AL amyloidosis: ten year single institutional experience. This book.
3.   Lewis WD: Liver transplantation: an effective treatment for familial ATTR amyloidosis. Amyloid: the Journal of Protein Folding Disorders. 9: 201-202, 2002.

Author Index

Subject Index

A

A2M, 58
AA amyloid
    autofluorescence, 223–224
    bovine, 201, 206–207
    cathepsin B degradation, 191–193
    deposition, 179, 281–283. *see also*
        Amyloid deposits
    detection in fat tissue, 287–288
    ovine, 206–207
    porcine, 206–207
    specific antibodies, 218–220
AA amyloidogenesis
    cell culture model, 185–187
    murine model, 6–8, 237
AA amyloidosis
    anemia in, 289–291
    bovine, 201, 206–207
    carbonyl/oxidative stress in, 269–271
    causes of, 517–518
    chemotherapy. *see* Chemotherapy, AA
        amyloidosis
    with concurrent AL amyloidosis, 284–286
    consequences, 266
    and Crohn's disease, 97, 173–175, 215, 272
    diagnosis of, 260–262, 525
    emerging clinical practices, 515–516
    experimental induction in mice, 231–233,
        235–237
    and familial Mediterranean fever (FMF),
        272, 524
    Fibrillex™, 492–493
    Fibrillex™ Amyloidosis Secondary Trial
        (FAST), 179–181
    heparan sulfate in, 491
    and hereditary periodic fevers, 182, 215
    incidence of, 517
    induction by naturally occurring fibrillar
        proteins, 225–226
    and infectious diseases, 517–519
    and inflammatory bowel disease, 524–525
    and inflammatory disorders, 524
    inhibitory effect of monensin, 250–252
    and juvenile rheumatoid arthritis, 272–274
    and leprosy, 524

    management of, 524–526
    in the marmoset, 247–249
    patient information, 266–268
    perlecan in, 491–492
    prevalence of, 516, 517
    and rheumatic diseases, 518–519
    and rheumatoid arthritis, 97, 215, 272,
        281–283
    SAA1 gene analysis, 176–178
    subclinical, 519
    treatment guidelines, 525
    and tuberculosis, 97, 272
    underlying conditions
        inflammatory, 173–175, 215, 266,
            272
        uncommon, 521–523
AApoAII amyloidosis, transmission in mice,
    457–459, 460–462
Ab initio approach, protein structure, 3–4
AB loop, transthyretin, 326–328
Activation relaxation technique (ART), amyloid
    forming peptides, 379–381
Acute-phase serum amyloid A (A-SAA), syn-
    ovial cell migration, 212–214
Acyl-CoA:cholesterol acyl transferase (ACAT),
    SAA2.1 suppression of, 241–243,
    244–246
Acylphosphatase (AcP), disaggregation model,
    18–20
Adrenalin, 238–240
Advanced glycation end products (AGEPs),
    269–271
    $\beta2$ microglobulin amyloidosis, 415–416
    receptor, 289–291
Advanced oxidation protein products (AOPP),
    269–271
African Americans
    congestive heart failure in, 342
    TTR Val122Ile amyloid mutation,
        337–338, 342–344
Aggressiveness, disease, 55
Agrin, 221–222, 491–492
AL amyloidosis
    aggressiveness of and protein folding,
        55–57
    antigen selection in, 43–45

Milton Keynes UK
Ingram Content Group UK Ltd.
UKHW052027071024
449327UK00027B/2461